Thermodynamics and Applications in Hydrocarbon Energy Production

Thermodynamics and Applications in Hydrocarbon Energy Production

Abbas Firoozabadi

McGraw Hill Education

New York Chicago San Francisco Athens
London Madrid Mexico City Milan
New Delhi Singapore Sydney Toronto

McGraw-Hill Education books are available at special quantity discounts to use as premiums and sales promotions or for use in corporate training programs. To contact a representative, please visit the Contact Us page at www.mhprofessional.com.

Thermodynamics and Applications in Hydrocarbon Energy Production

1 2 3 4 5 6 7 8 9 0 LPI/LPI 1 2 1 0 9 8 7 6 5

ISBN 978-0-07-184325-6
MHID 0-07-184325-6

This book is printed on acid-free paper.

Sponsoring Editor
Amanda Quinn

Editing Supervisor
Stephen M. Smith

Production Supervisor
Lynn M. Messina

Project Manager
Shruti Awasthi,
Cenveo® Publisher Services

Copy Editor
Nitin Goswami,
Cenveo Publisher Services

Proofreader
Amitha Karkera,
Cenveo Publisher Services

Art Director, Cover
Jeff Weeks

Composition
Cenveo Publisher Services

Dedicated to my wife Ghashang
and to my mother Azra

About the Author

Abbas Firoozabadi is the director of the Reservoir Engineering Research Institute (RERI) and a professor at Yale University. He is the author of *Thermodynamics of Hydrocarbon Reservoirs* and has published some 200 papers. More than 70 of the papers have appeared in Society of Petroleum Engineers (SPE) journals. Firoozabadi is a member of the National Academy of Engineering (USA).

Contents

Preface

A theory is the more impressive the greater the simplicity of its premises, the more different kinds of things it relates, and the more extended its area of applicability. Therefore, the deep impression that classical thermodynamics made upon me. It is the only physical theory of universal content which I am convinced will never be overthrown, within the framework of applicability of its basic concepts.—Albert Einstein

Thermodynamics shapes the understanding and formulation of diverse groups of problems in energy production, hard materials, soft matter, nanoparticles, climate change, Alzheimer's disease, stars, black holes, and many others. In thermodynamics, one examines bulk phases and the interface between the bulk phases as well as heterogeneous state in confinement. This book offers a unified approach to bulk-phase equilibrium thermodynamics, bulk-phase irreversible thermodynamics, and interfacial thermodynamics and thermodynamics of thin films in the framework of classical thermodynamics. Concepts of stability, criticality, and the effects of curvature, gravity, and electrical charge on equilibrium are covered in multicomponent systems. The choice of the Gibbs free energy function covers applications to a broad range of problems in chemical, environmental, biomedical, and petroleum engineering. The grand potential is also introduced to study equilibrium in open systems and in inhomogeneous nanopore domains. A number of concepts and problems related to phase change and interface energy, thin liquid films, and premelting and nonequilibrium effects are discussed in different chapters.

Molecular dynamics simulations, based on concepts from statistical thermodynamics, are being used increasingly in the study of molecular structures in complex fluid and other states of matter. Molecular dynamics and other molecular simulation techniques can also provide prediction and understanding of both complex and simple fluids. In molecular simulations, use of concepts from classical thermodynamics can greatly facilitate computations and guide simulations. An example is provided in Chap. 4 on the use of stability and criticality in molecular dynamics simulations.

No comprehensive thermodynamics text is available that covers bulk-phase equilibrium thermodynamics, bulk-phase irreversible thermodynamics, and interfacial and thin-film thermodynamics. This book presents a unified approach in the study of various problems in the domains of equilibrium, nonequilibrium, and the interfaces.

Chapter 1 covers basic concepts of bulk-phase equilibrium thermodynamics. Thermodynamic functions such as Gibbs free energy and grand potential are introduced. Many examples including thermodynamics of atmosphere and reverse melting are included to demonstrate the wide range of problems that can be solved from simple concepts.

Chapter 2 presents a general theory for bulk-phase equilibrium when gravity, interface curvature, and electric fields are accounted for. Examples help with clear understanding of topics. The effect of interface curvature on equilibrium both in single components and mixtures in ideal and nonideal systems is covered. Some of the examples relate to shale gas.

Chapter 3 provides the basic expressions for computation of the relation between pressure, temperature, and volume and presents with clarity how to account for association of polar molecules. The working equations allow for engineering computations in all the chapters. This chapter also contains examples for shale gas.

Chapter 4 provides a deep but simple mathematical approach to thermodynamic stability and criticality for both pure components and mixtures. One of the examples is devoted to derivations related to complete miscibility of solvents and polymers. Another example demonstrates the usefulness of classical thermodynamics in decreasing the computational time of molecular dynamics simulations of hydrate nucleation by orders of magnitude.

Chapter 5 offers efficient numerical solution of algebraic expressions of multicomponent and multiphase bulk mixtures. This chapter is self-contained for those who are interested in efficient and robust code development in fluid-phase equilbria.

Chapter 6 presents a new perspective on diffusion of multicomponents based on rigorous derivations of entropy production. Past climate change computations and species distributions in hydrocarbon reservoirs are covered in the *Examples and Theory Extension* section. Explanation is provided for unusual cases in which a heavy fluid can float on top of a light fluid in a hydrocarbon reservoir.

Chapter 7 covers thermodynamics of interfaces in a unified approach. There is no comparable material in current thermodynamics texts in English to the level and depth of this chapter. One of the examples relates to the effect of electric charge on vaporization of droplets and the unique features of Gibbs free energy change. Interestingly, Lord Kelvin solved the same problem in a different approach. Another example covers thermodynamics of the weightless state. The concept of line tension is presented in the last example of Chap. 7. Basics of classical nucleation are covered in this chapter.

Chapter 8 covers thermodynamics of thin liquid films. This is the shortest chapter of the book with only one example and several problems. Concepts, such as disjoining pressure, are clearly introduced. The key example of this chapter is calculation of the melting point of gold nanoparticles in relation to consideration of thin liquid films. This chapter is partly based on examination of thin-film thermodynamics and consistent thermodynamics derivations.

Abbas Firoozabadi

Acknowledgments

Thermodynamics and Applications in Hydrocarbon Energy Production has evolved from 10 years of teaching graduate thermodynamics in the Department of Chemical and Environmental Engineering at Yale University from 2003 to 2012. The material from my teaching and some parts of my earlier book, *Thermodynamics of Hydrocarbon Reservoirs*, constitute the new book.

The 10 years of teaching at Yale have been the most pleasant experience. I could not have enjoyed myself more and had more fun than teaching thermodynamics to eager graduate students. One of my most fond memories was correcting students' assignments as I, too, learned from their smart approaches to solving problems.

I would like to thank the following former graduate students who were my teaching assistants in the last 4 years of teaching and helped me with class duties: Dr. Livia Moreira (Max Planck Institute), Dr. Xiaokai Li (Apple), Dr. Xiaoming Wang (BASF), and Dr. Seyla Azoz (Yale University).

Three colleagues at Yale provided the opportunity for my teaching graduate thermodynamics. Prof. Daniel Rosner set the stage for my teaching, as I co-taught a special graduate course on *Advanced Topics in Transport and Thermodynamics* in 2002. Dan has more of a fascination with the Soret effect in flames and the thermodynamics of irreversible processes than anyone else I know in the United States. The common interest in the Soret effect evolved into a friendship and a genuine interest in graduate teaching at Yale. Profs. Menachem (Meny) Elimelech and Lisa Pfefferle provided much support to my teaching during their respective tenures as department chairs. Meny also facilitated the pleasant work with my Ph.D. students and postdocs.

Much of the material in Chaps. 5 through 8 is based on the research work performed by my Ph.D. students and postdocs at the Reservoir Engineering Research Institute (RERI), which was further refined by my interest in these special topics. The material in these chapters is based on the work of Dr. Hussein Hoteit (Chevron), Dr. Kjetil Haugen (ExxonMobil), Dr. Alana Leahy-Dios (ExxonMobil), Dr. Kassem Ghorayeb (Schlumberger), Dr. Wambui Mutoru (ConocoPhillips), Dr. Joachim Moortgat (Ohio State University), Dr. Hadi Nasrabadi (Texas A&M University), Dr. Philip Myint (Lawrence Livermore National Laboratory), Dr. Erick Santiso (North Carolina State University), Dr. Jiri Mikyska (Czech Technical University in Prague), and Dr. Boris Lukanov (RERI). Specific contributions to the book were also made by the following coworkers: Hussein read Chaps. 4 through 6; Philip read Chaps. 3 and 8, and I would like to acknowledge his work on the examples on invariant eigenvalues of the Fickian diffusion coefficients in multicomponent mixtures in Chap. 6 and the weightless state in Chap. 7; Erick solved the partial differential equations that lead to the augmented Young-Laplace equation of capillarity in Chap. 7; Dr. Seyma Aslan prepared two of the figures in Chap. 3; Kjetil prepared two examples on past climate changes from irreversible thermodynamics in Chap. 6; Alana read Chap. 1; Dr. Atefeh Khoshnood (RERI) read Chaps. 1, 2, 4, and 6 and helped with one example in Chap. 6; Wambui helped with reading Chap. 3; Hadi perused the notes of this book before the final stage of preparation and also read Chaps. 2 and 3; Dr. Zhehui Jin (RERI) helped with an example in Chap. 3; Dr. Nima Rezai (Kuwait University) did some early derivations in the CPA equation of state; Joachim read Chaps. 2 and 6; Jiri provided one example for Chap. 1 and also read a few chapters and gave me his suggestions; and Boris read Chap. 7 and helped with the climate change description in Chap. 6, and also solved many problems of different chapters (these solutions will be made available when this book is finalized).

Mrs. Sirma Pandev, the former RERI office manager, did most of the typing and some of the illustrations. Miss Anna Rowland, the former RERI assistant, enhanced most of the figures. Miss Sasha Rojas, the former RERI office manager, helped with final typing.

Dr. L. Kent Thomas, a RERI board member, helped with reading Chap. 7. He, along with Dr. Bret Beckner (ExxonMobil), Prof. Clay Radke (University of California, Berkeley), and other board members of RERI, has played a key role in advising and counseling the institute in order to allow me the opportunity to write the book.

My sincere thanks and gratitude go to all my colleagues at RERI and Yale University; and to my family—my wife Ghashang, my son Reza, my daughter Rozie, and my daughter-in-law Suzette—who created the atmosphere of love and support that allowed me the endless hours I needed to write the book. I will try to compensate for what I have missed.

Notation

All the terms and symbols used in the book are listed here, except for those that appear infrequently, or only in one section.

Terms

A	Helmholtz free energy
$\mathcal{A}$	(aP/R^2T^2)
$\mathcal{A}$	surface area
a	energy parameter of the EOS
a	sonic velocity
a	specific Helmholtz free energy
a	radius
$\mathcal{B}$	(bP/RT)
b	hard core parameter of the EOS
$\mathbf{B}$	matrix with elements of the Maxwell-Stefan diffusion coefficients
C_S	isentropic compressibility
C_T	isothermal compressibility
c	molar density
c	volume-translation parameter in the EOS
c_P	molar heat capacity at constant pressure
c_V	molar heat capacity at constant volume
$\mathbf{D}$	Fickian diffusion coefficient matrix
D	Fickian diffusion coefficient
$\mathcal{D}$	Maxwell-Stefan diffusion coefficient
d	diameter of droplet
d	mass density
E	extensive property of the phase, such as V, U, or G
$\bar{E}_i$	partial molar quantity of component i
$\vec{E}$	electric field vector
e	charge of an electron
e	thermal expansivity
F	force
F	mols of feed

F	degree of freedom
F	force
$\boldsymbol{F}$	matrix with elements $(\partial \ln f_i/\partial x_j)_{T,P,x_j}$
f	fugacity
G	gas gravity
G	Gibbs free energy
$\bar{G}_i$	partial molar Gibbs free energy of component i
g	acceleration of gravity
g	molar Gibbs free energy
g	radial distribution function
H	enthalpy
$\bar{H}_i$	partial molar enthalpy of component i
$\tilde{H}$	magnetic field strength
h	height
h	liquid-film thickness
h	molar enthalpy
$\boldsymbol{H}$	Hussian matrix
$\boldsymbol{J}$	Jacobian matrix
$\vec{J}$	diffusion flux
$\vec{j}$	diffusion flux vector based on the mass averaged velocity frame
$\vec{J}$	magnetic polarizability vector
$\vec{J}^s$	entropy flux
K	thermal conductivity
K_i	equilibrium ratio of component i
k	Boltzmann's constant (1.3805×10^{-23} J/K)
k	permeability
k_{ij}	interaction coefficient between components i and j
k_{iT}	thermal diffusion coefficient of component i
L	latent heat
L	length
L	mol fraction of liquid phase
L	phenomenological coefficient
$\mathcal{L}$	phenomenological coefficient
l	length
M	molecular weight
M	number of association sites
$\bar{M}$	average molecular weight
m	mass
N	number of mols
N	total number of mols (without index)
N_A	Avogadro's number (6.02217×10^{23})
n	displacement
n	total number of mols (without index)
n_c	number of components
n_i	number of mols of component i; also N_i

$\underline{\boldsymbol{n}}$	$(n_1, n_2, \ldots, n_c)$; also $\underline{\boldsymbol{N}}$
$\boldsymbol{n}_i$	$(n_1, n_2, \ldots, n_{i-1}, n_{i+1}, \ldots, n_c)$ also $\boldsymbol{N}_i$
P	parachore
P	pressure
P_b	bubblepoint pressure
P_c	capillary pressure
P_d	dew point pressure
P_e	equilibrium pressure
P_r	reduced pressure
P_n	pressure in the cluster
$\vec{\vec{P}}$	pressure tensor
p	number of phases, number of subsystems
Q	charge
Q	heat transfer across the interface
q	particle charge
$\underline{\boldsymbol{q}}$	$(q_1, q_2, \ldots, q_{n_c-1})$
$\vec{q}^{\,cond}$	energy flux by conduction
$\vec{q}^{\,conv}$	energy flux by convection
R	universal gas constant
r	number of segments in a polymer molecule
r	radius of bubble or droplet, or radius
$\mathfrak{r}$	radius of a droplet base
r_{ij}	distance between point i and point j
$\hat{r}$	unit outward normal
S	displacement
S	entropy
S	saturation
$\bar{S}_i$	partial molar entropy of component i
s	molar entropy; also specific entropy
T_b	boiling point temperature
T_c	critical temperature
T_e	equilibrium temperature
t	time
U	internal energy
$\bar{U}_i$	partial molar internal energy of component i
u	molar internal energy
V	mol fraction of gas phase
V	volume
V_n	volume of a cluster
$\bar{V}_i$	partial molar volume of component i
v	specific volume
$\bar{v}$	average molar volume
$\vec{v}$	velocity vector
v_o	molecular volume
v_r	reduced molar volume
v_x	velocity in the x-direction

W	width
W	work
w	mass fraction
x	mol fraction (often of the liquid phase)
x	x-axis
$\underline{x}$	$(x_1, x_2, \ldots, x_{n_c})$
$\mathbf{x}$	$(x_1, x_2, \ldots, x_{n_c-1})$
y	mole fraction (often of the gas phase)
$\underline{y}$	$(y_1, y_2, \ldots, y_{n_c})$
$\mathbf{y}$	$(y_1, y_2, \ldots, y_{n_c-1})$
Z	compressibility factor
z	overall compostion
z	valence
z	vertical direction, positive upward
$\underline{z}$	$(z_1, z_2, \ldots, z_{n_c})$
$\mathbf{z}$	$(z_1, z_2, \ldots, z_{n_c-1})$
α	temperature-dependent parameter of a in the EOS
α_T	thermal diffusion factor
β	mol fraction of vapor phase
Γ	adsorption density
γ	activity coefficient
γ	surface energy density
δ	solubility parameter
Δ	change in the variable from one state to another state
Δh^f	enthalpy of fusion
ε	liquid dielectric constant
ε_o	gas dielectric constant; also vacuum dielectric constant
θ	contact angle
κ	line tension
μ	chemical potential, per mole, or per molecule
μ	Joule-Thomson coefficient
χ	mol fraction not associated
Π	disjoining pressure
Π	osmotic pressure
π	3.14159...
ρ	charge density
ρ	fluid mass density
σ	entropy production per unit time per unit volume
σ	interfacial tension
$\vec{\vec{\tau}}_{ij}$	stress in the ij plane
Φ	electric potential
Φ	volume fraction
ϕ	porosity, often fraction
ϕ	surface energy density
$\vec{\phi}$	flux vector
φ	fugacity coefficient

ψ	stream function, $v_x = -\partial\psi/\partial z$
ψ_m	modified stream function, $\rho v_x = -\partial\psi_m/\partial z$
Ω	grand potential
Ω	thermal diffusion sensitivity parameter
Ω_a^o	constant of a parameter of the EOS
Ω_b^o	constant of b parameter of the EOS
ω	Pitzer's acentric factor
ω	solid angle
θ	contact angle

Superscripts

$'$	phase index
$''$	phase index
∞	flat interface
∞	infinite dilution
b	bubble
E	excess functions
G	gas
GL	gas-liquid
GS	gas-solid
(k)	kth Legendre transformation
L	liquid
LS	liquid-solid
S	solid
V	vapor
x	phase index
y	phase index
z	phase index
α	phase index
β	phase index
p	phase index
j	phase index
s	surface index

Subscripts

α	phase index
F	film
hg	hydrate-gas
hs	hydrate-solid
i	axis index: as an example, x-axis
i	component index
j	component index

j	axis index: as an example, y-axis
j	component index
j	phase index
k	derivative with respect to variable number k
kk	second derivative with respect to variable number k
r	reduced, relative to critical value
s	substrate
sg	solid-gas
ss	solution-solid
tp	two-phase
x	derivative with respect to variable x; x can be V, T, or any other variable
1	phase index, such as an L_1 (liquid phase 1)
2	phase index, such as an L_2 (liquid phase 2)

Symbols

$\rightharpoonup$	vector
$\rightrightarrows$	tensor
∇	gradient
$\nabla.$	divergence
$\nabla\times$	curl
$\cdot$	dot product
:	double dot product
( )	vector
| |	determinant
‖ ‖	absolute value
Bold	column vector; also matrix
—	average
-	partial molar quantity
*	critical size

Thermodynamics and Applications in Hydrocarbon Energy Production

CHAPTER 1

Review of Basic Concepts in Bulk Phase Equilibrium Thermodynamics

This review chapter presents basic concepts of bulk phase equilibrium thermodynamics when only the work of compression and expansion is taken into account. We define entropy mathematically and then establish the criteria of equilibrium. Simple powerful expressions such as $\mu_i' = \mu_i''$ and $(d\mu_i = \bar{V}_i dP)_{T,\underline{n}}$ allow solving a large number of problems. We also define ideal and nonideal mixtures and introduce two mathematical transformations. The Legendre and Jacobian transformation facilitate mathematical manipulation in thermodynamics. The twelve examples at the end of the chapter include problems from thermodynamics of atmosphere, and unusual pressure profiles in a porous semipermeable membrane. The thirty-four problems at the end of the chapter include both conceptual derivations and practical thermodynamic problems.

Broadly speaking, thermodynamics can be divided into two main branches: classical and statistical. The strength and powerful features of classical thermodynamics are that without knowledge of molecular and atomic interactions, a large class of problems in physical sciences can be solved. In statistical thermodynamics, the interactions between molecules and atoms are taken into account and the force field of these species constitutes the basis.

Both classical and statistical thermodynamics can be divided into equilibrium and nonequilibrium processes. Equilibrium thermodynamics is best understood among the two classifications. According to Tisza (1966), equilibrium thermodynamics can be further subdivided into Gibbsian thermodynamics and the early thermodynamics of Clausius and Kelvin. The latter considered the thermodynamic system as a black box, and all the relevant information was then derived from the energy absorbed and the work done by the system. The concepts of internal energy, U, and entropy, S, from the observable quantities are then established. In Gibbsian thermodynamics, the concepts of internal energy and entropy are assumed to be known and are used to provide a detailed description of the subsystems in equilibrium (we will soon define some of the terms used above).

In this chapter, we will mainly consider the Gibbsian thermodynamics of bulk phase equilibria and use the concepts to solve practical problems. The thermodynamics of equilibrium processes also provide the framework for nonequilibrium and irreversible thermodynamics. It is our intention that the material covered in this book should be self-contained. The postulational approach introduced by Callen (1985) and Tisza (1966) is, therefore, adopted to make brief the presentation of basic concepts and equations.

Equilibrium thermodynamics consists essentially of the first, second, and third laws. However, to obtain useful results, these laws should be combined with an equation of state (EOS) to provide knowledge of the fluid properties at any point in the system. Since gravity, capillary forces, and electrical field are important in a large number of problems, the concepts of phase-equilibria thermodynamics developed in Chap. 2 will include these forces. We exclude chemical reaction in our derivations.

Before addressing various concepts, a limited number of terms often used in thermodynamics will be defined.

Mol number Actual number of mols of each type of species (i.e., number of molecules, if there is no association between molecules divided by Avogadro's number, $N_A = 6.02217 \times 10^{23}$). Alternatively, mol number is the amount of mass of each type of species divided by its molecular mass. The mol number of species is denoted by n_i or N_i.

Mol fraction n_i/n_T, where $n_T = \sum_{i=1}^{n_c} n_i$ is the total mol number and n_c is the number of different species in the mixture; the subscript T is often dropped from n_T.

Molar quantity The ratio of a particular property of a phase divided by the total number of mols, n_T. As an example, the molar volume is defined by $v \equiv V/n_T$, where V is the total volume.

Wall, enclosure, partition A physical system idealized as a surface forming the common boundary of two different systems. The walls that completely enclose a system are called enclosures. Walls separating the subsystems of a composite system are called partitions.

Impermeable wall A wall that is restrictive of matter.

Semipermeable wall A wall that is restrictive to only certain chemical species and nonrestrictive to others.

Permeable wall A wall that is nonrestrictive of matter and energy.

Adiabatic wall A wall that is restrictive of matter and energy.

Rigid wall A wall that does not move or deform.

Diathermal wall A wall that is restrictive of matter but nonrestrictive of energy.

Isolated system A system enclosed in a completely restrictive enclosure.

Closed system A system in an impermeable wall.

Open system A system with a wall that is nonrestrictive to matter.

Heat A transfer quantity associated with energy transferred across a rigid diathermic wall, $Q = \Delta U$.

Composite system A conjunction of spatially disjoint simple systems (subsystems or phases).

Subsystems or phases Simple systems or phases that form a composite system. A phase may be homogeneous throughout or may have a continuous variation of properties.

Extensive variables Variables that depend on the total quantity of matter in the system or subsystem. Examples are volume, V, and mol numbers, n_i.

Intensive variables Variables that have point values and are independent of the size of the system or subsystem. Examples are temperature T and pressure P.

Equilibrium state The state in which the properties are determined by intrinsic properties that are time independent. A more precise definition is deferred to later.

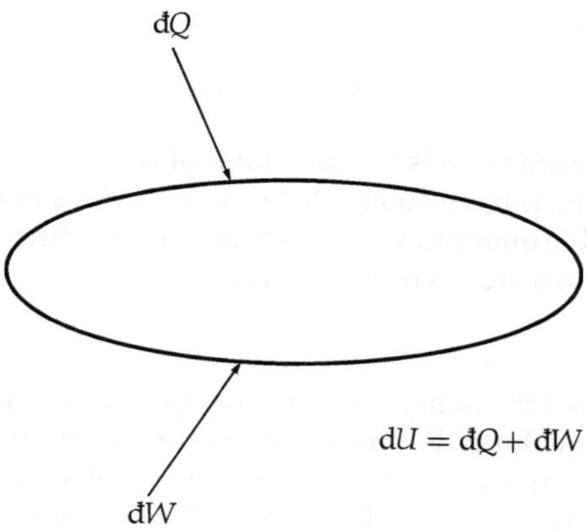

FIGURE 1.1 Sketch defining the first law of thermodynamics (the system is closed).

Next we present the five postulates of thermodynamics. These postulates are mainly adopted from Callen (1985).

Postulate I

The conservation of the internal energy, U, of a system, is postulated in the first law of thermodynamics. The internal energy of a closed system is defined in the form as

$$dU = đQ + đW \tag{1.1}$$

where dU is the differential of the internal energy, U, and $đQ$ and $đW$ are the infinitesimal amounts of heat absorbed by the system, and of the work done on the system across the system boundary, respectively (Fig. 1.1). In Eq. (1.1), dU is the differential of the state variable U. Q and W are not functions of state and do not have, in general, differentials $đQ$ and $đW$ in the state-space. Alternatively, it is said that U has an exact differential, and Q and W do not. (We will soon define an exact differential.) Note that Q and W depend on the interactions between the system and its surrounding. For the above system, if the enclosure is rigid and the system is not displaced in the vertical direction,

$$\Delta U = Q \tag{1.2}$$

where Q is the heat added to the system at constant volume and ΔU is the change in the internal energy as a result of heat absorbed by the system.

On the other hand, if the enclosure is adiabatic, but not rigid, then

$$\Delta U = W \tag{1.3}$$

where W is the work done on the system. As an example, if the volume of the system increases by dV and the system is subjected to a uniform external pressure P, we have

$$đW = -P dV \tag{1.4}$$

or at constant pressure

$$W = -P\Delta V \tag{1.5}$$

provided the work of expansion is the only kind of work.

Unlike Q and W, which have an absolute value, U does not have an absolute value; Q and W are measurable quantities, U is measured only with respect to a reference state. Internal energy, U, is also an extensive variable.

Postulate II

The equilibrium states of the subsystems of a composite system are fully defined by the internal energy U, the volume V, and the mol numbers $n_1, n_2, \ldots, n_c$ of the composite system. Note that $U, V, n_1, n_2, \ldots, n_c$ are extensive variables and are, therefore, additive properties. ($U = \sum_{j=1}^{p} U_j, V = \sum_{j=1}^{p} V_j$, and $n_i = \sum_{j=1}^{p} n_{ij}$, where p is the total number of subsystems and n_{ij} is the number of mols of component i in subsystem j.)

Postulate III

There exists an additive variable S (called entropy) that is a function of extensive variables $U, V, n_1, n_2, \ldots, n_{n_c}$ of the composite system and has the characteristic that at equilibrium is a maximum with respect to all of the extensive variables $U, V, n_1, n_2, \ldots, n_{n_c}$ of the subsystems.

From Postulates II and III, the entropy S of the composite system consisting of primed and double-primed subsystems, is a maximum at equilibrium with respect to $U', V', n_1', n_2', \ldots, n_{n_c}'$, and $U'', V'', n_1'', n_2'', \ldots, n_{n_c}''$.

Postulate IV

The entropy, S, is continuous and differentiable and is a monotonically increasing function of the internal energy, U, of the composite system.

From Postulates II and III, we can write the following expressions:

$$S = S' + S'' \tag{1.6}$$

$$S' = S'\left(U', V', n_1', \ldots, n_{n_c}'\right) \tag{1.7}$$

$$S'' = S''\left(U'', V'', n_1'', \ldots, n_{n_c}''\right) \tag{1.8}$$

where the composite system is assumed to be composed of subsystems primed and double-primed. Equations (1.6) to (1.8) are based on the additivity property of the entropy. S' and S'' are first-order homogeneous functions of the extensive properties $U', V', n_1', \ldots, n_{n_c}'$, and $U'', V'', n_1'', \ldots, n_{n_c}''$, respectively, that is,

$$S'\left(\lambda U', \lambda V', \lambda n_1', \ldots, \lambda n_{n_c}'\right) = \lambda S'\left(U', V', n', \ldots, n_{n_c}'\right) \tag{1.9}$$

where extensive variables are multiplied by a constant λ. The definition of a homogeneous function will be given soon.

The mathematical expression for the monotonically increasing property of S with respect to U can be written as

$$\left(\frac{\partial S}{\partial U}\right)_{V,n_1,\ldots,n_{n_c}} > 0 \tag{1.10}$$

where all the variables are for the composite system. The thermodynamic temperature T is defined as

$$T = \left(\frac{\partial U}{\partial S}\right)_{V,n_1,\ldots,n_{nc}} \tag{1.11}$$

Equation (1.11) implies that T cannot be negative. However, as we will see in Chap. 3, there is no such restriction on pressure; it can be negative.

The continuity, differentiability, and monotonic property imply that

$$S = S(U, V, n_1, \ldots, n_{n_c}) \tag{1.12}$$

can be uniquely solved for U given by

$$U = U(S, V, n_1, n_2, \ldots, n_{n_c}) \tag{1.13}$$

Postulates II, III, and IV define the second law of thermodynamics.

Finally, the third law of thermodynamics is defined on the postulational basis.

Postulate V

The entropy, S, vanishes as $T \rightarrow 0$.

Therefore, similar to V and n_i, and unlike U,S has a uniquely defined zero.

Prior to the presentation on equilibrium, we will discuss exact differentials and homogeneous functions.

Exact Differentials

Let $F(x_1, x_2, x_3, \ldots)$ be a continuous function of independent variables $x_1, x_2, x_3, \ldots$ except at certain isolated singular points. The total differential of F is

$$\mathrm{d}F = \left(\frac{\partial F}{\partial x_1}\right)_{x_2,x_3,\ldots} \mathrm{d}x_1 + \left(\frac{\partial F}{\partial x_2}\right)_{x_1,x_3,\ldots} \mathrm{d}x_2 + \cdots \tag{1.14}$$

which is called an exact differential. Since F is continuous, it possesses continuous higher-order derivatives except at isolated singular points. If F is integrated over some path connecting two points A and B in the $x_1, x_2, \ldots$ space, then F (at B) − F (at A) depends on the points A and B and not on the path connecting A and B.

Now consider the differential

$$\mathrm{d}\phi = X_1 \mathrm{d}x_1 + X_2 \mathrm{d}x_2 + \ldots \tag{1.15}$$

where $X_1, X_2, \ldots$ are the functions of $x_1, x_2, \ldots$. The above differential is an exact differential if $X_i = (\partial\phi/\partial x_i)_{x_1,\ldots,x_{i-1},x_{i+1},\ldots}$.

Homogeneous Functions

A function of several variables is said to be homogeneous of degree l if multiplying each variable by a constant, k, is the same as multiplying the original function by k^l. As an example, consider $F(x, y)$. This function is said to be a homogeneous function of degree l if $F(kx, ky) = k^l F(x, y)$. If $l = 1$, then F is said to be a first-order homogeneous function of x and y.

1.1 Conditions for Equilibrium

The fundamental equation for U in an open system is

$$U = U(S, V, n_1, n_2, \ldots, n_{n_c}) \tag{1.16}$$

which in the differential form can be written as

$$dU = \left(\frac{\partial U}{\partial S}\right)_{V,\underline{n}} dS + \left(\frac{\partial U}{\partial V}\right)_{S,\underline{n}} dV + \sum_{i=1}^{n_c} \left(\frac{\partial U}{\partial n_i}\right)_{S,V,\boldsymbol{n}_i} dn_i \tag{1.17}$$

where $\underline{n} \equiv (n_1, n_2, \ldots, n_{n_c})$ and $\boldsymbol{n}_i \equiv (n_1, \ldots, n_{i-1}, n_{i+1}, \ldots, n_{n_c})$. Therefore, the $\boldsymbol{n}_i$ symbol denotes that all the variables $n_1, n_2, \ldots, n_c$ are held constant except for n_i.

The first and second partial derivatives on the left side are given by

$$\left(\frac{\partial U}{\partial S}\right)_{V,\underline{n}} = T \tag{1.18}$$

$$\left(\frac{\partial U}{\partial V}\right)_{S,\underline{n}} = -P \tag{1.19}$$

Equation (1.18) was established earlier; Eq. (1.19) can be implied from $dU = TdS + đW$ and $đW = -PdV$ (for a closed system). The coefficient derivatives of the last term in Eq. (1.7) are defined by

$$\mu_i = \left(\frac{\partial U}{\partial n_i}\right)_{S,V,\boldsymbol{n}_i} \tag{1.20}$$

and are called chemical potentials. The chemical potential has a function analogous to temperature and pressure. The temperature difference causes heat flow, and pressure difference in the absence of gravity may result in bulk phase displacement. The chemical potential difference results in diffusion from the region of higher to the region of lower chemical potential in the absence of gravity effect (see Example 1.1). The chemical potential concept was introduced by Gibbs (1957). This concept, as we will see later, facilitates the description of the phase behavior of open systems. Denbigh (1971) presents the interpretation that the last term in Eq. (1.17), that is, $\sum_{i=1}^{n_c} \mu_i dn_i$, as a form of work (chemical work) which can be done by a system at constant volume due to change of composition.

Combining Eqs. (1.17) to (1.20), the expression for dU is given by

$$dU = TdS - PdV + \sum_{i=1}^{n_c} \mu_i dn_i \tag{1.21}$$

Note that the coefficients of the independent extensive variables in Eq. (1.21)—T, P, and μ_i—are intensive variables. Upon rearrangement,

$$dS = \frac{dU}{T} + \frac{P}{T}dV - \sum_{i=1}^{n_c} \frac{\mu_i}{T} dn_i \tag{1.22}$$

In the following, we will use Eq. (1.22), $S = S(U, V, n_1, \ldots, n_{n_c})$, and the entropy maximum principle from Postulate III to derive the conditions for various types of equilibria.

1.1.1 Thermal Equilibrium

Consider an isolated composite system consisting of two simple subsystems separated by a rigid diathermic partition. Therefore, $\mathrm{d}V' = 0, \mathrm{d}V'' = 0, \mathrm{d}n_i' = 0, \mathrm{d}n_i'' = 0; i = 1, \ldots, n_c$, and $\mathrm{d}(U' + U'') = 0$ (Fig. 1.2).

Mathematically, the entropy maximum principle can be written as

$$\mathrm{d}S = 0 \tag{1.23}$$

where $S = S' + S''$. (As we will see in detail in Chap. 4, to ensure the maxima, another requirement has to be met.) Writing Eq. (1.23) for the two subsystems,

$$\mathrm{d}S' = \frac{\mathrm{d}U'}{T'} + \frac{P'}{T'}\mathrm{d}V' - \sum_{i=1}^{n_c}\left(\frac{\mu_i'}{T'}\right)\mathrm{d}n_i' \tag{1.24}$$

$$\mathrm{d}S'' = \frac{\mathrm{d}U''}{T''} + \frac{P''}{T''}\mathrm{d}V'' - \sum_{i=1}^{n_c}\left(\frac{\mu_i''}{T''}\right)\mathrm{d}n_i' \tag{1.25}$$

and from $\mathrm{d}V' = \mathrm{d}V'' = \mathrm{d}n_i' = \mathrm{d}n_i'' = 0$,

$$\mathrm{d}S = \mathrm{d}S' + \mathrm{d}S'' = \frac{\mathrm{d}U'}{T'} + \frac{\mathrm{d}U''}{T''} \tag{1.26}$$

and since $\mathrm{d}U' = -\,\mathrm{d}U''$, then

$$\mathrm{d}S = \left(\frac{1}{T'} - \frac{1}{T''}\right)\mathrm{d}U' \tag{1.27}$$

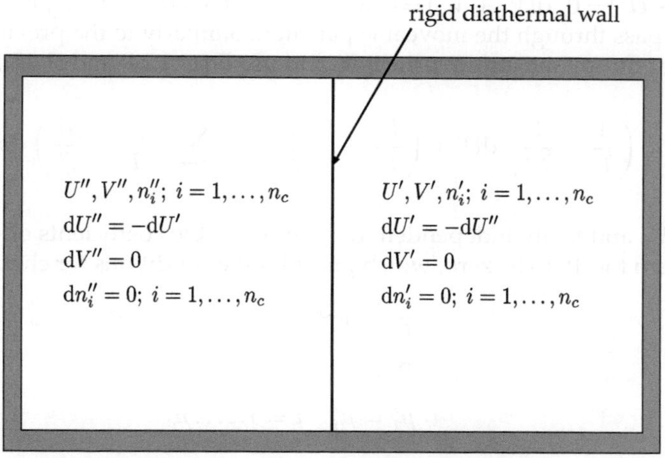

FIGURE 1.2 Isolated composite system with rigid diathermal wall.

From Eqs. (1.23) and (1.27),

$$T' = T'' \tag{1.28}$$

which states that the temperatures of the two subsystems should be equal at equilibrium for a diathermal wall.

1.1.2 Mechanical Equilibrium

Now we consider an isolated system consisting of two simple subsystems separated by a moveable diathermic partition that does not change shape. Therefore, $dV' + dV'' = 0$, $dU' + dU'' = 0$, and $dn_i' = dn_i'' = 0$. Invoking the entropy maximum principle [Eq. (1.23)] and writing Eqs. (1.24) and (1.25) with the above constraints,

$$dS = \left(\frac{1}{T'} - \frac{1}{T''}\right) dU' + \left(\frac{P'}{T'} - \frac{P''}{T''}\right) dV' = 0 \tag{1.29}$$

Since U' and V' are independent variables, each of the two coefficients, $(1/T' - 1/T'')$ and $(P'/T' - P''/T'')$ must be zero. Therefore,

$$T' = T''$$
$$P' = P'' \tag{1.30}$$

Note that the equality of pressures can also be derived from the force balance. The equilibrium with a moveable adiabatic wall is an indeterminate problem.

1.1.3 Chemical Equilibrium

In this case, the partition of the isolated composite system is permeable to some or all of the species and the partition can be either rigid or moveable. The constraint equations are $d(U' + U'') = 0$, $d(V' + V'') = 0$, and $d(n_i' + n_i'') = 0$; $i = 1, \ldots, n_c$ when all species can freely pass through the moveable partition. Similarly to the previous two cases, we invoke the entropy maximum principle, and use Eqs. (1.24) and (1.25),

$$dS = \left(\frac{1}{T'} - \frac{1}{T''}\right) dU' + \left(\frac{P'}{T'} - \frac{P''}{T''}\right) dV' - \sum_{i=1}^{n_c} \left(\frac{\mu_i'}{T'} - \frac{\mu_i''}{T''}\right) dn_i' = 0 \tag{1.31}$$

Since U', V', and n_i' are independent of each other, the coefficients of dU', dV', and dn_i' must be zero for dS to be zero, which provides the conditions for chemical equilibrium,

$$T' = T''$$
$$P' = P''$$
$$\mu_i' = \mu_i'' \quad i = 1, \ldots, n_c \tag{1.32a}$$

The flow of species between the subsystems results in the equality of chemical potentials at equilibrium, and therefore the equilibrium is classified as chemical equilibrium.

For a rigid partition, if some of the components can pass freely between the two subsystems and other components cannot go through the partition, one can readily show that

$$T' = T'' \tag{1.32b}$$

$$\mu_i'(T', P', n_1', n_2', \ldots, n_{n_c}') = \mu_i''(T', P'', n_1'', n_2'', \ldots, n_{n_c}'') \quad i = 1, \ldots, l \quad l \le n_c$$

The difference $(P' - P'')$ is the osmotic pressure. In Eq. (1.32b), l is the number of components that can pass freely between the two subsystems.

1.2 Mathematical Properties of *U* and *S*

The additivity property of entropy and internal energy of subsystems demands that both S and U of the subsystems be first-order homogeneous functions of extensive properties which define the subsystem. From the definition of the first-order homogeneous property for U, one can write

$$U(\lambda S, \lambda V, \lambda n_1, \lambda n_2, \ldots, \lambda n_{n_c}) = \lambda U(S, V, n_1, n_2, \ldots, n_{n_c}) \tag{1.33}$$

where λ is a parameter.

Differentiating Eq. (1.33) with respect to λ,

$$\left(\frac{\partial U}{\partial \lambda S}\right)_{\lambda V, \lambda \underline{n}} \left(\frac{d\lambda S}{d\lambda}\right) + \left(\frac{\partial U}{\partial \lambda V}\right)_{\lambda S, \lambda \underline{n}} \left(\frac{d\lambda V}{d\lambda}\right) + \sum_{i=1}^{n_c} \left(\frac{\partial U}{\partial \lambda n_i}\right)_{\lambda S, \lambda V, \lambda n_j} \left(\frac{d\lambda n_i}{d\lambda}\right)$$
$$= U(S, V, n_1, \ldots, n_{n_c}) \tag{1.34}$$

or

$$\left(\frac{\partial U}{\partial \lambda S}\right) S + \left(\frac{\partial U}{\partial \lambda V}\right) V + \sum_{i=1}^{n_c} \left(\frac{\partial U}{\partial \lambda n_i}\right) n_i = U(S, V, n_1, \ldots n_{n_c}) \tag{1.35}$$

where for brevity we have dropped the parameters that are held constant.

For $\lambda = 1$, Eq. (1.35) becomes

$$\left(\frac{\partial U}{\partial S}\right) S + \left(\frac{\partial U}{\partial V}\right) V + \sum_{i=1}^{n_c} \left(\frac{\partial U}{\partial n_i}\right) n_i = U(S, V, n_1, \ldots, n_{n_c}) \tag{1.36}$$

The partial derivatives of the first, second, and the third terms on the left side are T, $-P$, and μ_i, respectively. Therefore, the expression for U is

$$\boxed{U = TS - PV + \sum_{i=1}^{n_c} \mu_i n_i} \tag{1.37}$$

Of all the parameters on the right side of Eq. (1.37), only μ_i does not have an absolute value, which, as expected, makes U also not have an absolute value. The chemical potential's relative value will be discussed later.

From Eq. (1.37), one can derive the expression for S,

$$S = \frac{U}{T} + \left(\frac{P}{T}\right)V - \sum_{i=1}^{n_c} \frac{\mu_i}{T} n_i \tag{1.38}$$

1.3 Gibbs-Duhem Equation

The intensive properties of a phase (temperature, T, pressure, P, and the chemical potentials, μ_i) are not independent of each other. The Gibbs-Duhem equation provides the relationship between these variables. Let us write Eq. (1.37) in differential form:

$$dU = TdS + SdT - PdV - VdP + \sum_{i=1}^{n_c} \mu_i dn_i + \sum_{i=1}^{n_c} n_i d\mu_i \tag{1.39}$$

Comparison of the above equation with Eq. (1.21) provides the Gibbs-Duhem equation,

$$SdT - VdP + \sum_{i=1}^{n_c} n_i d\mu_i = 0 \tag{1.40}$$

Equation (1.40) demonstrates that T, P, and μ_i of a phase are not independent and provides the relation between them. We will use this equation extensively in various chapters.

For a single-component system, the Gibbs-Duhem equation is written as

$$SdT - VdP + nd\mu = 0 \tag{1.41}$$

or

$$d\mu = -sdT + vdP \tag{1.42}$$

where s and v are molar entropy and molar volume, respectively.

At constant temperature, Eq. (1.42) becomes

$$d\mu = vdP \tag{1.43}$$

Integrating Eq. (1.43) from pressure P' to a pressure P'',

$$\mu(P'', T) = \mu(P', T) + \left(\int_{P'}^{P''} vdP\right)_T \tag{1.44}$$

Now, for the first time we observe the relation between the chemical potential and pressure and volume. This implies that the change in the chemical potential can be determined from an equation of state, as we will discuss in Chap. 3.

1.4 Other Fundamental Equations

The enthalpy, H; Helmholtz free energy, A; and the Gibbs free energy, G, are, respectively, defined as

$$H = U + PV \tag{1.45}$$

$$A = U - TS \tag{1.46}$$

$$G = A + PV = U - TS + PV = H - TS \tag{1.47}$$

Later we will see the usefulness of the above fundamental equations.

Combining Eqs. (1.45), (1.46), (1.47), and (1.37), one obtains

$$\boxed{H = TS + \sum_{i=1}^{n_c} n_i \mu_i} \tag{1.48}$$

$$\boxed{A = -PV + \sum_{i=1}^{n_c} n_i \mu_i} \tag{1.49}$$

$$\boxed{G = \sum_{i=1}^{n_c} n_i \mu_i} \tag{1.50}$$

where μ_i is given by Eq. (1.20). Note that Eqs. (1.45) to (1.50) apply to both closed and open systems. Let us write the differential form of Eq. (1.45):

$$\mathrm{d}H = \mathrm{d}U + P\mathrm{d}V + V\mathrm{d}P \tag{1.51}$$

Combining Eqs. (1.51) and (1.21) provides

$$\mathrm{d}H = T\mathrm{d}S + V\mathrm{d}P + \sum_{i=1}^{n_c} \mu_i \mathrm{d}n_i \tag{1.52}$$

from which one may infer that

$$H = H(S, P, n_1, n_2, \ldots, n_{n_c}) \tag{1.53}$$

Equation (1.53) in differential form is written as

$$\mathrm{d}H = \left(\frac{\partial H}{\partial S}\right)_{P,\underline{n}} \mathrm{d}S + \left(\frac{\partial H}{\partial P}\right)_{S,\underline{n}} \mathrm{d}P + \sum_{i=1}^{n_c} \left(\frac{\partial H}{\partial n_i}\right)_{S,P,n_j} \mathrm{d}n_i \tag{1.54}$$

Comparison of Eqs. (1.52) and (1.54) leads to

$$\mu_i = \left(\frac{\partial H}{\partial n_i}\right)_{S,P,n_j} \tag{1.55}$$

Similarly, we can write the differential forms of Eqs. (1.46) and (1.47) and combine the results with Eq. (1.21) to obtain

$$\boxed{dA = -SdT - PdV + \sum_{i=1}^{n_c} \mu_i dn_i} \tag{1.56}$$

$$\boxed{dG = -SdT + VdP + \sum_{i=1}^{n_c} \mu_i dn_i} \tag{1.57}$$

which imply that

$$A = A(T, V, n_1, n_2, \ldots, n_{n_c}) \tag{1.58}$$

$$G = G(T, P, n_1, n_2, \ldots, n_{n_c}) \tag{1.59}$$

Note that A is a function of $T, V, n_1, n_2, \ldots, n_{n_c}$ and G is a function of $T, P, n_1, \ldots, n_{n_c}$. Such dependency in terms of measurable variables makes A and G the natural choices for the processes at constant T and V, and at constant T and P, respectively. In particular, G is best suited for describing the state of a fluid because it has the most convenient independent variables—all easily measurable.

In Example 1.9 we will present another important thermodynamic function, the Grand potential, which has wide applications in statistical thermodynamics.

Writing differential forms of Eqs. (1.58) and (1.59) and comparing the results with Eqs. (1.56) and (1.57) leads to

$$\mu_i = \left(\frac{\partial U}{\partial n_i}\right)_{S,V,n_j} = \left(\frac{\partial A}{\partial n_i}\right)_{T,V,n_j} = \left(\frac{\partial G}{\partial n_i}\right)_{T,P,n_j} \tag{1.60}$$

1.5 Internal Energy Minimum Principle

The entropy maximum principle can be stated equivalently in terms of the internal energy minimum principle. The proof is straightforward as seen in Prob 1.9. In the following, we present the mathematical derivations from Callen (1986).

The entropy maximum principle can be mathematically stated as

$$\left(\frac{\partial S}{\partial \underline{\mathbf{X}}}\right)_{U,V,\underline{n}} = 0 \tag{1.61}$$

and

$$\left(\frac{\partial^2 S}{\partial \underline{\mathbf{X}}^2}\right)_{U,V,\underline{n}} < 0 \tag{1.62}$$

where the negativity of the second derivative shows that the extremum is a maximum. Note that the expression given by Eq. (1.62) is not rigorous. In the above equations, S, U, V, $\underline{n}$ are extensive variables of the composite system, and $\underline{\mathbf{X}}$ corresponds to the extensive variables of the subsystems that maximize the entropy, S, at equilibrium. Note that for a composite system composed of two subsystems primed and double-primed,

$$\underline{\mathbf{X}} = (U', V', n'_1, n'_2, \ldots, n'_{n_c}; U'', V'', n''_1, n''_2, \ldots, n''_{n_c}) \tag{1.63}$$

and

$$\underline{\boldsymbol{n}} = (n_1, n_2, \ldots, n_{n_c}) \tag{1.64}$$

Since $U = U(S, \underline{\mathbf{X}})$, then

$$\mathrm{d}U = \left(\frac{\partial U}{\partial S}\right)_{\underline{\mathbf{X}}} \mathrm{d}S + \left(\frac{\partial U}{\partial \underline{\mathbf{X}}}\right)_S \mathrm{d}\underline{\mathbf{X}} \tag{1.65}$$

According to the partial derivatives rule (to be discussed shortly),

$$\left(\frac{\partial U}{\partial \underline{\mathbf{X}}}\right)_S = -\frac{\left(\dfrac{\partial S}{\partial \underline{\mathbf{X}}}\right)_U}{\left(\dfrac{\partial S}{\partial U}\right)_{\underline{\mathbf{X}}}} \tag{1.66}$$

Since $(\partial S/\partial U)_{\underline{\mathbf{X}}} = 1/T$ and at equilibrium $(\partial S/\partial \underline{\mathbf{X}})_U = 0$ [see Eq. (1.61)],

$$\left(\frac{\partial U}{\partial \underline{\mathbf{X}}}\right)_S = 0 \tag{1.67}$$

which states that U has an extremum at equilibrium (in the above and subsequent equations, V and $\underline{\boldsymbol{n}}$ are kept constant). We have to prove that the extremum is a minimum. Let $\eta = (\partial U/\partial \underline{\mathbf{X}})_S$. Since $U = U(S, \underline{\mathbf{X}})$, then

$$\eta = \eta(U, \underline{\mathbf{X}}) \tag{1.68}$$

Writing Eq. (1.68) in differential form,

$$\mathrm{d}\eta = \left(\frac{\partial \eta}{\partial U}\right)_{\underline{\mathbf{X}}} \mathrm{d}U + \left(\frac{\partial \eta}{\partial \underline{\mathbf{X}}}\right)_U \mathrm{d}\underline{\mathbf{X}} \tag{1.69}$$

and dividing it by $\mathrm{d}\underline{\mathbf{X}}$ and holding S constant,

$$\left(\frac{\partial \eta}{\partial \underline{\mathbf{X}}}\right)_S = \left(\frac{\partial \eta}{\partial U}\right)_{\underline{\mathbf{X}}} \left(\frac{\partial U}{\partial \underline{\mathbf{X}}}\right)_S + \left(\frac{\partial \eta}{\partial \underline{\mathbf{X}}}\right)_U \tag{1.70}$$

Using $\eta = (\partial U/\partial \underline{\mathbf{X}})_S$, $(\partial \eta/\partial \underline{\mathbf{X}})_U$ can be written as

$$\left(\frac{\partial \eta}{\partial \underline{\mathbf{X}}}\right)_U = \frac{\partial}{\partial \underline{\mathbf{X}}}\left[\left(\frac{\partial U}{\partial \underline{\mathbf{X}}}\right)_S\right]_U \tag{1.71}$$

Eq. (1.11) can be written as,

$$\left(\frac{\partial U}{\partial S}\right)_{\underline{\mathbf{X}}} = T \tag{1.72}$$

Then we can combine Eqs. (1.72) and (1.73) to obtain

$$\left(\frac{\partial^2 U}{\partial \underline{\mathbf{X}}^2}\right)_S = \eta \left(\frac{\partial \eta}{\partial U}\right)_{\underline{\mathbf{X}}} - T\left(\frac{\partial^2 S}{\partial \underline{\mathbf{X}}^2}\right)_U + T^2\left(\frac{\partial S}{\partial \underline{\mathbf{X}}}\right)\left(\frac{\partial^2 S}{\partial \underline{\mathbf{X}} \partial U}\right) \tag{1.73}$$

At equilibrium, both η and $(\partial S/\partial \underline{\mathbf{X}})_U$ are zero and $(\partial^2 S/\partial \underline{\mathbf{X}}^2)$ is negative [see Eq. (1.62)]. Therefore,

$$\left(\frac{\partial^2 U}{\partial \underline{\mathbf{X}}^2}\right)_S = -T\left(\frac{\partial^2 S}{\partial \underline{\mathbf{X}}^2}\right)_U \tag{1.74}$$

$$\left(\frac{\partial^2 U}{\partial \underline{\mathbf{X}}^2}\right)_S > 0 \tag{1.75}$$

Equation (1.75) together with Eq. (1.67) provide the internal energy minimum principle. In other words at equilibrium, the internal energy U must be a minimum at constant S, V, and n_i; $i = 1, \ldots, n_c$. The equilibrium criteria in terms of the H, A, and G can be shown to be the following:

1. The enthalpy, H, must be a minimum at constant S, P, and n_i; $i = 1, \ldots, n_c$.
2. The Helmholtz free energy, A, must be a minimum at constant T, V, and n_i; $i = 1, \ldots, n_c$.
3. The Gibbs free energy, G, must be a minimum at constant T, P, and n_i; $i = 1, \ldots, n_c$.

The mathematical proof for the H, A, and G minima principle is straightforward. Note that the Gibbs free energy minimum principle is very convenient to apply because pressure, temperature, and the total mol numbers of species i are held constant.

Next, we derive the relation between partial derivatives that was used in the mathematical derivation of the internal energy minimum principle.

1.5.1 Relation between Partial Derivatives of Implicit Functions

Let η be a continuous differentiable function of three variables x, y, and z:

$$\eta = \eta(x, y, z) \tag{1.76}$$

$$d\eta = \left(\frac{\partial \eta}{\partial x}\right)_{y,z} dx + \left(\frac{\partial \eta}{\partial y}\right)_{x,z} dy + \left(\frac{\partial \eta}{\partial z}\right)_{x,y} dz \tag{1.77}$$

Let us keep η and z constant and divide the remaining two terms by dx, then

$$\left(\frac{\partial \eta}{\partial x}\right)_{y,z} + \left(\frac{\partial \eta}{\partial y}\right)_{x,z}\left(\frac{\partial y}{\partial x}\right)_{\eta,z} = 0 \tag{1.78}$$

Equation (1.78) can be rearranged as

$$\boxed{\left(\frac{\partial y}{\partial x}\right)_{\eta,z} = -\frac{\left(\frac{\partial \eta}{\partial x}\right)_{y,z}}{\left(\frac{\partial \eta}{\partial y}\right)_{x,z}}} \tag{1.79}$$

The change of variables given by Eq. (1.79) is very useful in thermodynamics. We have already used it in Eqs. (1.66) and (1.72). Next we derive another relationship between the partial derivatives.

1.5.2 Reciprocity Relation

The reciprocity relation or the cross-differentiation identity is very useful in thermodynamic derivations. Let us represent the partial derivatives in Eq. (1.77) by

$$\eta_x = \left(\frac{\partial \eta}{\partial x}\right)_{y,z}, \quad \eta_y = \left(\frac{\partial \eta}{\partial y}\right)_{x,z}, \quad \eta_z = \left(\frac{\partial \eta}{\partial z}\right)_{x,y} \tag{1.80}$$

Then

$$\boxed{\frac{\partial}{\partial y}(\eta_x)_{x,z} = \frac{\partial}{\partial x}(\eta_y)_{y,z}} \tag{1.81}$$

1.6 Chemical Potential of a Component in a Mixture

We can use the above reciprocity relation to derive the expression for the evaluation of μ_i. Consider the expression for dG given by Eq. (1.57) and apply the reciprocity relation to the coefficients of dP and dn_i terms to obtain

$$\left(\frac{\partial V}{\partial n_i}\right)_{T,P,n_j} = \left(\frac{\partial \mu_i}{\partial P}\right)_{T,\underline{n}} \tag{1.82}$$

and then define the partial molar volume as

$$\bar{V}_i = \left(\frac{\partial V}{\partial n_i}\right)_{T,P,n_j} \tag{1.83}$$

Combining Eqs. (1.82) and (1.83) provides

$$\boxed{(d\mu_i = \bar{V}_i dP)_{T,\underline{n}}} \tag{1.84}$$

Equation (1.84) is an important expression in the thermodynamics of phase equilibrium. It relates the chemical potential of component i in the mixture to the measurable properties of pressure, temperature, composition, and volume. For a one-component system Eq. (1.84) simplifies to $(d\mu = v dP)_T$, which was derived earlier [see Eq. (1.43)]. The significance and usefulness of partial molar volume and other partial quantities are discussed next.

1.7 Partial Molar Quantities

Consider a single phase composed of various components at constant temperature and pressure. Let E represent any extensive property of the phase such as V, U, or G. The partial molar quantity of any component i within the phase is defined as

$$\bar{E}_i = \left(\frac{\partial E}{\partial n_i}\right)_{T,P,n_j} \tag{1.85}$$

which gives the change in the extensive property of the phase due to a small change in the amount of component i while temperature, T, pressure, P, and the amount of all other components are held constant. As an example, let us add a small amount of component, i Δn_i, at constant T and P to the container holding the phase. Then the partial molar volume, $\bar{V}_i$, of component i is given by

$$\bar{V}_i \approx \left(\frac{\Delta V}{\Delta n_i}\right)_{T,P,n_j} \tag{1.86}$$

Figure 1.3 shows the process for the determination of $\bar{V}_i$. Figure 1.3a sketches the initial system, then Δn_i mols of component i are added to the system. The result may be the sketches in Fig. 1.3b or Fig. 1.3c. In Fig. 1.3b, after adding Δn_i mols, the volume decreases by ΔV. In Fig. 1.3c, after addition of Δn_i mols, the volume increases by ΔV. Either could happen in hydrocarbon mixtures, in water-alcohol mixtures, and in some other fluid systems. For the system in Fig. 1.3b, $\bar{V}_i = -\Delta V/\Delta n_i$ and for the system shown in Fig. 1.3c, $\bar{V}_i = +\Delta V/\Delta n_i$.

Similarly the partial molar enthalpy is defined as

$$\bar{H}_i \approx \left(\frac{\Delta H}{\Delta n_i}\right)_{T,P,n_j} \tag{1.87}$$

Since V, U, S, H, A, and G are all extensive properties, then the extensive property, E, when expressed as a function of T, P, and n_i is a first-degree homogeneous function in $\underline{n}$

$$E\left(T, P, \lambda n_1, \lambda n_2, \ldots, \lambda n_{n_c}\right) = \lambda E\left(T, P, n_1, n_2, \ldots, n_{n_c}\right) \tag{1.88}$$

FIGURE 1.3 The addition of Δn_i mols to a system at constant T and P.

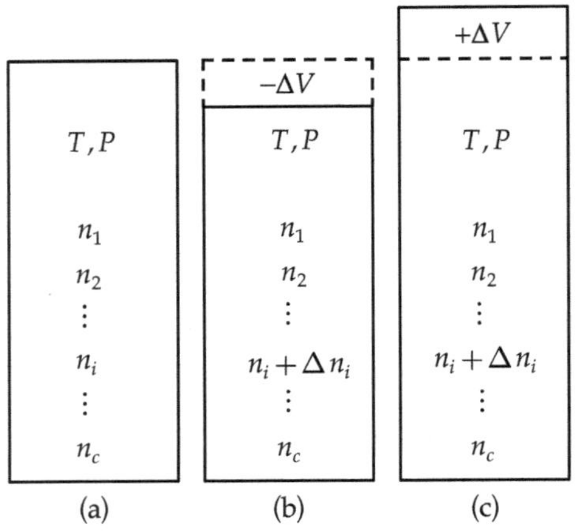

Differentiating the above equation with respect to λ at constant T and P,

$$\sum_{i=1}^{n_c}\left(\frac{\partial E}{\partial \lambda n_i}\right)_{T,P,n_j}\left(\frac{\mathrm{d}\lambda n_i}{\mathrm{d}\lambda}\right) = E\left(T, P, n_1, n_2, \ldots, n_{n_c}\right) \tag{1.89}$$

For $\lambda = 1$

$$E\left(T, P, n_1, \ldots, n_c\right) = \sum_{i=1}^{n_c} n_i \left(\frac{\partial E}{\partial n_i}\right)_{T,P,n_j} \tag{1.90}$$

Using the definition given by Eq. (1.85), then

$$\boxed{E = \sum_{i=1}^{n_c} n_i \bar{E}_i} \tag{1.91}$$

Therefore, any extensive property can be calculated from its partial molar quantities; $V = \sum_{i=1}^{n_c} n_i \bar{V}_i$, $G = \sum_{i=1}^{n_c} n_i \bar{G}_i$, etc. The partial molar Gibbs free energy, $\bar{G}_i = (\partial G/\partial n_i)_{T,P,n_j}$ has a unique feature. From Eq. (1.60) and the definition of partial molar Gibbs free energy,

$$\boxed{\mu_i = \bar{G}_i} \tag{1.92}$$

which can also be obtained from Eq. (1.50).

The partial molar quantities could be either positive or negative which is particularly important in regards to partial molar volumes. The physical meaning of negative partial molar volume of component i is that its addition at constant T and P results in a decrease in volume. Figure 1.4 shows the plot of measured partial molar volumes of C_2 and n-C_7 in

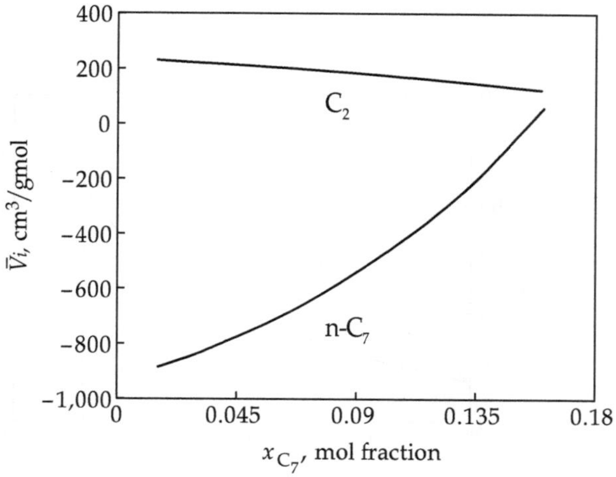

FIGURE 1.4 Measured partial molar volumes of C_2 and n-C_7 at 80°C and $P = 74.5$ atm (data from Wu and Ehrlich, 1973).

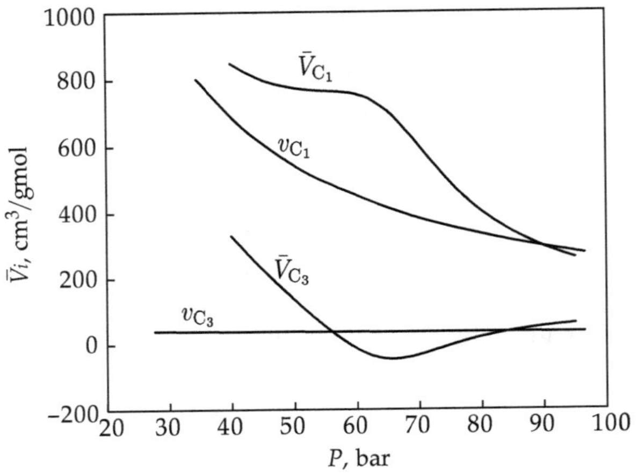

FIGURE 1.5 Partial molar and pure component molar volumes of the C_1/C_3 system at 346 K: $x_{C_1} = 0.34$ (data from Sage and Lacy, 1949).

the C_2/n-C_7 mixture from Wu and Ehrlich (1973). Note that the partial molar volume of the heavier hydrocarbon n-C_7 is negative. Negative partial molar volume implies that if a small amount of n-C_7 is added to the system at constant T and P, the volume decreases.

Figure 1.5 shows the partial molar volumes of C_1 and C_3 vs. pressure in a mixture of C_1/C_3 with $x_{C_3} = 0.34$ (mol fraction) at $T = 346$ K. The mixture in the entire range of pressure is in the gas state (Fig. 1.6). The molar volumes of pure C_1 and C_3 are also graphed in Fig. 1.5 at $T = 346$ K and different pressures. The partial molar volumes can be calculated from the Peng-Robinson equation of state (1976), which will be discussed in Chap. 3. The pure component molar volumes are from Starling (1973). The large difference between

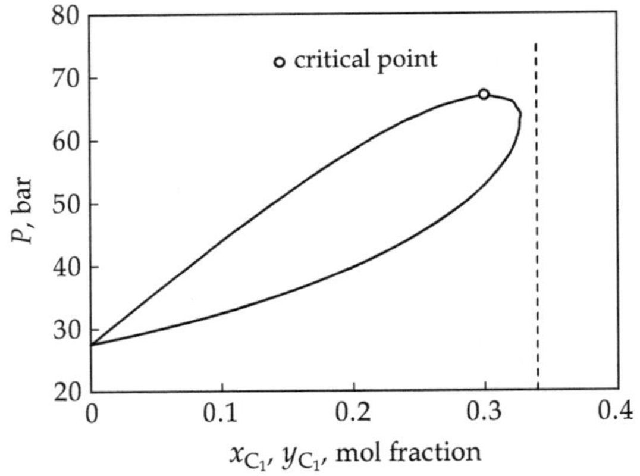

FIGURE 1.6 Phase diagram of the C_1/C_3 system at 346 K.

the partial molar volume and pure component molar volume provides strong evidence of the effect of mixing on the density.

Partial molar volumes take into account the volume change due to mixing, and there is no need to assume that the partial molar volume is equal to the molar volume of the pure component at the same temperature and pressure. In fact, the partial molar volume of some light components of reservoir fluids can be two times that of the pure component molar volume, and for some heavy components it can be negative. A significant change in volume due to mixing is a distinct feature of reservoir fluid systems. The change in volume due to mixing makes the use of equations of state very appropriate for reservoir fluids because such use readily takes into account the volume change. It is important to note that there is no relation between partial molar volume and the volume fraction of the component in the mixture. The error of assuming, partial molar volume = volume fraction, is sometimes seen in the literature.

Figure 1.7 presents partial molar volume of several alcohols in the water-alcohol mixtures at $T = 25°C$ and $P = 1$ atm. Note that up to mol fraction of about 0.40, the alcohols have negative partial molar volume. The alcohol *t*-butanol has the most pronounced negative partial molar volume. The negative partial molar volume arises from the dense packing of molecules.

The relationship between partial molar quantities is analogous to the relationships between the extensive variables. As an example,

$$\bar{G}_i = \mu_i = \bar{H}_i - T\bar{S}_i \tag{1.93}$$

This equation can be obtained by differentiating Eq. (1.47) with respect to n_i at constant temperature, pressure, and n_j. Next, we will define the fugacity, which has all the features of chemical potential, but unlike the chemical potential has an absolute value. The introduction of fugacity facilitates the phase-split calculations to be discussed in Chap. 5.

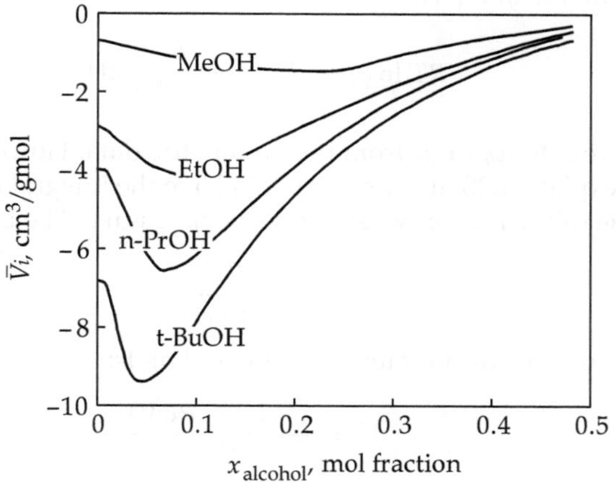

FIGURE 1.7 Partial molar volumes of alcohols in alcohol-water mixtures vs. alcohol concentration (adapted from Franks, 1983).

1.8 Fugacity

The fugacity, often represented by the symbol f, has the unit of pressure. It is defined from the following relationship:

$$(\mathrm{d}\mu_i = RT\,\mathrm{d}\ln f_i)_{T,\underline{n}} \quad i = 1, \ldots, n_c \tag{1.94}$$

Another relationship is needed to complete the definition of fugacity:

$$\lim_{P\to 0}\left(\frac{f_i}{x_i P}\right) = 1 \tag{1.95}$$

where x_i is the mol fraction and f_i is the fugacity of component i in the mixture. The ratio $(f_i/x_i P)$ is called the fugacity coefficient, φ_i:

$$\varphi_i = \frac{f_i}{x_i P} \tag{1.96}$$

At low pressures, as $P \to 0$, $\varphi_i = 1$, and it is said that the fluid has an ideal behavior.

Let us derive the expression for calculating φ_i or f_i. Subtract $RT\,\mathrm{d}\ln x_i P$ from both sides of Eq. (1.94):

$$(RT\,\mathrm{d}\ln f_i - RT\,\mathrm{d}\ln x_i P = \mathrm{d}\mu_i - RT\,\mathrm{d}\ln x_i P)_{T,\underline{n}} \quad i = 1, \ldots, n_c \tag{1.97}$$

or

$$(RT\,\mathrm{d}\ln\varphi_i = \mathrm{d}\mu_i - RT\,\mathrm{d}\ln P)_{T,\underline{n}} \quad i = 1, \ldots, n_c \tag{1.98}$$

Note that on the right side of Eq. (1.98), the term $RT\,\mathrm{d}\ln x_i$ is dropped since composition is held constant. Combining Eqs. (1.84) and (1.98), one writes

$$(RT\,\mathrm{d}\ln\varphi_i = \bar{V}_i\mathrm{d}P - RT\,\mathrm{d}\ln P)_{T,\underline{n}} \quad i = 1, \ldots, n_c \tag{1.99}$$

Drop the subscripts T and $\underline{n}$ for brevity and integrate Eq. (1.99) from 0 to pressure P and combine the results with Eq. (1.95):

$$RT\ln\varphi_i = \int_0^P \left(\bar{V}_i - \frac{RT}{P}\right)\mathrm{d}P \tag{1.100}$$

This equation provides φ_i or f_i from the volumetric data. However, since we will later use a pressure-explicit EOS, it is preferable to have the integral in terms of volume. For this purpose, the following derivations may be necessary. The EOS in the form

$$Z = \frac{PV}{nRT} \tag{1.101}$$

can be differentiated at constant temperature and mols:

$$\mathrm{d}Z = \frac{(V\mathrm{d}P + P\mathrm{d}V)}{nRT} \tag{1.102}$$

Multiplying both sides by $(nRT)/(PV)$ results in

$$\frac{\mathrm{d}Z}{Z} = \frac{\mathrm{d}P}{P} + \frac{\mathrm{d}V}{V} \tag{1.103}$$

Now consider the expression for dA given by Eq. (1.56). From the reciprocity relation,

$$-\left(\frac{\partial P}{\partial n_i}\right)_{T,V,\underline{n}} = \left(\frac{\partial \mu_i}{\partial V}\right)_{T,\underline{n}} \tag{1.104}$$

Combining Eqs. (1.98) and (1.103),

$$RT\,\mathrm{d}\ln\varphi_i = \mathrm{d}\mu_i - RT\left[\frac{\mathrm{d}Z}{Z} - \frac{\mathrm{d}V}{V}\right] \tag{1.105}$$

Dividing Eq. (1.105) by dV and once more showing the parameters that are held constant for clarity, one writes

$$RT\left(\frac{\partial \ln\varphi_i}{\partial V}\right)_{T,\underline{n}} = \left(\frac{\partial \mu_i}{\partial V}\right)_{T,\underline{n}} - \frac{RT}{Z}\left(\frac{\partial Z}{\partial V}\right)_{T,\underline{n}} + \frac{RT}{V} \tag{1.106}$$

Combining Eqs. (1.106) and (1.104) results in

$$\mathrm{d}\ln\varphi_i = \left[-\frac{1}{RT}\left(\frac{\partial P}{\partial n_i}\right)_{T,V,n_j} + \frac{1}{V}\right]\mathrm{d}V - \mathrm{d}\ln Z \tag{1.107}$$

Integration of Eq. (1.107) provides

$$\ln\varphi_i = \int_0^P \left[\frac{1}{V} - \frac{1}{RT}\left(\frac{\partial P}{\partial n_i}\right)_{T,V,n_j}\right]\mathrm{d}V - \ln Z \tag{1.108}$$

The integration limits in terms of volume are

$$\boxed{\ln\varphi_i = \int_V^\infty \left[\frac{1}{RT}\left(\frac{\partial P}{\partial n_i}\right)_{T,V,n_j} - \frac{1}{V}\right]\mathrm{d}V - \ln Z} \tag{1.109}$$

The above equation provides φ_i or f_i in terms of volumetric properties to be discussed in Chap. 3.

Now let us examine the usefulness of Eq. (1.109) for vapor-liquid equilibria calculations. First, we will establish that at equilibrium, instead of the equality of chemical potentials, one may equally write

$$f_i'(T, P, \boldsymbol{x}') = f_i''(T, P, \boldsymbol{x}'') \tag{1.110}$$

where $\boldsymbol{x}' = (x_1', x_2', \ldots, x_{n_c-1}')$ and $\boldsymbol{x}'' = (x_1'', x_2'', \ldots, x_{n_c-1}'')$. From Eq. (1.94) (see Prob. 1.33),

$$\mu_i'(T, P, \boldsymbol{x}') = \mu_i^0(T, P) + RT\ln\frac{f_i'(T, P, \boldsymbol{x}')}{f_i^0(T, P)} \tag{1.111}$$

and

$$\mu_i''(T, P, \boldsymbol{x}'') = \mu_i^0(T, P) + RT\ln\frac{f_i''(T, P, \boldsymbol{x}'')}{f_i^0(T, P)} \tag{1.112}$$

At equilibrium, $\mu_i'(T, P, x') = \mu_i''(T, P, x'')$; then this equation together with Eqs. (1.111) and (1.112), provide Eq. (1.110). From the definition of the fugacity coefficient according to Eq. (1.96) and using Eq. (1.110), one can derive

$$\varphi_i' x_i' = \varphi_i'' x_i'' \quad i = 1, \ldots, n_c \tag{1.113}$$

Now if we represent the gas phase by prime and the liquid phase by double-primes, and the composition of the gas and liquid phases by y_i and x_i, respectively,

$$K_i = \frac{y_i}{x_i} = \frac{\varphi_i^L}{\varphi_i^V} \tag{1.114}$$

Equation (1.114) relates the vapor-liquid equilibrium ratio, K_i, to the ratio of fugacity coefficients. The fugacity coefficients can be obtained from the volumetric properties given by an EOS. However, as Eq. (1.109) demands, the volumetric data are required from zero pressure to pressure P of the system at constant temperature and composition. Therefore, the EOS should represent the volumetric behavior over the whole range.

Next, we define ideal and nonideal fluids and the representation of the corresponding chemical potentials.

1.9 Ideal and Nonideal Fluids

1.9.1 Ideal Gas

An ideal gas is defined as the fluid that obeys the equation

$$PV = nRT \tag{1.115}$$

For a multicomponent system, n is given by

$$n = n_1 + n_2 + \cdots + n_{n_c} \tag{1.116}$$

Note that the partial molar volume, $\bar{V}_i$, of component i in an ideal gas mixture is simply

$$\bar{V}_i = \left(\frac{\partial V}{\partial n_i}\right)_{T,P,n_j} = \frac{RT}{P} \tag{1.117}$$

From Eq. (1.109), we notice that the fugacity coefficients of all the components of an ideal gas mixture is $\varphi_i = 1$, and, therefore, the fugacity of component i in an ideal gas mixture is equal to its partial pressure, P_i,

$$f_i = y_i P = P_i \quad i = 1, \ldots, n_c \tag{1.118}$$

Using Eqs. (1.94) and (1.118), one obtains

$$\mu_i(T, P, y) = \mu_i^0(T, P^0) + RT \ln \frac{P_i}{P^0} \tag{1.119}$$

where $\mu_i^0(T, P^0)$ is the chemical potential of pure component i at the reference state of pressure P^0 and temperature T. Note that for pure-component ideal gas at temperature T and pressure P, $f(T, P) = P$ from Eq. (1.118).

1.9.2 Ideal Solution

An ideal solution is defined by the following relation

$$\boxed{\mu_i(T, P, x) = \mu_i^0(T, P) + RT \ln x_i} \quad i = 1, \ldots, n_c \tag{1.120}$$

From the above relation one can obtain (see Prob. 1.20)

$$V(T, P, \underline{n}) = \sum_{i=1}^{n_c} n_i v_i(T, P) \tag{1.121}$$

which implies that when n_i mols of component i at temperature T, and pressure P are mixed, there is no volume change in mixing for ideal mixtures.

Many hydrocarbon mixtures at high pressure, when the volatility of components is different, are highly nonideal and one cannot assume

$$\bar{V}_i(T, P, x) = v_i(T, P) \tag{1.122}$$

Substitution of Eqs. (1.120) and (1.111) results in

$$f_i(T, P, x) = x_i f_i^0(T, P) \tag{1.123}$$

Now let us examine the heat of mixing of an ideal solution. In other words, is there any heat of mixing when several different species at constant temperature and pressure are mixed? In order to answer this question, divide Eq. (1.120) by T and take the derivative with respect to T while holding P and $\underline{n}$ constant:

$$\frac{\partial}{\partial T}\left(\frac{\mu_i}{T}\right)_{P,\underline{n}} = \frac{\partial}{\partial T}\left(\frac{\mu_i^0}{T}\right)_{P,\underline{n}} \tag{1.124}$$

Using the reciprocity relation on the coefficients of the first and third terms on the right side of Eq. (1.57) provides

$$-\left(\frac{\partial S}{\partial n_i}\right)_{T,P,n_j} = \left(\frac{\partial \mu_i}{\partial T}\right)_{P,\underline{n}} = -\bar{S}_i \tag{1.125}$$

Combining the above equation with Eq. (1.93) gives

$$\mu_i = \bar{H}_i + T\left(\frac{\partial \mu_i}{\partial T}\right)_{P,\underline{n}} \tag{1.126}$$

Equation (1.126) can be written as

$$\frac{\partial}{\partial T}\left(\frac{\mu_i}{T}\right) = -\bar{H}_i/T^2 \tag{1.127}$$

For a pure substance at temperature T and pressure P,

$$\frac{\partial}{\partial T}\left(\frac{\mu_i^0}{T}\right) = -\frac{h_i}{T^2} \tag{1.128}$$

Eqs. (1.124), (1.127), and (1.128) give

$$H(T, P, \underline{n}) = \sum_{i=1}^{n_c} n_i \bar{H}_i = \sum_{i=1}^{n_c} n_i h_i \tag{1.129}$$

which implies that there is no enthalpy of mixing at temperature T, and pressure P. Since $\Delta V_{mix} = 0$, and since the mixing process is carried out at constant pressure, then

$$\Delta U_{mix} = \Delta H_{mix} = Q = 0 \tag{1.130}$$

Therefore, for an ideal fluid, there is no heat of mixing. Due to nonideality, mixing of hydrocarbons at constant temperature and pressure may, however, require heating or cooling.

1.9.3 Nonideal Solution

In the understanding of a nonideal solution, excess functions are often defined. Excess functions are thermodynamic properties of solutions that are in excess of those of an ideal solution at the same pressure, temperature, and composition. For an ideal solution, therefore, all excess functions are zero. Let us show all thermodynamic properties by E and all excess functions by E^E; then

$$E^E = E(\text{ real solution at } T, P, \underline{n}) - E(\text{ ideal solution } T, P, \underline{n}) \tag{1.131}$$

The excess Gibbs free energy, excess enthalpy, and excess volume are defined as

$$G^E = G\ (\text{real solution at } T, P, \underline{n}) - G\ (\text{ideal solution at } T, P, \underline{n}) \tag{1.132}$$

$$H^E = H\ (\text{real solution at } T, P, \underline{n}) - H\ (\text{ideal solution at } T, P, \underline{n}) \tag{1.133}$$

$$V^E = V\ (\text{real solution at } T, P, \underline{n}) - V\ (\text{ideal solution at } T, P, \underline{n}) \tag{1.134}$$

The relations between excess functions are exactly the same as those of total thermodynamic functions,

$$G^E = H^E - TS^E \tag{1.135}$$

$$H^E = U^E + PV^E \tag{1.136}$$

Similarly, partial molar excess functions are defined in the same manner as the partial molar quantities,

$$\bar{E}_i^E = \left(\frac{\partial E^E}{\partial n_i}\right)_{T,P,n_j} \tag{1.137}$$

Since E^E is a first-degree homogeneous function in mol numbers, one can readily show that,

$$E^E = \sum_{i=1}^{n_c} n_i \bar{E}_i^E \tag{1.138}$$

1.10 Activity Coefficient

In the expression for the chemical potential of an ideal solution given by Eq. (1.120), a modification may be necessary to describe nonideal solutions, or real solutions. The modified expression is

$$\mu_i(T, P, x) = \mu_i^0(T, P) + RT \ln \gamma_i(T, P, x)x_i \tag{1.139}$$

where γ_i is called the activity coefficient which is generally a function of temperature, T, composition, x, and pressure, P. The magnitude of γ_i depends on the magnitude of $\mu_i^0(T, P)$, which is also unknown. To complete the definition of γ_i, we define the condition under which γ_i becomes equal to unity. There are two conventions:

$$\text{(I)} \qquad \mu_i(T, P, x) = \mu_i^0(T, P) + RT \ln \gamma_i(T, P, x)x_i \tag{1.140}$$

$$\gamma_i \to 1 \quad \text{as} \quad x_i \to 1 \qquad i = 1, \ldots, n_c$$

which means all of the components of a real solution approach ideal behavior as $x_i \to 1$. This implies that all the components in the pure state are in the same state as the mixture (i.e., if the mixture is liquid, components in the pure state at temperature, T, and pressure, P, are also in the liquid state). The activity coefficients obeying the above relationship are known as symmetrically organized.

$$\text{(II)} \qquad \mu_i(T, P, x) = \mu_i^*(T, P) + RT \ln \gamma_i^*(T, P, x)x_i \tag{1.141}$$

$$\gamma_i^* \to 1 \text{ as } x_i \to 0 \qquad i = 1, \ldots, a \quad a < n_c$$

For $i = a+1, \ldots, n_c$, Eq. (1.140) applies, and the components are identified as solvents. The components belonging to the indexing of $i = 1, \ldots, a$ are known as solutes. When the solutes are not in the same state as the solution, the second convention becomes useful. This is often the case for liquid solutions of noncondensable gases. Activity coefficients defined by convention II are unsymmetrical. The relationship between γ_i and γ_i^* is the subject of Prob. 1.34. Note that as $x_i \to 0$ the corresponding γ_i which is shown by γ_i^∞ has a defined value; as $x_i \to 0$, the corresponding γ_i^* which is shown by $\gamma_i^{*,\infty} \to 1$, as defined above.

1.11 Relation between γ_i and G^E

Let us differentiate Eq. (1.132) with respect to n_i while holding T, P, and $\underline{n}_i$ constant. The result is,

$$\bar{G}_i^E = \mu_i(\text{real solution at } T, P, x) - \mu_i(\text{ideal solution at } T, P, x) \tag{1.142}$$

From $(d\mu_i = RT d \ln f_i)_{T,\underline{n}}$, and Eqs. (1.120) and (1.142), one may derive

$$\bar{G}_i^E = RT \ln \frac{f_i(T, P, x)}{x_i f_i^0(T, P)} \tag{1.143}$$

By comparing Eq. (1.111) (without the prime superscript) and Eq. (1.139),

$$\boxed{f_i(T, P, x) = \gamma_i(T, P, x) x_i f_i^0(T, P)} \tag{1.144}$$

Combining Eqs. (1.143) and (1.144) results in

$$\bar{G}_i^E = RT \ln \gamma_i \tag{1.145}$$

and from Eq. (1.138)

$$\boxed{G^E = RT \sum_{i=1}^{n_c} n_i \ln \gamma_i} \tag{1.146}$$

Equation (1.146) shows that the excess Gibbs free energy and the activity coefficients are related. An alternative form of Eq. (1.146) is given by

$$\boxed{g^E = RT \sum_{i=1}^{n_c} x_i \ln \gamma_i} \tag{1.147}$$

where g^E is the excess molar Gibbs free energy. Next, we will examine the effect of pressure and temperature on γ_i.

1.11.1 Pressure and Temperature Derivative of γ_i

Let us divide Eq. (1.139) by T and take the derivative of the resulting expression with respect to temperature while holding P and x constant.

$$\frac{\partial}{\partial T}\left(\frac{\mu_i(T, P, x)}{T}\right)_{P,x} = \frac{\partial}{\partial T}\left(\frac{\mu_i^0(T, P)}{T}\right)_P + R\frac{\partial}{\partial T}(\ln \gamma_i)_{P,x} \tag{1.148}$$

Combining the above equation with Eqs. (1.127), and (1.128),

$$\left(\frac{\partial \ln \gamma_i}{\partial T}\right)_{P,x} = -\frac{\bar{H}_i - h_i}{RT^2} = -\frac{\bar{H}_i^E}{RT^2} \tag{1.149}$$

The above equation provides the effect of temperature on the activity coefficients. The effect of pressure on γ_i can be obtained by taking the derivative of Eq. (1.139) with respect to pressure at constant T and x:

$$\bar{V}_i = v_i + RT\left(\frac{\partial \ln \gamma_i}{\partial P}\right)_{T,x} \tag{1.150}$$

Rearrangement of Eq. (1.150) results in

$$\left(\frac{\partial \ln \gamma_i}{\partial P}\right)_{T,x} = \frac{\bar{V}_i - v_i}{RT} = \frac{\bar{V}_i^E}{RT} \tag{1.151}$$

Figure 1.5 shows $\bar{V}_i$ and v_i for the C_1/C_3 vs. pressure at 346 K. This figure implies that the effect of the pressure on γ_i may be very pronounced.

Next, we will briefly discuss the activity coefficient models.

1.12 Activity Coefficient Models

Prausnitz, Lichtenthaler, and de Azevedo (1998) provide a comprehensive presentation of activity coefficient models and theories of solutions in Chaps. 6 and 7 of their book. Here, we briefly review the basic concepts beyond various activity coefficient models.

1.12.1 Margules Activity Coefficient Equations

The expressions for the activity coefficients for a binary system using this model are

$$\ln \gamma_1 = \frac{A}{RT} x_2^2 \tag{1.152}$$

$$\ln \gamma_2 = \frac{A}{RT} x_1^2 \tag{1.153}$$

where A is a function of temperature. Consequently, $\bar{V}_1^E = 0$ and $\bar{V}_2^E = 0$, and the volume change on mixing is zero, that is, $V^E = 0$.

1.12.2 Van Laar Activity Coefficient Equations

The van Laar activity coefficient model is based on the assumption that $V^E = 0$ and $S^E = 0$. Therefore, $G^E = U^E$. In other words, the model allows for heat of mixing but does not allow for volume change on mixing. Van Laar used the van der Waals EOS to calculate G^E and then Eq. (1.145) can be used to obtain the activity coefficients. The activity coefficients for a binary system are given by,

$$\ln \gamma_1 = \frac{A^1}{\left[1 + \frac{A^1}{B^1}\frac{x_1}{x_2}\right]^2} \tag{1.154}$$

and

$$\ln \gamma_2 = \frac{B^1}{\left[1 + \frac{B^1}{A^1}\frac{x_2}{x_1}\right]^2} \tag{1.155}$$

where parameters A^1 and B^1 are functions of temperature and properties of pure components. This model does not account for the effect of pressure on γ.

1.12.3 Scatchard-Hildebrand Regular-Solution Activity Coefficients

Hildebrand (1929) defined a regular solution as the mixture in which components mix with no excess entropy provided there is no volume change on mixing. Scatchard in an independent work arrived at the same conclusion. The definition of regular solutions (Hildebrand and Scott, 1950) is in line with van Laar's assumption that the excess entropy and the excess volume of mixing are negligible. Scatchard and Hildebrand used an approach different from van Laar's to calculate G^E. They defined parameter C as

$$C \equiv \frac{\Delta u^G}{v^L} \tag{1.156}$$

where Δu^G is the molar internal energy change upon isothermal vaporization of the saturated liquid to the ideal-gas state, and v^L is the saturated liquid molar volume. Note that $\Delta u^G = u^G(T, P^0) - u^L(T, P)$ where P^0 may be zero pressure. Then the solubility parameter δ, is defined as

$$\delta = C^{1/2} \tag{1.157}$$

The solubility parameter has a clear physical meaning; when the difference between the solubility parameters of two substances is small, one can dissolve the other appreciably. The difference in solubility parameters is a measure of the solubility power. As an example, the solubility parameters of asphaltenes, heptane, and toluene at 20°C are, 9.5, 7.5, 8.9 $(\text{cal/cm}^3)^{0.5}$, respectively. The solubility of asphaltenes in heptane is very low, and toluene dissolves the asphaltenes. In Chap. 3, the expression for the solubility parameter from the equation of state will be presented (see Prob. 3.10).

Scatchard and Hildebrand obtained U^E from Eq. (1.156) for binary mixtures. The final results are

$$\ln \gamma_1 = \frac{v_1 \Phi_2^2}{RT}[\delta_1 - \delta_2]^2 \tag{1.158}$$

and

$$\ln \gamma_2 = \frac{v_2 \Phi_1^2}{RT}[\delta_1 - \delta_2]^2 \tag{1.159}$$

where the subscript L is dropped from the v_1 and v_2 quantities. In Eqs. (1.158) and (1.159), Φ_1 and Φ_2 are the volume fraction of components 1 and 2, respectively. The volume fractions are defined by

$$\Phi_1 = \frac{x_1 v_1}{x_1 v_1 + x_2 v_2} \tag{1.160}$$

$$\Phi_2 = \frac{x_2 v_2}{x_1 v_1 + x_2 v_2} \tag{1.161}$$

Unlike mass and mol fractions, the meaning of which is very clear, volume fractions do not have a clear physical meaning when there is volume change on mixing.

For three- and higher-component mixtures, the equation for the activity coefficients from the regular-solution theory is

$$\ln \gamma_i = \frac{v_i}{RT}(\delta_i - \delta)^2 \tag{1.162}$$

where

$$\delta = \sum_{i=1}^{n_c} \Phi_i \delta_i \quad \text{and} \quad \Phi_i = \frac{x_i v_i}{\sum_{i=1}^{n_c} x_i v_i} \tag{1.163}$$

1.12.4 Flory-Huggins Polymer-Solution Activity Coefficients

When the molecules of one component are much larger than the molecules of the other components in the mixture (i.e., polymers), the assumption of $S^E = 0$ is not appropriate. For such systems it is found that H^E may be assumed to be zero since V^E is assumed to be zero (i.e., $U^E = 0$). From $G^E = H^E - TS^E$, then $G^E = -TS^E$ assuming $H^E = 0$. On this basis, activity coefficient models are proposed (Flory, 1953). The activity-coefficient model from the Flory-Huggins polymer solution theory has been used in the petroleum industry for asphaltene precipitation. Prausnitz et al. (1998) provide details of that model.

1.13 Legendre Transformation

Let η be a continuously differentiable function of variables $X_1, X_2, \ldots, X_{n_c}, X_{n_c+1}$ and X_{n_c+2}; $\eta = \eta(X_1, X_2, \ldots, X_{n_c}, X_{n_c+1}, X_{n_c+2})$. Then $d\eta$ can be written as

$$d\eta = \sum_{i=1}^{n_c+2} \left(\frac{\partial \eta}{\partial X_i}\right)_{X_i} dX_i = \sum_{i=1}^{n_c+2} C_i dX_i \tag{1.164}$$

where the coefficients $C_i = C_i(\underline{X})$. Let us define ζ as

$$\zeta = -\sum_{i=1}^{n_c+2} C_i X_i + \eta \tag{1.165}$$

The differential of the above equation results in

$$d\zeta = -\sum_{i=1}^{n_c+2} X_i dC_i \tag{1.166}$$

Therefore,

$$\zeta = \zeta(C_1, C_2, \ldots, C_{n_c+2}) = \zeta\left(\left(\frac{\partial \eta}{\partial X_1}\right)_{X_1}, \ldots, \left(\frac{\partial \eta}{\partial X_{n_c+2}}\right)_{X_{n_c+2}}\right) \tag{1.167}$$

Note that in the above transformation, the variables $X_1, X_2, \ldots, X_{n_c+2}$ are transformed into $(\partial \eta/\partial X_1)_{X_1}, (\partial \eta/\partial X_2)_{X_2}, \ldots$ and $(\partial \eta/\partial X_{n_c+2})_{X_{n_c+2}}$, respectively. This kind of transformation which is very useful in transforming some nonlinear differential equations to linear differential equations (Courant and Hilbert, 1962) is called the Legendre transformation. Geometrically, the surface in the $X_1, X_2, \ldots, X_{n_c+2}, \eta$-space, which is a point set, is transformed into tangent-plane coordinates. For example, in the case of a curve in a

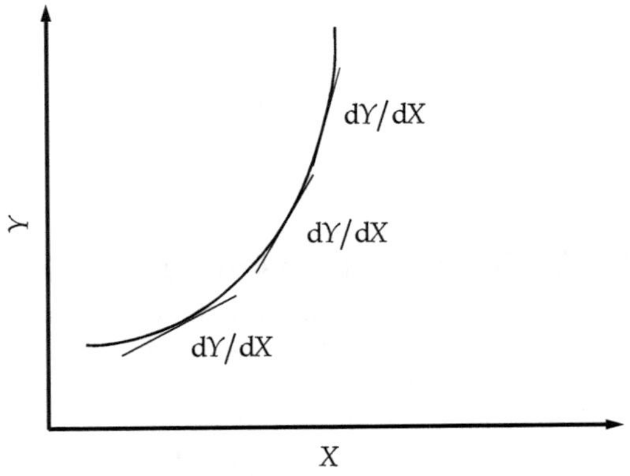

FIGURE 1.8 Depiction of Y vs. X and dY/dX.

2-D space, instead of $\eta(X)$ representation, one could have $\eta(d\eta/dX)$, where instead of X, $(d\eta/dX)$, the tangent, is the independent variable. Figure 1.8 shows $\eta(X)$ and $\eta(d\eta/dX)$ in the point and tangent spaces. In the above transformations, there is no need to transform all the independent variables into their derivative variables. One could change only one or two variables in the group of $(\partial C_{k-1}/\partial X_k)_{C_1,C_2,\ldots,C_{k-3},C_{k-2},X_{k-1},X_{k+1},\ldots,X_{n_c+2}} = y^{(k-2)}_{k-1,k}$ variables. In this case, the transformation is called the first and the second Legendre transformation with respect to the particular variables.

Let us go back to the internal energy and perform Legendre transformation on its independent variables, and write

$$U = U(X_1, X_2, \ldots, X_{n_c+2}) \tag{1.168}$$

The first Legendre transformation on the first variable is written as

$$U^{(1)} = U^{(1)}(C_1, X_2, \ldots, X_{n_c+2}) = U - C_1X_1 \tag{1.169}$$

The second Legendre transformation with respect to the first and second variables is

$$U^{(2)} = U^{(2)}(C_1, C_2, X_3, \ldots, X_{n_c+2}) = U - C_1X_1 - C_2X_2 \tag{1.170}$$

Finally, the total Legendre transformation of U is

$$U^{(n_c+2)} = U^{(n_c+2)}(C_1, C_2, \ldots, C_{n_c+2}) = U - \sum_{i=1}^{n_c+2} C_iX_i \tag{1.171}$$

The kth Legendre transformations of U and its differential are

$$U^{(k)} = U - \sum_{i=1}^{k} C_iX_i \tag{1.172}$$

$$dU^{(k)} = -\sum_{i=1}^{k} X_i dC_i + \sum_{i=k+1}^{n_c+2} C_i dX_i \tag{1.173}$$

Let us write the first and second Legendre transformations for $U = U(S, V, n_1, \ldots, n_{n_c})$.

$$U^{(1)} = U - TS = A \tag{1.174}$$

$$U^{(2)} = U - TS - (-P)V = U - TS + PV = G \tag{1.175}$$

$$dU^{(1)} = -SdT - PdV + \sum_{i=1}^{n_c} \mu_1 dn_i = dA \tag{1.176}$$

$$dU^{(2)} = -SdT + VdP + \sum_{i=1}^{n_c} \mu_i dn_i = dG \tag{1.177}$$

Therefore, A and G are the first and second partial Legendre transformation of U. The total Legendre transformation of U is

$$U^{(n_c+2)} = U - TS - (-P)V - \sum_{i=1}^{n_c} \mu_i n_i = U - TS + PV - \sum_{i=1}^{n_c} \mu_i n_i \tag{1.178}$$

$$dU^{(n_c+2)} = -SdT + VdP - \sum_{i=1}^{n_c} n_i d\mu_i = 0 \tag{1.179}$$

In Eq. (1.179) the Gibbs-Duhem equation, Eq. (1.40), is implemented.

Let us denote the kth Legendre transformation for a thermodynamic function $y^{(0)}$ by $y^{(k)}$; $y^{(k)}$ could represent the kth Legendre transformation of any of the four thermodynamic functions U, H, A, and G that we have defined. We can then write

$$y^{(0)} = y^{(0)}\left(X_1, X_2, \ldots, X_{n_c+2}\right) \tag{1.180}$$

$$y^{(k)} = y^{(k)}\left(C_1, C_2, \ldots, C_k, X_{k+1}, \ldots, X_{n_c+2}\right) \tag{1.181}$$

$$dy^{(k)} = -\sum_{i=1}^{k} X_i dC_i + \sum_{i=k+1}^{n_c+2} C_i dX_i \tag{1.182}$$

Note that Eq. (1.182) is the same as Eq. (1.173).

From Eq. (1.182), for $i \geq k+1$,

$$C_i = \left(\frac{\partial y^{(k)}}{\partial X_i}\right)_{C_1,\ldots,C_k, X_{k+1},\ldots, X_{i-1}, X_{i+1},\ldots, X_{n_c+2}} \tag{1.183}$$

Writing the expressions for $dy^{(k-1)}$, $dy^{(k-2)}$, $dy^{(k-3)}$, $dy^{(1)}$, and $dy^{(0)}$, one readily establishes

$$\begin{aligned} C_k &= \left(\frac{\partial y^{(k-1)}}{\partial X_k}\right)_{C_1,\ldots,C_{k-1}, X_{k+1},\ldots, X_{n_c+2}} = \left(\frac{\partial y^{(k-2)}}{\partial X_k}\right)_{C_1,\ldots,C_{k-2}, X_{k-1}, X_{k+1},\ldots, X_{n_c+2}} \\ &= \left(\frac{\partial y^{(1)}}{\partial X_k}\right)_{C_1, X_2,\ldots, X_{k-1}, X_{k+1},\ldots, X_{n_c+2}} = \left(\frac{\partial y^{(0)}}{\partial X_k}\right)_{X_1,\ldots, X_{k-1}, X_{k+1},\ldots, X_{n_c+2}} \end{aligned} \tag{1.184}$$

The above set of equations will be used frequently in the manipulations of Legendre transformations in Chap. 4.

In Eq. (1.183), when $i = n_c + 2$ and $k = n_c + 1$,

$$C_{n_c+2} = \left(\frac{\partial y^{(n_c+1)}}{\partial X_{n_c+2}}\right)_{C_1,\ldots,C_{n_c+1}} \tag{1.185}$$

Let us assume that $y^{(0)} = U(S, V, n_1, \ldots, n_{n_c})$. Then from the dU expression, $C_1 = T$, $C_2 = -P$, $C_3 = \mu_1, \ldots,$ and $C_{n_c+1} = \mu_{n_c-1}$. For a single phase according to the Gibbs-Duhem expression [see Eq. (1.40)], $n_c + 1$ intensive variables $T, P, \mu_1, \ldots, \mu_{n_c-1}$ define the system. Therefore C_{n_c+2} is fixed, and there could be no variation of C_{n_c+2} with respect to X_{n_c+2}; that is,

$$C_{n_c+2,n_c+2} = \left(\frac{\partial^2 y^{(c+1)}}{\partial^2 X_{n_c+2}}\right)_{C_1,\ldots,C_{n_c+1}} = 0 \tag{1.186}$$

The above relationship will be used in establishing the criteria of phase stability in Chap. 4.

Next we will discuss Jacobian transformations.

1.14 Jacobian Transformation

Jacobians, which are simply functional determinants, have long been used to simplify derivatives of functions in thermodynamics (Crawford, 1950; Carroll, 1965). Our interest in Jacobian transformation is for its simplicity and usefulness in transformation of the various derivatives of the Legendre transformation of thermodynamic functions. Without the Jacobian transformation, the interrelation between the derivatives of Legendre transformation becomes very complicated.

Let us define the general $(n_c + 2)$ dimensional Jacobian J as

$$J = \frac{\partial(Z_1, Z_2, \ldots, Z_{n_c+2})}{\partial(X_1, X_2, \ldots, X_{n_c+2})} = \begin{vmatrix} \left(\frac{\partial Z_1}{\partial X_1}\right) & \left(\frac{\partial Z_1}{\partial X_2}\right) & \cdots\cdots & \left(\frac{\partial Z_1}{\partial X_{n_c+2}}\right) \\ \vdots & \vdots & \vdots & \vdots \\ \vdots & \vdots & \vdots & \vdots \\ \left(\frac{\partial Z_{n_c+2}}{\partial X_1}\right) & \left(\frac{\partial Z_{n_c+2}}{\partial X_2}\right) & \cdots\cdots & \left(\frac{\partial Z_{n_c+2}}{\partial X_{n_c+2}}\right) \end{vmatrix} \tag{1.187}$$

where dependent variables $Z_1, \ldots, Z_{n_c+2}$ are functions of independent variables $X_1, \ldots, X_{n_c+2}$. In the above determinant, we have dropped the variables that are held constant in the derivatives. As an example, $(\partial Z_1/\partial X_1)$ should be written as $(\partial Z_1/\partial X_1)_{X_2,\ldots,X_{n_c+2}}$, where variables X_2 to X_{n_c+2} are kept constant. There are well-known rules in the manipulation of determinants. These include: (1) sign rule, (2) reduction properties, and (3) transformation properties. As an example, the sign of J will change when a neighboring pair of Z's or X's are interchanged. The reduction property is useful for the type of derivatives we have in mind. Whenever a common variable occurs between Z's and X's, a reduction takes place. As an example if $Z_1 = X_1$ and $Z_2 = X_2$, then from Eq. (1.187),

$$\frac{\partial(Z_1, Z_2, Z_3, \ldots, Z_{n_c+2})}{\partial(X_1, X_2, X_3, \ldots, X_{n_c+2})} = \frac{\partial(X_1, X_2, Z_3, \ldots, Z_{n_c+2})}{\partial(X_1, X_2, X_3, \ldots, X_{n_c+2})} = \left.\frac{\partial(Z_3, \ldots, Z_{n_c+2})}{\partial(X_3, \ldots, X_{n_c+2})}\right|_{X_1, X_2} \tag{1.188}$$

Note that in Eq. (1.188), X_1 and X_2 must remain constant. In Eq. (1.188), the order of the Jacobian has been reduced from $(n_c + 2)$ to n_c. The above process can be reversed, going from order n_c to $n_c + 2$, which is also true from Eq. (1.188). The transformation properties of Jacobians is useful in the manipulation of Legendre transformation derivatives and can be utilized by the introduction of a new set of independent variables $\underline{Y} = (Y_1, Y_2, \ldots, Y_{n_c+2})$,

$$\frac{\partial(Z_1, Z_2, \ldots, Z_{n_c+2})}{\partial(X_1, X_2, \ldots, X_{n_c+2})} = \frac{\dfrac{\partial(Z_1, Z_2, \ldots, Z_{n_c+2})}{\partial(Y_1, Y_2, \ldots, Y_{n_c+2})}}{\dfrac{\partial(X_1, X_2, \ldots, X_{n_c+2})}{\partial(Y_1, Y_2, \ldots, Y_{n_c+2})}} \tag{1.189}$$

which is simply the ratio of the two Jacobians.

Now we will use the Jacobian transformation to provide a relationship between the derivatives of higher-order Legendre transformations to the derivatives of a lower-order one. Let us denote the derivative of $y^{(k-1)}$ with respect to X_k while holding $C_1, C_2, \ldots, C_{k-1}, X_{k+1}, \ldots, X_{n_c+2}$ constant by $y_k^{(k-1)}$; that is, $y_k^{(k-1)} = (\partial y^{(k-1)}/\partial X_k)_{C_1,\ldots,C_{k-1},X_{k+1},\ldots,X_{n_c+2}}$. The second derivative with respect to X_k is shown by $y_{kk}^{(k-1)}$, which is $(\partial^2 y^{(k-1)}/\partial X_k^2)_{C_1,\ldots,C_{k-1},X_{k+1},\ldots,X_{n_c+2}}$. From Eq. (1.184),

$$y_{kk}^{(k-1)} = \left(\frac{\partial C_k}{\partial X_k}\right)_{C_1,\ldots,C_{k-1},X_{k+1},\ldots,X_{n_c+2}} \tag{1.190}$$

From Eq. (1.188), and the sign rule of determinants,

$$y_{kk}^{(k-1)} = \frac{\partial(C_1, C_2, \ldots, C_{k-1}, C_k, X_{k+1}, \ldots, X_{n_c+2})}{\partial(C_1, C_2, \ldots, C_{k-1}, X_k, X_{k+1}, \ldots, X_{n_c+2})} \tag{1.191}$$

which can be also written as

$$y_{kk}^{(k-1)} = \left.\frac{\partial(C_{k-2}, C_{k-1}, C_k)}{\partial(C_{k-2}, C_{k-1}, X_k)}\right|_{C_1,C_2,\ldots,C_{k-3},X_{k+1},\ldots,X_{n_c+2}} \tag{1.192}$$

According to Eq. (1.189),

$$y_{kk}^{(k-1)} = \left. \frac{\dfrac{\partial(C_{k-2}, C_{k-1}, C_k)}{\partial(C_{k-2}, X_{k-1}, X_k)}}{\dfrac{\partial(C_{k-2}, C_{k-1}, X_k)}{\partial(C_{k-2}, X_{k-1}, X_k)}} \right|_{C_1,C_2,\ldots,C_{k-3},X_{k+1},\ldots,X_{n_c+2}} \tag{1.193}$$

Equation (1.193) can now be written in a determinant form,

$$y_{kk}^{(k-1)} = \frac{\begin{vmatrix} \left(\dfrac{\partial C_{k-2}}{\partial C_{k-2}}\right)_{X_{k-1},X_k} & \left(\dfrac{\partial C_{k-2}}{\partial X_{k-1}}\right)_{C_{k-2},X_k} & \left(\dfrac{\partial C_{k-2}}{\partial X_k}\right)_{C_{k-2},X_{k-1}} \\ \left(\dfrac{\partial C_{k-1}}{\partial C_{k-2}}\right)_{X_{k-1},X_k} & \left(\dfrac{\partial C_{k-1}}{\partial X_{k-1}}\right)_{C_{k-2},X_k} & \left(\dfrac{\partial C_{k-1}}{\partial X_k}\right)_{C_{k-2},X_{k-1}} \\ \left(\dfrac{\partial C_k}{\partial C_{k-2}}\right)_{X_{k-1},X_k} & \left(\dfrac{\partial C_k}{\partial X_{k-1}}\right)_{C_{k-2},X_k} & \left(\dfrac{\partial C_k}{\partial X_k}\right)_{C_{k-2},X_{k-1}} \end{vmatrix}}{\begin{vmatrix} \left(\dfrac{\partial C_{k-2}}{\partial C_{k-2}}\right)_{X_{k-1},X_k} & \left(\dfrac{\partial C_{k-2}}{\partial X_{k-1}}\right)_{C_{k-2},X_k} & \left(\dfrac{\partial C_{k-2}}{\partial X_k}\right)_{C_{k-2},X_{k-1}} \\ \left(\dfrac{\partial C_{k-1}}{\partial C_{k-2}}\right)_{X_{k-1},X_k} & \left(\dfrac{\partial C_{k-1}}{\partial X_{k-1}}\right)_{C_{k-2},X_k} & \left(\dfrac{\partial C_{k-1}}{\partial X_k}\right)_{C_{k-2},X_{k-1}} \\ \left(\dfrac{\partial X_k}{\partial C_{k-2}}\right)_{X_{k-1},X_k} & \left(\dfrac{\partial X_k}{\partial X_{k-1}}\right)_{C_{k-2},X_k} & \left(\dfrac{\partial X_k}{\partial X_k}\right)_{C_{k-2},X_{k-1}} \end{vmatrix}} \tag{1.194}$$

On the right side we have dropped the variables $C_1, C_2, \ldots, C_{k-3}, X_{k+1}, \ldots, X_{n_c+2}$ which are held constant. One could readily establish the values of various elements of the two determinants. We will first write those that are either 0 or 1.

$$(\partial C_{k-2}/\partial C_{k-2}) = 1, (\partial C_{k-2}/\partial X_{k-1})_{C_{k-2},X_k} = (\partial C_{k-2}/\partial X_k)_{C_{k-2},X_{k-1}} = 0, (\partial X_k/\partial X_k) = 1,$$

$$(\partial X_k/\partial C_{k-2})_{X_{k-1},X_k} = (\partial C_{k-2}/\partial X_{k-1})_{C_{k-2},X_k} = (\partial C_{k-2}/\partial X_k)_{C_{k-2},X_{k-1}}$$
$$= (\partial X_k/\partial X_{k-1})_{C_{k-2},X_k} = 0$$

The derivative of a function with respect to itself is 1 and the derivative of a function while the function itself is held constant is zero. With the above entries, $y_{kk}^{(k-1)}$ simplifies to

$$y_{kk}^{(k-1)} = \left. \frac{\begin{vmatrix} \left(\dfrac{\partial C_{k-1}}{\partial X_{k-1}}\right)_{C_{k-2},X_k} & \left(\dfrac{\partial C_{k-1}}{\partial X_k}\right)_{C_{k-2},X_{k-1}} \\ \left(\dfrac{\partial C_k}{\partial X_{k-1}}\right)_{C_{k-2},X_k} & \left(\dfrac{\partial C_k}{\partial X_k}\right)_{C_{k-2},X_{k-1}} \end{vmatrix}}{\left(\dfrac{\partial C_{k-1}}{\partial X_{k-1}}\right)_{C_{k-2},X_k}} \right|_{C_1,C_2,\ldots,C_{k-3},X_{k+1},\ldots,X_{n_c+2}} \tag{1.195}$$

Note that we have shown the parameters that are held constant on the right side of Eq. (1.195). Now we can use the expression for $dy^{(k-2)}$ to obtain the elements of the above equation. The reason for using $dy^{(k-2)}$ is that $C_1, \ldots, C_{k-3}$ are held constant for all four elements [see Eq. (1.184)]

$$\begin{aligned}
(\partial C_{k-1}/\partial X_{k-1})_{C_1,C_2,\ldots C_{k-3},C_{k-2}X_k,\ldots X_{n_c+2}} &= y_{k-1,k-1}^{(k-2)} \\
(\partial C_k/\partial X_{k-1})_{C_1,C_2,\ldots,C_{k-3},C_{k-2},X_k\ldots,X_{n_c+2}} &= y_{k,k-1}^{(k-2)} \\
(\partial C_k/\partial X_{k-1})_{C_1,C_2,\ldots,C_{k-3},C_{k-2},X_k,\ldots,X_{n_c+2}} &= y_{k,k-1}^{(k-2)} \\
(\partial C_k/\partial X_k)_{C_1,C_2,\ldots,C_{k-3},C_{k-2},X_{k-1},X_{k+1},\ldots,X_{n_c+2}} &= y_{k,k}^{(k-2)}
\end{aligned} \tag{1.196}$$

From the reciprocity relationship [see Eq. (1.81)], $y_{k,k-1}^{(k-2)} = y_{k-1,k}^{(k-2)}$. Substitution of the above results in Eq. (1.195) leads to

$$y_{kk}^{(k-1)} = y_{kk}^{(k-2)} - \frac{\left(y_{k,k-1}^{(k-2)}\right)^2}{y_{k-1,k-1}^{(k-2)}} \tag{1.197}$$

In Chap. 4, we will provide an alternative derivation of the above equation, and discuss the fact that $y_{kk}^{(k-1)}$ goes to zero before $y_{kk}^{(k-2)}$ goes to zero. Eq. (1.197) is a useful relationship between the derivatives of $y^{(k-1)}$ and the derivatives of $y^{(k-2)}$. For $k=3$, then

$$y_{33}^{(2)} = y_{33}^{(1)} - \frac{\left(y_{32}^{(1)}\right)^2}{y_{22}^{(1)}} \tag{1.198}$$

If $y^{(0)} = U$, then $y^{(1)} = A$ and $y^{(2)} = G$. Eq. (1.198) provides the relationship between the derivatives of Gibbs free energy and the Helmholtz free energy.

1.15 Maxwell's Relations

Maxwell's relations can be readily obtained from the Jacobian transforms (see Prob. 1.25). However, we will obtain them by applying the reciprocity relations to the differential expressions of dU, dH, dA, and dG, given by Eqs. (1.21), (1.52), (1.56), and (1.57). The full set of Maxwell's relations is:

$$\left(\frac{\partial T}{\partial V}\right)_{S,\underline{n}} = -\left(\frac{\partial P}{\partial S}\right)_{V,\underline{n}} \tag{1.199}$$

$$-\left(\frac{\partial S}{\partial P}\right)_{T,\underline{n}} = \left(\frac{\partial V}{\partial T}\right)_{P,\underline{n}} \tag{1.200}$$

$$\left(\frac{\partial T}{\partial P}\right)_{S,\underline{n}} = \left(\frac{\partial V}{\partial S}\right)_{P,\underline{n}} \tag{1.201}$$

$$\left(\frac{\partial S}{\partial V}\right)_{T,\underline{n}} = \left(\frac{\partial P}{\partial T}\right)_{V,\underline{n}} \tag{1.202}$$

Note that in Eqs. (1.200) and (1.202), on the right side the variables are only P, T, V, and $\underline{n}$, all directly measurable. We will use these relationships in Chap. 3 in the derivation of the expressions for specific heat quantities c_P and c_V.

1.16 Examples and Theory Extension

1.1: Driving Force for Diffusion Prove that a substance in a mixture at constant temperature and pressure tends to pass from the region of higher to the region of lower chemical potential. Assume the effect of gravity to be negligible.

Solution Consider a composite system (shown in Fig. 1.9) consisting of two subsystems primed and double-primed. The partition could be either rigid or moveable and permeable to component i. The temperature and the pressure are the same in both subsystems.

The change of the entropy of the composite system, dS, is given by

$$dS = dS' + dS''$$

From Eq. (1.22),

$$dS = -\sum_{i=1}^{n_c} \left(\frac{1}{T}\right) (\mu_i' - \mu_i'')dn_i'$$

If the system has not reached equilibrium, then $dS > 0$. Assume $dn_i' < 0$ then $dn_i'' > 0$ which implies that component i is flowing from the primed subsystem to the double-primed subsystem. Since $dS > 0$, and $dn_i < 0$, then $(\mu_i' - \mu_i'') > 0$; $\mu_i' > \mu_i''$. Therefore, diffusion of component i will be in the direction of $\mu_i' > \mu''$. Note that we need to account for the Gibbs-Duhem equation. The other issue is related to the summation of positive and negative combination of various terms. Only in the case of binary mixtures, the diffusion flux is always from higher to lower chemical potentials.

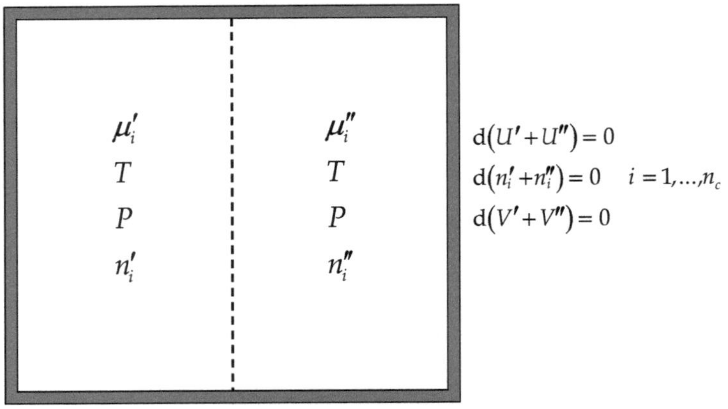

FIGURE 1.9 Isothermal System composed of two different mixtures at constant pressure.

1.2: Thermodynamic Functions in Terms of Total Number of Mols It is sometimes useful to write the expressions for thermodynamic functions in terms of the variable set $(n_1, n_2, \ldots, n_{n_c-1}, n)$ instead of $(n_1, n_2, \ldots, n_{n_c-1}, n_{n_c})$ where $n = \sum_{i=1}^{n_c} n_i$. In the new variable set,

$$dU = TdS - PdV + \sum_{i=1}^{n_c-1} (\mu_i - \mu_{n_c})dn_i + \mu_{n_c}dn$$

and

$$dG = -SdT + VdP + \sum_{i=1}^{n_c-1} (\mu_i - \mu_{n_c})dn_i + \mu_{n_c}dn$$

From the above two equations, derive the expression for du and dg.

Solution For molar quantities, we simply fix $n = 1$; therefore,

$$du = Tds - Pdv + \sum_{i=1}^{n_c-1} (\mu_i - \mu_{n_c})dx_i$$

and

$$dg = -sdT + vdP + \sum_{i=1}^{n_c-1} (\mu_i - \mu_{n_c})dx_i$$

where $x_i = n_i/n$ is the mol fraction of component i, and therefore $\sum_{i=1}^{n_c} x_i = 1$, that is, $x_{n_c} = 1 - \sum_{i=1}^{n_c-1} x_i$.

1.3: Gibbs Free Energy and Entropy of Mixing at Constant Temperature and Pressure for Ideal Gas Derive the following expressions for the entropy and Gibbs free energy of mixing at constant temperature and pressure for an ideal fluid mixture of mol fractions y_i:

$$\Delta G_{mix} = +RT \sum_{i=1}^{n_c} n_i \ln y_i$$

$$\Delta S_{mix} = -R \sum_{i=1}^{n_c} n_i \ln y_i$$

Solution The expression for the ΔG_{mix} is

$$\Delta G_{mix} = G(T, P, \underline{n}) - \sum_{i=1}^{n_c} n_i g_i(T, P) = \sum_{i=1}^{n_c} n_i \mu_i(T, P, y) - \sum_{i=1}^{n_c} n_i g_i(T, P)$$

where $g_i(T, P)$ is the molar Gibbs free energy of pure component i at temperature T and pressure P.

The chemical potential of component i in an ideal gas mixture and in an ideal pure gas can be written as

$$\mu_i(T, P, \mathbf{y}) = \mu_i^0(T, P^0) + RT \ln y_i P - RT \ln P^0$$
$$\mu_i(T, P) = \mu_i^0(T, P^0) + RT \ln P - RT \ln P^0$$

Combining the above equations,

$$\boxed{\Delta G_{mix} = +RT \sum_{i=1}^{n_c} n_i \ln y_i}$$

Note that since $\ln y_i < 0$, then ΔG_{mix} is negative. We can relate ΔG_{mix} to ΔS_{mix} through

$$G(T, P, \underline{n}) = H(T, P, \underline{n}) - TS(T, P, \underline{n})$$
$$G(T, P) = H(T, P) - TS(T, P) = \sum_{i=1}^{n_c} n_i h_i(T, P) - T \sum_{i=1}^{n_c} n_i s_i(T, P)$$

where $G(T, P)$ is the total Gibbs free energy of various components before mixing and $G(T, P, \underline{n})$ is the Gibbs free energy after mixing.

Subtracting the above two equations,

$$\Delta G_{mix} = \Delta H_{mix} - T \Delta S_{mix}$$

or

$$\Delta S_{mix} = \frac{(\Delta H_{mix} - \Delta G_{mix})}{T}$$

From Eq. (1.130), $\Delta H_{mix} = 0$; therefore,

$$\boxed{\Delta S_{mix} = -R \sum_{i=1}^{n_c} n_i \ln y_i}$$

Note that the entropy of mixing in an ideal gas mixture is positive since $\ln y_i$ is negative.

1.4: Expression for Helmholtz Free Energy of Mixtures The Helmhotz free energy of a mixture at temperature T and pressure P can be expressed as

$$A = \int_V^\infty \left(P - \frac{nRT}{V}\right) dV - RT \sum_{i=1}^{n_c} n_i \ln \frac{V}{n_i RT} + \sum_{i=1}^{n_c} n_i \left(u_i^0 - Ts_i^0\right)$$

where u_i^0 is the ideal gas molar internal energy at temperature T and s_i^0 is the molar entropy of pure component i at temperature T and pressure $P^0 = 1$ atm. Derive the above expression.

Solution The expression for dA at constant temperature and composition [see Eq. (1.56)] is given by $(\mathrm{d}A = -P\mathrm{d}V)_{T,\underline{n}}$. Let A represent the Helmhotz free energy of the mixture at T, P, and $\underline{n}$. Then

$$A - A^0 = -\int_{P^0}^{P} P\mathrm{d}V = -\int_{V^0}^{V} P\mathrm{d}V$$

The above equation can be written as

$$A - A^0 = -\int_{V^0}^{V'} P\mathrm{d}V - \int_{V'}^{V} P\mathrm{d}V$$

Note that from V^0 to V' (i.e., from P^0 to a lower pressure), the fluid is an ideal gas. Therefore, the first term on the right side can be computed for an ideal gas. Let us add and subtract $\int_{V'}^{V}(nRT/V)\mathrm{d}V$ to and from the above equation:

$$A - A^0 = -\int_{V^0}^{V'} P\mathrm{d}V - \int_{V}^{V'}\left(\frac{nRT}{V}\right)\mathrm{d}V + \int_{V}^{V'}\left(\frac{nRT}{V}\right)\mathrm{d}V + \int_{V}^{V'} P\mathrm{d}V$$

The term $\int_{V}^{V'}(nRT/V)\mathrm{d}V = \int_{V}^{V^0}(nRT/V)\mathrm{d}V + \int_{V^0}^{V'}(nRT/V)\mathrm{d}V$ and, therefore,

$$A - A^0 = nRT\ln\frac{V^0}{V} - \int_{V'}^{V}\left(P - \frac{nRT}{V}\right)\mathrm{d}V$$

$A^0 = U^0 - TS^0$; $U^0 = \sum_{i=1}^{n_c} n_i u_i^0$, and $S^0 = \Delta S_{mix}^0 + \sum_{i=1}^{n_c} n_i s_i^0$ (see Example 1.3), where u_i^0 and s_i^0 are the ideal gas molar internal energy and entropy of pure component i at T and P^0, respectively. From Example 1.3,

$$S^0 = -R\sum_{i=1}^{n_c} n_i \ln y_i + \sum_{i=1}^{n_c} n_i s_i^0$$

The next step is the derivation of an expression for $nRT\ln(V^0/V)$. At total pressure of P^0, the partial pressure of component i is P_i^0; $P_i^0 V^0 = n_i RT$.

$$RT\sum_{i=1}^{n_c} n_i \ln\frac{V}{(n_i RT)} = RT\sum_{i-1}^{n_c} n_i \ln\frac{V}{P_i^0 V^0}$$

From the above equations,

$$A = \int_{V}^{V'}\left(P - \frac{nRT}{V}\right)\mathrm{d}V - RT\sum_{i=1}^{n_c} n_i \ln\frac{V}{(n_i RT)}$$
$$+ \sum_{i=1}^{n_c} n_i\left(u_i^0 - Ts_i^0\right) + RT\sum_{i=1}^{n_c} n_i \ln y_i - RT\sum_{i=1}^{n_c} n_i \ln P_i^0$$

The last two terms on the right side are $RT\sum_{i=1}^{n_c} n_i \ln P^0$ since $P_i^0/y_i = P^0$. For $P^0 = 1$ atm, and $V' \to \infty$

$$A = \int_V^\infty \left(P - \frac{nRT}{V}\right) dV - RT\sum_{i=1}^{n_c} n_i \ln \frac{V}{(n_i RT)} + \sum_{i=1}^{n_c} n_i\left(u_i^0 - Ts_i^0\right)$$

Note that the reference pressure for the ideal gas entropy is, therefore, 1 atm.

1.5: Expression for Mixture Entropy Derive the expression for the mixture entropy, S, given by

$$S = \int_V^\infty \left[\left(\frac{nR}{V}\right) - \left(\frac{\partial P}{\partial T}\right)_{V,\underline{n}}\right] dV + R\sum_{i=1}^{n_c} n_i \ln \frac{V}{(n_i RT)} + \sum_{i=1}^{n_c} n_i s_i^0$$

Solution From Eq. (1.202), $[dS = (\partial P/\partial T)_{V,\underline{n}} dV]_{T,\underline{n}}$ and, therefore, $S(T, P, \underline{n}) - S^0(T, P^0, \underline{n}) = \int_{P^0}^{P} (\partial P/\partial T)_{V,\underline{n}} dV = \int_{V^0}^{V} (\partial P/\partial T)_{V,\underline{n}} dV$. The integral on the right side can be written as $S - S^0 = \int_{V^0}^{V'} (\partial P/\partial T)_{V,\underline{n}} dV + \int_{V'}^{V} (\partial P/\partial T)_{V,\underline{n}} dV$. Adding and subtracting $\int_V^{V'} (nR/V)dV$ to and from the right side, $S - S^0 = \int_V^{V'} [(nR/V) - (\partial P/\partial T)_{V,\underline{n}}]dV - \int_V^{V'} (nR/V)dV + \int_V^{V'} (\partial P/\partial T)_{V,\underline{n}} dV$, but $\int_V^{V'} (nR/V)dV = \int_V^{V^0} (nR/V)dV + \int_{V^0}^{V'} (nR/V)dV$; therefore, $S - S^0 = \int_V^{V'} [(nR/V) - (\partial P/\partial T)_{V,\underline{n}}]dV + \int_{V^0}^{V'} [(\partial P/\partial T)_{V,\underline{n}} - (nR/V)]dV - \int_V^{V^0} (nR/V)dV$.

In the above equation, the second term of the right side is zero, and the third term is $-nR\ln V^0/V$. The second term is zero because from volume V^0 to V', the ideal gas describes the fluid; $PV = nRT$, and $(\partial P/\partial T)_{V,\underline{n}} = nR/V(V' \to \infty)$.

$$S - S^0 = \int_V^\infty \left[\left(\frac{nR}{V}\right) - \left(\frac{\partial P}{\partial T}\right)_{V,\underline{n}}\right] dV - nR\ln\frac{V^0}{V}$$

From Example 1.4, $nR\ln V^0/V = -R\sum_{i=1}^{n_c} n_i \ln P_i^0 - R\sum_{i=1}^{n_c} n_i \ln V/(n_i RT)$ and from Example 1.3, $S^0 = \sum_{i=1}^{n_c} n_i s_i^0 - R\sum_{i=1}^{n_c} n_i \ln y_i$. Combining the above three equations and assuming $P^0 = 1$ atm, the sought expression is obtained. Note that here, similarly to the previous example, $P^0 = 1$ atm corresponds only to the ideal gas state for entropy.

1.6: Derivation of Raoult's Law Suppose a multicomponent mixture is in the gas-liquid equilibrium state. Under what assumptions can one write

$$x_i P_i^{sat}(T) = y_i\, P$$

In the above equation, x_i and y_i are the liquid and vapor phase mol fractions, respectively, and P is the pressure. $P_i^{sat}(T)$ is the vapor pressure at temperature T for pure component i.

Solution At equilibrium, one can write

$$f_i^L(T, P, \mathbf{x}) = f_i^V(T, P, \mathbf{y})$$

For an ideal solution $f_i^L(T, P, \mathbf{x}) = x_i f_{pure\,i}^L(T, P)$ and $f_i^V(T, P, \mathbf{y}) = y_i f_{pure\,i}^V(T, P)$. For an ideal gas $f_{pure\,i}^V(T, P) = P$, since $\varphi_{pure\,i}(T, P) = 1$. Now assume that the liquid phase of pure component i at T has vapor pressure $P_i^{sat}(T)$; at $P_i^{sat}(T)$, $f_{pure\,i}^L(T, P_i^{sat}) = f_{pure\,i}^V(T, P_i^{sat})$ and if the gas phase is ideal, then $f_{pure\,i}^V(T, P_i^{sat}) = P_i^{sat}(T)$. If we assume that the effect of pressure on $f_{pure\,i}^L$ is negligible, then $f_{pure\,i}^L(T, P) = P_i^{sat}(T)$. These assumptions lead to the derivation of Raoult's Law.

1.7: Gibbs Phase Rule (Flat Interface) Derive the phase rule $F = n_c + 2 - p$, where F is the number of degrees of freedom, n_c is the number of components, and p is the number of phases. Assume the interface between the phases is flat.

Solution The criteria of chemical equilibrium of multicomponent systems with n_c components and p phases are [see Eq. (1.32a)]

$$T^{(1)} = T^{(2)} = \ldots = T^{(p)}$$
$$P^{(1)} = P^{(2)} = \ldots = P^{(p)}$$
$$\mu_1^{(1)} = \mu_1^{(2)} = \ldots = \mu_1^{(p)}$$
$$\vdots$$
$$\mu_{n_c}^{(1)} = \mu_{n_c}^{(2)} = \ldots = \mu_{n_c}^{(p)}$$

There are $(p-1)$ temperature and $(p-1)$ pressure equations. The number of equations for the chemical potentials are $n_c(p-1)$. Therefore, the total number of equations is $(n_c+2)(p-1)$. For each phase the intensive variables are related by the Gibbs-Duhem equations [see Eq. (1.40)]. Therefore, the number of independent intensive variables of each phase are (n_c+1); T, P, and $\mu_1, \mu_2, \ldots, \mu_{n_c-1}$. The degrees of freedom, F, is defined as the difference between the number of independent intensive variables of the system [which is $p(n_c+1)$] and the number of equations between these variables [which is $(n_c+2)(p-1)$].

Therefore,

$$F = p(n_c+1) - (p-1)(n_c+2) = n_c + 2 - p$$

The above relationship can be used to describe the state of a particular phase of a composite system; it provides the number of independent intensive variables for specifying the system. Since chemical potential of a phase is a function of temperature, pressure, and mol fraction of components 1 to $(n_c - 1)$, then one may use $(n_c - 1)$ mol fractions instead of n_c chemical potentials for every phase.

Let us now give two examples of the phase rule. In the first example, consider a two-component system consisting of three fluid phases; therefore, $F = 2 + 2 - 3 = 1$. If we specify the temperature, the system is then defined; we are not free to specify both temperature and pressure of a three-phase two-component composite system. In the second example, we consider a two-phase two-component system; $F = 2$. If we specify the temperature and pressure, each of the phases are specified. In other words, if we change the amount of each of the two components, as long as the system is in two phase, the composition of each phase remains unchanged.

The phase rule presented above is valid if all the components are present in all phases. It should be modified when some or all components are absent in one or more phases. We will see in Chap. 2 that the above phase rule relationship should be modified when the interface between the phases is curved. A modification is also needed when there is influence of gravity on equilibrium.

1.8: Derivatives with Respect to Mol Numbers and Mol Fractions Derivatives with respect to mol numbers and mol fractions and the relation between them often result in confusion. This example is designed to avoid such confusion.

a. Consider the simple derivative $(\partial w_1/\partial x_1)$ for binary and ternary systems (w is the weight fraction, x is the mol fraction, and n is the number of mols). Derive the expressions for $(\partial w_1/\partial x_1)$ and $(\partial w_1/\partial n_1)$.
b. What is the relationship between $(\partial f_i/\partial n_j)_{T,P,n_1,\ldots,n_{j-1},n_{j+1},\ldots,n_{nc}}$, $(\partial f_i/\partial n_j)_{T,P,n_1,\ldots,n_{j-1},n_{j+1},\ldots,n_{nc-1},n}$, and $(\partial f_i/\partial x_j)_{T,P,x_1,\ldots,x_{j-1},x_{j+1},\ldots,x_{nc-1}}$? In our notation $n = \sum_{i=1}^{nc} n_i$.

Solution

a. Let us consider first the binary system.

$$w_1 = \frac{x_1 M_1}{x_1 M_1 + x_2 M_2} = \frac{x_1 M_1}{\bar{M}}$$

where M_1 and M_2 are the molecular weights of components 1 and 2, respectively and $\bar{M}$ is the average molecular weight.

$$\frac{\partial w_1}{\partial x_1} = \frac{M_1 \bar{M} - x_1 M_1 \left(\dfrac{\partial \bar{M}}{\partial x_1}\right)}{\bar{M}^2} \tag{1.203}$$

The key term is the derivative $(\partial \bar{M}/\partial x_1)$; from $\bar{M} = x_1 M_1 + x_2 M_2 = x_1 M_1 + (1 - x_1) M_2$

$$\frac{\partial \bar{M}}{\partial x_1} = M_1 - M_2$$

Note that in the above derivative x_2 cannot be held constant since $x_1 + x_2 = 1$. Combining the expression for $(\partial w_1/\partial x_1)$ and $(\partial \bar{M}/\partial x_1)$,

$$\frac{\partial w_1}{\partial x_1} = \frac{M_1 \bar{M} - x_1 M_1 (M_1 - M_2)}{\bar{M}^2} = \frac{M_1 M_2}{\bar{M}^2}$$

Now let us calculate $(\partial w_1/\partial n_1)$. Here $\partial w_1/\partial n_1|_{n_2}$ and $\partial w_1/\partial n_1|_n$ (where, $n = n_1 + n_2$) are not the same. The latter derivative is related to $(\partial w_1/\partial x_1)$, but the former is not.

From

$$n\bar{M} = n_1 M_1 + n_2 M_2$$

the derivative with respect to n_1 at constant n_2 is

$$n \left.\frac{\partial \bar{M}}{\partial n_1}\right|_{n_2} + \bar{M} = M_1$$

and, therefore,

$$\left.\frac{\partial \bar{M}}{\partial n_1}\right|_{n_2} = \frac{M_1 - \bar{M}}{n}$$

Note that since n is not constant, we cannot write $\partial \bar{M}/\partial x_1|_{n_2} = M_1 - \bar{M}$. The derivative of $n\bar{M}$ at constant n is given by

$$n\left.\frac{\partial \bar{M}}{\partial n_1}\right|_n = M_1 + M_2\left.\frac{\partial n_2}{\partial n_1}\right|_n$$

Since $n = n_1 + n_2$, then $(\partial n_2/\partial n_1)_n = -1$, and, therefore,

$$n\left.\frac{\partial \bar{M}}{\partial n_1}\right|_n = M_1 - M_2$$

Since n is constant, then $n(\partial \bar{M}/\partial n_1)|_n = \partial \bar{M}/\partial x_1$.

Now let us consider a ternary system. One can first write

$$w_1 = \frac{x_1 M_1}{\bar{M}}$$

where $\bar{M} = x_1 M_1 + x_2 M_2 + x_3 M_3$. The expression for $\partial w_1/\partial x_1$ is given by Eq. (1.203). The main task is finding the derivative $\partial \bar{M}/\partial x_1$. In a ternary system $x_1 + x_2 + x_3 = 1$; therefore, $x_3 (= 1 - x_2 - x_1)$ is not an independent variable. Hence, $\partial \bar{M}/\partial x_1|_{x_2} = M_1 - M_3$

$$\left.\frac{\partial w_1}{\partial x_1}\right|_{x_2} = \frac{M_1 \bar{M} - x_1 M_1 (M_1 - M_3)}{\bar{M}^2} = \frac{(1 - x_2) M_1 M_3 + x_2 M_1 M_2}{\bar{M}^2} \tag{1.204}$$

The calculation of $(\partial w_1/\partial n_1)_{n,n_2}$ is based on $n_3 (= n - n_1 - n_2)$ not being an independent variable. From

$$n\bar{M} = n_1 M_1 + n_2 M_2 + n_3 M_3 = n_1 M_1 + n_2 M_2 + (n - n_1 - n_2) M_3$$

$$n\left.\frac{\partial \bar{M}}{\partial n_1}\right|_{n,n_2} = M_1 - M_3$$

or

$$\left.\frac{\partial \bar{M}}{\partial x_1}\right|_{x_2} = M_1 - M_3$$

The expression for $(\partial w_1/\partial n_1)_{n_2,n_3}$ and $(\partial w_1/\partial n_1)_{n,n_2}$ are readily calculated.

b. The fugacity of component i in a mixture can be expressed by,

$$f_i = f_i\left(T, P, x_1, x_2, \ldots, x_{n_c - 1}\right)$$

The differential of f_i at constant T and P is

$$\mathrm{d}f_i = \sum_{k=1}^{n_c-1} \left.\frac{\partial f_i}{\partial x_k}\right|_{\boldsymbol{x}_k} \mathrm{d}x_k \tag{1.205}$$

where $\boldsymbol{x}_k = (x_1, x_2, \ldots, x_{k-1}, x_{k+1}, \ldots, x_{n_c-1})$. The mol fraction of component k is a function of mol numbers of all the components

$$x_k = x_k(n_1, n_2, \ldots, n_{n_c})$$

The differential of x_k is expressed by

$$\mathrm{d}x_k = \sum_{l=1}^{n_c} \left.\frac{\partial x_k}{\partial n_l}\right|_{\boldsymbol{n}_l} \mathrm{d}n_l$$

where $\boldsymbol{n}_l = (n_1, n_2, \ldots, n_{l-1}, n_{l+1}, \ldots, n_{n_c})$. Substituting $\mathrm{d}x_k$ from the above expression into the expression for $\mathrm{d}f_i$,

$$\mathrm{d}f_i = \sum_{k=1}^{n_c-1} \left.\frac{\partial f_i}{\partial x_k}\right|_{x_k} \sum_{l=1}^{n_c} \left.\frac{\partial x_k}{\partial n_l}\right|_{\boldsymbol{n}_l} \mathrm{d}n_l$$

Now divide the above equations by ∂n_j while holding $(n_1, n_2, \ldots, n_{j-1}, n_{j+1}, \ldots, n_{n_c})$ constant

$$\left(\frac{\partial f_i}{\partial n_j}\right)_{\boldsymbol{n}_j} = \sum_{k=1}^{n_c-1} \left.\frac{\partial f_i}{\partial x_k}\right|_{x_k} \left(\frac{\partial x_k}{\partial n_j}\right)_{\boldsymbol{n}_j} \tag{1.206}$$

The derivative $(\partial x_k/\partial n_j)_{n_j}$ can be readily calculated from

$$x_k = \frac{n_k}{n_1 + n_2 + \cdots + n_{n_c}} \qquad k = 1, \ldots, n_{n_c}$$

$$\left.\frac{\partial x_k}{\partial n_j}\right|_{\boldsymbol{n}_j} = -\frac{n_k}{n^2} \quad \text{for } j \neq k \tag{1.207}$$

$$\left.\frac{\partial x_k}{\partial n_j}\right|_{\boldsymbol{n}_j} = \frac{n - n_k}{n^2} \quad \text{for } j = k \tag{1.208}$$

Combining Eqs. (1.206) to (1.208) provides the relation between mol-number and mol-fraction derivatives of fugacity of component i

$$\left.\frac{\partial f_i}{\partial n_j}\right|_{\boldsymbol{n}_j} = -\sum_{k=1}^{n_c-1} \frac{x_k}{n} \left.\frac{\partial f_i}{\partial x_k}\right|_{x_k} + \left.\frac{1}{n}\frac{\partial f_i}{\partial x_j}\right|_{x_j} \tag{1.209}$$

In a similar manner, by writing $x_k = x_k(n_1, n_2, \ldots, n_{n_c-1}, n)$ in differential form, and further manipulations, we obtain,

$$\left(\frac{\partial f_i}{\partial n_j}\right)_{n_1, n_2, \ldots, n_{j-1}, n_{j+1}, \ldots, n} = \frac{1}{n}\left(\frac{\partial f_i}{\partial x_j}\right)_{x_j}$$

Note that according to Eq.(1.209), we need an array of fugacity derivatives with respect to mol fractions to calculate a single mol-number derivative.

1.9: Grand Potential Ω Use Legendre transformation to derive the expression for a thermodynamic function with variables T, V, and $\mu_1, \ldots, \mu_{n_c}$.

Solution Let us use the Helmholtz free energy A and rearrange the variables:

$$A = A(n_1, n_2, \ldots, n_{n_c}, T, V)$$

$$\mathrm{d}A = \sum_{i=1}^{n_c} \mu_i dn_i - S\mathrm{d}T - P\mathrm{d}V$$

Using the notation $X_i = n_i$, $X_{n_c+1} = T, X_{n_c+2} = V$, $C_i = \mu_i$, $C_{n_c+1} = -S$, $C_{n_c+2} = -P$

$$A^{(n_c)} = A - \sum_{i=1}^{n_c} C_i X_i = A - \sum_{i=1}^{n_c} n_i \mu_i$$

$$A^{(n_c)} = U - TS - \sum_{i=1}^{n_c} n_i \mu_i = TS - PV + \sum_{i=1}^{n_c} n_i \mu_i - TS - \sum_{i=1}^{n_c} n_i \mu_i = -PV$$

$$\mathrm{d}A^{(n_c)} = -\sum_{i=1}^{n_c} X_i \mathrm{d}C_i + \sum_{i=n_c+1}^{n_c+2} C_i \mathrm{d}X_i$$

$$\mathrm{d}A^{(n_c)} = -\sum_{i-1}^{n_c} n_i \mathrm{d}\mu_i - S\mathrm{d}T - P\mathrm{d}V$$

Therefore,

$$A^{(n_c)} = \Omega = A^{(n_c)}(\mu_1, \ldots, \mu_i, \ldots, \mu_{n_c}, T, V)$$

$$A^{(n_c)} = \Omega(\mu_1, \ldots, \mu_i, \ldots, \mu_{n_c}, T, V)$$
$$\Omega = -PV \tag{1.210}$$

The above thermodynamic function is called grand potential Ω. Unlike U, H, A, and G which are at minimum for closed systems at equilibrium, the grand potential Ω is the minimum for an open system at equilibrium. The grand potential has wide applications in statistical thermodynamics. One major application relates to phase behavior in nanopores such as those of shale gas (Li, Jin, and Firoozabadi, 2014). One important aspect of Ω is that its use may alleviate the need for pressure calculation in nanopores with adsorption.

1.10: Thermodynamics of Atmosphere The temperature of the atmosphere decreases with height in the lower layer called the troposphere (near the poles the height is 6 km, and at the equator it raises to 18 km). Find the pressure and temperature vs. height. Assume the process to be adiabatic (due to low thermal conductivity of air) and ideal gas behavior. One can readily relate P and T by $PT^{-(\gamma/(\gamma-1))} = constant$ for an ideal gas in an adiabatic process (where $\gamma = c_P/c_V$).

a. Derive the relation

$$\frac{dT}{dh} = -\frac{(\gamma-1)gM}{\gamma R}$$

where h is the height.

b. Use $\gamma = 1.40$ to find $dT/dh = -9.9°C/km$. In reality, due to condensation of H_2O, the temperature gradient is less than $-9.9°C/km$; it is around $-6.5°C/km$.

c. Show that $P = P^0\{1 - [(\gamma-1)/(\gamma RT_0)]Mgh\}^{\gamma/\gamma-1}$, where P is the pressure at height h; P^0 and T^0 are the pressure and temperature at the Earth's surface, respectively. Note that $(P/P^0) = (T/T^0)^{\gamma/\gamma-1}$.

d. Plot (P/P^0) and (T/T^0) vs. h (from 0 to 6 km). The pressure data provides the height (basis for altimeter).

Solution

a. The differential of $PT^{-(\gamma/(\gamma-1))} = constant$ is given by

$$d\left(PT^{-\left(\frac{\gamma}{(\gamma-1)}\right)}\right) = T^{-\left(\frac{\gamma}{(\gamma-1)}\right)}dP - P\frac{\gamma}{(\gamma-1)}\frac{T^{-\left(\frac{\gamma}{(\gamma-1)}\right)}}{T}dT = 0$$

Therefore,

$$\frac{dP}{P} = \frac{\gamma}{(\gamma-1)}\frac{dT}{T} \tag{1.211}$$

From $dP = -\rho g dh$ (Hydrostatic head) and $P = nRT/V = \rho RT/M$ (Ideal gas law) and Eq. (1.211)

$$\boxed{\frac{dT}{dh} = -\frac{(\gamma-1)gM}{\gamma R}} \tag{1.212}$$

b. Using $\gamma = 1.40$, $R = 8.314$ J/mol·K, $g = 9.81$ m/s^2, and $M = 0.029$ kg/mol for dry air, and substituting into Eq. (1.212):

$$\frac{dT}{dh} = -9.8°C/km$$

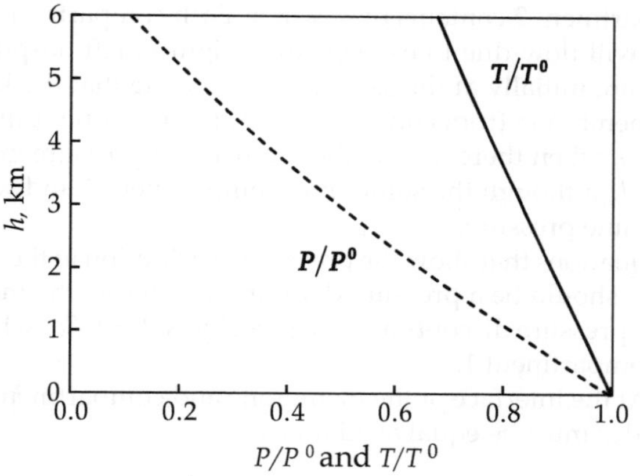

FIGURE 1.10 Variation of pressure and temperature with height.

c. Integrating Eq. (1.212):

$$\int_{T^0}^{T} \mathrm{d}T = -\int_0^h \frac{(\gamma-1)Mg}{\gamma R}\mathrm{d}h$$

$$T = T^0\left[1 - \frac{(\gamma-1)}{\gamma R T^0}Mgh\right]$$

and using $P = P^0(T/T^0)^{\gamma/\gamma-1}$,

$$P = P^0\left[1 - \frac{(\gamma-1)}{\gamma R T^0}Mgh\right]^{\frac{\gamma}{\gamma-1}}$$

d. We plot the above equations in Fig. 1.10 using $\gamma = 1.40$, $R = 8.314$ J/mol · K, $g = 9.81$ m/s^2, $M = 0.029$ kg/mol, and $T^0 = 300$ K.

1.11: Pressure, Chemical Potential, and Concentration across a Semipermeable Porous Membrane
Consider a porous membrane consisting of a large number of water-filled circular capillary tubes of length L (see Fig. 1.11). Compartment 1 contains a solution of water and

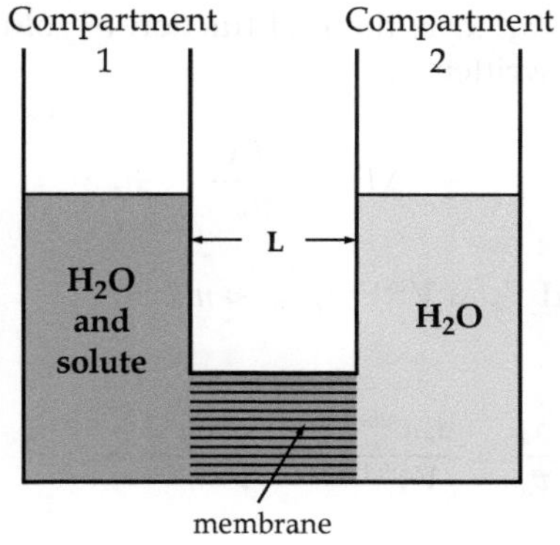

FIGURE 1.11 Semipermeable membrane at initial condition.

a solute. Compartment 2 contains pure water. Both compartments are at the same pressure P. Water will flow due to osmosis from right to left despite the fact that the two compartments are initially at the same pressure. Note that the impermeant solute cannot enter the membrane from compartment 1. Because only pure water is in the pores (capillary tubes), when there is flow there must be a pressure gradient across capillary tubes of length L, although the solution in compartment 1 and water in compartment 2 may be at the same pressure.

Derive the equations that show the pressure distribution in the membrane and demonstrate that there should be a pressure discontinuity across the membrane-compartment 1 interface. The pressure discontinuity is given by $\Delta P = RTc_s$ where c_s is the solute concentration in compartment 1.

Solution At the interface of the membrane and compartment 1, the chemical potentials on both sides must be equal at all times:

$$\mu_w^{solution}(T, P, x_w) = \mu_w^{pure}(T, P - \Delta P) \tag{1.213}$$

where ΔP is the pressure drop across the capillary tubes. It is also the pressure discounting at the interface of the membrane and compartment 1.

From Eq. (1.84), one may write

$$\mu_w^{pure}(T, P - \Delta P) = \mu_w^{pure}(T, P) + \int_P^{P-\Delta P} v_w \mathrm{d}P$$

$$\mu_w^{pure}(T, P - \Delta P) = \mu_w^{pure}(T, P) - \bar{v}_w \Delta P$$

where $\bar{v}_w$ is the average molar volume of water in the pressure range P and $P - \Delta P$.

For an ideal liquid solution:

$$\mu_w^{solution}(T, P, x_w) = \mu_w^{pure}(T, P) + RT \ln x_w$$

Combining the above expressions with the condition for equilibrium in Eq. (1.213), one obtains:

$$\bar{v}_w \Delta P = -RT \ln x_w \tag{1.214}$$

From $\ln x_w = \ln(1 - x_s) \approx -x_s$; x_s is the mol fraction of solute, assumed to be small. Therefore, Eq. (1.214) can be written as

$$\Delta P = \frac{RTx_s}{\bar{v}_w}$$

Since $x_s/\bar{v}_w = x_s/(V_w/n_w)$ and $V_w \approx V^{solution}$, $n_w \approx n^{solution}$, then

$$\frac{x_s}{\bar{v}_w} = \frac{x_s n^{solution}}{V^{solution}} = \frac{n_s}{V^{solution}} = c_s$$

Therefore,

$$\boxed{\Delta P = RTc_s}$$

where c_s is the solute concentration in compartment 1.

1.12: Inverse Melting (Freezing from Isobaric Heating of a Liquid) Consider the following unusual $P - T$ plot (Fig. 1.12). In certain regions, there is crystallization from heating a liquid at constant pressure (inverse melting). Use the Clausius-Clapeyron equation:

$$\frac{\mathrm{d}P}{\mathrm{d}T} = \frac{h^{\alpha}(T,P) - h^{\beta}(T,P)}{T\left(v^{\alpha}(T,P) - v^{\beta}(T,P)\right)}$$

to show that points B, D, and E (where $\mathrm{d}P/\mathrm{d}T = 0$) are associated with vanishing entropy change. α represents the liquid phase and β represents the crystalline phase. Do you expect liquid molar enthalpy to be lower than the crystal molar entropy in certain regions for phase transformation? Stillinger and Debenedetti (2003) discuss the inverse melting in detail.

Solution From Prob. 1.23

$$\frac{\mathrm{d}P}{\mathrm{d}T} = \frac{h^{\alpha}(T,P) - h^{\beta}(T,P)}{T\,[v^{\alpha}(T,P) - v^{\beta}(T,P)]} = \frac{s^{\alpha}(T,P) - s^{\beta}(T,P)}{v^{\alpha}(T,P) - v^{\beta}(T,P)}$$

Thus, when $\mathrm{d}P/\mathrm{d}T = 0$, $s^{\alpha}(T,P) - s^{\beta}(T,P) = 0$ and $S^{\alpha} - S^{\beta} = n(s^{\alpha} - s^{\beta}) = 0$. Therefore, points B, D, and E (where $\mathrm{d}P/\mathrm{d}T = 0$) are associated with vanishing entropy.

Let α be a liquid phase and β be the crystalline phase. Assuming that $v^{\alpha} > v^{\beta}$, which is true for most substances (water being a notable exception), the liquid molar enthalpy will be lower than the crystal molar enthalpy ($h^{\alpha} < h^{\beta}$) in regions where $\mathrm{d}P/\mathrm{d}T < 0$ (regions A-B, C-D, and E-F). Conversely, the liquid molar enthalpy will be higher than the crystal molar enthalpy ($h^{\alpha} > h^{\beta}$) in regions where $\mathrm{d}P/\mathrm{d}T > 0$ (regions B-C and D-E).

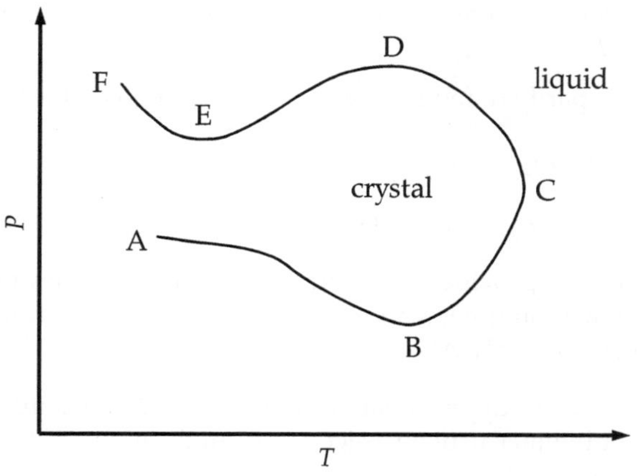

FIGURE 1.12 Inverse melting.

1.17 Problems

1.1 Derive the critieria of: (1) thermal equilibrium, (2) mechanical equilibrium, and (3) chemical equilibrium from the internal energy minimum principle. These criteria were derived from the entropy maximum principle at the beginning of this chapter.

1.2 Show that the heat added to a closed system at constant volume results in an increase of the internal energy, U, while the heat added at constant pressure results in an increase in the enthalpy.

1.3 Demonstrate that T, P, and μ_i are intensive variables. You may use a simple reasoning by using the following expression for T:

$$T = \left(\frac{\partial U}{\partial S}\right)_{V,\underline{N}}$$

1.4 Consider an isolated system composed of subsystems $'$ and $''$ similar to the sketch in Fig. 1.2. Show that the heat always flows from the subsystem at higher temperature to the subsystem at lower temperature using $\mathrm{d}S > 0$, where $\mathrm{d}S$ is the differential entropy change from time t to time $t + \mathrm{d}t$.

1.5 Consider the chemical potential of component i in a n_c-component mixture. Based on zero-order homogeneity of the chemical potential, one may write

$$\mu_j(\lambda V, T, \lambda N_1, \ldots \lambda N_{n_c}) = \mu_j(V, T, N_1, \ldots N_{n_c}) \quad j = 1, \ldots n_c$$

where λ is a constant. Show that

$$\sum_{i=1}^{n_c} \left(\frac{\partial \mu_j}{\partial N_i}\right) N_i + \left(\frac{\partial \mu_j}{\partial V}\right) V = 0 \qquad i = 1, \ldots, n_c$$

The chemical potentials in terms of the above variables are discussed by Mikyška and Firoozabadi (2012).

1.6 Suppose ϕ is partly homogeneous function of degree unity, that is,

$$\phi(\lambda x_1, \lambda x_2, \ldots .\lambda x_r, x_{r+1}, \ldots, x_{n_c}) = \lambda \phi(x_1, x_2, \ldots \ldots \ldots, x_{n_c})$$

Show that $\sum_{i=1}^{r} (\partial \phi / \partial x_i) x_i = \phi$. Relate the above expression to G and A.

1.7 The table below provides the thermodynamic properties of water for the liquid to solid phase transition (phase transition of first-order) at 1 atm pressure at the given temperature. Note that only $\Delta G = 0$ at 0°C.

- Is $\Delta U = \Delta H$ an approximation or an exact relationship? If an approximation, how would you calculate exact ΔH from the given ΔU?
- What is the significance of $\Delta G = 0$? What is your interpretation of $\Delta G > 0$ and $\Delta G < 0$?
- Would ΔU, ΔH, or ΔS at constant T and P provide information on equilibrium?

Thermodynamic property changes for the liquid to solid phase transition of water at 1 atm pressure at three different temperatures.

T (°C)	ΔU (J/mol)	ΔH (J/mol)	ΔS (J/mol K)	$-T\Delta S$ (J/mol)	ΔG (J/mol)
−10	−5619	−5619	−21	5406	−213
0	−6008	−6008	−22	6008	0
+10	−6397	−6397	−23	6623	+226

1.8 Show that (P, V, N) do not uniquely characterize the equilibrium state of a single-component system. On the other hand, (P, T, N) do characterize the equilibrium state of a single-component system. You may give an example. The symbols have their usual meanings. Figures 1.13 and 1.14 present one example.

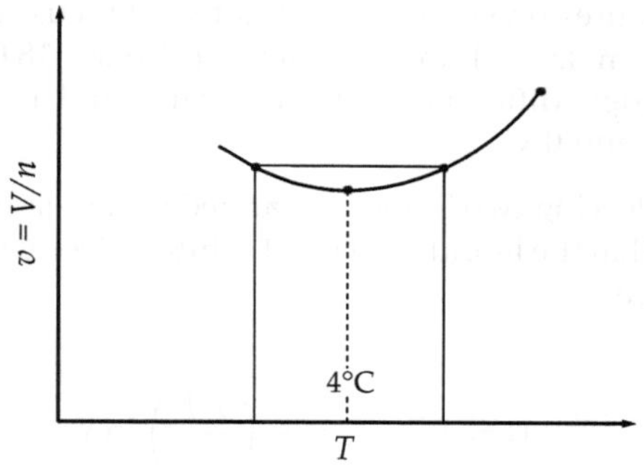

FIGURE 1.13 Molar volume of water as a function of T at $P = 1$ atm.

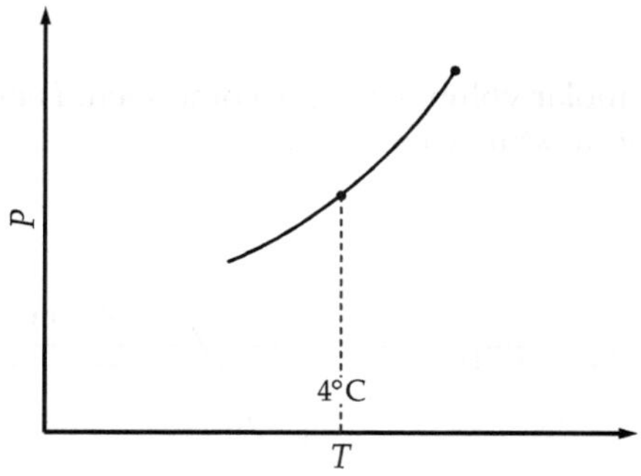

FIGURE 1.14 Vapor pressure of water as a function of T.

1.9 For a nonequilibrium process in a closed system, one can write

$$TdS > dU + PdV$$

where the only form of work is due to expansion/compression. Derive the criteria of equilibrium in terms of thermodynamic functions U, H, A, and G.

1.10 The following table shows the melting-point temperature of ice as a function of pressure. Calculate the melting-point temperature and compare your results with data.

P, bar	T, °C
907	−7.96
614	−5.10
356	−2.8
49	−0.36
−48	0.34
−115	0.86
−228	1.66

Data from Henderson and Speedy (1987)

Note that at negative pressures, the melting temperature of ice is higher than 0°C (hot ice). At very high pressures one may also get hot ice. Discuss the negative pressure for the melting of ice. The molar volume of water and ice are 18.018 and 19.652 cm^3/mol, respectively. The entropy difference between water and ice is 22.0 J/K.mol. These properties are at 1 atm and 0°C.

1.11 Consider the following two liquids separated by a membrane that is permeable to component i. Assume that the liquid is an ideal solution. Derive the following expression for the osmotic pressure

$$\Pi = P' - P = -\left(\frac{RT}{\bar{v}_i}\right) \ln x_i$$

where $\bar{v}_i$ is the average molar volume of pure component i and x_i is the mol fraction of component i in the liquid mixture (see Fig. 1.15).

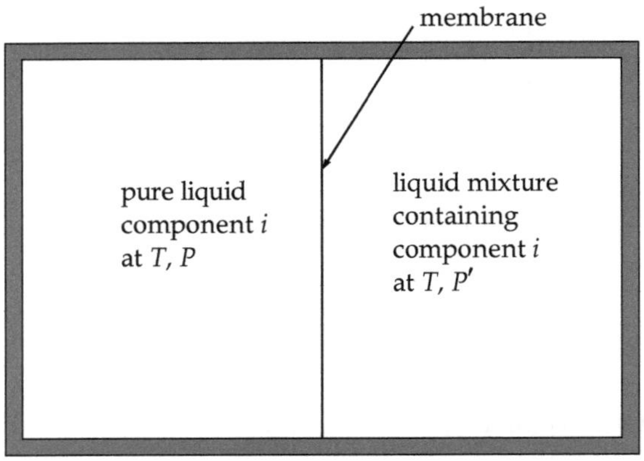

FIGURE 1.15

1.12 Consider a mixture of C_1/C_3 with a composition of $x_{C_1} = 0.34$ and $x_{C_3} = 0.66$ (mol fractions). Calculate the volume of 1 kg of this mixture at 346 K and 60 bar using: (1) the partial molar volumes of Fig. 1.5, and (2) pure component data also of Fig. 1.5 and $v_{mix} = \sum_{i=1}^{n_c} x_i v_{pure\,i}$. What is the percentage difference between the two and which is accurate?

1.13 At constant T and P, the Gibbs-Duhem equation simplifies to $\sum_{i=1}^{n_c} n_i d\mu_i = 0$. Analogous to chemical potentials μ_i are partial molar properties such as $\bar{V}_i$, $\bar{H}_i$, and $\bar{S}_i$. Show that for all partial molar quantities such as $\bar{V}_i$, $\bar{H}_i$, and $\bar{S}_i$, at constant T and P one can also write $\sum_{i=1}^{n_c} n_i d\bar{V}_i = 0$, $\sum_{i=1}^{n_c} n_i d\bar{H}_i = 0$, and $\sum_{i=1}^{n_c} n_i d\bar{S}_i = 0$.

1.14 The partial molar volumes can be calculated from the experimental data of $V(x_1)$ of binary mixtures from

$$\bar{V}_1 = v - x_2 \left(\frac{\partial v}{\partial x_2}\right)_{T,P}$$

$$\bar{V}_2 = v - x_1 \left(\frac{\partial v}{\partial x_1}\right)_{T,P}$$

where v is the molar volume of the mixture. Derive the above relationships.

1.15 Show that the partial molar enthalpy in a binary mixture can be calculated from

$$\bar{H}_1 = h - x_2 \left(\frac{\partial h}{\partial x_2}\right)_{T,P}$$

$$\bar{H}_2 = h - x_1 \left(\frac{\partial h}{\partial x_i}\right)_{T,P}$$

where h is the molar enthalpy of the mixture.

1.16 A student was preparing a cylinder of natural gas for his use in the laboratory. In the course of preparing the gas cylinder, he observed some unusual behavior. The cylinder free from liquid at 75°F, and 2005 psia, was cooled to 32°F. As a result of lower temperature, condensation took place and he removed about 500 cm^3 of liquid from the cylinder. He then brought back the temperature of the cylinder to 75°F, the initial temperature. But the pressure increased to 2060 psia, 55 psi higher than before liquid removal. How would you explain this unusual observation?

1.17 Show that in an adiabatic process for an ideal gas

$$PT^{-\frac{\gamma}{(\gamma-1)}} = constant$$

where $\gamma = c_P/c_V$; c_P and c_V are heat capacities at constant pressure and volume, respectively.

1.18 Liquefied petroleum gas (LPG) is a mixture of propane (C_3) and n-butane (n-C_4). Use Raoult's law to determine the bubble point and dew point pressures of a mixture of 50 percent C_3 and 50 percent n-C_4 (equimolar mixture) at 150°F. The vapor pressure data are $P_{C_3}^{sat}(150°F) = 350$ psia, $P_{n\text{-}C_4}^{sat}(150°F) \approx 105$ psia.

1.19 Derive the following expressions:

$$U = \int_V^\infty \left[P - T\left(\frac{\partial P}{\partial T}\right)_{V,\underline{n}}\right] dV + \sum_{i=1}^{n_c} n_i u_i^0$$

$$H = \int_V^\infty \left[P - T\left(\frac{\partial P}{\partial T}\right)_{V,\underline{n}}\right] dV + PV + \sum_{i=1}^{n_c} n_i u_i^0$$

$$\mu_i = \int_V^\infty \left[\left(\frac{\partial P}{\partial n_i}\right)_{T,V,n_j} - RT/V\right] dV - RT \ln\frac{V}{(n_i RT)} + RT + u_i^0 - Ts_i^0$$

Note: in the derivation of μ_i one can use $\mu_i - \mu_i^0 = [\partial(A - A^0)/\partial n_i]_{T,V,n_j}$.

1.20 Use the expression for ideal solution enthalpy, and the ideal solution chemical potential

$$\mu_i(T, P, \underline{n}) = \mu_i(T, P) + RT \ln x_i$$

to show that there is no volume change on mixing. The above expression provides the most valid definition of an ideal solution. Demonstrate that no volume change in mixing does not lead to the derivation of the above expression.

1.21 Find the enthalpy and entropy of mixing when 1 mol of pure benzene is mixed with 2 mols of cyclohexane at constant T and P. The activity coefficient of benzene can be expressed by

$$RT \ln \gamma_B = (a - bT)(1 - x_B)^2$$

where a and b are constant parameters and x_B is the mol fraction of benzene. What is the expression for heat released or heat absorbed?

1.22 The activity coefficient at infinite dilution for n-butane in n-C_{36} is around 0.6 at 100°C and 1 bar. What is your interpretation of a large deviation from one?

1.23 The effect of pressure on equilibrium temperature in vapor-liquid and liquid-solid phase transition of pure substances can be determined by the Clausius-Clapeyron equation. Here you are asked to derive the Clausius-Clapeyron equation.

Use

$$d\mu^L(T, P) = d\mu^V(T, P)$$

and

$$d\mu^L(T, P) = d\mu^S(T, P)$$

to derive

$$\frac{dP}{dT} = \frac{h^\alpha(T, P) - h^\beta(T, P)}{T[v^\alpha(T, P) - v^\beta(T, P)]}$$

where v is the molar volume and h is the molar enthalpy. The superscripts α, β, V (vapor), L (liquid), and S (solid) denote phase identity.

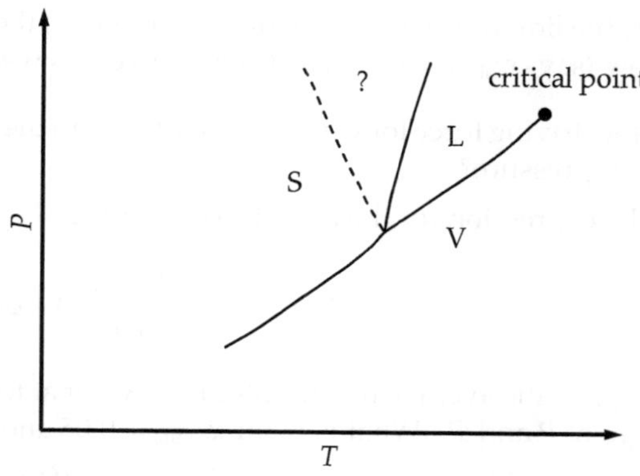

FIGURE 1.16

Can (dP/dT) be negative for equilibrium liquid-solid phase transition (see Fig. 1.16)? If it can, give an example.

1.24 A topic of general importance in biochemical engineering is osmosis; it is an equilibrium phenomenon that involves a semi-permeable membrane. The membrane is permeable to solvents (such as water) and impermeable to solutes (such as hemoglobin). The motivation for part b below is an appreciation of the fate of a red blood cell placed in pure water or in liquid solutions. Red blood cells are full of impermeant solute particles. When placed in pure water, a red cell experiences a very large osmotic pressure of about 8 atm or higher. The plasma membrane of a red cell is not strong enough to withstand such a high pressure and breaks.

Consider the following sketch (Fig. 1.17) when a container of a solvent and solute liquid mixture at pressure P and T is immersed in pure solvent at the same T and P. The wall of the container is made of rigid semi-permeable material.

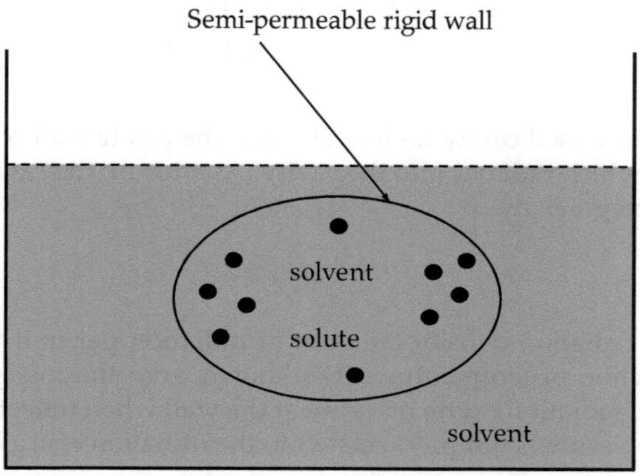

FIGURE 1.17

Assume that: (1) the liquid mixture is an ideal solution, (2) the concentration of solute in the solvent is low (say, $x_{\text{solute}} < 0.1$), and (3) the solvent is water.

(a) What is the driving force for water transfer through the membrane from a lower to a higher pressure?

(b) Derive the expression for the osmotic pressure Π,

$$\Pi = P' - P = -\frac{RT}{\bar{v}_{\text{water}}} \ln x_{\text{water}}$$

where $\bar{v}_{\text{water}}$ is the average molar volume of water at temperature T in the pressure range of P and P'. What is Π for $x_{\text{solute}} = 0.05$ and $T = 310$ K?

(c) Use the relationship $\ln(1 + x) = x - x^2/2 + x^3/3$ (for $-1 < x < 1$) to derive

$$\Pi = \frac{RT c_{\text{solute}}}{M_{\text{solute}}}$$

where c_{solute} is the solute concentration. This is the van't Hoff law of osmotic pressure for ideal dilute solutions.

(d) Derive the expression in part "b"for a nonideal liquid mixture.

1.25 Use the Jacobian transformation to obtain the Maxwell's relations (assume PV to be the only type of work).

1.26 Use the basic function

$$y^{(0)} = U = U(S, V, n_1, n_2, \ldots n_{n_c})$$

to derive,

$$y_{1,4}^{(2)} = \frac{\begin{vmatrix} U_{VV} & U_{V,n_2} \\ U_{SV} & U_{S,n_2} \end{vmatrix}}{\begin{vmatrix} U_{SS} & U_{SV} \\ U_{SV} & U_{VV} \end{vmatrix}}$$

1.27 Consider a closed cavity under vacuum. The inside wall is a perfect blackbody at temperature T which radiates into the cavity. As a result, the cavity has internal energy U from radiation given by

$$U = uV \tag{1.215}$$

where u is the radiation density (amount of radiation per unit volume of the cavity). u is only a function of temperature. It is known from thermodynamics that isotropic electromagnetic radiation exerts pressure at the wall when reflects or absorbs radiation. The radiation pressure is uniquely related to the radiation energy by

$$P = \frac{u}{3} \tag{1.216}$$

Use the above relations to derive

$$u = \alpha T^4 \tag{1.217}$$

where α is a constant. Eq. (1.217) which gives radiation density of a perfect blackbody is the Stefan–Boltzmann law. In this problem, radiation in a cavity is derived from thermodynamics.

1.28 From $\mu_1 = (\partial G/\partial n_1)_{T,P,n_2\ldots,n_{n_c}}$, use the Jacobian transformation to show that

$$\mu_1 - \mu_{n_c} = \left(\frac{\partial G}{\partial n_1}\right)_{T,P,n_2,\ldots,n_{n_c-1},n} = \left(\frac{\partial g}{\partial x_1}\right)_{T,P,x_2,\ldots,x_{n_c-1}}$$

where $n = \sum_{i=1}^{n_c} n_i$.

1.29 Show that along an isotherm,

$$\int_0^P \bar{V}_i \mathrm{d}P = \int_V^\infty \left(\frac{\partial P}{\partial n_i}\right)_{T,V,n_j} \mathrm{d}V$$

Hint: Use $\bar{V}_i = (\partial V/\partial n_i)_{T,P,n_j} = -[(\partial P/\partial n_i)_{T,V,n_j}]/[(\partial P/\partial V)_{T,P,\underline{n}}]$ to derive the above relationship.

1.30 Derive the Gibbs-Duhem equation in the following forms,

$$-S\mathrm{d}T + V\mathrm{d}P - \sum_{i=1}^{n_c-1} n_i \mathrm{d}(\mu_i - \mu_{n_c}) - n\mathrm{d}\mu_{n_c} = 0$$

$$-s\mathrm{d}T + v\mathrm{d}P - \sum_{i=1}^{n_c-1} x_i \mathrm{d}(\mu_i - \mu_{n_c}) - \mathrm{d}\mu_{n_c} = 0$$

where $n = \sum_{i=1}^{n_c} n_i$.

1.31

(a) Show that for excess properties, the Gibbs-Duhem equation takes the following form,

$$-S^E\mathrm{d}T + V^E\mathrm{d}P - \sum_{i=1}^{n_c} n_i \mathrm{d}\bar{G}_i^E = 0$$

(b) Show that at constant T and P, the activity coefficients are not independent from each other,

$$\left(\sum_{i=1}^{n_c} n_i d \ln \gamma_i = 0\right)_{T,P}$$

1.32 Derive the following expression for the activity coefficient of component i at infinite dilution γ_i^∞ (that is, γ_i as $x_i \rightarrow 0$),

$$\gamma_i^\infty = \frac{\varphi_i^\infty}{\varphi_{pure\ i}}$$

where both φ_i^∞, and $\varphi_{pure\ i}$ are at temperature T, and pressure P.

1.33 Derive the following expression

$$\mu_i(T, P, \underline{\mathbf{n}}) = \mu_i(T, P) + RT \ln \frac{f_i(T, P, \underline{\mathbf{n}})}{f_i(T, P)}$$

Hint: You may combine the following equations:

$$\mu_i(T, P, \underline{\mathbf{n}}) = \mu_i(T, P^o, \underline{\mathbf{n}}) + RT \ln \frac{f_i(T, P, \underline{\mathbf{n}})}{f(T, P^o, \underline{\mathbf{n}})}$$

$$f_i(T, P^o, \underline{\mathbf{n}}) = x_i f_i(T, P^o)$$

$$\mu_i(T, P^o, \underline{\mathbf{n}}) = \mu_i(T, P^o) + RT \ln x_i$$

$$\mu_i(T, P) = \mu_i(T, P^o) + RT \ln \frac{f_i(T, P)}{f_i(T, P^o)}$$

in your derivation. Note that P^o is chosen to be a low enough pressure at which the fluid has an ideal behavior.

1.34 Derive the relation between γ_i and γ_i^*.

1.18 References

Callen, H. B., *Thermodynamics and an Introdunction to Thermostatistics*. 2nd ed. 1985, New York: John Wiley & Sons.

Carroll, B., On the use of Jacobians in thermodynamics. *Journal of Chemical Education*, 1965. 42(4): p. 218.

Courant, R., and D. Hilbert, *Methods of Mathematical Physics, Vol II: Partial Differential Equations*. 1962, New York: John Wiley & Sons.

Crawford, F. H., Thermodynamic relations in n-variable systems in jacobian form: part I, general theory and application to unrestricted systems. *Proceedings of the American Academy of Arts and Sciences*, 1950. 78: pp. 165–184.

Denbigh, K., *The Principles of Chemical Equilibrium*. 3rd ed. 1971, London: Cambridge University Press.

Denbigh, K., *The Thermodynamics of Steady State*. 1951, New York: John Wiley & Sons.

Franks, F., *Water a Comprehensive Treatise, Vol II: Water in Crystalline Hydrates Aqueous Solutions of Simple Nonelectrolytes*. 1983, London: Springer Science and Business Media.

Flory, P. J., *Principles of Polymer Chemistry*. 1953, Ithaca, NY: Cornell University Press.

Gibbs, J. W., *Collected Works, Vol I: Thermodynamics*. 1957, New Haven, Connecticut: Yale University Press.

Henderson, S. J., and R. J. Speedy, Melting temperature of ice at positive and negative pressures. *Journal of Physical Chemistry*, 1987. 91(11): pp. 3069–3072.

Hildebrand, J. H., Solubility. XII. Regular solutions. *Journal of the American Chemical Society*, 1929. 51(1): pp. 66–80.

Hildebrand, J. H., and R. L. Scott, *The Solubility of Nonelectrolytes*. 3rd ed. 1950, New York: Reinhold Publishing Corporation.

Katz, D. L., Cornell, R., Kobayashi, R., Poettmann, F. H., Vary, J. A., Elenblass, J. R., and Weinaug, C. G., *Handbook of Natural Gas Engineering*. 1959, New York: McGraw-Hill Book Company.

Li, Z., Z., Jin, and A. Firoozabadi, Phase behavior and adsorption of pure substances and mixtures and characterization in nanopore structures by density functional theory. *SPE Journal*, 2014. p. 1096–1109.

Mikyška, J., and A. Firoozabadi, Investigation of mixture stability at given volume, temperature, and number of moles. *Fluid Phase Equilibria*, 2012. 321: pp. 1–9.

Peng, D.-Y., and D. B. Robinson, A new two-constant equation of state. *Industrial & Engineering Chemistry Fundamentals*, 1976. 15(1): pp. 59–64.

Prausnitz, J. M., R. N. Lichtenhaler, and E. SG. de Azevedo, *Molecular Thermodynamics of Fluid-Phase Equilibria*. 1998, Englewood Cliffs, NJ: Prentice-Hall Inc.

Sage, B. H., and W. N. Lacey, *Volumetric and Phase Behavior of Hydrocarbons*, Chapter VIII. 1949, Stanford, CA: Stanford University Press.

Starling, K. E., *Fluid Thermodynamic Properties for Light Petroleum Systems*. 1973, Gulf Publishing Company.

Stillinger, F. H., and P. G. Debenedetti, Phase transitions, kauzmann curves, and inverse melting. *Biophysical Chemistry*, 2003. 105(2–3): pp. 211–220.

Tisza, L., *Generalized Thermodynamics*. 1966, Cambridge, MA: M.I.T. Press.

Wu, P. C., and P. Ehrlich, Volumetric properties of supercritical ethane-n-heptane mixtures: Molar volumes and partial molar volumes. *AIChE Journal*, 1973. 19(3): pp. 533–540.

CHAPTER 2

General Theory of Bulk Phase Equilibria

The first law of thermodynamics of a closed system is expressed as $dU = đQ + đW$; $đW$ is the infinitesimal amount of work done on the system. In addition to the work of expansion (compression), which is the work against (from) external pressure, there are many other types of work. Sychev (1973) presents different forms of work from the magnetic field to the gravity field. In this chapter we will discuss three different types of work due to: (1) gravity, (2) interfacial phenomena, and (3) electric field. The effect of gravity on equilibrium segregation of mixtures, and the effect of interface curvature on equilibrium both in pure components and mixtures will be formulated. We will also derive basic equations for the effect of the electric field. Interestingly, the effect of gravity and electric fields on equilibrium and the expressions for chemical potentials have the same form. In the *Examples and Theory Extension* and *Problems* sections other types of work will be introduced. The effect of interface curvature on equilibrium will be discussed in Chap. 7 when the interface will be assigned a phase identity for a more general formulation. In this chapter, the interfacial formulation is simplified. Even with simplification, we will be able to demonstrate that the effect of curvature on saturation pressure in single- and multicomponent mixtures can be drastically different. In multicomponent mixtures, due to curvature, the dewpoint pressure can increase in the critical region. Away from the critical region, there is a decrease in dewpoint and bubblepoint pressures. For single-component systems, there is always a decrease of vapor pressure. All these changes only apply when the liquid is the wetting phase. Many of these features are directly applicable to thermodynamics of fluids in shale reservoirs.

2.1 Equilibrium Condition under the Influence of Gravity

The work term in Eq. (1.4) includes only the expansion or compression contribution; $đW = -PdV$ represents the work of expansion (compression) done by the system on the surroundings. Consider a mass m undergoing either expansion (or compression), and change in position in the vertical direction (Fig. 2.1).

In order to raise mass m to height dz (positive upward), a certain amount of work must be done; $đW = mgdz$, provided there is no change in the volume of mass m. When there

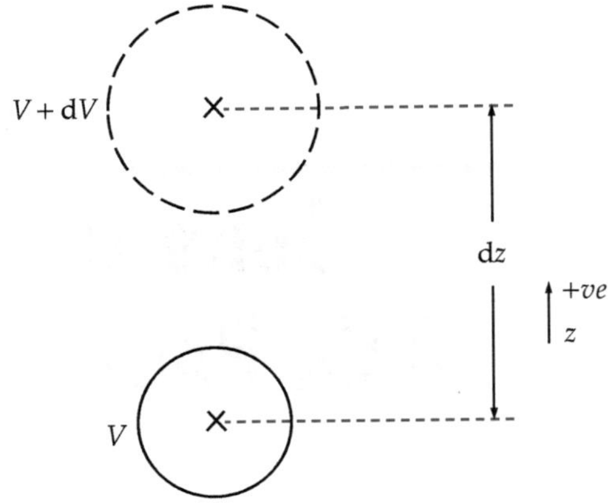

FIGURE 2.1 Schematic of work performed due to gravity in a single-component fluid.

are both changes in the volume of mass m and displacement in the vertical direction by the distance, dz, the work term is simply

$$đW = -P\mathrm{d}V + mg\mathrm{d}z \tag{2.1}$$

The expression for dU for a closed system is given by

$$\mathrm{d}U = T\mathrm{d}S - P\mathrm{d}V + mg\mathrm{d}z \tag{2.2}$$

One may derive expressions for dH, dA, and dG in a gravity field as

$$\mathrm{d}H = T\mathrm{d}S + V\mathrm{d}P + mg\mathrm{d}z \tag{2.3}$$

$$\mathrm{d}A = -S\mathrm{d}T - P\mathrm{d}V + mg\mathrm{d}z \tag{2.4}$$

$$\mathrm{d}G = -S\mathrm{d}T + V\mathrm{d}P + mg\mathrm{d}z \tag{2.5}$$

Now let us apply the Gibbs criterion of equilibrium: dG must vanish at equilibrium. As a result, the independent terms on the right side of Eq. (2.5) must vanish. Pressure (P) and vertical position (z) are not independent. Therefore, at equilibrium,

$$\mathrm{d}T = 0 \tag{2.6}$$

and

$$V\mathrm{d}P + mg\mathrm{d}z = 0 \tag{2.7}$$

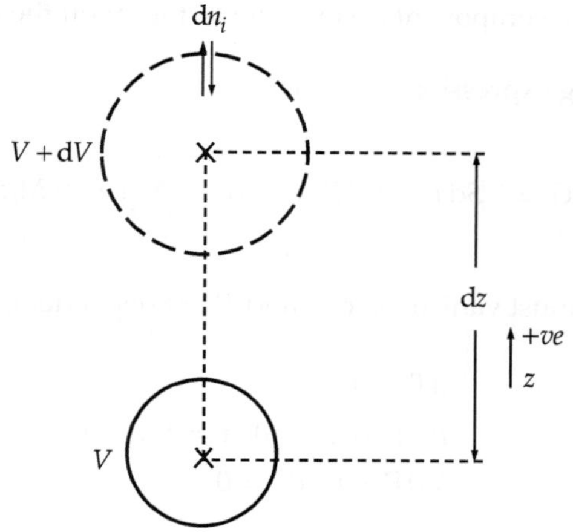

FIGURE 2.2 Schematic of work performed due to gravity in a multicomponent fluid.

Equation (2.6) states that the temperature T must be the same everywhere in the system. Since $\rho = m/V$, then Eq. (2.7) becomes

$$dP = -\rho g dz, \tag{2.8}$$

which is the expression for the hydrostatic head.

In the above derivations, we defined our system to be closed, a definition that is applicable to single-component systems. For a multicomponent fluid, when the mass m is displaced from z to $z + dz$, at the new position we expect transfer of components across the wall of volume of fixed mass m, and we should account for the chemical work discussed in Chap. 1. We will first move mass m from z to $z+dz$; because of the change in pressure, there will be a change in the volume of mass m. The corresponding work terms are similar to those of a single-component system: $-PdV$ and $mgdz$. At $z+dz$, because of transfer of various components across the system, there is the chemical work (see Chap. 1), which is given by $\mu_i dn_i$ for each component i (Fig. 2.2). The work associated with the corresponding mass transfer, dm_i, for component i is $zgdm_i$. Since $m_i = n_i M_i$, where M_i is the molecular weight of component i; $zdm_i = M_i z dn_i$. For all the components, the work when the system is brought to $z + dz$ is

$$đW' = \sum_{i=1}^{n_c} (\mu_i + M_i zg) dn_i \tag{2.9}$$

Adding Eq. (2.9) to Eq. (2.2) provides the expression for dU when mass m is displaced from z to $z + dz$ for a multicomponent fluid:

$$dU = TdS - PdV + mgdz + \sum_{i=1}^{n_c} (\mu_i + M_i zg) dn_i \tag{2.10}$$

Note that for a single-component system, the last term on the right side of Eq. (2.10) is absent [see Eq. (2.2)].

The corresponding expression for dG is

$$dG = -SdT + VdP + mgdz + \sum_{i=1}^{n_c} (\mu_i + M_i zg) dn_i \tag{2.11}$$

At equilibrium, dG must vanish. Since z and P are dependent, then

$$\begin{aligned} dT &= 0 \\ \mu_i + M_i gz &= 0 \quad i = 1, \ldots, n_c \\ VdP + mgdz &= 0 \end{aligned} \tag{2.12}$$

The first and third expressions in Eq. (2.12) provide the same results as previously for the single-component system. The second equation provides the Gibbs sedimentation expression,

$$\boxed{(d\mu_i = -M_i g dz)_T} \quad i = 1, \ldots, n_c \tag{2.13}$$

The equilibrium concept of gravity segregation leads to the expression that, in a multi-component fluid column under isothermal conditions, the chemical potential of the i-th component, μ_i, is a function of position, z, according to the above differential equation. In Chap. 6, Eq. (2.13) will be derived from the concept of entropy production. Equation (2.13) provides both composition and pressure as a function of depth, as we will see next. From $(d\mu_i = RT d \ln f_i)_{T,\underline{n}}$ and Eqs. (1.111) without primes and (2.13),

$$(RT d \ln f_i = -M_i g dz)_{T,\underline{n}} \quad i = 1, \ldots, n_c \tag{2.14}$$

Equation (2.14) can be derived by an alternative approach by minimizing the total Helmholtz free energy of the system when it has a continuous variation (Aavatsmark, 1995; Wheaton, 1991).

Integrating Eq. (2.14) from a reference depth of zero to z,

$$\boxed{f_i = f_i^o \exp\left(-\frac{M_i}{RT} gz\right)} \quad i = 1, \ldots, n_c \tag{2.15}$$

Equation (2.15) provides the fugacity of component i in a given phase as a function of position; given composition and pressure at the reference point, one can calculate both the composition and pressure at any point in the vertical direction.

$$\begin{aligned} F_i = {} & f_i(T, P, y_1, y_2, \ldots, y_{n_c-1}) - f_i^o\left(T, P^o, y_1^o, y_2^o, \ldots, y_{n_c-1}^o\right) \\ & \times \exp\left(-\frac{M_i}{RT} gz\right) = 0 \quad i = 1, \ldots, n_c \end{aligned} \tag{2.16}$$

subject to the constraint equation,

$$\sum_{i=1}^{n_c} y_i = 1 \tag{2.17}$$

The unknowns are pressure and composition at the desired depth. Newton's method can be used to solve the above system of nonlinear equations. Note that in Eqs. (2.16) and (2.17), there is no need to write the expression for hydrostatic head. One may relate the pressure and composition at any point to those at the reference point from Eqs. (2.16) and (2.17).

2.1.1 Conditions for Pronounced Compositional Variation

Let us write $d\mu_i$ from Eq. (2.13) in terms of the independent variables, P, $\mathbf{y}(y_1, \ldots, y_{n_c-1})$:

$$\left[d\mu_i = \left(\frac{\partial \mu_i}{\partial P}\right)_{\mathbf{y}} dP + \sum_{j=1}^{n_c-1} \left(\frac{\partial \mu_i}{\partial y_j}\right)_{P, \mathbf{y}_j} dy_j \right]_T \tag{2.18}$$

Combining $(d\mu_i = \bar{V}_i dP)_{T,\mathbf{y}}$, Eqs. (2.13) and (2.18) and the hydrostatic expression given by Eq. (2.8), one obtains

$$\sum_{j=1}^{n_c-1} \left(\frac{\partial \mu_i}{\partial y_j}\right)_{P,T,\mathbf{y}_j} \left(\frac{dy_j}{dz}\right) = (\rho \bar{V}_i - M_i) g \quad i = 1, \ldots, n_c - 1 \tag{2.19a}$$

Note that in Eq. (2.19a), $i = 1, \ldots, n_c - 1$ and not $i = 1, \ldots, n_c$ as in Eq. (2.13). Equation (2.13) gives $\mathbf{y}_j$ and P; Eq. (2.19a) gives only (dy_j/dz) but not (dP/dz). The above equation can also be written as

$$\begin{pmatrix} \dfrac{\partial \mu_1}{\partial y_1} & \dfrac{\partial \mu_1}{\partial y_2} & \cdots & \dfrac{\partial \mu_1}{\partial y_{n_c-1}} \\ \dfrac{\partial \mu_2}{\partial y_1} & \dfrac{\partial \mu_2}{\partial y_2} & \cdots & \dfrac{\partial \mu_2}{\partial y_{n_c-1}} \\ \vdots & \vdots & \cdots & \vdots \\ \dfrac{\partial \mu_{n_c-1}}{\partial y_1} & \dfrac{\partial \mu_{n_c-1}}{\partial y_2} & \cdots & \dfrac{\partial \mu_{n_c-1}}{\partial y_{n_c-1}} \end{pmatrix} \begin{pmatrix} \dfrac{dy_1}{dz} \\ \dfrac{dy_2}{dz} \\ \vdots \\ \dfrac{dy_{n_c-1}}{dz} \end{pmatrix} = g \begin{pmatrix} \rho \bar{V}_1 - M_1 \\ \rho \bar{V}_2 - M_2 \\ \vdots \\ \rho \bar{V}_{n_c-1} - M_{n_c-1} \end{pmatrix} \tag{2.19b}$$

For the sake of brevity we have dropped the subscripts of T, P, and $\mathbf{y}_i$ from the elements of the matrix above. According to Eq. (2.19b), compositional variation in a multicomponent column could be pronounced when: (1) the term $(\rho \bar{V}_i - M_i)$ is large for one of the species, or (2) the determinant of the matrix above is small. The term $(\rho \bar{V}_i - M_i)$ for asphaltene aggregates in a crude oil could be very large; the aggregate molecular weight could be of the order of several thousand and even over 20,000. The molecular weight of other species, especially lighter components, range from 16 to perhaps 500. The determinant of the matrix in Eq. (2.19b) becomes very small near the critical point, and becomes zero at the critical point as we will show in Chap. 6. Therefore, compositional segregation and saturation-pressure variation in an oil column can be very pronounced when

FIGURE 2.3
Schematic of depth vs. pressure and saturation pressure in the gas cap and in oil column with a distinct gas-oil contact (GOC).

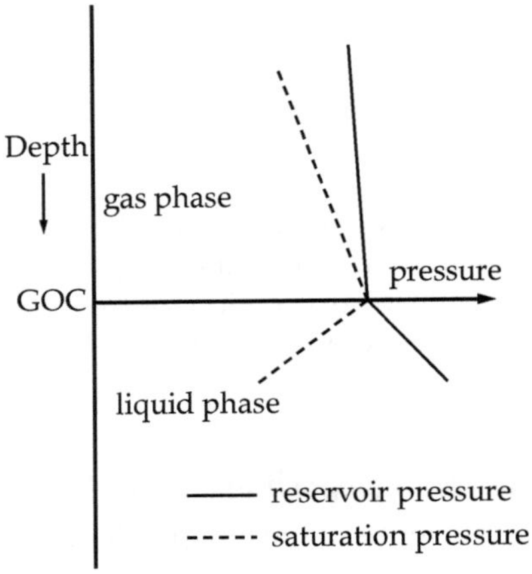

asphaltene materials are present. Compositional variation in a gas column is enhanced when a gas condensate fluid is in the critical region.

Figure 2.3 shows schematically the variation of pressure and saturation pressure with depth in both the oil leg and the gas cap. Saturation pressure variation of Fig. 2.3 is due to compositional grading. Kingston and Niko (1975) report considerable bubblepoint-pressure variation for the Brent and Statford reservoirs of the North Sea. The bubblepoint pressure gradient in the oil zone is reported to be 3.6 and 4.0 psi/ft, respectively, for these reservoirs. Some field data show even a greater bubblepoint pressure decrease with depth, greater than 5 psi/ft.

Figure 2.4 shows another schematic of the variation of pressure and saturation pressure with depth. Note that there is no GOC. At the top, there is the gas phase, and at the

FIGURE 2.4
Schematic of depth vs. pressure and saturation pressure for a near-critical multicomponent fluid.

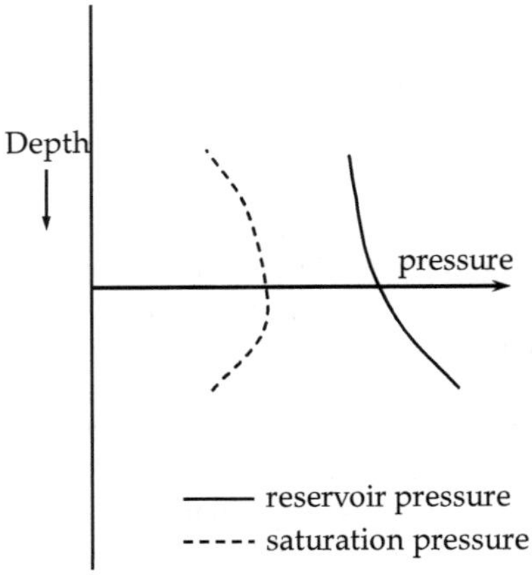

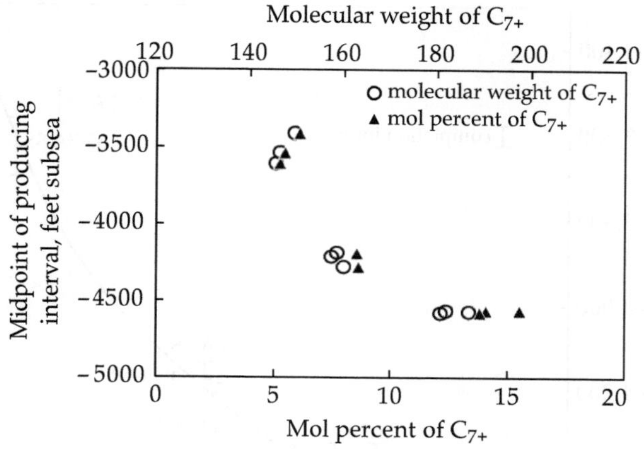

FIGURE 2.5 Variation of mol percent and molecular weight of C_{7+} with depth in the East Painter field (adapted from Creek and Schrader, 1985).

bottom, there is the liquid phase. In such a case, the critical temperature is less than the reservoir temperature at the top and more than the reservoir temperature at the bottom; the reservoir pressure is higher than the dewpoint pressure at the top and it is also higher than the bubblepoint pressure at the bottom. Oil-field examples of such a behavior are given by Neveux and Sathikumar (1988) and by Espach and Fry (1951).

Figure 2.5 shows the variation of the molecular weight and the mol percent of C_{7+} of the East Painter Reservoir (Creek and Schrader, 1985). Figure 2.6 depicts the variation of the amount of C_1 as a function of depth. Figure 2.7 shows the variation of total producing gas-oil ratio (GOR) as a function of depth.

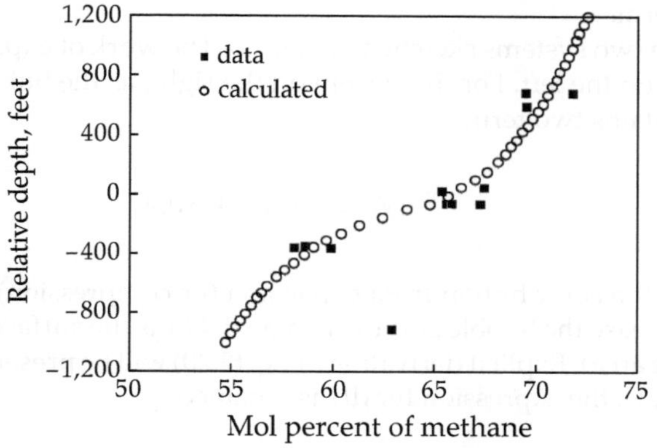

FIGURE 2.6 Variation of mol percent of methane with depth in the East Painter field (adapted from Creek and Schrader, 1985).

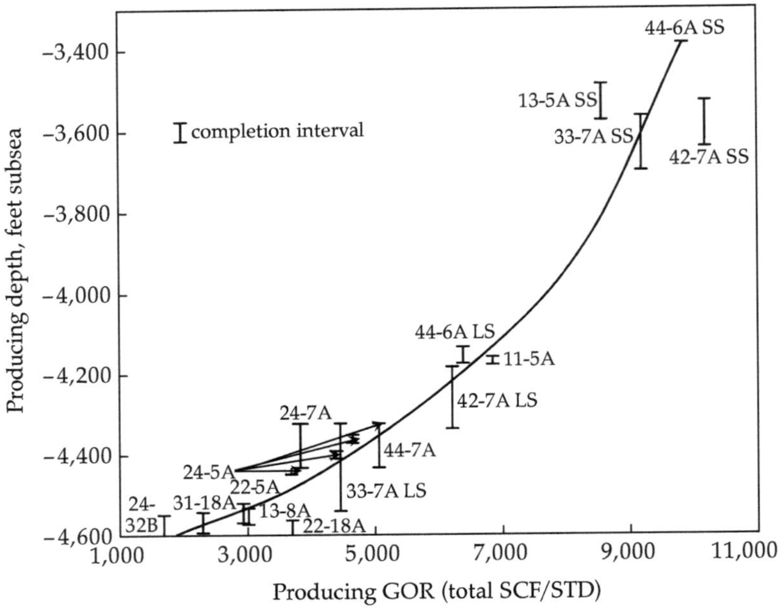

FIGURE 2.7 Variation of total producing GOR with depth in the East Painter field (adapted from Creek and Schrader, 1985).

2.2 Equilibrium Condition for Curved Interfaces

The interface between different phases (e.g., gas and oil) may not be flat. As an example, in a capillary tube, the interface between phases may be curved. In the derivation of equilibrium condition, the effect of interface curvature should be taken into account. Our approach to account for the effect of interface curvature is to modify the expression for the work term.

Consider the two systems sketched in Fig. 2.8. The work of expansion is $đW = -PdV$ for the system on the left. For the system on the right, as the bubble expands, the work expression contains two terms:

$$đW = -PdV + \sigma d\mathcal{A} \tag{2.20}$$

where PdV is the contribution from expansion (or compression) and $\sigma d\mathcal{A}$ is the work required to increase the bubble surface area by $d\mathcal{A}$ (σ is the surface or interfacial tension; $\mathcal{A}$ is the surface area). Explicit derivation of Eq. (2.20) will be presented shortly. Including $\sigma d\mathcal{A}$ in Eq. (1.21), the expression for dU is obtained:

$$dU = TdS - PdV + \sum_{i=1}^{n_c} \mu_i dn_i + \sigma d\mathcal{A} \tag{2.21}$$

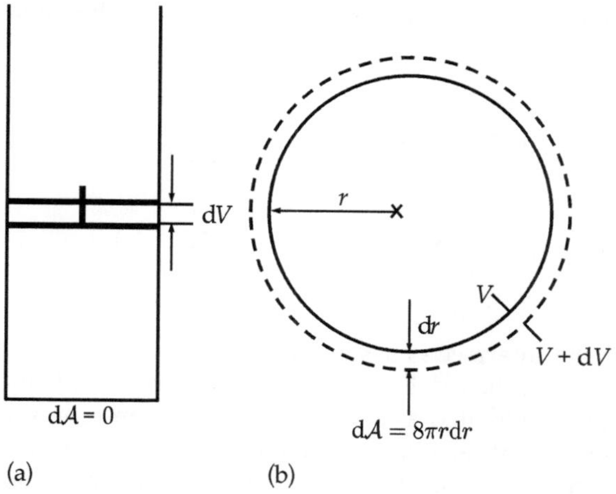

FIGURE 2.8 Change of interfacial surface due to expansion.

Similar expressions for $\mathrm{d}H$, $\mathrm{d}A$, and $\mathrm{d}G$ are given below.

$$\mathrm{d}H = T\mathrm{d}S - V\mathrm{d}P + \sum_{i=1}^{n_c} \mu_i \mathrm{d}n_i + \sigma \mathrm{d}\mathcal{A} \tag{2.22}$$

$$\mathrm{d}A = -S\mathrm{d}T - P\mathrm{d}V + \sum_{i=1}^{n_c} \mu_i \mathrm{d}n_i + \sigma \mathrm{d}\mathcal{A} \tag{2.23}$$

$$\mathrm{d}G = -S\mathrm{d}T + V\mathrm{d}P + \sum_{i=1}^{n_c} \mu_i \mathrm{d}n_i + \sigma \mathrm{d}\mathcal{A} \tag{2.24}$$

Now consider a composite system consisting of a gas bubble (shown by superscript prime or by b superscript) and the surrounding liquid phase (shown by superscript double-primes or L) sketched in Fig. 2.9. The expression for $\mathrm{d}U$ of each phase

FIGURE 2.9 Gas bubble.

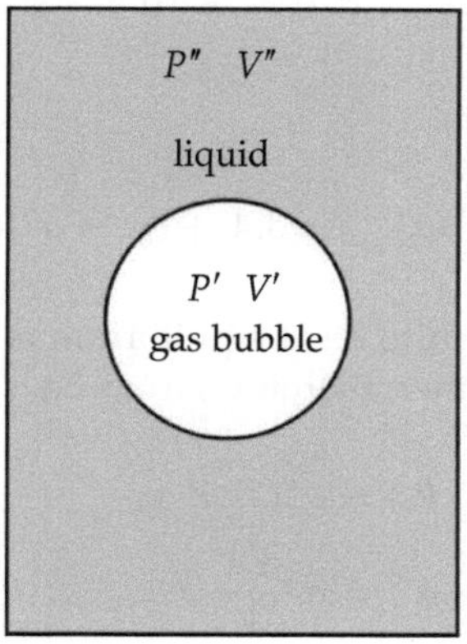

depends on how one assigns the interface. If we assign the interface to the gas bubble, then

$$dU^b = T^b dS^b - P^b dV^b + \sum_{i=1}^{n_c} \mu_i^b dn_i^b + \sigma d\mathcal{A}^b \tag{2.25}$$

$$dU^L = T^L dS^L - P^L dV^L + \sum_{i=1}^{n_c} \mu_i^L dn_i^L \tag{2.26}$$

Alternatively, if the interface is assigned to the surrounding liquid phase and not as part of the gas bubble, the $\sigma d\mathcal{A}^b$ term should be part of dU^L and not part of dU^b. The interface could also be treated as a distinct entity; it can represent the surface phase or even a line phase when three phases with curved interfaces share a common line. However, for convenience and simplicity, we regard the interface as part of either the gas bubble or the surrounding liquid. Such a simplification is justified as long as the thickness of the interfacial region between the bulk phases is negligible in comparison to the radii of the interface. Physically, the boundary between the two bulk phases is a heterogeneous region with a thickness of a few molecular diameters away from the critical region. Other considerations should be taken into account for highly curved interfaces as we will discuss in Chap. 7.

In order to establish the criteria of equilibrium when the interface between the phases is curved, let us consider the system of Fig. 2.9 at constant volume and constant temperature. The pressure inside the gas bubble is P' and the surrounding liquid pressure is P''. The volumes of the gas bubble and the surrounding liquid phase are V' and V'', respectively. The expressions for the differential Helmhotz free energy of the bubble and surrounding liquid for a single-component system are

$$dA' = -P'dV' + \sigma d\mathcal{A}' + \mu' dn' \tag{2.27}$$

$$dA'' = -P''dV'' + \mu'' dn'' \tag{2.28}$$

and the differential Helmhotz free energy of the total system, dA, then is

$$dA = dA' + dA'' = -P'dV' - P''dV'' + \sigma d\mathcal{A}' + \mu' dn' + \mu'' dn'' \tag{2.29}$$

Since $dV' = -dV''$, and $dn' = -dn''$,

$$dA = -(P' - P'')dV' + \sigma d\mathcal{A}' + (\mu' - \mu'')dn' \tag{2.30}$$

The necessary condition for the system to be at equilibrium is that dA must vanish. V' and $\mathcal{A}'$ are not independent of each other; therefore, from Eq. (2.30),

$$\mu'(T, P') = \mu''(T, P'') \tag{2.31a}$$

$$P' - P'' = \sigma \frac{d\mathcal{A}'}{dV'} \tag{2.31b}$$

If we assume that the gas bubble has a spherical shape of radius r,

$$\mathrm{d}\mathcal{A}' = 8\pi r\,\mathrm{d}r \tag{2.32}$$

$$\mathrm{d}V' = 4\pi r^2\mathrm{d}r \tag{2.33}$$

Combining Eqs. (2.31b) through (2.33),

$$\boxed{P' - P'' = P_c = \frac{2\sigma}{r}} \tag{2.34}$$

Equation (2.34) is the well-known Young-Laplace equation of capillarity, which provides the condition for mechanical equilibrium of a curved interface.

For a multicomponent system at constant temperature,

$$\mathrm{d}A' = -P'\mathrm{d}V' + \sum_{i=1}^{n_c} \mu_i'\mathrm{d}n_i' + \sigma\mathrm{d}\mathcal{A}' \tag{2.35}$$

$$\mathrm{d}A'' = -P''\mathrm{d}V'' + \sum_{i=1}^{n_c} \mu_i''\mathrm{d}n_i'' \tag{2.36}$$

and with $V' + V'' =$ constant and $n_i' + n_i'' = constant$ $(i = 1, \ldots, n_c)$,

$$\mathrm{d}A = \mathrm{d}A' + \mathrm{d}A'' = -(P' - P'')\mathrm{d}V' + \sigma\mathrm{d}\mathcal{A}' + \sum_{i=1}^{n_c}(\mu_i' - \mu_i'')\mathrm{d}n_i'$$

Again, the necessary condition for the system to be at equilibrium is that $\mathrm{d}A$ must vanish. From the first two terms, Eq. (2.33) can be obtained, and from the last term,

$$\mu_i'(T, P', \underline{n}') = \mu_i''(T, P'', \underline{n}''). \tag{2.37}$$

Equation (2.37) shows that for a curved interface, the chemical potentials of each component on both sides of the interface should be equal to achieve equilibrium. The chemical potentials, however, are evaluated at different pressures; the pressures on different sides of the interface are related by Eq. (2.34). In Eq. (2.20), the work required to increase the surface area of the bubble was expressed by $\sigma\mathrm{d}\mathcal{A}$. Defay and Prigogine (1966) derive this work term in a straightforward manner. Consider the system shown in Fig. 2.10,

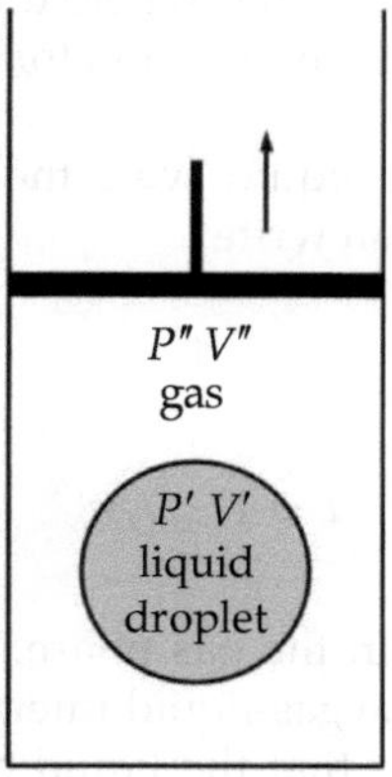

FIGURE 2.10 Liquid droplet.

where a spherical liquid droplet of volume V' and surface area $\mathcal{A}'$ is surrounded by its vapor. The total volume of the system, V, is equal to $V' + V''$, where V'' is the volume of the vapor surrounding the liquid droplet. Suppose part of the droplet vaporizes and the piston moves upward to expand the system to volume $V + \mathrm{d}V$. The work done by the system is, therefore,

$$đW = -P''\mathrm{d}V = -P''\mathrm{d}(V' + V'') = -P''\mathrm{d}V' - P''\mathrm{d}V'' \quad (2.38)$$

Let us add and subtract $P'\mathrm{d}V'$ from Eq. (2.38) and rearrange:

$$đW = -(P'' - P')\mathrm{d}V' - P''\mathrm{d}V'' - P'\mathrm{d}V' \quad (2.39)$$

By using the Young-Laplace equation of capillarity, $-(P'' - P') = 2\sigma/r$, the expression for $đW$ becomes

$$đW = -P'\mathrm{d}V' - P''\mathrm{d}V'' + \sigma\mathrm{d}\mathcal{A} \quad (2.40)$$

where $\mathrm{d}\mathcal{A} = \mathrm{d}V'/2r$. The first two terms on the right side of Eq. (2.40) are the work of expansion of the droplet and the surrounding liquid. The third term, $\sigma\mathrm{d}\mathcal{A}$, is the work required to increase the surface area by $\mathrm{d}\mathcal{A}$.

2.2.1 Effect of Curvature on Saturation Pressure: Condensation and Vaporization in Porous Media

Equations (2.34) and (2.37) reveal that the interface curvature affects the equilibrium. We wish to derive an explicit expression for the effect of curvature on the saturation pressure in porous media. Consider the simple system sketched in Fig. 2.11a. In this figure, the diameter of capillary tube 1, d_1, is less than the diameter of capillary tube 2, d_2. The volume of the space above the capillary tubes where the piston is located is very large compared to the volume of capillary tubes 1 and 2. Suppose a superheated gas at temperature T and pressure P is charged to the system. The saturation pressure of the fluid at temperature T with a flat interface is P^{∞}. If the pressure is raised by isothermal compression, the fluid can condense. The question is where the liquid would form first? To answer this question, we need to derive the expression for the effect of curvature on saturation pressure.

Consider a gas-liquid system where the liquid wets the solid surface. Based on mechanical and chemical equilibrium, one can write

$$P^G - P^L = \frac{2\sigma}{r} \quad (2.41)$$

$$\mu_i^G = \mu_i^L \qquad i = 1, \ldots, n_c \quad (2.42)$$

In the above equations, P^G is the pressure in the gas phase, P^L is the pressure in the liquid phase, σ is the interfacial tension at the gas-liquid interface, and r is the interface radius. Note that in Eq. (2.41), if we assume that the liquid completely wets the solid

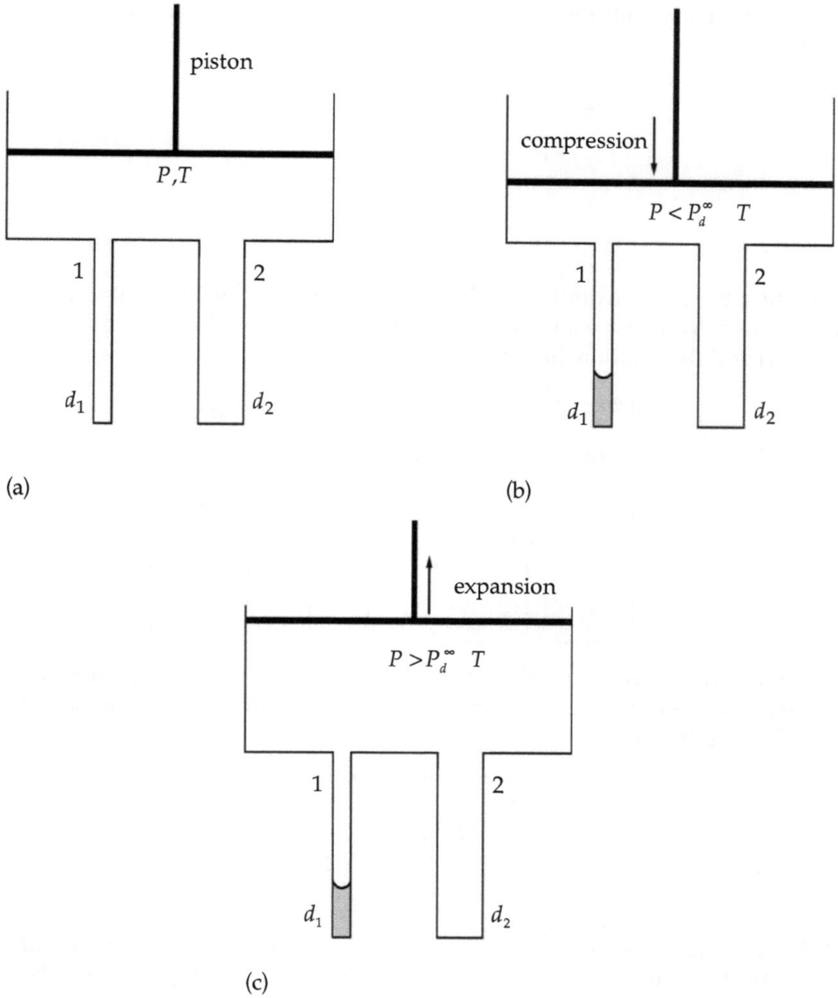

FIGURE 2.11 The influence of curvature on the vapor pressure of a pure substance and the dewpoint pressure of a hydrocarbon mixture: a) system in the single-phase gaseous state, b) for a pure substance, as pressure increases, gas may condense and the condensation will occur first in tube 1, c) for a hydrocarbon gas mixture with retrograde condensation behavior, as pressure decreases, liquid may form also in tube 1 (liquid wets the substrate).

substrate, that is, contact angle $\theta = 0$, then $r = d/2$. For $\theta > 0$, $r = d/(2\cos\theta)$(d is the tube diameter).

For an equilibrium displacement by change of r, Eqs. (2.41) and (2.42) are written as

$$d(P^G - P^L) = d\left(\frac{2\sigma}{r}\right) \tag{2.43}$$

$$d\mu_i^G = d\mu_i^L \quad i = 1, \ldots, n_c \tag{2.44}$$

Under isothermal conditions, the Gibbs-Duhem equation for gas and liquid phases can be written as

$$-v^L \mathrm{d}P^L + \sum_{i=1}^{n_c} x_i \mathrm{d}\mu_i^L = 0 \tag{2.45}$$

$$-v^G \mathrm{d}P^G + \sum_{i=1}^{n_c} y_i \mathrm{d}\mu_i^G = 0 \tag{2.46}$$

Subtracting Eq. (2.46) from Eq. (2.45), assuming that for the gas phase the composition y_i is constant, which is valid at the dewpoint (i.e., $\mathrm{d}\mu_i^G = \bar{V}_i^G \mathrm{d}P^G$), and using Eqs. (2.43) and (2.44) and the relationship $v^G = \sum_{i=1}^{n_c} y_i \bar{V}_i^G$,

$$\boxed{\mathrm{d}P^G = \frac{2\mathrm{d}\left(\dfrac{\sigma}{r}\right)}{\left[1 - \dfrac{\sum_{i=1}^{n_c} x_i \bar{V}_i^G}{v^L}\right]} = \frac{2\mathrm{d}\left(\dfrac{\sigma}{r}\right)}{\left[1 - \dfrac{\sum_{i=1}^{n_c} x_i \bar{V}_i^G}{\sum_{i=1}^{n_c} x_i \bar{V}_i^L}\right]}} \tag{2.47}$$

The above equation provides the effect of interface curvature on the dewpoint pressure at constant temperature. Let us now turn our attention to a pure substance, for which Eq. (2.47) simplifies to

$$\boxed{\mathrm{d}P^G = \frac{2\mathrm{d}\left(\dfrac{\sigma}{r}\right)}{1 - \left(\dfrac{v^G}{v^L}\right)}} \tag{2.48}$$

The above simple equation reveals in very clear terms that the vapor pressure decreases as the interface curvature $(1/r)$ increases, provided the liquid wets the solid substrate. The denominator is negative since for pure substances $v^G > v^L$, and $\mathrm{d}(\sigma/r) = (\sigma/r)|_{r=R} - (\sigma/r)|_{r=\infty} = \sigma/R$ is positive (for a flat interface, $r = \infty$). Therefore, for the process of condensation sketched in Fig. 2.11a, since $\mathrm{d}P^G < 0$, the condensation for a pure substance will first occur in tube 1 with $d_1 < d_2$, as shown in Fig. 2.11b.

Now let us examine Eq. (2.47). For hydrocarbon mixtures in the critical region, the denominator can be positive, $(1 - \sum_{i=1}^{n_c} x_i \bar{V}_i^G / v^L) > 0$, and, therefore, $\mathrm{d}P^G > 0$. Consequently, the dewpoint pressure may increase as the interface curvature increases. For the system sketched in Fig. 2.11a, if a multicomponent natural gas is introduced into the system at $T > T_c$ (where T_c is the critical temperature of the natural gas) in the course of expansion at constant temperature, the liquid might also first form in tube 1 (see Fig. 2.11c).

For a pure substance, the dewpoint and bubblepoint pressures are the same when the gas-liquid interface is flat, and therefore Eq. (2.48) can be used to study the effect of curvature on the saturation pressure, whether we approach the saturation pressure from the gas or liquid side (provided there is no hysteresis in contact angle). The substrate

FIGURE 2.12 The influence of curvature on the bubblepoint of a near-critical hydrocarbon mixture.

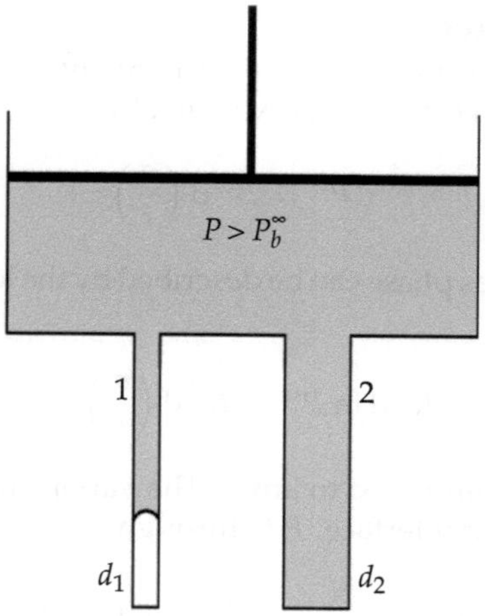

is assumed to wet the liquid, and the pressure in the liquid phase is less than that in the gas phase. For mixtures, dewpoint and bubblepoint (for a given composition) are different for a flat interface between the gas and liquid phases. Equation (2.47) applies to the dewpoint point only. The expression for the change of bubblepoint pressure, that is, vaporization, can be obtained from

$$\boxed{dP^G = \frac{2d\left(\frac{\sigma}{r}\right)}{\left[1 - \frac{v^G}{\sum_{i=1}^{n_c} y\bar{V}_i^L}\right]} = \frac{2d\left(\frac{\sigma}{r}\right)}{\left[1 - \frac{\sum_{i=1}^{n_c} y_i\bar{V}_i^G}{\sum_{i=1}^{n_c} y_i\bar{V}_i^L}\right]}} \tag{2.49}$$

This equation is obtained by holding the composition of the liquid phase x_i constant (i.e., $d\mu_i^L = \bar{V}_i^L dP^L$). Since for a hydrocarbon mixture in the critical region, v^G might be smaller than $\sum_{i=1}^{n_c} y_i\bar{V}_i^L$, dP^L can be positive, which implies that the bubblepoint pressure can increase with an increase in curvature. Based on the above, when we lower the pressure of an undersaturated near-critical oil mixture in a porous medium, the tight pores may provide nucleation sites where gas will form first. In Fig. 2.12, when the pressure of a compressed near-critical liquid is lowered, the gas bubble will first appear in tube 1 and then in tube 2 at a lower pressure, provided the liquid wets the solid; for $\theta = 0$, the gas should form spherical bubbles.

Next, we will derive simple expressions for the vapor pressure of pure substances as a function of interface curvature.

Vapor Pressure of Pure Substances

For a pure substance at low and moderate pressures away from the critical point, often $v^G >> v^L$. Therefore, Eq. (2.48) can be approximated by

$$-v^G \mathrm{d}P^G = 2v^L \mathrm{d}\left(\frac{\sigma}{r}\right) \tag{2.50}$$

If we further assume that the gas phase can be described by the ideal gas law, $P^G v^G = RT$, then

$$-RT \mathrm{d}\ln P^G = 2v^L \mathrm{d}\left(\frac{\sigma}{r}\right) \tag{2.51}$$

We can integrate Eq. (2.51) from $r = \infty$ to any r. The vapor pressure at r, P^G, is related to the vapor pressure of the flat interface, P^∞, through

$$RT \ln\left(\frac{P^\infty}{P^G}\right) = 2\int_{r=\infty}^{r} v^L \mathrm{d}\left(\frac{\sigma}{r}\right) \tag{2.52}$$

If we further assume that the liquid molar volume does not change with pressure,

$$\boxed{P^G = P^\infty \exp\left[-\frac{2\sigma}{r}\frac{v^L}{RT}\right]} \tag{2.53}$$

Equation (2.53), known as the Kelvin equation, reveals that the vapor pressure, P^G, decreases with increasing interface curvature. An alternative derivation of Eq. (2.53) will be presented in Example 7.8. So far, we have assumed that the substrate is liquid-wet or that the new phase forms as a bubble. For a droplet or when gas is the wetting phase, the effect of curvature on saturation pressure is formulated shortly. When the radius of a bubble or droplet becomes very small (say, $r < 10^{-5}$ cm, for a pure substance), the interfacial tension may become a function of the radius (see Chap. 7). However, the derivation of Eq. (2.53) is not based on the assumption of the interfacial tension being independent of r.

Figure 2.13 provides a schematic of liquid- and gas-wet systems. Bubble and droplet formations are also sketched in parallel analogy with liquid-wet and gas-wet systems, respectively.

Effect of Wettability

Our attention to the effect of curvature on the saturation pressure was directed in the case of a bubble or, equivalently, when the liquid wets the solid surface. This is often true in hydrocarbon reservoirs in which gas is the nonwetting phase and oil is the wetting phase. However, in some other systems, gas may be the wetting phase and oil (or the liquid) may be the nonwetting phase. Contact angle in the gas-liquid-rock system can be altered (Fahes and Firoozabadi, 2007). In some reservoir rocks, water may form a contact angle of 30 to 40 degrees in the rock-gas-water system (Fahes and Firoozabadi, 2007). Even in rocks, liquid mercury is the nonwetting phase and air is the wetting phase. In such cases, the Young-Laplace equation of capillarity in a tube should be written as

$$P_c = P^G - P^L = \frac{2\sigma}{r} = \frac{\sigma\cos\theta}{d} \tag{2.54}$$

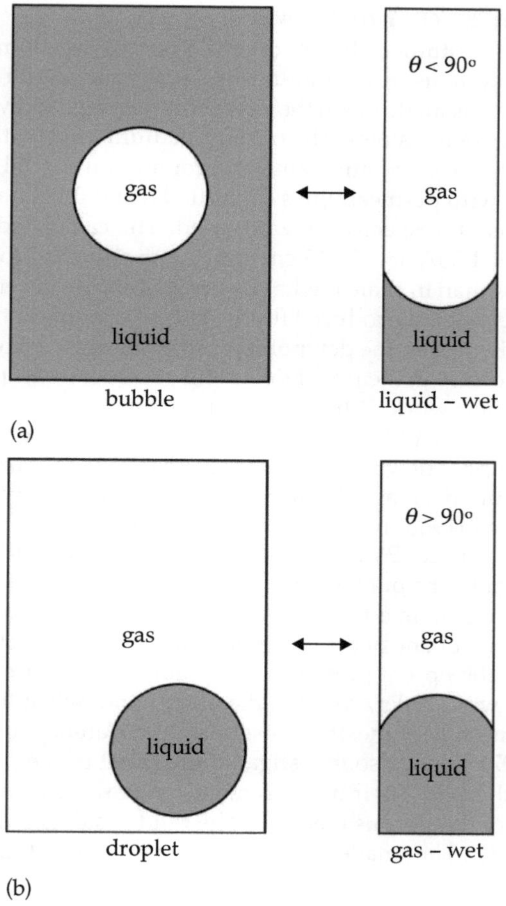

FIGURE 2.13 Parallel analogy between wettability, a) bubble, and b) droplet.

where d is the tube diameter and θ is the contact angle. Note that when gas wets the substrate or for a droplet, gas pressure is less than the liquid pressure because when $\theta > 90°$, $\cos\theta < 0$. For a droplet, or when $\theta = 180°$,

$$dP^G = \frac{2d\left(\frac{\sigma}{r}\right)}{\left(\frac{v^G}{v^L}\right) - 1} \tag{2.55}$$

where r is the radius of the droplet. Equation (2.55) indicates that the vapor pressure of a pure substance increases as the interface curvature increases. Note that in reservoir fluids, according to Eqs. (2.47) and (2.49), decrease or increase in saturation pressure may be different from single-component systems.

Effect of Porous Medium on Phase Behavior

Several authors have studied the effect of the porous medium on the phase behavior of reservoir fluid systems. Russian authors Trebin and Zadora (1968) report a strong influence of the porous medium on the dewpoint pressure and vapor-liquid equilibrium (VLE) of gas condensate systems. The porous medium used by these authors was a silica sand mixture (0.300- to 0.215-mm diameter) ground by a special cutter-pulverizer. Three different packings with permeabilities of 5.6, 0.612, and 0.111 darcy and porosities of 34, 31.4, and 29.8 percent, respectively, were used. The calculated surface areas for these packings were 563, 1,307, and 3,415 cm^2/cm^3, respectively. Note that the three surface areas are much less than in shale media. Figure 2.14 shows the effect of the porous media on the liquid content of the produced fluids. This figure shows that as the surface area of the porous media increases, the dewpoint pressure increases. Trebin and Zadora report a 10 to 15 percent increase in the upper dewpoint pressure in the porous media of the type that was used in their work. When these authors increased the temperature, the effect of the porous media on VLE was decreased. In Chap. 3, we will present examples to show that: (1) the upper dewpoint pressure increases due to curvature effect, and (2) the bubblepoint pressure decreases from the curvature effect away from the critical point.

Tindy and Raynal (1966) measured the bubblepoint pressure of two reservoir crude oils in both an open space (PVT cell) and a porous medium with grain sizes in the range of 160 to 200 microns. The bubblepoint pressures of these two crude oils were higher in the porous medium than in a PVT cell by 7 and 4 kg/cm^2, respectively. Specifically, the bubblepoint pressure of one of the two crude oils measured at 80°C in a PVT cell was 121 kg/cm^2 and the bubblepoint pressure at the same temperature in a porous medium of 160 to 20 microns was 128 kg/cm^2. On the other hand, when these authors used a mixture of methane and n-heptane, they observed no differences in the saturation pressure. Sigmund et al (1973) have also investigated the effect of the porous medium on phase behavior of model fluids. Their measurements on dewpoint and bubblepoint pressures showed no effect of the porous medium. The fluid systems used by these authors were $C_1/n\text{-}C_4$ and $C_1/n\text{-}C_5$. The smallest bead size used was 30 to 40 U. S. mesh. In Example 2.3

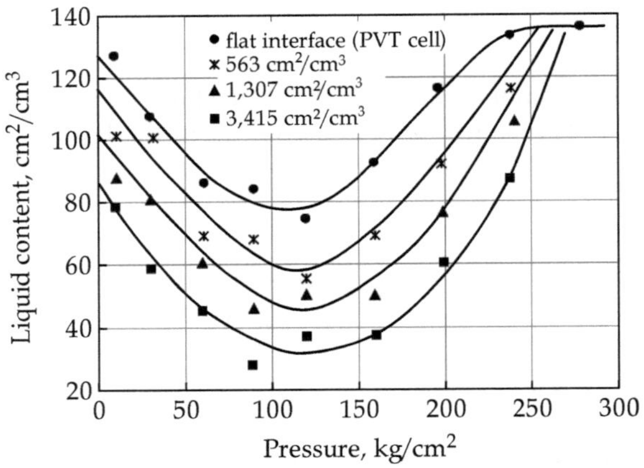

FIGURE 2.14 Effect of surface area of the porous medium on produced liquid content at different surface areas (cm^2/cm^3); (adapted from Trebin and Zadora, 1968).

presented at the end of this chapter, the effect of interface curvature on dewpoint pressure and equilibrium phase composition will be examined. We have two assumptions in relation to the effect of curvature on phase behavior. We have neglected adsorption onto the substrate. We have also assumed that the phase in the pores or capillary tubes is homogeneous. When the pores become small (less than, say, 10 nm), the effect of walls on the confined fluids may make the fluids heterogeneous. A molecular simulation approach will be required for such conditions (see Li, Jin, and Firoozabadi, 2014).

2.3 Equilibrium Condition for Charged Systems

The general expression for differential of work is given by

$$đW = -\vec{F} \cdot d\vec{S} \tag{2.56}$$

where $\vec{F}$ is the force vector, and $d\vec{S}$ is the differential displacement (also a vector) along the path. The minus sign indicates the work is done by the system. For two charged particles 1 and 2 at rest, the force between them $\vec{F}$ is given by Coulomb's law,

$$\vec{F} = \frac{1}{4\pi\varepsilon_0}\frac{q_1 q_2}{r_{12}^2}\vec{e}_{12} \tag{2.57}$$

where q_1 and q_2 are the charge of particles 1 and 2, respectively; r_{12} is the distance between particles 1 and 2, and $\vec{e}_{12}$ is the unit vector. $(1/4\pi\varepsilon_0)$ is the constant of proportionality; it is equal to 9.0×10^9 Nm^2/C^2, or 9.0×10^9 Vm/C (V is the volt; C is the Coulomb). ε_0 is the permeabitivity of vacuum. Let $\vec{F}$ represent the force on charge q_1. Then the work required to move q_1 from point a to point b (Fig. 2.15) is given by

$$W = -\int_a^b \vec{F}\cdot d\vec{S} = \frac{q_1 q_2}{4\pi\varepsilon_0}\left(\frac{1}{r_2} - \frac{1}{r_1}\right) \tag{2.58}$$

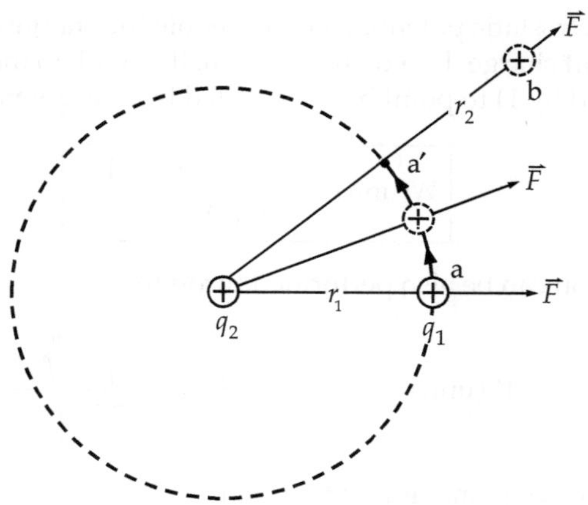

FIGURE 2.15 Displacement of charge q_1 from point a to point b through point a′.

Figure 2.15 shows the schematic to compute the work. Here the work is independent of the path. When we move charge q_1 from point a to point b (see Fig. 2.15), we can go along the arc of the circle with radius r_1. There is no work done from point a to point a′; the force on test charge is at right angle to the direction of motion.

Let us consider the force on the unit charge at point a which is numerically the same as the electric field $\vec{E}$. The field $\vec{E}(1)$ is the force per unit charge of q_1, from charge q_2. Based on Eq. (2.57),

$$\vec{E}(1) = \frac{1}{4\pi\varepsilon_0}\frac{q_2}{r_{12}^2}\vec{e}_{12} \tag{2.59}$$

When there are various charges present, the field at point a is the sum of the contributions of all the other charges. Let q_j represent the magnitude of charge at point j and $\vec{r}_{1j}$ the distance from point j to point a. Then the electrostatic field at point a is given by

$$\vec{E}(1) = \sum_j \frac{1}{4\pi\varepsilon_0}\frac{q_j}{r_{1j}^2}\vec{e}_{1j} \tag{2.60}$$

In a similar way, the force on charge q_1 from all the other charges is given by

$$\vec{F}(1) = \sum_j \frac{1}{4\pi\varepsilon_0}\frac{q_1 q_j}{r_{1j}^2}\vec{e}_{1j} \tag{2.61}$$

When the charges are denoted by a charge distribution, the charge distribution is given by

$$\rho(2) = \frac{\Delta q_2}{\Delta V_2} \tag{2.62}$$

where Δq_2 is the charge in volume increment ΔV_2. In this case the electrostatic field at point a is given by

$$\vec{E}(1) = \frac{1}{4\pi\varepsilon_0}\int \frac{\rho(2)\vec{e}_{12}\mathrm{d}V_2}{r_{12}^2} \tag{2.63}$$

We now define electrostatic potential. Let us denote the charge q_2 by q and the charge at point a to be the unit charge. Based on Eq. (2.58), the work to move the unit charge from point a (represented by 1) to point b (represented by 2) is given by,

$$\boxed{\mathrm{W(unit)} = \frac{q}{4\pi\varepsilon_0}\left(\frac{1}{r_2} - \frac{1}{r_1}\right)} \tag{2.64}$$

The above derivation can be also performed through

$$\mathrm{W(unit)} = -\int_a^b \vec{E}\cdot \mathrm{d}\vec{S} = -\frac{q}{4\pi\varepsilon_0}\int_a^b \frac{dr}{r^2} \tag{2.65}$$

Note that at point a $r = r_1$, and at point b $r = r_2$.

Now let us introduce the electrostatic potential by $\Phi = \Phi(x, y, z)$ and define it by

$$-\int_a^b \vec{E} \cdot d\vec{S} = \Phi(b) - \Phi(a) \tag{2.66}$$

Comparison of Eqs. (2.65) and (2.66) provides,

$$\boxed{\Phi(x, y, z) = \frac{1}{4\pi \varepsilon_0} \frac{q}{r}} \tag{2.67}$$

Equation (2.67) implies that the electrostatic potential at infinity is zero. Therefore, the potential can be interpreted as the work done to carry a unit charge from infinity to r. We can also write the expression for the electrostatic potential due to several charges, and for a charge density of $\rho(2)$

$$\Phi = \sum_j \frac{1}{4\pi_0} \frac{q_j}{r_{1j}} \tag{2.68}$$

$$\Phi = \frac{1}{4\pi_0} \int \frac{\rho(2) d V_2}{r} \tag{2.69}$$

respectively.

2.3.1 Electrostatic Energy of a Sphere with Uniform Charge

Suppose charges q_1 and q_2 are separated by a distance r_{12}. The work done in bringing them from long distance to a distance of r_{12} is given by [based on Eq. (2.58)].

$$W = \frac{q_1 q_2}{4\pi \varepsilon_0 r_{12}} \tag{2.70}$$

The change in the electrostatic energy of the system of two charges by bringing from a long distance to a distance of r_{12} is, therefore, $\Delta U = W$ or simply

$$\boxed{\Delta U = \frac{q_1 q_2}{4\pi \varepsilon_0 r_{12}}} \tag{2.71}$$

Now we can write down, using the principle of superposition, the total energy for many charged particles. If q_i and q_j are any one of the two particles and r_{ij} is the distance between them, the energy of the particle pair is $q_i q_j/(4\pi \varepsilon_0 r_{ij})$. The total electrostatic energy ΔU is the sum of the energies of all particle pairs,

$$\Delta U = \sum_{\textit{all pairs}} \frac{q_i q_j}{4\pi \varepsilon_0 r_{ij}} \tag{2.72}$$

For a distribution of charges described by charge density ρ, Eq. (2.72) is replaced by an appropriate integral.

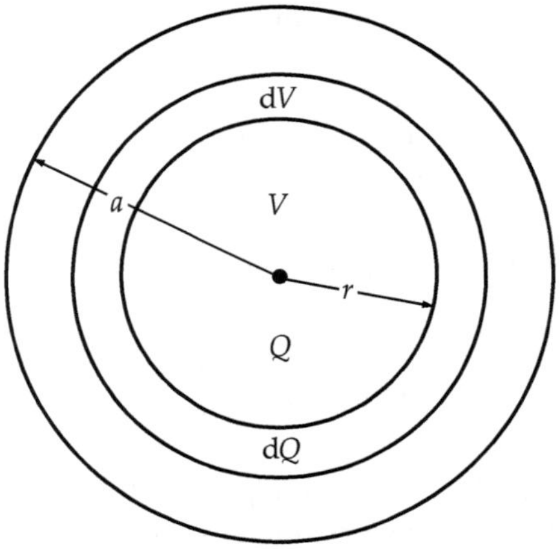

FIGURE 2.16 Sketch of assembling a droplet of uniform charge.

Now let us compute the electrostatic energy required to assemble a sphere of uniform charge. Let us consider bringing a charge dQ from infinity to position r as shown in Fig. 2.16. The work for the process is given by,

$$đW = \frac{QdQ}{4\pi\varepsilon_0 r} \tag{2.73}$$

The charge density ρ in the sphere of radius r is given by

$$Q = \left(\frac{4\pi}{3}\right) r^3 \rho \tag{2.74}$$

Assuming ρ to be constant, we write

$$dQ = \rho(4\pi r^2 dr) \tag{2.75}$$

Substituting Eqs. (2.74) and (2.75) in Eq. (2.73), one obtains

$$đW = \frac{4\pi\rho^2 r^4 dr}{3\varepsilon_0} \tag{2.76}$$

The electrostatic energy, or the work of forming a sphere with radius a, is given by

$$W = \int_0^a \frac{4\pi\rho^2 r^4 dr}{3\varepsilon_0} = \frac{4\pi\rho^2 a^5}{15\varepsilon_0} \tag{2.77}$$

From $\rho = 3Q/(4\pi a^3)$ we obtain

$$\boxed{\Delta U = \left(\frac{3}{5}\right)\frac{Q^2}{4\pi\varepsilon_0 a}} \tag{2.78}$$

If the sphere or a droplet is assembled in a medium of dielectric constant ε, then

$$\boxed{W = \Delta U = \left(\frac{3}{5}\right)\frac{Q^2}{4\pi\varepsilon_0\varepsilon a}} \tag{2.79}$$

We will use a modified form of the above equation to derive the expression for the effect of charge on vapor pressure of a charged droplet in Chap. 7 (see Example 7.8).

2.3.2 Criteria of Equilibrium for a Charged System

In Eq. (2.73), $Q/(4\pi\varepsilon_0 r)$ is the electrostatic potential. Therefore, the expression for work can be also written as

$$\boxed{\text{đ}W = \Phi \mathrm{d}Q} \tag{2.80}$$

Based on Eq. (2.80) we can write the general expression for $\mathrm{d}U$ in a single-component system with charge effect

$$\mathrm{d}U = T\mathrm{d}S - P\mathrm{d}V + \Phi\mathrm{d}Q + \mu\mathrm{d}n \tag{2.81}$$

The charge in the system is described by,

$$Q = zen \tag{2.82}$$

where z is the valence, e is the unit charge, and n is the number of mols. Taking the differential of Q in Eq. (2.82), and combining it with Eq. (2.81), one obtains

$$\mathrm{d}U = T\mathrm{d}S - P\mathrm{d}V + (\mu + ze\Phi)\mathrm{d}n \tag{2.83}$$

The sum of the two terms in the parenthesis is called the electrochemical potential,

$$\mu' = \mu + ze\Phi \tag{2.84}$$

In charged systems the electrochemical potential replaces the chemical potential for equilibrium considerations. The expression for $\mathrm{d}G$ is given by

$$\mathrm{d}G = -S\mathrm{d}T - V\mathrm{d}P + (\mu + ze\Phi)\mathrm{d}n$$

At equilibrium,

$$\boxed{\mathrm{d}\mu = -ze\mathrm{d}\Phi} \tag{2.85}$$

Note the similarity between Eqs. (2.13) and (2.85).

2.4 Examples and Theory Extension

2.1: Gravity Segregation in an Isothermal Ideal Liquid and Gas Column

(a) Consider an isothermal column of an ideal liquid mixture of thickness h. Suppose the composition at the bottom of the liquid column is $\boldsymbol{x}^0 = (x_1^0, x_2^0, \ldots, x_{n_c-1}^0)$. Derive the following expression for the composition at the top of the liquid column:

$$x_i = x_i^0 \exp\left[\frac{(\rho\bar{v}_i - M_i)}{RT} gh\right] \quad i = 1, \ldots, (n_c - 1)$$

where x_i and x_i^0 are the mol fractions at the top and bottom of the column, respectively, $\overline{\rho v}_i$ is the average of the product of the mass density and the molar volume of component i along the liquid column, and M_i is the molecular weight of component i.

(b) Consider an isothermal column of an ideal gas mixture of thickness h. Suppose the composition at the bottom of the gas column is $\boldsymbol{y}^0 = (y_1^0, y_2^0, \ldots, y_{n_c-1}^0)$. Derive the following expression for the composition at the top of the gas column:

$$y_i = y_i^0 \exp\left[\frac{(\bar{M} - M_i)}{RT} gh\right] \quad i = 1, \ldots, (n_c - 1)$$

where $\bar{M}$ is the average molecular weight of the gas mixture along the gas column.

(c) Calculate the mol fraction of C_3 at $h = 1{,}000$ and $5{,}000$ ft for a mixture of C_3 and n-C_7. At $z = 0$, $x_{C_3}^0 = 0.50$ and $P^0 = 2{,}000$ psia. Assume $T = 100\,^\circ$F.

(d) Calculate the mol fraction of C_1 at $h = 1{,}000$ and $5{,}000$ ft for a mixture of C_1 and N_2. At $z = 0$, $y_{C_1}^0 = 0.50$ and $P^0 = 2{,}000$ psia. Assume $T = 100\,^\circ$F.

Solution

(a) For an ideal-liquid solution,

$$\mu_i(T, P, \boldsymbol{x}) = \mu_i^0(T, P) + RT \ln x_i \qquad i = 1, \ldots, n_c.$$

The derivatives of μ_i with respect to x_i and x_j, $j \neq i$, are

$$\left(\frac{\partial \mu_i}{\partial x_i}\right) = \frac{RT}{x_i} \quad i = 1, \ldots, n_c$$

$$\left(\frac{\partial \mu_i}{\partial x_j}\right) = 0 \quad i \neq j$$

Now let us write Eq. (2.19a) for an ideal-liquid solution,

$$\sum_{j=1}^{n_c-1} \left(\frac{\partial \mu_i}{\partial x_j}\right)_{P, x_j} \left(\frac{dx_j}{dz}\right) = (\rho v_i - M_i) g \quad i = 1, \ldots, n_c - 1$$

In the above equation, v_i is the molar volume of component i at temperature T and pressure P (that is, $\bar{V}_i = v_i$ for an ideal-liquid solution).

Combining the above three equations,

$$\left(\frac{RT}{x_i}\right)\left(\frac{dx_i}{dz}\right) = (\rho v_i - M_i) g \quad i = 1, \ldots, n_c - 1$$

Integrating the above equation from $z=0$ to $z=h$,

$$\boxed{x_i = x_i^0 \exp\left[\frac{\overline{\rho v_i} - M_i}{RT} gh\right]} \qquad i = 1, \ldots, n_c - 1$$

(b) For an ideal gas mixture,

$$\mu_i(T, P, y) = \mu_i^0(T, P^0) + RT \ln P y_i - RT \ln P^0$$

Taking the derivatives of μ_i with respect to y_i and $y_j, i \neq j$,

$$\left(\frac{\partial \mu_i}{\partial y_i}\right) = \frac{RT}{y_i} \quad i = 1, \ldots, n_c - 1$$

$$\left(\frac{\partial \mu_i}{\partial y_i}\right) = 0 \quad i \neq j$$

For an ideal gas mixture, $\bar{V}_i = RT/P$ and $\rho = (PM/RT)$. Combining Eq. (2.19a) with the above equations,

$$\left(\frac{RT}{y_i}\right)\left(\frac{\mathrm{d}y_i}{\mathrm{d}z}\right) = (M - M_i)g \qquad i = 1, \ldots, n_c - 1$$

Integrating the above equation from $z=0$ to $z=h$,

$$y_i = y_i^0 \exp\left[\frac{(\bar{M} - M_i)}{RT} gh\right] \qquad i = 1, \ldots, n_c - 1$$

(c) The specific volumes of C_3 and n-C_7 at 100°F and 2,000 psia are (Starling, 1973)

$$v_{C_3}(100\,°\mathrm{F},\ 2{,}000\ \mathrm{psia}) = 0.03157\ \mathrm{ft}^3/\mathrm{lb_m}$$

$$v_{\text{n-}C_7}(100\,°\mathrm{F},\ 2{,}000\ \mathrm{psia}) = 0.02283\ \mathrm{ft}^3/\mathrm{lb_m}$$

The mass density of the mixture of C_3 and n-C_5 ($x_{C_3} = 0.5$) is then $\rho_{mix} = 39.212\ \mathrm{lb_m/ft^3}$. Since $R = 1545(\mathrm{ft \cdot lb_f/R \cdot lb\ mol})$ and $g_c = 1(\mathrm{lb_f/lb_m})$, then

$$\frac{(\rho v_{C_3} - M_{C_3})}{RT}\left(\frac{g}{g_c}\right) h = 1.2 \times 10^{-5} h(\mathrm{ft})$$

Therefore, for $h = 1{,}000$ ft, $x_{C_3} = 0.506$ and for $h = 5{,}000$ ft, $x_{C_3} = 0.53$.

(d) The molecular weight of the gas mixture at the bottom of the column is $M = 22\ \mathrm{lb_m/lb\ mol}$. We set $\bar{M} = M$. Therefore, $[(\bar{M} - M_{C_1})/RT](g/g_c)h = 6.9 \times 10^{-6} h$ (ft).

At $h = 1{,}000$ ft, $y_{C_1} = 0.503$ and at $h = 5{,}000$ ft, $y_{C_1} = 0.517$. Note that in the above two examples, there is very little variation of composition with height.

2.2: Gravity Segregation in a Nonideal Isothermal Liquid and Gas Column Consider a mixture of methane and normal butane in a gravitational field. Use the PR-EOS to compute the composition of the C_1/n-C_4 system at intervals of 1,000, 2,000, 3,000, 4,000, 5,000, and 7,000 ft below the reference depth for the following cases.

(a) At the reference depth, C_1 and n-C_4 compositions are 27.27 and 72.73 mol percent, respectively. Pressure at the reference depth is 1,300 psia. Temperature throughout the liquid column is assumed to be 220 °F.

(b) At a given reference depth, C_1 and n-C_4 compositions are 88.88 and 11.12 mol percent, respectively. Pressure at the reference depth is 514 psia. Temperature throughout the gas column is assumed to be 160 °F.

Solution From Eq. (2.15),

$$f_1(P, y_1) = f_1^0\left(P^0, y_1^0\right) \exp\left[-\frac{M_1}{RT} gz\right]$$

$$f_2(P, y_1) = f_2^0\left(P^0, y_1^0\right) \exp\left[-\frac{M_2}{RT} gz\right]$$

where the subscripts 1 and 2 represent methane and normal butane, respectively. We use the expression for fugacity of component i in a mixture from Eq. (3.32) for the PR-EOS. In the above equations, the right side is known and the two unknowns P and y_1, on the left are to be found from the two nonlinear equations. One can use Newton's method to solve the two equations. The procedure is to calculate the right side at every position z and then obtain P and y_1. The results of the calculation are shown in Figs. 2.17 and 2.18.

2.3: Effect of Interface Curvature on Dewpoint Pressure Consider a binary system of C_1/n-C_{10} with the following composition at 100 °F: $x_{C_1} = 99.894$ mol percent and $x_{n\text{-}C_{10}} = 0.106$ mol percent.

(a) Compute the dewpoint pressure of the above system in a PVT cell (measured $P_d \approx 1{,}450$ psia).

(b) What would be the dewpoint pressure of the above fluid system for droplets of mean radii of 10, 1, 0.1, and 0.01 microns?

(c) What are the equilibrium liquid-phase compositions of the above cases?

Data: At 100 °F and 1,500 psia $\sigma_{C_1/n\text{-}C_{10}} = 9.76$ dyne/cm. Assume the interfacial tension to be independent of the interface curvature and composition in this problem.

Solution At the dewpoint,

$$f_i^L\left(T, P^L, x_1\right) = f_i^G\left(T, P_d^G, z_1\right) \qquad i = 1, 2$$

$$P_d^G - P^L = \frac{2\sigma}{r}$$

$$\sum_{i=1}^{2} x_i = 1$$

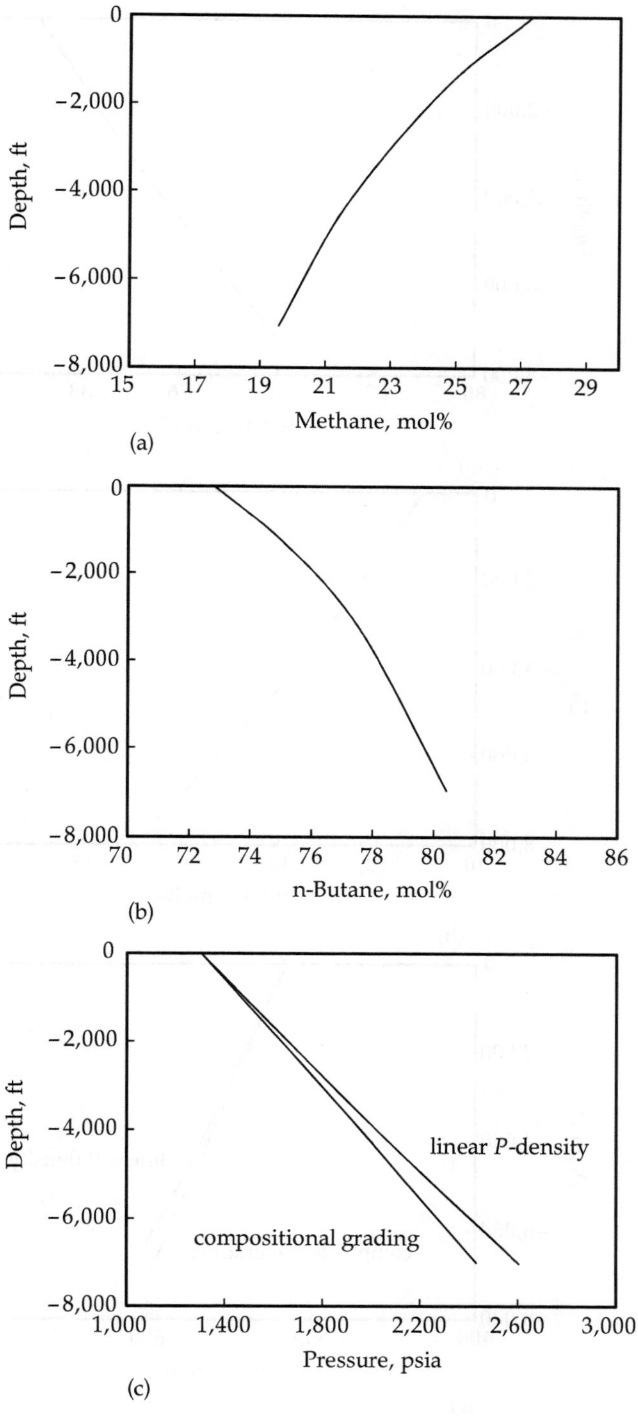

FIGURE 2.17 Compositional grading and pressure: $C_1/n\text{-}C_4$ (liquid), $T = 220°F$.

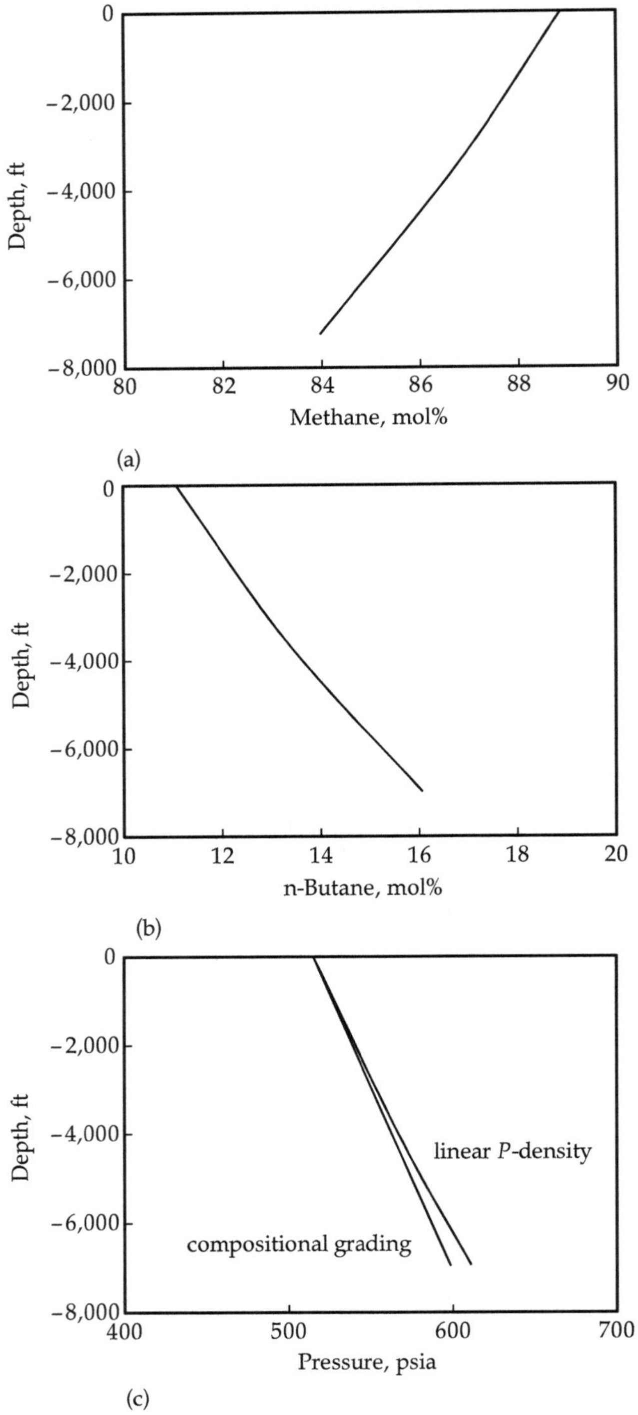

FIGURE 2.18 Compositional grading and pressure: C_1/n -C_4 (gas), $T = 160°F$.

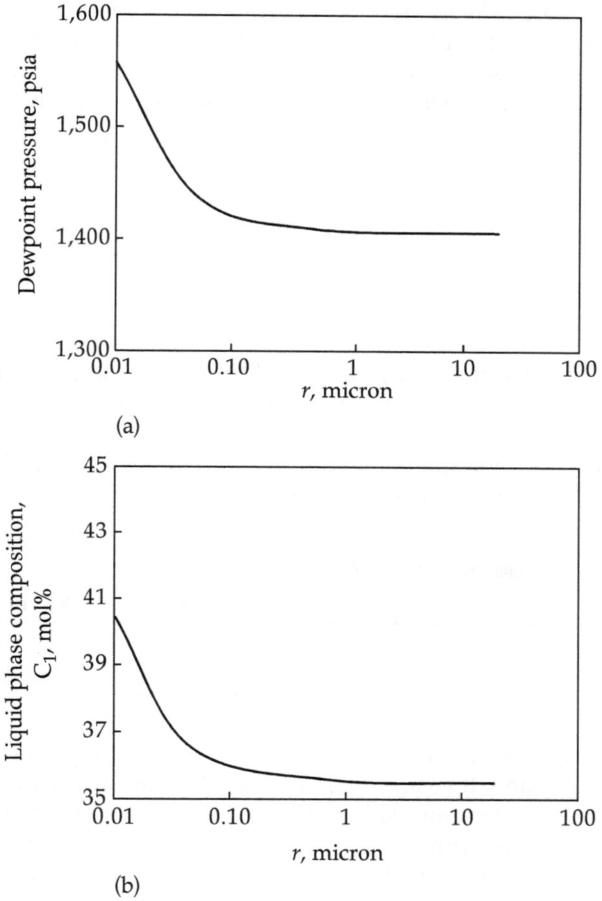

FIGURE 2.19 Effect of pore size on (a) dewpoint pressure and (b) liquid phase composition of the C_1/n-C_{10} system (99.98 mol C_1) at 100°F.

The fugacity expression [see Eq. (3.32)] is used to calculate the fugacities. Newton's method can be used to solve the above system of nonlinear equations to obtain P^L, P_d^G, and x_1 at a given T and z_1. Numerical results from the problem are shown in Fig. 2.19. Note that there is hardly any increase in dewpoint pressure when $r > 0.1$ micron. When $r < 0.1$ micron, dewpoint pressure increases with an increase in curvature. Similarly, the effect of interface curvature on the liquid phase composition is negligible when $r > 0.1$ micron. The results presented in Fig. 2.19 suggest that the porous medium, unless it is very tight (that is, shale), may not have a significant effect on equilibrium. The implication from this problem is that when condensation takes place, the liquid phase may first form in smaller pores in porous media. Another implication is that the curvature increases the solubility of methane in the liquid phase.

Note that when $r < 8$ nm, one cannot use a bulk homogeneous phase description. Adsorption may also have a significant effect (Li, Jin, and Firoozabadi, 2014).

2.4: Effect of Curvature on Vapor Pressure in a Bubble and a Droplet Use Eq. (2.53) to calculate the vapor pressure of a bubble and a droplet of water for radii of 10^{-4}, 10^{-5}, 5×10^{-6}, and 2×10^{-6} cm at 20°C. Assume the interfacial tension of water-vapor to be 72 dyne/cm.

Solution The expressions for the vapor pressure of a pure substance for bubble and droplet are

$$P^G = P^\infty \exp\left[-\frac{2\sigma}{r}\frac{v^L}{RT}\right]$$

$$P^G = P^\infty \exp\left[+\frac{2\sigma}{r}\frac{v^L}{RT}\right]$$

respectively. Note that the curvature of a bubble or a droplet is related to its diameter by $1/r = 2\cos\theta/d$ where θ is the contact angle. For a bubble, $\theta = 0$ and $\cos\theta = 1$. For a droplet, $\theta = 180°$ and $\cos\theta = -1$.

Using the above two equations with $v^L = 18$ cm³/gmol and $R = 82.06$ (atm · cm³)/(gmol · K), the following results are obtained:

Vapor Pressure Ratio P^G/P^∞ of Water at 20°C					
***r*, cm**	**∞**	**10^{-4}**	**10^{-5}**	**5×10^{-6}**	**2×10^{-6}**
Bubble	1	0.999	0.989	0.979	0.948
Droplet	1	1.001	1.010	1.022	1.055

Note that the vapor pressure of a bubble is less than the vapor pressure for a flat interface. On the other hand, the vapor pressure of a droplet is more than the vapor pressure for a flat interface. In the above solution, we have assumed that the interfacial tension is independent of the curvature, even at $r < 10^{-5}$ cm. The interfacial tension changes with the curvature (Defay and Prigogine, 1966; Santiso and Firoozabadi, 2006). However, this trend is only true for pure components; for mixtures, the effect of curvature change on the interfacial tension may be different (see Example 3.6).

2.5: Work of Creating a Gas Bubble The work to create a gas bubble of radius r for a pure substance is given by

$$W = \left(\frac{1}{3}\right)\sigma \mathcal{A}^b$$

where W is the work, σ is the interfacial tension, and $\mathcal{A}^b$ is the surface area of the spherical bubble. Derive the above equation.

Solution In order to create a gas bubble in a liquid phase, one needs: (1) to displace the liquid by the gas bubble, and (2) to create the interfacial area of the gas bubble [see Eq. (2.20) and Fig. 2.20a]. Therefore, the work required to create a gas bubble consists of two parts:

$$W = -\int_{V^b} \Delta P dV + \sigma \mathcal{A}^b$$

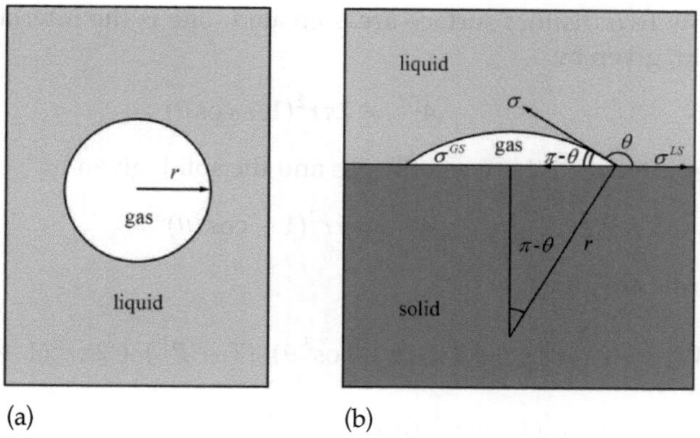

FIGURE 2.20 (a) Homogeneous and (b) heterogeneous bubble nucleation.

where ΔP is the difference in pressure in the gas bubble, P^b, and the liquid surrounding the gas bubble, P^L; $\Delta P = P^b - P^L$. Assuming that $P^b - P^L$ is constant,

$$W = -(P^b - P^L)V^b + \sigma \mathcal{A}^b$$

where V^b is the volume of the gas bubble. Since for a spherical gas bubble, $P^b - P^L = 2\sigma/r$ [see Eq. (3.34)] and $V^b = (1/3)r\mathcal{A}^b$, then

$$W = \left(\frac{1}{3}\right)\sigma \mathcal{A}^b$$

The above fundamental equation is of importance in nucleation theory. Nucleation is a phenomenon of interest in many engineering applications, including metallurgical processes and solution-gas drive in porous media for oil production (Firoozabadi and Kashchiev, 1996). In Chap. 7, a more rigorous approach will be used in the derivation of $W = (1/3)\sigma\mathcal{A}^b$.

2.6: Work of Forming a Segment of a Spherical Bubble Consider the creation of a new gas phase that consists of a segment of a spherical bubble as shown in Fig. 2.20b. Derive the expression for the work required to create the new gas phase, given by

$$W_{het} = W\varphi_{het}$$

and

$$\varphi_{het} = \frac{1}{4}(2 - \cos\theta)(1 + \cos\theta)^2$$

where θ is the contact angle measured through the liquid phase.

Solution The solution to this problem is similar to Example 2.5, with the difference that the presence of the solid should be considered (see Fig. 2.20b).

The work required to displace the liquid is the volume of the spherical segment, V^b, multiplied by $-(P^b - P^L)$. The volume of the spherical segment shown in Fig. 2.20b is

$$V^b = \frac{1}{3}\pi r^3(2 + 3\cos\theta - \cos^3\theta)$$

There are now two distinct surface areas created; one is the interface between the gas and the liquid, given by

$$\mathcal{A}^{GL} = 2\pi r^2(1 + \cos\theta)$$

and the other interface is between the gas and the solid, given by

$$\mathcal{A}^{GS} = \pi r^2(1 - \cos^2\theta)$$

The work terms are given by

$$W_{het} = -\frac{1}{3}\pi r^3(2 + 3\cos\theta - \cos^3\theta)(P^b - P^L) + 2\pi r^2(1 + \cos\theta)\sigma$$
$$+ \pi r^2(1 - \cos^2\theta)(\sigma^{GS} - \sigma^{LS})$$

In the above expression, the first term on the right represents the work of liquid displacement by gas, the second term represents the work of creating the surface area between the gas and the liquid, and the last term represents the work of creating the interface between the gas and the solid. σ is the gas-liquid interfacial tension, and σ^{GL} and σ^{LS} are the interfacial tensions between the gas and solid and liquid and solid, respectively. The gas-solid and liquid-solid interfacial tensions from a simple force balance are related through

$$\sigma^{GS} - \sigma^{LS} = \sigma\cos\theta$$

(we will derive this expression in Chap. 7).

Combining the above equations and the results from Example 2.5,

$$\boxed{W_{het} = W\varphi_{het}}$$

where

$$\varphi_{het} = \frac{1}{4}(2 - \cos\theta)(1 + \cos\theta)^2$$

In nucleation theory, W_{het} is the heterogeneous nucleation work as compared with the homogeneous nucleation work W. The heterogeneous nucleation refers to the fact that the gas bubble forms at the interface of a liquid and a solid phase. Note that φ_{het} is simply W_{het}/W, which varies in the range of 0 and 1.

2.7: Thermodynamic Functions with Interface Effect Derive the following relationships for thermodynamic functions U, A, and G of an open system with a curved interface:

$$U = TS - PV + \sigma\mathcal{A} + \sum_{i=1}^{n_c} n_i\mu_i$$

$$A = -PV + \sigma\mathcal{A} + \sum_{i=1}^{n_c} n_i\mu_i$$

$$G = \sigma\mathcal{A} + \sum_{i=1}^{n_c} n_i\mu_i$$

In Chaps. 7 and 8, we will derive a more general expression for Gibbs free energy of the system. The term $\sigma\mathcal{A}$ in G will not appear in the derivation in Chap. 7. Show that the interfacial tension, $\sigma=(G-\sum_{i=1}^{n_c} n_i\mu_i)/\mathcal{A}$, is given by

$$\sigma=\left(\frac{\partial U}{\partial \mathcal{A}}\right)_{S,V,\underline{n}}=\left(\frac{\partial A}{\partial \mathcal{A}}\right)_{T,V,\underline{n}}=\left(\frac{\partial G}{\partial \mathcal{A}}\right)_{T,P,\underline{n}}$$

Solution The internal energy of an open system with a curved interface is given by [see Eq. (2.21)]

$$dU=TdS-PdV+\sum_{i=1}^{n_c}\mu_i dn_i+\sigma d\mathcal{A}$$

and, therefore, $U=U(S,V,n_1,n_2,\ldots,n_{n_c},\mathcal{A})$. Note that for an open system with a curved interface, the extensive variables S, V, $n_1,\ldots,n_{n_c}$ and $\mathcal{A}$ define the system. The bulk phase, which excludes the interface, is defined by S, V, and $n_1,\ldots,n_{n_c}$. The intensive variables T, P, μ_i, and σ are conjugates to the extensive variables S, V, n_1, $n_2,\ldots,n_{n_c}$ and $\mathcal{A}$. Writing the total differential of $U(S,V,n_1,n_2,\ldots,n_{n_c},\mathcal{A})$ and comparing the results with the above equation,

$$\sigma=\left(\frac{\partial U}{\partial \mathcal{A}}\right)_{S,V,\underline{n}}$$

which defines the interfacial tension as the change in internal energy when the interfacial area, $\mathcal{A}$, of a closed system is changed at constant S and V. Since σ is intrinsically positive, the internal energy U increases as the interfacial area increases.

From the property that U is a first-order homogeneous function (see Chap. 1), $\lambda U=(\lambda S,\lambda V,\lambda n_1,\ldots,\lambda n_{n_c},\lambda\mathcal{A})$, where λ is a parameter [see Eq. (1.33)]. Differentiating the expression for λU with respect to λ and performing the algebra similarly to Eqs (1.33) to (1.35), one obtains

$$\boxed{U=TS-PV+\sigma\mathcal{A}+\sum_{i=1}^{n_c}n_i\mu_i}$$

The expressions for A, G, and the corresponding expressions for σ are obtained in a similar manner.

2.8: Gibbs-Duhem Equation with Interface Effect Derive the following Gibbs-Duhem equation for an open phase with a curved interface:

$$d\sigma=-\left(\frac{S}{\mathcal{A}}\right)dT+\left(\frac{V}{\mathcal{A}}\right)dP-\sum_{i=1}^{n_c}\left(\frac{n_i}{\mathcal{A}}\right)d\mu_i$$

Note that the Gibbs-Duhem equation for a bulk phase does not depend on the interface; it is the same whether the interface is flat or curved. In other words, the Gibbs-Duhem equation of the bulk phase, b, is always

$$-S^b \mathrm{d}T + V^b \mathrm{d}P - \sum_{i=1}^{n_c} n_i^b \mathrm{d}\mu_i = 0$$

Solution The internal energy of an open phase with a curved interface, when the interface is counted as part of one of the phases, is given by (see Example 2.7)

$$U = TS - PV + \sigma\mathcal{A} + \sum_{i=1}^{n_c} n_i \mu_i$$

Writing the total differential of the above equation and comparing the results with the equation for $\mathrm{d}U$ in Example 2.7, one obtains

$$\boxed{\mathrm{d}\sigma = -\left(\frac{S}{\mathcal{A}}\right) dT + \left(\frac{V}{\mathcal{A}}\right) dP - \sum_{i=1}^{n_c} \left(\frac{n_i}{\mathcal{A}}\right) d\mu_i} \tag{2.86}$$

The above equation provides the relationship between the intensive variables σ, T, P, $\mu_1, \ldots, \mu_{n_c}$ of a phase with a curved interface when the interface is counted as part of that phase.

2.9: Derivation of Effect of Curvature on Saturation Temperature at Constant External Pressure Derive the expression for the effect of curvature on saturation temperature at constant external pressure (in a single component system) for both a droplet and a bubble. We derived the expressions at constant temperature in the text.

Solution Let us designate the gas phase by α and the liquid phase by β. We start with droplet.

Droplet Constant external pressure implies $\mathrm{d}P^\alpha = 0$

Equilibrium conditions provide

$$\mathrm{d}\mu^\alpha = \mathrm{d}\mu^\beta$$

$$\mathrm{d}P^\alpha - \mathrm{d}P^\beta = -2\mathrm{d}\left(\frac{\sigma}{r}\right) \text{ or } \mathrm{d}P^\beta = 2\mathrm{d}\left(\frac{\sigma}{r}\right)$$

Gibbs-Duhem equations for phases α and β reads:

$$S^\alpha \mathrm{d}T + n^\alpha \mathrm{d}\mu^\alpha = 0 \text{ or } \mathrm{d}\mu^\alpha = -s^\alpha \mathrm{d}T$$

$$S^\beta \mathrm{d}T - V^\beta \mathrm{d}P^\beta + n^\beta \mathrm{d}\mu^\beta = 0 \text{ or } \mathrm{d}\mu^\beta = -s^\beta \mathrm{d}T + v^\beta \mathrm{d}P^\beta$$

The above two expressions and $\mathrm{d}\mu^\alpha = \mathrm{d}\mu^\beta$ give

$$(s^\alpha - s^\beta)\mathrm{d}T + v^\beta \mathrm{d}P^\beta = 0$$

Using the relation $\Delta h_e = T(s^\alpha - s^\beta)$, where Δh_e is the molar latent heat of evaporation, and $\mathrm{d}P^\beta = 2\mathrm{d}(\sigma/r)$, we get:

$$\left(\frac{\Delta h_e}{T}\right) \mathrm{d}T + v^\beta 2d\left(\frac{\sigma}{r}\right) = 0 \quad \text{or} \quad \frac{\mathrm{d}T}{T} = -\frac{2v^\beta}{\Delta h_e}\mathrm{d}\left(\frac{\sigma}{r}\right)$$

Making the assumption that Δh_e and σ are independent of temperature and integrating from $r=\infty$ to finite r:

$$\int_{T^\infty}^{T} \frac{\mathrm{d}T}{T} = -\frac{2v^\beta}{\Delta h_e}\int_{\infty}^{r} \mathrm{d}\left(\frac{\sigma}{r}\right)$$

Integrating the above expression,

$$\boxed{\ln\frac{T}{T^\infty} = -\frac{2v^\beta}{\Delta h_e}\frac{\sigma}{r}} \tag{2.87}$$

Bubble For a bubble, $\mathrm{d}P^\beta=0$. In a similar approach, we obtain

$$(s^\alpha - s^\beta)\mathrm{d}T - v^\alpha \mathrm{d}P^\alpha = 0$$

Using the relation $\Delta h_c = T(s^\beta - s^\alpha)$, where Δh_c is the molar latent heat of condensation, and assuming the ideal gas behavior, we obtain:

$$-\frac{\Delta h_c}{T^2}\mathrm{d}T = \frac{R}{P^\alpha}\mathrm{d}P^\alpha$$

Combining the above expression with the Young-Laplace equation $P^\alpha = P^\beta + 2\sigma/r$:

$$-\frac{\Delta h_c}{T^2}\mathrm{d}T = \frac{R}{\left(P^\beta + \dfrac{2\sigma}{r}\right)}\mathrm{d}\left(P^\beta + \frac{2\sigma}{r}\right)$$

Assuming Δh_c and σ to be independent of temperature and integrating from $r=\infty$ to finite r:

$$-\int_{T^\infty}^{T} \frac{\Delta h_c}{T^2}\mathrm{d}T = \int_{\infty}^{r} \frac{R}{\left(P^\beta + \dfrac{2\sigma}{r}\right)}\mathrm{d}\left(P^\beta + \frac{2\sigma}{r}\right)$$

Integration of the above expression provides

$$\boxed{\Delta h_c\left(\frac{1}{T^\infty} - \frac{1}{T}\right) = R\ln\left(\frac{P^\beta + \dfrac{2\sigma}{r}}{P^\beta}\right)} \tag{2.88}$$

Note that the effect of curvature in a droplet and in a bubble is not the same. Equations (2.87) and (2.88) can be used to show the drastic difference in limits of condensation and vaporization (see Prob. 2.23).

2.10: Gibbs Phase Rule for Curved Interfaces Derive the phase rule for a composite system of p phases and n_c components with curved interfaces; $F=n_c+1$ where F is the number of degrees of freedom. If some of the interfaces are flat, then $F=n_c+1-I$ where I, is the number of flat interfaces between the bulk phases.

Solution The Gibbs Phase Rule for a flat interface between the phases established in Example 1.9, is based on the assumption that the PdV work is the only mode of

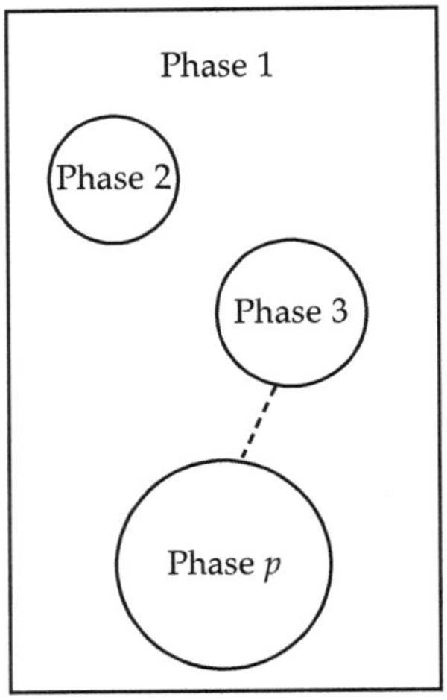

FIGURE 2.21 Sketch of phases with curved interfaces.

work. As we have seen in this chapter, the equilibrium conditions for systems with curved interfaces and under the influence of gravity are different from the equilibrium conditions of systems with flat interfaces and negligible gravity. For systems with curved interfaces and also with gravity effect, the Gibbs Phase Rule should be modified. In this example, we will only consider the effect of the curved interface.

The criteria of equilibrium of multicomponent systems with n_c components and p phases with curved interface between phase 1 and the other phases are (Fig. 2.21)

$$
\begin{aligned}
T^{(1)} = T^{(2)} &= \ldots = T^{(p)} \\
P^{(1)} - P^{(2)} &= \sigma^{(1,2)} J^{(1,2)} \\
&\vdots \\
P^{(1)} - P^{(p)} &= \sigma^{(1,p)} J^{(1,p)} \\
\mu_1^{(1)}\left(T^{(1)}, P^{(1)}, x_1^{(1)}, \ldots, x_{n_c-1}^{(1)}\right) &= \ldots = \mu_1^{(p)}\left(T^{(p)}, P^{(p)}, x_1^{(p)}, \ldots, x_{n_c-1}^{(p)}\right) \\
&\vdots \\
\mu_{n_c}^{(1)}\left(T^{(1)}, P^{(1)}, x_1^{(1)}, \ldots, x_{n_c-1}^{(1)}\right) &= \ldots = \mu_{n_c}^{(p)}\left(T^{(p)}, P^{(p)}, x_1^{(p)}, \ldots, x_{n_c-1}^{(p)}\right)
\end{aligned}
$$

where $\sigma^{(1,j)}$ and $J^{(1,j)}$ are the surface tension and the mean curvature of the interfaces between phase 1 and phase $j(j=2, \ldots, p)$, respectively. There are $(p-1)$ equations for temperature, $(p-1)$ equations for pressure relationships, and $n_c(p-1)$ equations for chemical equilibria. The total number of equations are, therefore, $(p-1)(n_c+2)$. The intensive variables for phase $j(j=2, \ldots, p)$ are $T^{(j)}$, $P^{(j)}$, $\mu_i^{(j)}$ $(i=1, \ldots, n_c)$, and $\sigma^{(1,j)}$. For phase 1, the intensive variables are $T^{(1)}$, $P^{(1)}$, and $\mu_i^{(1)}$ $(i=1, \ldots, n_c)$. Note that we have assigned the interfaces to phases $j=2, \ldots, p$, which is consistent with our derivation of Eqs. (2.25) and (2.26). In phase 1, only (n_c+1) intensive variables among a total of (n_c+2) intensive variables are independent [see the Gibbs-Duhem Eq. (1.40)]. Of the (n_c+3) intensive variables $T^{(j)}$, $P^{(j)}$, $\mu_i^{(j)}$, and $\sigma^{(1,j)}$ for phases $j=2, \ldots, p$, only (n_c+2) intensive variables are independent (see Example 2.8). Therefore, the total number of independent intensive variables of the composite system is: $(p-1)(n_c+2)+(n_c+1)$. The number of degrees of freedom is given by the expression $F =$ number of independent intensive variables — number of independent equations. After substitution in the above expression,

$$\boxed{F = n_c + 1}$$

If there are I flat interfaces between the phases, one can readily show that

$$\boxed{F = n_c + 1 - I}$$

In the above two equations, the intensive variables are temperature, pressure, chemical potentials, and interfacial tensions. As we saw in Example 1.9, instead of the chemical potentials, one may use the mol fractions as intensive variables. It is also more convenient and meaningful to use the interface curvature instead of the interfacial tension as the intensive variable. For a pure substance, as the interface curvature changes, the interfacial tension also changes. For an isothermal process of a single-component system when the curvature of the bubble in a bulk liquid phase at constant pressure changes, the interfacial tension also changes because of: (1) curvature change, and (2) gas-phase pressure change. Similarly, for a bubble in a multicomponent system at constant temperature, when the bulk liquid phase is held at constant pressure and composition, if the interface curvature changes, the interfacial tension may also change because of: (1) curvature change, (2) gas-phase pressure change, and (3) gas-phase composition change. For a pure substance, the interfacial tension is a weak function of curvature when say $r > 10^{-5}$ cm (Santiso and Firoozabadi, 2006). For mixtures, the effect of curvature change on interfacial-tension change becomes more complicated because of both curvature and composition effects (see Example 3.6). In the context of the Phase Rule, if we specify the interface curvature, the interfacial tension is also specified because other intensive variables such as pressure, temperature, and composition are among the variables of the Phase Rule.

For systems in which there is a three-phase contact line between the phases as a result of a solid phase, the concept of contact angle is introduced. For such systems, the Phase Rule remains the same (Li et al., 1989). For highly curved interfaces where the thickness of the heterogeneous region between the phases is not small compared to r, there are other considerations in the derivation of the Phase Rule (Li et al., 1989; Li, 1994).

Let us give two examples for the use of the Phase Rule for curved interfaces. First, consider a single component gas-liquid system with a curved interface between the gas and liquid phases; for this system $F = 2$. Therefore, we can fix two intensive variables, say temperature and pressure, of the vapor phase. Then the system is fully defined. We can also specify temperature and curvature. Note that for a single-component gas-liquid system with a flat interface, $F = 1$. If we fix the temperature, the vapor pressure is fixed. In the second example, we consider a two-component two-phase system with a curved interface between the gas and liquid phases, $F = 3$. Unlike the system with a flat interface, specifying the temperature and pressure of the gas phase does not specify the system. We also need to specify the interface curvature.

2.11: Expression for dU in Multicomponent Mixtures with Electrical Effect When a system is composed of different charged species, one has to account for chemical forces and for electrical forces. Let us consider a multicomponent system of n_i species ($i = 1, \ldots, n_c$) with charges $q_i = (i = 1, \ldots, n_c)$. Derive the following expression for differential of internal energy,

$$dU = TdS - PdV + \sum_{i=1}^{n_c} (\mu_i + ez_i\Phi)dn_i$$

where z_i is the valence of component of i and Φ is the electrostatic potential. You may compare the above equation with

$$dU = TdS - PdV + mgdz + \sum_{i=1}^{n_c} (\mu_i + M_i gz)dn_i$$

where the gravity effect is included in the expression for dU (z is the vertical height; it is positive upward). Suggest a general expression for the effect of gravity potential and electrostatic potential. Write the criteria of equilibrium based on the above two equations.

Solution When a system is composed of different charged species, one has to account for chemical forces and for electrical forces. Therefore, the work includes two terms:

$$đW = đW_Q - PdV$$

For a system with charged species, the electrostatic work is given by:

$$đW_Q = \frac{QdQ}{4\pi\varepsilon_0 r} = \Phi dQ$$

where Φ is the electrostatic potential. The charge in the system is described by $Q = \sum_{i=1}^{n_c} z_i e n_i$ and the charge in each species in the system is

$$Q_i = z_i e n_i$$

where z_i is the valance of species i, and n_i is the number of mols of species i. Taking the differential, we have

$$dQ_i = z_i e dn_i$$

Therefore,

$$đW_Q = \sum_{i=1}^{n_c} \Phi e z_i dn_i$$

The work term $đW_Q$ can be expressed as:

$$đW = -PdV + \sum_{i=1}^{n_c} \Phi e z_i dn_i$$

The expression for the internal energy is:

$$dU = TdS + đW + \sum_{i=1}^{n_c} \mu_i dn_i$$

Combining the above two equations, the expression for dU is

$$dU = TdS - PdV + \sum_{i=1}^{n_c} (\mu_i + \Phi e z_i) dn_i$$

The expressions for dG which account for electrostatic and gravitational potential are

$$dG = -SdT + VdP + \sum_{i=1}^{n_c} (\mu_i + \Phi e z_i) dn_i$$

and

$$dG = -SdT + VdP + mgdz + \sum_{i=1}^{n_c} (\mu_i + zgM_i) dn_i$$

respectively. At equilibrium, dG must vanish. Therefore, the criteria for equilibrium with electrostatic are $dT = 0$, $dP = 0$, and $\mu_i + \Phi e z_i = 0$.

$$(d\mu_i = e z_i d\Phi)_{T,P} \qquad i = 1, \ldots, n_c \tag{2.89}$$

Note that the above expression has the same form as the Gibbs sedimentation expression in Eq. (2.13).

2.12: Effect of Spring Force on Work Consider the system in Fig. 2.22 where a spring is placed between two flat plates. Show that the change in Gibbs free energy is given by,

$$dG = -SdT + VdP + \sum_{i=1}^{n_c} \mu_i dn_i + 2F dh$$

where F is the force exerted on the plates and h is the distance between the plates. What are the extensive and intensive variable in the above equation? The steric repulsion between two particles is modeled by the use of the above expression.

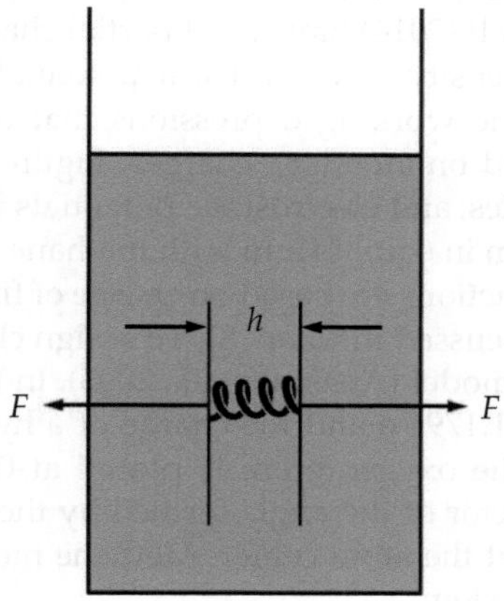

FIGURE 2.22 Effect of spring force on work.

Solution The Gibbs free energy G is defined as $G = U + PV - TS$
The total differential of G is

$$dG = dU + PdV + VdP - TdS - SdT \tag{2.90}$$

The internal energy of a closed system is (Postulate I):

$$dU = đQ + đW$$

For a reversible process,

$$đQ = TdS$$

The mechanical work in this system can be the work of expansion and the elastic work done by the force F over an infinitesimal displacement h:

$$đW = -PdV + 2F\,dh$$

Therefore,

$$dU = TdS - PdV + 2F\,dh$$

Considering a multicomponent system, the internal energy change becomes:

$$dU = TdS - PdV + \sum_{i=1}^{n_c} \mu_i dn_i + 2F\,dh$$

Substituting the internal energy into Eq. (2.90) above:

$$dG = -SdT + VdP + \sum_{i=1}^{n_c} \mu_i dn_i + 2F\,dh$$

Extensive variables are $S,\ V,\ n_i$, and F. Intensive variables are $T,\ P,\ \mu_i$, and h.

2.13: Computation of Electrostatic Field, Surface Charge Density, and Electrical Potential in Thin Film Jiménez-Ángeles and Firoozabadi (2014) have used partial charge of atoms to compute electric field, surface charge density, and electrical potential of thin liquid films in hydrates (Fig. 2.23). Provide the working expressions that can be used to compute charge-related parameters based on atomistic charges. Figure 2.24 shows the number densities, surface charge densities, and electrostatic potentials in methane hydrates and methane hydrate thin liquid film in equilibrium with methane at 100 bar.

Solution All the electric functions are based on charge of fixed atoms in the hydrate and thin film of water (to be discussed in Chap. 8). To assign charges to O and H atoms of water, we use the TIP4P-ice model (Abascal et al., 2005). In this model for water, the charge of an oxygen atom is −1.1794 e and the charge of a hydrogen atom is 0.5897 e (see Chap. 3). The charge of the oxygen atom is placed at 0.1577 Å from the center of the oxygen atom on the bisector of the angle formed by the O-H bonds. The charge of hydrogen atoms are placed at the atom center. Methane molecules are not assigned charge in hydrates and in the system.

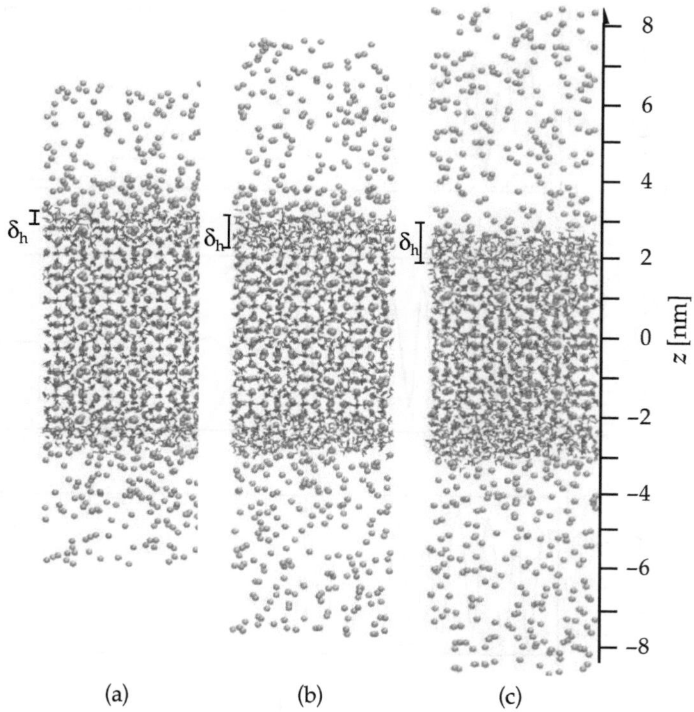

FIGURE 2.23 Snapshots of the hydrate/methane gas setups at (a) $T_m - T = 17K$, (b) $T_m - T = 7$ K, and (c) $T_m - T = 1$ K. The pressure is $P = 100$ bar; δ_h is the water film thickness. At this pressure the computed melting temperatures is $T_m = 287$ K. Oxygen and hydrogen atoms of water are in black and methane molecules are shown in light color (from Jiménez-Ángeles and Firoozabadi, 2014).

The number density of O and H atoms are computed from,

$$\rho_i(z) = \frac{N_i(z)}{l_x l_y \Delta h}$$

where $N_i(z)$ is the number of O, H, and methane particles in the slab of $l_x l_y \Delta h$. l_x and l_y are the x and y direction size of the box in Fig. 2.23; $l_x = 3.5$ nm, $l_y = 3.5$ nm, and $\Delta h = 0.2$ Å. The charge density is given by,

$$\rho_{el}(z) = q_O \rho_O(z) + q_H \rho_H(z)$$

$q_O = -1.1794$ e and $q_H = 0.5897$ e as mentioned above. Next, we calculate the electric field along the z-direction.

Based on Eq. (2.66) in the z-direction,

$$E_z(z) = -\frac{d\Phi(z)}{dz}$$

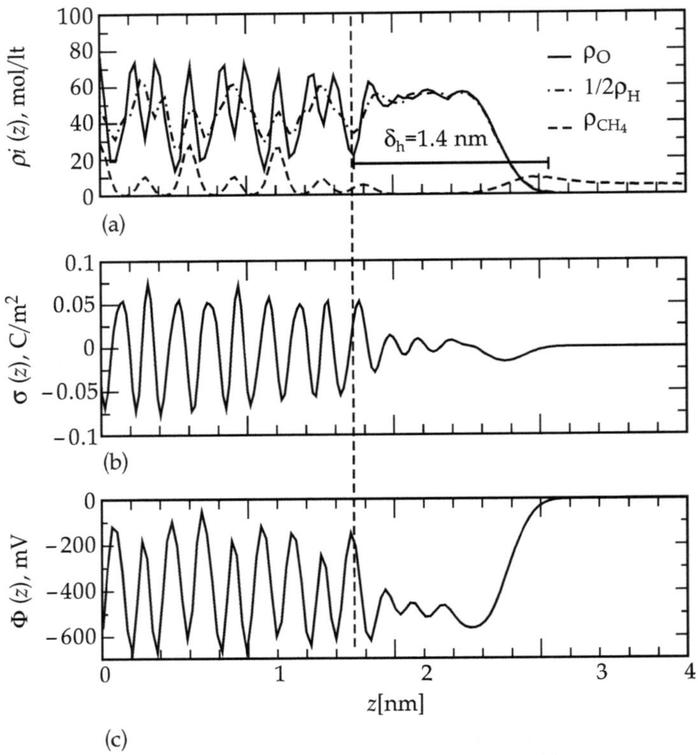

FIGURE 2.24 Hydrate/methane gas interface at 1 K subcooling ($T_m - T = 1$ K) and $P = 100$ bar. (a) Number density profiles of oxygen (solid black line) and methane (dashed line) and half of the density profile of hydrogen (dot-dash line). (b) Charge density profiles. (c) Mean electrostatic potential profile. Simulation temperature is $T = 286$ K and the melting temperature is $T_m = 287$ K. The hydrate crystal is centered at $z = 0$. For clarity only the interval for $z > 0$ is shown. The vertical dashed line is the hydrate-liquid layer boundary (adopted from Jiménez-Ángeles and Firoozabadi, 2014).

where Φ is the electrostatic potential. From the Poisson equation, $d^2\Phi(z)/dz^2 = -\rho_{el}(z)/\varepsilon_0$, one readily obtains,

$$E_z(z) = \frac{1}{\varepsilon_0}\int_{\infty}^{z} \rho_{el}(z)dz$$

In deriving the above equation, $E_z(z \to \infty) = 0$ is invoked. The $z \to \pm\infty$ is the boundary. The local surface charge density is simply

$$\sigma(z) = \rho_{el}(z)\Delta h$$

which is charge/area. The electrostatic potential is readily computed from

$$\Phi(z) = -\int_{\infty}^{z} E_z(z)dz$$

which requires the use of the boundary condition $\Phi(z \to \infty) = 0$. The number density profiles of O and H atoms, local surface charge density, and mean electrostatic potentials are computed along the z-axis.

We have not covered the Poisson equation, $\nabla^2 \Phi = -\rho_{el}(x, y, z)/\varepsilon_0$ in the chapter. Poisson's equation allows the calculation of the electrostatic potential from a distribution of fixed electric charges.

2.5 Problems

2.1 In a gravity field, for a closed reversible process,

$$dH = TdS$$

When the effect of gravity is neglected,

$$dH = TdS + VdP$$

Derive the first equation.

2.2 Derive the following barometric formula which describes the isothermal change of pressure with increasing altitude z:

$$P = P^o \exp\left[1 - \frac{Mg}{RT} z\right]$$

P^o is the pressure at $z = 0$.

2.3 Show that the Phase Rule when the gravity effect is not negligible should be modified to the following form:

$$F = n_c + 3 - p$$

Note that in the above equation, the interface between the phases is assumed to be flat.

2.4 Consider capillary condensation in Fig. 2.25 where $r_1 = 0.2\ \mu m$, $r_2 = 10\ \mu m$, and $r_3 = 100\ \mu m$. The system is initially saturated with pure n-C_5 vapor at 100°F at very low pressure. Calculate the pressure at which condensation occurs in each of the capillary tubes for different contact angles.

FIGURE 2.25

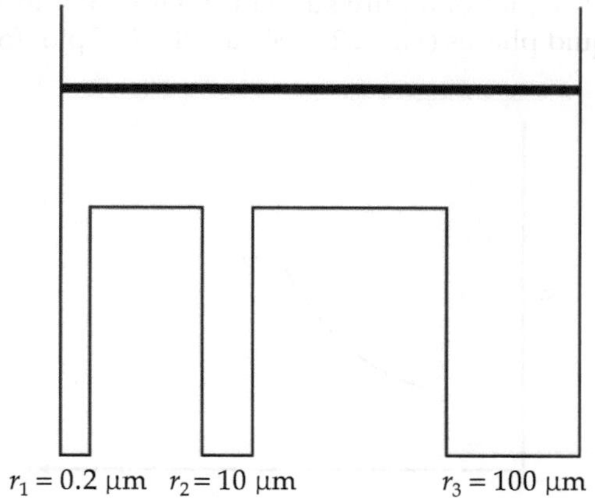

(a) $\theta = 0$ (i.e., liquid completely wets the substrate)

(b) $\theta = 30°$

(c) $\theta = 150°$

Pertinent data are as follows:

Vapor pressure of n-C_5 for a flat interface at 100°F = 15.57 psia,

Surface tension for n-C_5 at 100°F = 14 dynes/cm.

2.5 Consider a gas bubble and a liquid droplet sketched in Fig. 2.26. The fluid is n-C_5 and the temperature is 100°F. Calculate pressure P' for both systems for the following radii: r = 0.05, 0.1, 10 μm.

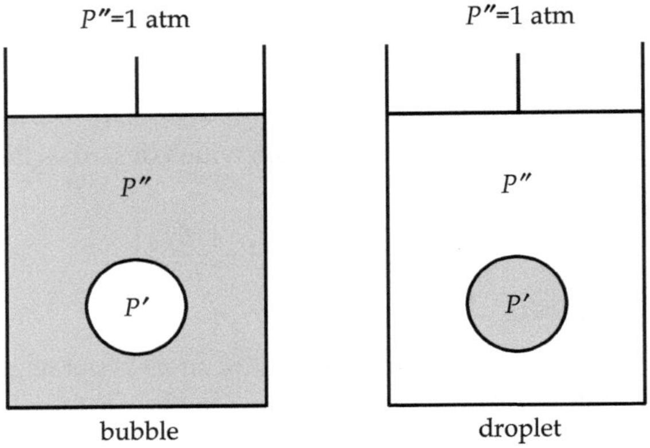

FIGURE 2.26

2.6 Show that the work required to generate a gas bubble can be expressed as

$$W = \frac{16\pi\sigma^3}{[3(P^b - P^L)^2]}$$

2.7 Consider the P-T plot of a pure substance with a flat interface between the equilibrium gas and liquid phases (Fig. 2.27). Sketch the P-T plot for the same system when

FIGURE 2.27

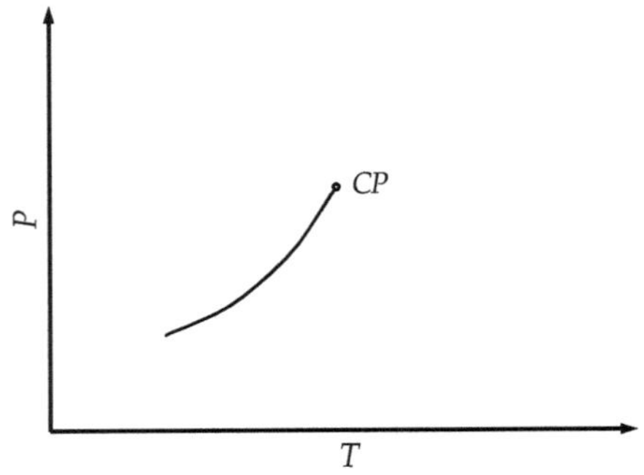

the interface between the equilibrium gas and liquid phases has a constant curvature. Assume liquid is the wetting phase and gas is the nonwetting phase. What would the results be if gas were the wetting phase?

2.8 The effect of interface curvature on the saturation pressure of a pure substance is given by [see Eq. (2.48)]

$$\mathrm{d}P^G = \frac{2\mathrm{d}\left(\frac{\sigma}{r}\right)}{\left(1-\frac{v^G}{v^L}\right)} = \frac{2\mathrm{d}\left(\frac{\sigma}{r}\right)}{\left(1-\frac{\rho^L}{\rho^G}\right)}$$

However, as the critical point is approached, both the numerator and denominator of the above equation for a fixed curvature approach zero. Show that for a fixed curvature, approaching the critical point, $\mathrm{d}P^G \to 0$ and at the critical point, $\mathrm{d}P^G = 0$.

Hint: At the critical point $\sigma = 0$, but the rates at which σ and $(1 - v^G/v^L)$ approach zero are not the same. From the Macleod-Sugdon equation (Macleod, 1923; Sugden, 1924; Fowler, 1937; Firoozabadi et al., 1988),

$$\sigma^{\frac{1}{4}} = \left(\frac{P}{M}\right)(\rho^L - \rho^G)$$

The interfacial tension and phase densities are related. In the above equation, P is the parachor of the pure substance (which can be obtained from pure-component interfacial tension measurements), M is the molecular weight, and ρ^L and ρ^G are mass densities of the equilibrium liquid and vapor phases, respectively. Note that in the Macleod-Sugden equation above, the effect of interface curvature on the interfacial tension of a pure substance is neglected.

2.9 Consider a circular ring of radius r inserted in a liquid horizontally. The force on the ring required to break the liquid-vapor interface is measured (F). Why is the interfacial tension given by

$$\sigma = \frac{F}{2(2\pi r)}$$

2.10 Derive the Gibbs-Duhem equation for an open phase with a curved interface [see Eq. (2.86) in Example 2.8] by taking the Legendre transform of $U = U(S, V, n_1, \ldots, n_{n_c}, \mathcal{A})$ with respect to all the extensive variables $S, V, n_1, \ldots, n_{n_c}$, and $\mathcal{A}$.

2.11 Consider the gas bubble shown in the cavity in Fig. 2.28. Derive the expression for the work of creating the new gas phase that consists of the top spherical segment and the cone:

$$W_{het} = W\varphi_{het}$$

FIGURE 2.28

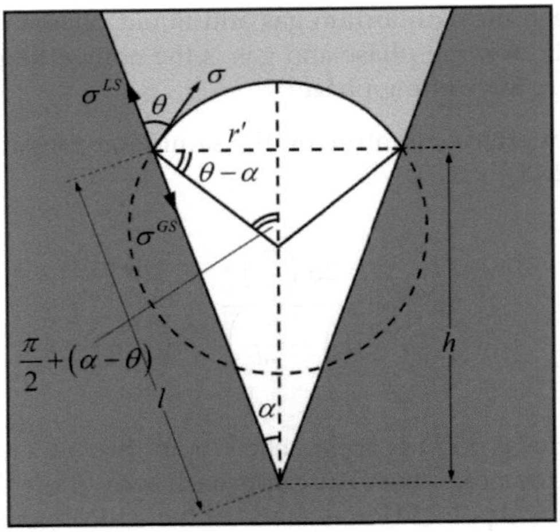

where W is the work given in Example 2.5,

$$\varphi_{het} = \frac{3}{2}(1+\sin\gamma) + \frac{3}{4}\frac{(1-\sin^2\gamma)\cos\theta}{\sin\theta} - \frac{1}{2}\left[2+3\sin\gamma - \sin^3\gamma + \frac{\cos^3\gamma}{\tan\alpha}\right]$$

and

$$\gamma = \alpha - \theta$$

2.12 Derive Eq. (2.49) of the text.

2.13 Derive the following simple relations for the gas and liquid pressures corresponding to the bubble and droplet systems:

$$P^G - P^L = \frac{2\sigma}{r}$$

$$P^G - P^L = -\frac{2\sigma}{r}$$

where r is the radius of the bubble and the droplet, respectively. Calculate the pressure in the liquid phase for the bubble and droplet of Example 2.4 for $P^G = 0.2$ atm. Note that negative pressure of liquid phase inside the bubble indicates that the liquid is under tension (see Chap. 3).

2.14 Consider the phases sketched in Fig. 2.29. In spite of the difference between the arrangement of the phases in the figure below and in Fig. 2.21, show that the Phase Rule remains unchanged: $F = n_c + 1$.

FIGURE 2.29

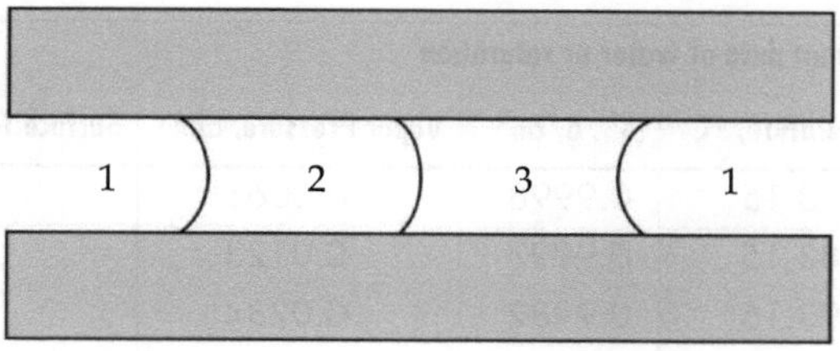

Hint: You may need to establish the following relationship from the criterion of mechanical equilibrium in your derivations:

$$\left(\frac{\sigma^{(1,2)}}{r^{(1,2)}}\right) + \left(\frac{\sigma^{(2,3)}}{r^{(2,3)}}\right) = \left(\frac{\sigma^{(1,3)}}{r^{(1,3)}}\right)$$

2.15 The Kelvin equation is often thought to be valid up to a pore size of about 8 nm (Walton and Quirke, 1989; Fisher and Israelachvili, 1983). Use the attached data for vapor pressure, density, and interfacial tension to compute the vapor pressure, and pressure in the liquid phase vs. temperature for a droplet and bubble of diameter = 8 nm. Which one of your results relates to capillary condensation in porous media (capillary tube) with contact angle of zero? Plot the results of your vapor pressure computations on the same figure together with vapor pressure data of the table. Compare your results with Fig 2.30.

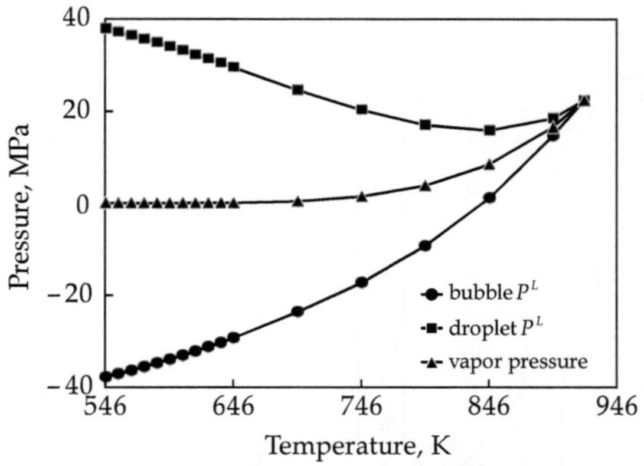

FIGURE 2.30

Relevant data of water at saturation			
Temperature, °C	ρ^L, g/cm³	Vapor Pressure, atm	Surface Tension, dyne/cm
273.15	0.9998	0.0061	75.7
283.15	0.9997	0.0123	74.2
293.15	0.9982	0.0234	72.8
303.15	0.9956	0.0425	71.2
313.15	0.9922	0.0738	69.6
323.15	0.9880	0.123	67.9
333.15	0.9832	0.199	66.2
343.15	0.9778	0.312	64.4
353.15	0.9718	0.474	62.6
363.15	0.9653	0.702	60.8
373.15	0.9584	1.013	58.8
423.15	0.9171	4.76	48.2
473.15	0.8647	15.544	37.6
523.15	0.7991	39.753	26.4
573.15	0.7124	85.903	14.7
623.15	0.5747	165.321	3.7
647.15	0.3560	220.640	0

2.16 Consider a circular capillary tube immersed partly in water (Fig. 2.31). The pressure outside is 1 atm and the ambient temperature is 20°C. The height of water in the tube is 14 m. The expression for the rise in a capillary tube is given by $h = 2\sigma/r\rho_w$, where σ is the interfacial tension, ρ_w is the water density, and r is the tube radius. Plot the pressure profile in water vs. height. Note that the pressure is negative above a height of about 10 m. Is negative pressure in contradiction with basic postulates of classical thermodynamics?

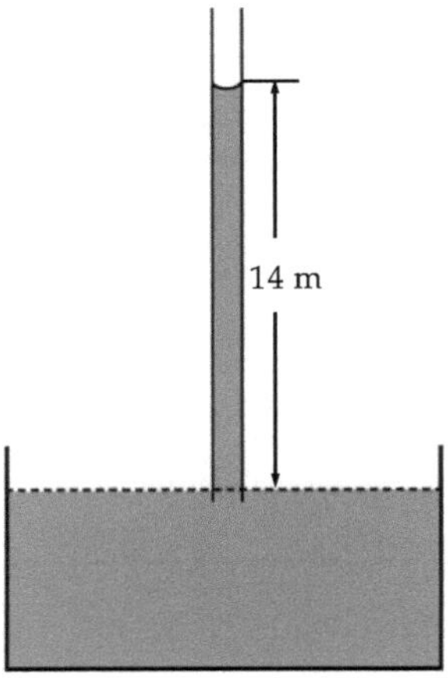

FIGURE 2.31 Water rise in a capillary tube.

2.17 Consider three configurations for a small water droplet (the sizes are all the same) on a substrate depicted in Fig. 2.32. We are interested to evaluate the equilibrium vapor pressure of the water droplets with respect to that of the flat surface. Will there be an increase or a decrease in vapor pressure for the three droplets?

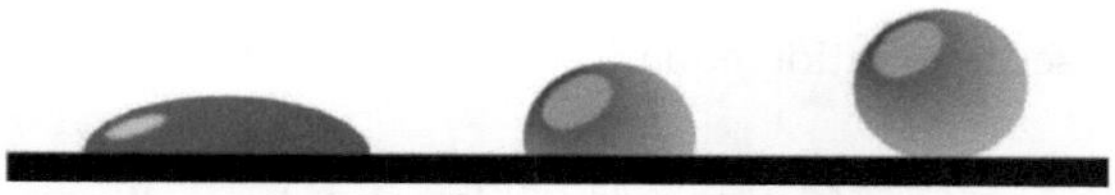

FIGURE 2.32

2.18 It is often believed that the mathematics and physics of long-ranged interactions are very different from that of short-ranged interactions. As an example, a particle that interacts through short-ranged interactions is affected only by its nearest neighbors in uncharged particles. But when interactions are long range, as in the case of charged particles, more distant neighbors contribute to the energies. Explain the reason that we need to add all the pairwise coulombic interactions for charged particles whereas for the uncharged particles we may need only the contributions from the neighboring pairs.

2.19 Suppose a system is composed of n_i species $(i = 1, \ldots, n_c)$ with charges $q_i (i = 1, \ldots, n_c)$. Then,

$$U = U(S, V, \underline{n}, \underline{q})$$

The electrostatic potential felt by charged species is Φ due to nearby charged surfaces, electrodes or constellation of charges. Derive the following expression for dU.

$$dU = TdS - PdV + \sum_{i=1_i}^{n_c} \mu_i dn_i + \sum_{i=1_i}^{n_c} \Phi dq_i$$

The total charge of species i is given by

$$q_i = z_i e n_i$$

where

z_i = valency (number of charges per ion),
e = unit charge of a proton, and
n_i = number of ions of type i.

Show that the expression for dG is given by

$$dG = -SdT + VdP + \sum_{i=1}^{n_c} (\mu_i + z_i e \Phi) dn_i$$

The quantity $(\mu_i + z_i e\Phi)$ defines the electrochemical potential

$$\mu_i' = \mu_i + z_i e\Phi$$

What is the condition of equilibrium for a system composed of two subsystems with charged particles?

2.20 The Gibbs sedimentation equation is given below

$$\boxed{(d\mu_i = -M_i g dz)_T} \quad i = 1, \ldots, n_c$$

Let us also write the differential of chemical potential at constant temperature.

$$\left[\mathrm{d}\mu_i = \left(\frac{\partial \mu_i}{\partial P}\right)_y \mathrm{d}P + \sum_{j=1}^{n_c-1} \left(\frac{\partial \mu_i}{\partial y_j}\right)_{P,y_j} \right]_T \quad i = 1, \ldots, n_c \tag{2.91}$$

From combing the above two equations and some preliminary steps we arrive at the following expression.

$$\sum_{j=1}^{n_c-1} \left(\frac{\partial \mu_i}{\partial y_j}\right)_{P,T,y_j} \left(\frac{\mathrm{d}y_j}{\mathrm{d}z}\right) = (\rho \bar{V}_i - M_i)g \quad i = 1, \ldots, n_c - 1 \tag{2.92}$$

Note that the number of equations has been reduced from n_c to $(n_c - 1)$. Explain the reason for the reduction from n_c in Eq. (2.91) to $(n_c - 1)$ in Eq. (2.92).

2.21 Consider a cylindrical capillary tube of radius r containing two immiscible fluids. The interface between the two fluids is characterized by a hemispherical surface ∂S that has a unit outward normal $\hat{r}$. Let the pressure on one side of the interface be $P_1\hat{r}$. The pressure on the other side of the interface is $-P_2\hat{r}$. What is the force on the interface due to these pressures?

2.22 The work in a magnetic field can be expressed by,

$$đW = -P\mathrm{d}V + \vec{H}\mathrm{d}(\vec{J}V)$$

where $\vec{H}$ is the magnetic field strength, and $\vec{J}$ is the magnetic polarization, P and V have their usual definitions. Note that in the above equation, the two contributions to work are the product of an intensive property and the differential of an extensive property. The change in internal energy of the open system in a magnetic field can be expressed as,

$$\mathrm{d}U = T\mathrm{d}S + \sum_{i=1}^{n_c} \tilde{\mu}_i \mathrm{d}n_i - P\mathrm{d}V + \vec{H}\mathrm{d}(\vec{J}V)$$

Use the Legendre transformation to show that

$$\mathrm{d}G = -S\mathrm{d}T + \sum_{i=1}^{n_c} \tilde{\mu}_i \mathrm{d}n_i + V\mathrm{d}P - (\vec{J}V)\mathrm{d}\vec{H}$$

Use the definition $(d\tilde{\mu}_i = RTd \ln \tilde{f}_i)_{T,\underline{n},\tilde{H}}$ to show that

$$\tilde{\mu}_i(T, P, \underline{n}, \vec{H}) - \mu_i(T, P, \underline{n}) = \frac{RT \ln \tilde{f}_i(T, P, \underline{n}, \vec{H})}{f_i(T, P, \underline{n})}$$

Note that $\tilde{\mu}_i = (\partial G/\partial n_i)_{T,P,n_j,\vec{H}}$ and $\mu_i = (\partial G/\partial n_i)_{T,P,n_j,\vec{H}=0}$. Similar definitions apply to $\tilde{f}_i$ and f_i.

2.23 In vaporization and in condensation, the degrees of superheating and supercooling are often very different. In a different way of stating the same problem, at constant temperature, when lowering the pressure of a liquid, the pressure can be reduced significantly below the saturation pressure. However, when increasing the pressure of the vapor from low pressure, the pressure increase may not be significant above the saturation pressure (see Fig. 3.6). Explain the difference between vaporization and condensation. As we will discuss in Chap. 8, crystallization and melting are also radically different.

2.24 Consider a small volume of a liquid in a capillary tube in Fig. 2.33. When the pressure on both sides of a liquid droplet is the same, the contact angle will be symmetric (a). The volume of the liquid in the vertical tube may also change the contact angle (d).

When the pressure on one side increases, the liquid droplet may not flow because of contact angle hysteresis. Based on the various sketches below, derive the following expressions:

(a) $(P_1 - P_2) \leq (2\sigma/r)(\cos\theta_R - \cos\theta_A)$ for sketches (b) and (c).

(b) $h \leq (2\sigma/\rho rg)(\cos\theta_R - \cos\theta_A)$ for sketches (d) and (e).

Use the sketches to discuss the end effects in (c) and (e).

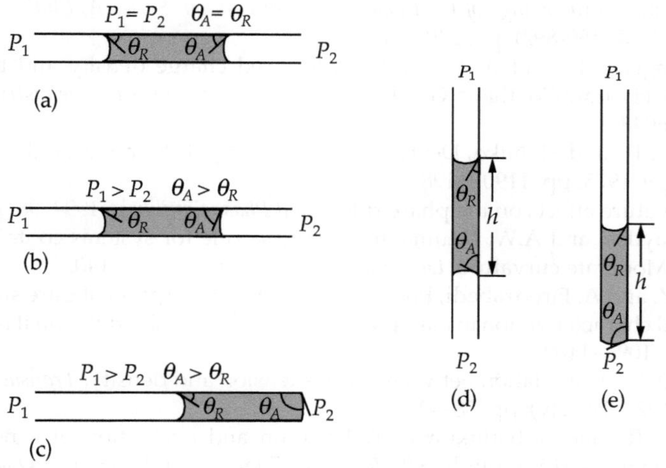

FIGURE 2.33

2.25 Equations (2.14) and (1.194) seem to be in contradiction. In Eq. (1.194), the composition is held constant. In Eq. (2.14), composition varies. Provide detailed derivation that the use of Eq. (1.194) leads to Eq. (2.14).

2.6 References

Aavatsmark, I., *Mathematiche Einfuhrug in die Thermodynamik der Gemische*. 1995, Berlin: Akademie Verlag.

Abascal, J. L. F., E. Sanz, R. Garcia Fernández, and C. Vega, A potential model for the study of ices and amorphous water: TIP4P/Ice. *Journal of Chemical Physics*, 2005. 122(23): p. 234511.

Acs, G., S. Dolescholl, and E. Farkas, General purpose compositional model. *SPE Journal*, 1985. 25(4): pp. 543–553.

Creek, J. L., and M. L. Schrader, East painter reservoir: an example of a compositional gradient from a Gravitational Field, in 60th Annual Technical Conference and Exhibition of the society of petroleum engineers. *SPE Journal:Las Vegas, Nevada*, 1985: p. 14411.

Defay, R., and I. Prigogine, *Surface Tension and Adsorption*. 1966, London: Longmans.

Espach, R. H., and J. Fry, Variable characteristics of the oil in the tensleep sandstone reservoir, elk Basin Field, wyoming and montana. *Transactions of the American Institute of Mining, Metallurgical and Petroleum Engineers*, 1951. 192: pp. 75–82.

Fahes, M., and A. Firoozabadi, Wettability alteration to intermediate gas-wetting in gas-condensate reservoirs at high temperatures. *SPE Journal*, 2007. 12(4): pp. 397–407.

Firoozabadi, A., and D. Kashchiev, Pressure and volume evolution during gas phase formation in solution gas drive process. *SPE Journal*, 1996. 1(3): pp. 219–227.

Firoozabadi, A., Katz, D. L., Soroosh, H. and Sajjadian, V.A., Surface tension of reservoir crude oil/gas systems recognizing the asphalt in the heavy fraction. *SPE Reservoir Engineering*, 1988. 3(1): pp. 265–272.

Fisher, L. R., and J. N. Israelachvili, Experimental studies on the applicability of the kelvin equation to highly curved concave menisci. *Journal of Colloid and Interface Science*, 1981. 80(2): pp. 528–541.

Fowler, R. H., A tentative statistical theory of macleod's equation for surface tension, and the parachor. *Proceedings of the Royal Society of London. Series A, Mathematical and Physical Sciences*, 1937. 159(897): pp. 229–246.

Jiménez-Ángeles, F., and A. Firoozabadi, Induced charge density and thin liquid film at Methane Hydrate/Methane Gas Interface, *Journal of Physical Chemistry C*, 2014. 118: pp. 26041–26048.

Kingston, P. E., and H. Niko, Development planning of the brent field. *Journal of Petroleum Technology*, 1975. pp. 1190–1198.

Li, D., Curvature effects on the phase rule. *Fluid Phase Equilibria*, 1994. 98: pp. 13–34.

Li, D., J. Gaydos, and A.W. Neumann, The phase rule for systems containing surfaces and lines. 1. Moderate curvature. *Langmuir*, 1989. 5(5): pp. 1133–1140.

Li, Z., Jin, Z. and A. Firoozabadi, Phase behavior and adsorption of pure substances and mixtures and characterization in nanopore structures by density functional theory. *SPE Journal*, 2014: pp. 1096–1109.

Macleod, D. B. On a relation between surface tension and density. *Transactions of the Faraday Society*, 1923. 19(July): pp. 38–41.

Neveux, A. R., and S. Sathikumar, Delineation and evaluation of a north sea reservoir containing near-critical fluids. *SPE Reservoir Engineering*, 1988: pp. 842–848.

Peng, D. Y., and D. B. Robinson, A new two-constant equation of state. *Industrial & Engineering Chemistry Fundamentals*, 1976. 15(1): pp. 59–64.

Santiso, E., and A. Firoozabadi, Curvature dependency of surface tension in multicomponent systems. *AIChE Journal*, 2006. 52(1): pp. 311–322.

Sigmund, P. M., P. M. Dranchuk, N. R. Morrow, and R. A. Purvis, Retrograde Condensation in Porous Media. *SPE Journal*, 1973. 13(2): pp. 93–104.

Starling, K. E., *Fluid Thermodynamic Properties for Light Petroleum Systems*. 1973, Gulf Publishing Company.

Sugden, S., VI.-The variation of surface tension with temperature and some related functions. *Journal of the Chemical Society, Transactions*, 1924. 125: pp. 32–41.

Sychev, V. V., *Complex Thermodynamic Systems*. 1973, New York: Consultant Bureau.

Tindy, R., and M. Raynal, Are test-cell saturation pressures accurate enough? *Oil and Gas Journal*, 1966. pp. 126–139.

Trebin, F. A., and G. I. Zadora, Experimental study of the effect of a porous media on phase changes in gas condensate systems. *Neft'i Gaz*, 1968. 81(8): pp. 37–41.

Walton, J. P. R. B., and N. Quirke, Capillary condensation: a molecular simulation study. *Molecular Simulation*, 1989. 2(4-6): pp. 361–391.

Wheaton, R., Treatment of variations of composition with depth in gas-condensate reservoirs. *SPE Reservoir Engineering*, 1991. 6(2): pp. 239–244.

CHAPTER 3

Equation-of-State Representation of Fluid Phase Behavior and Properties

The relation between pressure, temperature, and volume in pure substances in fluid state, and the additional composition variables in mixtures is one of the most fundamental aspects in phase behavior, physical, and thermal properties. The description of various fluid phases (gas and liquid) by cubic equations of state, when the aggregation effect of species is negligible, has been surprisingly successful. The cubic equations of state are deceptively simple. When there is association between species due to partial charge, we may resort to noncubic equations of state. In this chapter we will demonstrate powerful features of one of the most popular cubic equations of state, the Peng-Robinson equation of state, when neglect of association is a good approximation. When one needs to account for association, the cubic-plus-association (CPA) equation of state may become a method of choice.

This chapter starts with a simple analysis of the *P-V* space. In this connection we will include a short discussion of negative pressure which is present in many physical and natural processes. Then cubic equations of state will be presented in relation to phase behavior and density of reservoir fluids. Association of molecules and equations of state will be covered in relation to phase behavior of water, water-containing mixtures, and asphaltene precipitation. Working expressions for two-phase compressibility and sonic velocity and thermal properties will complete the chapter. In transition from single-phase to two-phase state, compressibility and sonic velocity may have a sharp discontinuity which implies lack of validity of averaging procedures. Some 15 examples and theory extension at the end of the chapter will provide details which are not covered in the main text. Among the examples, two are related to phase behavior in shale media when the pores are in the range of 10 to 100 nm. When pores are less than 10 nm, a fluid may become heterogeneous due to the interaction between the atoms/molecules of the pore walls and the fluid. In inhomogeneous fluids statistical thermodynamic models are required (Li, Jin, and Firoozabadi, 2014).

3.1 EOS Representation of Volumetric and Phase Behavior

Consider the plot of pressure versus total volume of a pure substance shown in Fig. 3.1. An equation of state (EOS) is desired to represent the volumetric behavior of the pure substance in the entire range of volume—both in liquid and in gaseous state. It is also desired that the EOS be a continuous function of volume.

Let us first examine the phase diagram of Fig. 3.1 at T_1, where $T_1 < T_c$; T_c is the critical temperature. At point S, the liquid is compressed. As the pressure is lowered, the volume increases. In Chap. 4 we will show that the volume increase is in line with the second law. At point A, the liquid is in the saturated and stable state. Toward the right of point A, as the pressure is lowered, the substance might follow one of two routes. It might follow the line AD, in which case point D represents the saturated and stable vapor, or it might follow curve AB, for which the fluid will be in a metastable condition. In this case, the limit of metastability is determined by the condition that $(\partial P/\partial V)_{T_1}$ vanishes (i.e., point B). Similarly, one can start from point R and observe that as the pressure increases, condensation may not occur up to point C, where $(\partial P/\partial V)_{T_1}$ will vanish at the limit of stability. Curve DC represents the locus of metastability and point C is the limit of metastability at T_1 for the vapor. The dashed line in Fig. 3.1 is the locus of the limit of metastability (i.e., the spinodal curve). Any point on this curve or curve AB, below zero, represents the liquid under the state of tension. The envelope shown by the solid thick line is the binodal curve representing the saturated equilibrium liquid and vapor loci.

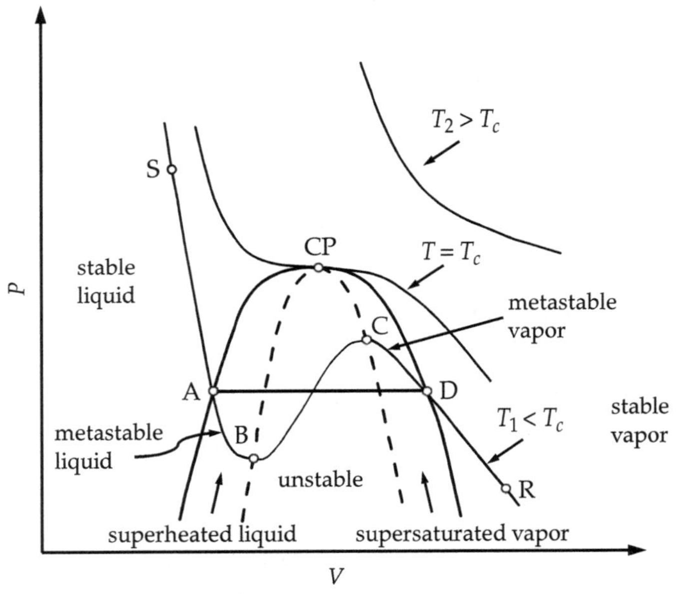

FIGURE 3.1 Pressure vs. volume of a pure substance. The sketch is abstract. Figure 3.6 represents the real behavior.

The region between the spinodal curve and the saturated liquid curve represents liquid in a superheated state. Similarly, the region between the saturated vapor curve and the spinodal curve represents the supersaturated vapor state. Inside the spinodal curve is the unstable region. In the next chapter, we will discuss the stability concept in detail and derive the corresponding mathematical expressions from the first principles.

Kenrick, Gilbert, and Wismer (1924) studied the superheating of a number of liquids at atmospheric pressure and found that there was a limiting temperature for each liquid, at or below which it always exploded. For example, the maximum value for the superheating of water at atmospheric pressure is around 270°C. Debenedetti (1996) presents metastability of water and other liquids in detail. In Example 3.11 we will calculate the maximum superheating of water at atmospheric pressure.

It was mentioned that a point on curve AB may be at negative pressure. In other words, a liquid could have negative pressure without vaporizing. Consider a simple experiment that was performed to examine the effect of negative pressure in a tube (Skuse, Firoozabadi, and Ramey, 1992). The apparatus of Fig. 3.2 was used to create negative pressure in a glass capillary. First, the bottle was partially filled to 2-cm depth with vacuum pump oil and a partial vacuum of 0.00038 psia (2.6 Pa) was created to remove air from the oil, then the vacuum was released. A capillary tube of 0.025-cm radius was placed in a vertical position inside the bottle, dipping into the vacuum pump oil. A capillary rise of 2.5 cm was observed. The capillary rise remained the same when

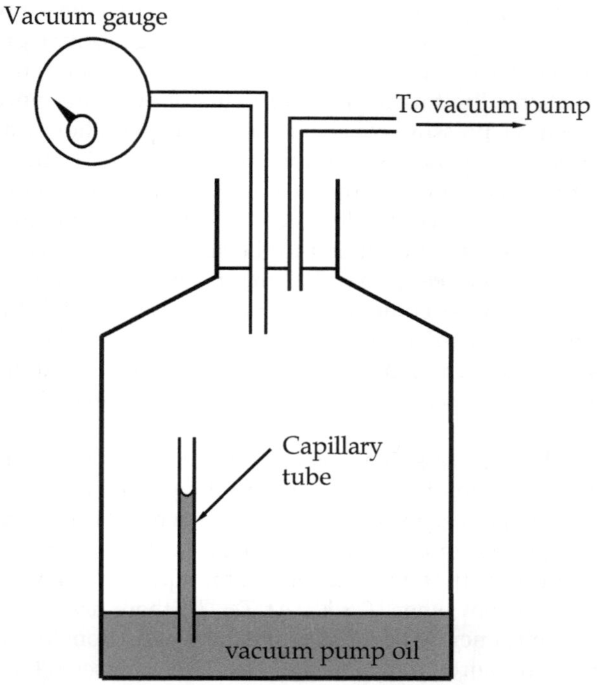

FIGURE 3.2 Apparatus used to establish negative pressure in a capillary tube (adapted from Skues, Firoozabadi, and Ramey, 1992).

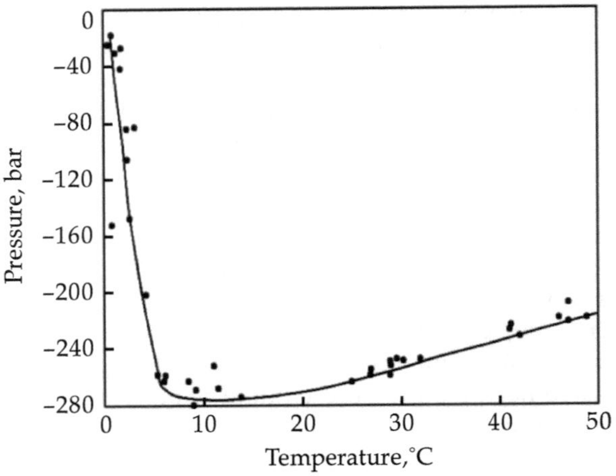

FIGURE 3.3 Limiting negative pressure of water between 0 and 50°C (adapted from Briggs, 1950).

the bottle was pumped to a vacuum of 0.00038 psia (2.6 Pa). For an oil density of 0.9 g/cm^3, the hydrostatic pressure at the top of the column was, therefore, about −0.0315 psia (−217 Pa). Nearly the entire column was at negative pressure. No vaporization was observed. Briggs (1950) measured the limiting negative pressure of water as a function of temperature using a centrifuge to generate negative pressure in a capillary tube. Figure 3.3 is a reproduction of Briggs' data. The limiting negative pressure of water from Fig. 3.3 is about −280 bar at a temperature of 9°C. The EOS and the stability criteria can be used to estimate the limiting negative pressure of pure substances and mixtures (see Chap. 4). The negative pressure concept is used to explain how water gets to the top of the trees. Tall trees may generate negative pressures of −11 atm when a tree is as tall as 112 m. Water is pulled through the tree by a tensible force generated from evaporation of water from leaf surfaces (Holbrook et al., 1995). In cosmology, negative pressure in empty space is discussed in relation to the accelerating universe (Steinhardt, 2003). When the gas-liquid interface is curved, there may be a permanent state of metastability inside the binodal curve. We could have negative pressures while the fluid is metastable; this can occur in a porous medium (Skuse, Firoozabadi, and Ramey, 1992).

It is also desired that an EOS predicts the limits of stability. Another desirable feature of an EOS is that it can describe the volumetric behavior even in the negative pressure region.

Next, we consider the pressure versus total volume plot shown in Fig. 3.4.

An EOS is an expression that can represent the phase behavior of the fluid, both in the two-phase envelope (i.e., inside the binodal curve), as was demanded above, on the two-phase envelope, and outside the binodal curve. The equations of state are divided into two main groups: cubic and noncubic. Cubic equations may have three roots when $T \leq T_c$ and only one root when $T > T_c$. At $T = T_c$, there are three equal roots. Figure 3.4 portrays the deficiency with most of the cubic equations of state. In this plot, the solid circles show measured data and the solid line represents the predictions from an EOS. The flatness around the critical point cannot be adequately described by most cubic equations. The liquid phase description is also not so good as the description of the gas phase. Later, we will discuss how the volume-translation technique can alleviate

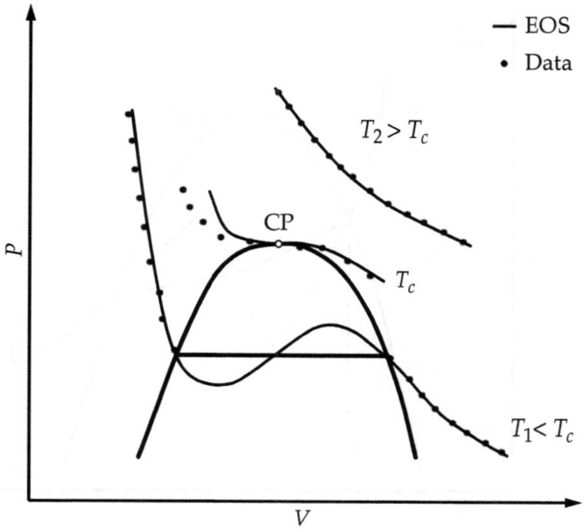

FIGURE 3.4 Equation-of-state representation of P vs. V for a pure substance. The sketch is abstract. Figure 3.6 represents a realistic behavior.

this deficiency. Noncubic equations can better describe the volumetric behavior of pure substances but may not be suitable for complex hydrocarbon mixtures, especially in the critical region.

For mixtures, the pressure-volume plot differs from that of pure substances; a schematic of a P-V plot for mixtures is shown in Fig. 3.5. The main differences between Figs. 3.4 and 3.5 are: (1) V^L and V^G of Fig. 3.5 do not represent the equilibrium states, and (2) the critical points have different features. For a pure substance, $(\partial P/\partial V) = (\partial^2 P/\partial V^2) = 0$ at the critical point. For a mixture, these two equations do not hold (see Fig. 3.5). The Z-factors in Fig. 3.4 of the equilibrium gas and liquid phases always meet the condition $Z^L < Z^G$. However, for mixtures, when gas and liquid phases are at equilibrium, Z^L might be smaller or larger than Z^G. At equilibrium, the mass density of the liquid phase is higher than the mass density of the gas phase, $\rho^L > \rho^G$. Then from

$$\rho = \frac{PM}{ZRT} \tag{3.1}$$

it follows that

$$\frac{M^L}{Z^L} > \frac{M^G}{Z^G} \tag{3.2}$$

when Z-factors are less than one, then Z^L could be smaller or larger than Z^G.

Similarly to pure substances, an equation of state should also represent the volumetric behavior of multicomponent fluids. The volumetric representation of mixtures is a more difficult task, especially around the critical point.

There are a large number of equations of state that have been proposed to represent the phase behavior of pure substances and mixtures in the gas and liquid states. In 1873,

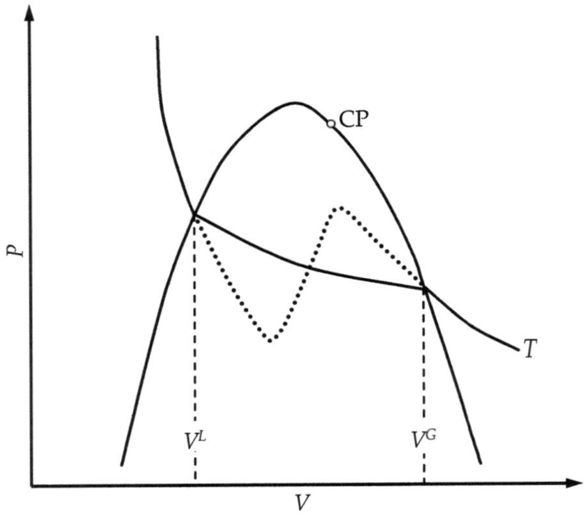

FIGURE 3.5 Representation of pressure vs. volume of a mixture. The sketch is abstract.

van der Waals introduced an EOS, which is known as the van der Waals equation of state:

$$\left(P + \frac{a}{v^2}\right)(v - b) = RT \tag{3.3}$$

In the above equation, $v = V/n$ is the molar volume. The parameters a and b have clear physical meaning. As an example, b is the so-called hard-core or covolume parameter, which is the state at which the fluid is completely packed with the molecules at infinite pressure. Therefore, v should be larger than b. Parameter a is energy related and accounts for interaction between particles, and will simply be referred to as the attraction parameter.

Equation (3.3) may be written as

$$\left[P_r + \frac{3}{v_r^2}\right](3v_r - 1) = 8T_r \tag{3.4}$$

where $P_r = P/P_c$, $T_r = T/T_c$, and $v_r = v/v_c$. Therefore Eq. (3.4) is a statement of "the principle of corresponding states," which means that at the same reduced pressure and reduced temperature, all substances have the same reduced volume. Generalized compressibility-factor charts for natural gases are based on the corresponding-states principle; that concept was also introduced by van der Waals in 1873.

In 1949, Redlich and Kwong (RK) made an important modification to the van der Waals EOS. They proposed

$$\left[P + \frac{a}{T^{\frac{1}{2}} v(v + b)}\right](v - b) = RT \tag{3.5}$$

Later Soave (1972) improved on the RK-EOS by replacing the term $a/T^{1/2}$ with a more general temperature-dependent term $a(T)$ and proposed a simple form for $a \equiv a(T_r, \omega)$ for all pure substances, taking advantage of the concept of the acentric factor of Pitzer (1939) and Pitzer et al. (1955). Pitzer's acentric factor, ω, was intended as an additional parameter for the improvement of the corresponding-states principle. The acentric factor is a measure of the difference in molecular structure between a given component and a gas with spherically symmetric molecules with $\omega = 0$ (such as argon).

Another important variation of the van der Waals EOS was introduced in 1976 by Peng and Robinson:

$$\left[P + \frac{a(T)}{v(v+b) + b(v-b)}\right](v - b) = RT \tag{3.6}$$

This equation improves the liquid density prediction, but still cannot describe volumetric behavior around the critical point because of a fundamental reason that will be discussed later. There are thousands of cubic equations of states, and many noncubic equations. The noncubic equations such as the Benedict-Webb-Rubin equation (1942) and its modification by Starling (1973) have a large number of constants; they describe accurately the volumetric behavior of pure substances. But for hydrocarbon mixtures and crude oils, because of mixing rule complexities, they may not be suitable (Katz and Firoozabadi, 1978). Cubic equations with more than two constants also may not improve the volumetric behavior prediction of complex reservoir fluids. In fact, most of the cubic equations have the same accuracy for phase-behavior prediction of complex hydrocarbon systems; the simpler equations often do better. Therefore, the discussion will be limited to the Peng-Robinson EOS, which according to the author's experience enjoys more simplicity and reliability than many other equations.

Before turning to the specifics of the PR-EOS, Maxwell's equal-area rule for pure substances will be derived for the van der Waals family of equations and the mathematical structure of these equations will be discussed. Maxwell's equal-area rule, which applies to the subcritical isotherm ($T < T_c$), is shown schematically in Fig. 3.6.

From the equality of chemical potentials or the Gibbs free energy at the saturation points A and E,

$$G^L = G^G \tag{3.7}$$

$$A^L + P^{sat}V^L = A^G + P^{sat}V^G \tag{3.8}$$

and, therefore,

$$A^G - A^L = -P^{sat}(V^G - V^L) \tag{3.9}$$

which is the area of the rectangle AEFG. An alternative expression for $A^L - A^G$ can be obtained by evaluating the integral of $\int \mathrm{d}A$ along the path ABCDE using $\mathrm{d}A = -P\mathrm{d}V$, since $\mathrm{d}T = 0$,

$$A^G - A^L = -\int P\mathrm{d}V \tag{3.10}$$

which is the area under the curve in Fig. 3.6. It then follows that the two shaded areas are equal. Note that Maxwell's equal-area rule on a *P-V* plot applies to pure substances only. In Prob. 4.4, we will address this concept for mixtures.

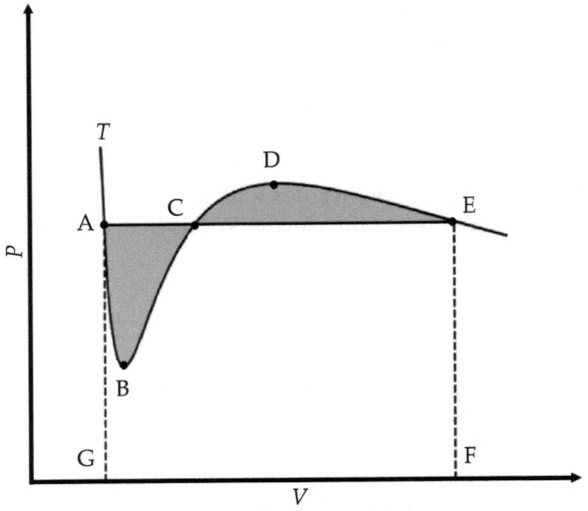

FIGURE 3.6 Maxwell's equal-area rule.

3.1.1 Algebraic Form of Cubic Equations

The cubic equations of state may have four branches in the P-V space. Examine the PR-EOS given by Eq. (3.6). A plot of Eq. (3.6) shown in Fig. 3.7 exhibits three vertical asymptotes:

$$v = b \tag{3.11}$$

$$v = +(\sqrt{2} - 1)b \tag{3.12}$$

$$v = -(\sqrt{2} + 1)b \tag{3.13}$$

The branches for which $v < b$ have no physical meaning. When $P > P_1$, there may be three roots: $-(\sqrt{2}+1)b < v < 0$, $0 < v < (\sqrt{2}-1)b$, and $v > b$. Only the root that is larger than b is identified as the molar volume of the liquid phase. The other two roots have no physical meaning. When $P_2 < P < P_1$, the only real root corresponds to the liquid phase. For $P_3 < P < P_2$, three real roots are obtained. The largest root corresponds to the vapor phase and the smallest root corresponds to the liquid phase. The intermediate root has no physical meaning. When $P < P_3$, two situations may arise for $v > b$: one shown in Fig. 3.7a and the other shown in Fig. 3.7b. In Fig. 3.7a, the root corresponds to the vapor phase, and in Fig. 3.7b, the root corresponds to the liquid phase, which might have a negative pressure.

3.1.2 Peng-Robinson Equation of State (PR-EOS)

The PR-EOS, Eq. (3.6), can be written as

$$Z^3 - (1 - \mathcal{B})Z^2 + (\mathcal{A} - 3\mathcal{B}^2 - 2\mathcal{B})Z - (\mathcal{A}\mathcal{B} - \mathcal{B}^2 - \mathcal{B}^3) = 0 \tag{3.14}$$

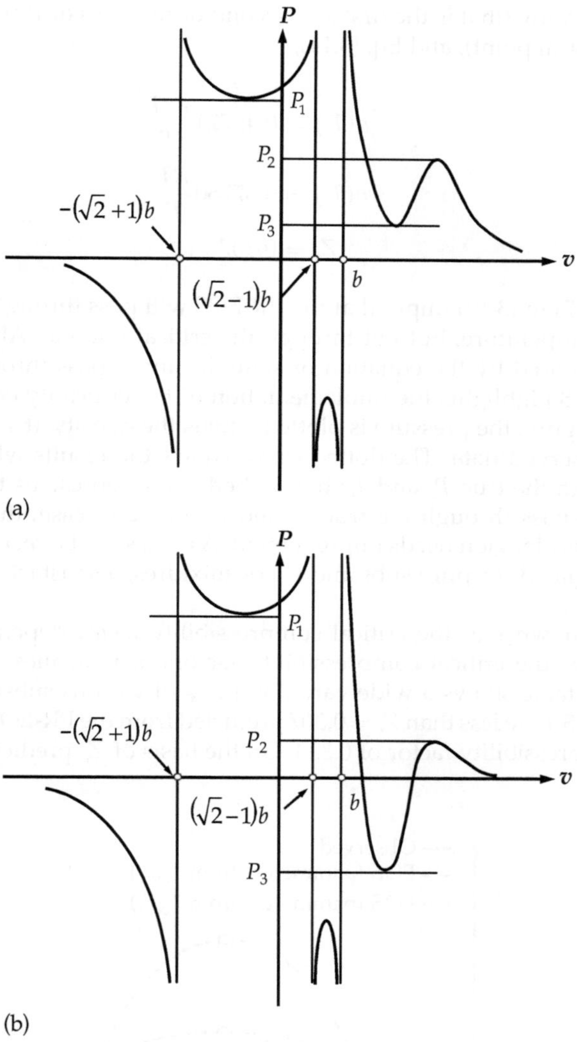

FIGURE 3.7 Various branches of the PR-EOS.

where

$$\mathcal{A} = \frac{aP}{R^2T^2} \tag{3.15}$$

$$\mathcal{B} = \frac{bP}{RT} \tag{3.16}$$

For pure substances, the first and the second derivatives of pressure with respect to volume at the critical point are equal to zero. (The condition of criticality will be derived in the next chapter.) Note that these derivatives may not be zero for mixtures. Using the criteria of criticality (that is the first and second derivatives of P with respect to v being zero at the critical point), and Eq. (3.14),

$$a(T_c) = 0.45724\frac{R^2T_c^2}{P_c} \tag{3.17}$$

$$b(T_c) = 0.07780\frac{RT_c}{P_c} \tag{3.18}$$

$$Z_c = 0.307 \tag{3.19}$$

Equations (3.17) to (3.19) imply that the PR-EOS will pass through the critical pressure and critical temperature, but not through the critical volume. Alternatively, if true T_c and v_c are honored by the equation of state, it cannot pass through the critical pressure. Figure 3.8 highlights the implementation of the criticality criteria in equations of state. In this figure, the pressure is plotted versus the density at $T = T_c$. The solid curve shows the observed data. The dotted curve shows the results when an EOS is forced to pass through the true P_c and T_c; the dashed curve represents the EOS results when it is forced to pass through the true T_c and v_c. In either case, the shortcoming of the EOS in the critical region needs improvement. As will soon be seen, the deficiency could be easily alleviated for pure substances. For mixtures, a satisfactory solution is not yet available.

It is well known that the critical compressibility factor depends on the substance. Table 3.1 gives the critical compressibility factors of n-alkanes and some nonhydrocarbons. This table shows a wide variation in Z_c of various substances. All the values listed in Table 3.1 are less than $Z_c = 0.307$ predicted from the PR-EOS. The SRK-EOS gives a critical compressibility factor of 0.333. On the basis of Z_c predictions, one expects the

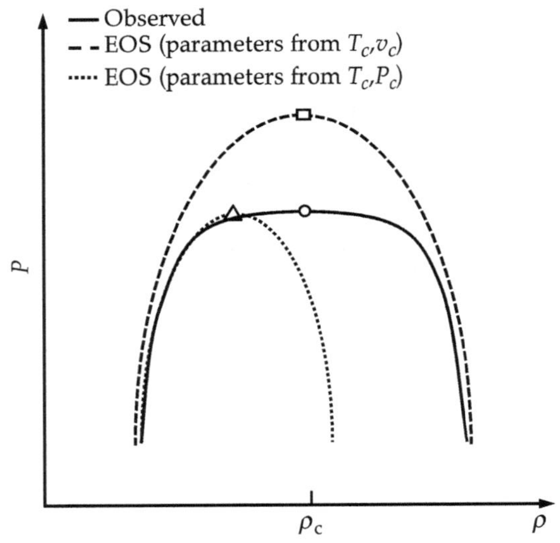

FIGURE 3.8 Coexistence curve in the critical region: EOS vs. observed (adapted from Chou and Prausnitz, 1989).

Comp.	T_c, K	P_c, MPa	ρ_c, g/cm³	T_b, K	Z_c
C_1	190.58	4.604	0.162	111.63	0.288
C_2	305.42	4.880	0.203	184.55	0.285
C_3	369.82	4.250	0.217	231.05	0.281
n-C_4	425.18	3.797	0.228	272.64	0.274
n-C_5	469.7	3.369	0.230	309.22	0.271
n-C_6	507.3	3.014	0.233	341.88	0.264
n-C_7	540.1	2.734	0.233	371.57	0.262
n-C_8	568.7	2.495	0.232	398.82	0.260
n-C_9	594.6	2.280	0.231	423.97	0.256
n-C_{10}	617.7	2.099	0.228	447.30	0.255
n-C_{11}	638.8	1.948	0.227	469.08	0.253
n-C_{12}	658.4	1.810	0.226	489.47	0.249
n-C_{13}	675.9	1.679	0.224	508.62	0.246
n-C_{14}	692.3	1.573	0.222	526.73	0.244
n-C_{15}	707.8	1.479	0.220	543.84	0.242
n-C_{16}	722.6	1.401	0.219	560.01	0.241
n-C_{17}	735.6	1.342	0.218	575.17	0.242
n-C_{18}	747.2	1.292	0.214	589.50	0.247
H_2O	647.4	22.104	0.400	373.30	0.229
CO_2	304.4	7.398	0.461	194.80	0.274
N_2	126.2	3.392	0.311	77.50	0.290

**n*-Alkanes data from Teja et al. (1990);
*Nonhydrocarbons data from *GPA Handbook* (1972)

TABLE 3.1 Critical Properties and Normal Boiling Points of *n*-Alkanes and Some Selected Nonhydrocarbons*

PR-EOS to predict pure component densities better than the SRK-EOS. Density predictions will be further discussed later.

At temperatures other than the critical temperature, the parameter a in Eq. (3.6) is given by

$$a(T) = a(T_c)\alpha(T_r, \omega) \tag{3.20}$$

The dimensionless parameter α is a function of T_r and the acentric factor, ω. Vapor pressure data are used to obtain α in the following manner.

At the boiling point (vapor-pressure condition),

$$f^L = f^G \tag{3.21}$$

Using Eq. (3.6), and Eq. (1.109),

$$\ln\left(\frac{f}{P}\right) = Z - 1 - \ln(Z - \mathcal{B}) - \frac{\mathcal{A}}{2\sqrt{2}\mathcal{B}} \ln\frac{Z + 2.414\mathcal{B}}{Z - 0.414\mathcal{B}} \tag{3.22}$$

Vapor-pressure data and Eqs. (3.14), (3.21), and (3.22) are used to estimate α for pure substances. The parameter α can be correlated by a simple expression:

$$\alpha = \left[1 + m\left(1 - T_r^{\frac{1}{2}}\right)\right]^2 \tag{3.23}$$

In Eq. (3.23), m is given by

$$m = 0.37464 + 1.54226\omega - 0.26992\omega^2 \tag{3.24}$$

Equation (3.24) is apparently based on vapor pressure data of hydrocarbons with $0 < \omega < 0.5$. The correlation was later expanded for ω in the range of $0.1 < \omega < 2.0$ (Robinson et al., 1985):

$$m = 0.3796 + 1.485\omega - 0.1644\omega^2 + 0.01667\omega^3 \tag{3.25}$$

For water and other polar substances, Mathias and Copeman (1983) suggest a different correlation which will be presented later.

The acentric factor in Eqs. (3.24) and (3.25) is defined as

$$\omega = -\log \left.\frac{P^{sat}}{P_c}\right|_{T_r=0.7} - 1 \tag{3.26}$$

If we assume the vapor-pressure data to be represented by

$$\log P^{sat} = \beta + \frac{\gamma}{T} \tag{3.27}$$

where β and γ are constants and P^{sat} is the saturation pressure at absolute temperature T, then

$$\omega = \frac{3}{7}\left[\frac{\log\frac{P_c}{14.695}}{\left(\frac{T_c}{T_b} - 1\right)}\right] - 1 \tag{3.28}$$

where P_c is in psia, and T_b is the normal boiling point with the same units as T_c. For mixtures, the a and b parameters are defined according to certain mixing rules. The following mixing rules for petroleum fluids have proved useful:

$$a = \sum_{i=1}^{n_c}\sum_{j=1}^{n_c} x_i x_j (a_{ij})^{1/2} \tag{3.29}$$

$$a_{ij}^{1/2} = (1 - \delta_{ij}) a_i^{1/2} a_j^{1/2} \tag{3.30}$$

$$b = \sum_{i=1}^{n_c} x_i b_i \tag{3.31}$$

where δ_{ij} (we also use the symbol k_{ij}) is the interaction parameter between component i and component j, and $\delta_{ij} = \delta_{ji}$. The interaction parameter is assumed to be independent of pressure and composition and generally independent of temperature. In the above equations, x_i represents the mol fraction and a_i and b_i represent the parameters of pure substance i.

From Eq. (1.109) and Eq. (3.6),

$$\ln\frac{f_i}{y_i P} = \frac{b_i}{b}(Z-1) - \ln(Z-\mathcal{B}) - \frac{\mathcal{A}}{2\sqrt{2}\mathcal{B}}\left[\frac{2\sum_{j=1}^{n_c} x_i a_{ij}}{a} - \frac{b_i}{b}\right]\ln\frac{Z+2.414\mathcal{B}}{Z-0.414\mathcal{B}} \tag{3.32}$$

This equation is very important in the thermodynamics of phase equilibria; the condition of the equality of the fugacity of equilibrium phases and the above equation provide the phase composition.

Next, we will present the use of the cubic equations in predicting: (1) the volumetric properties of pure components, (2) the phase behavior of multicomponent mixtures, and (3) the phase behavior of reservoir fluid systems.

Pure Substances

Generally, the PR-EOS and other similar cubic equations reliably represent the vapor pressure of pure substances since vapor pressure data are used to obtain the parameter α. The density prediction is the weak point and may need a modification. An exception is the ZJRK-EOS (Zudkevitch and Jaffe, 1970). In this equation the constants of the a and b parameters of the SRK-EOS, Ω_a^0 and Ω_b^0, are assumed to be temperature-dependent. These two parameters are given by

$$a = \Omega_a^0 \frac{R^2 T_c^2}{P_c} \tag{3.33}$$

$$b = \Omega_b^0 \frac{R T_c}{P_c} \tag{3.34}$$

For the SRK-EOS, the dimensionless constants Ω_a^0 and Ω_b^0 are 0.4274 and 0.0867, respectively. However, in the ZJRK-EOS, these two parameters are determined from saturated liquid density and the equality of the saturated liquid and vapor phase fugacities. At the critical temperature and above, these two parameters are assigned values of 0.4274 and 0.0867, respectively. In the PR and SRK equations, no parameter is adjusted for density. As a result, these two equations have a density-prediction deficiency. Figure 3.9 shows the deviation in liquid molar volume of selected substances at $T_r = 0.7$ versus ω. The SRK-EOS underestimates the liquid density of all substances that are shown in the figure. The PR-EOS overestimates the density to $\omega = 0.35$, and then underestimates the density of n-alkanes heavier than n-C_8. This figure clearly shows that at $T_r = 0.7$, the SRK-EOS is best suited for density prediction of pure hydrocarbons with $\omega \approx 0$, while the PR-EOS performs best for n-heptane and other hydrocarbons with $\omega \approx 0.35$. Figure 3.10 shows the deviation for the molar liquid volume of n-hexane as a function of reduced temperature. This figure reveals that the deviation is nearly constant up to a reduced temperature of 0.75. The volume-translation concept was introduced to take advantage of this feature

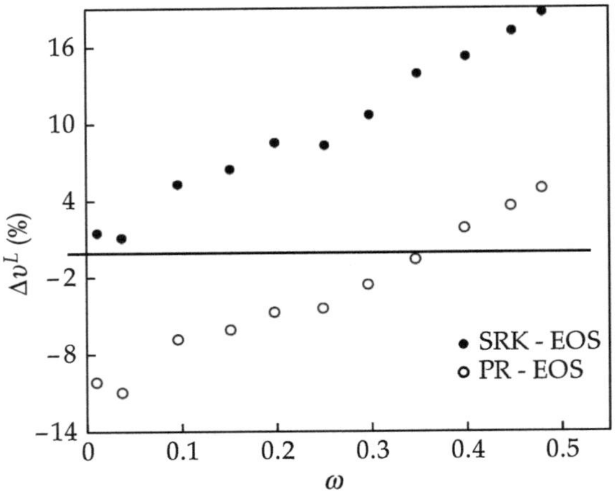

FIGURE 3.9 Percent deviation (normalized) in liquid molar volume at $T_r = 0.7$ as a function of acentric factor (adapted from Firoozabadi, 1988a).

(Péneloux et al., 1982; Jhaveri and Youngren, 1988). The volume-translation technique separates vapor-liquid-equilibria (VLE) from density calculations (see Example 3.3). The translation along the volume axis is given by

$$v^{true} = v^{EOS} + c \tag{3.35}$$

where c is the volume-translation parameter. Figure 3.10 shows that the volume translation may not improve volumetric prediction above $T_r = 0.7$; for the PR-EOS, it will make

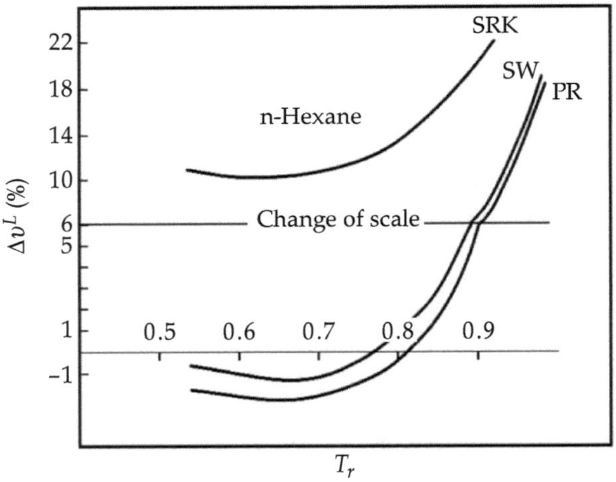

FIGURE 3.10 Percent deviation (normalized) in liquid molar volume for n-hexane as a function of reduced temperature (adapted from Firoozabadi, 1988a). SW represents the Schmidt-Wenzel EOS, 1980.

the predicted density worse. An additional correction term has been suggested by Mathias, Naheiri, and Oh (1989):

$$v^{true} = v^{EOS} + c + f_c \left[\frac{\lambda}{\lambda - \dfrac{v^2}{RT}\left(\dfrac{\partial P}{\partial v}\right)_T} \right] \tag{3.36}$$

In Eq. (3.36), $(-v^2/RT)(\partial P/\partial v)_T$ is a dimensionless quantity related to the inverse of the compressibility. This dimensionless quantity is zero at the critical point, and its value is relatively high at low reduced temperatures. Therefore, at conditions away from the critical point, the third term on the right side of Eq. (3.36) becomes negligible. At the critical point,

$$v_c^{true} = v_c^{EOS} + c + f_c \tag{3.37}$$

where f_c from Eq. (3.37) is

$$f_c = v_c^{true} - \left(c + v_c^{EOS}\right) \tag{3.38}$$

The modified volume translation forces the EOS to pass through T_c, P_c, and v_c. There is a universal value for λ for the PR-EOS, $\lambda = 0.41$. This value was determined by regressing data for many substances.

Figure 3.11 shows the predicted liquid densities for nitrogen and water when the modified volume translation is used. It is evident that the last term on the right side of Eq. (3.36) provides an important description for pure substances in the critical region.

Multicomponent Mixtures

The volume-translation concept is extended to mixtures according to the following mixing rule:

$$c = \sum_{i=1}^{n_c} x_i c_i \tag{3.39}$$

$$v_c = \sum_{i=1}^{n_c} x_i v_{c_i} \tag{3.40}$$

The application of the conventional volume-translation technique generally improves the volume predictions away from the critical region, although the improvement is not guaranteed. In the retrograde and near-critical regions, both volume and modified volume translations may fail to predict volumetric behavior accurately (Bjorlykke and Firoozabadi, 1992).

3.1.3 Phase Behavior of Mixtures with Well-Defined Components

The vapor pressure data for pure components are used to obtain the α parameter of the EOS; the density is then predicted. For mixtures, the EOS can be used to calculate not only the mixture density but also the phase behavior. In a limited sense, the phase behavior means the compositions and amounts of the equilibrium phases. This chapter presents

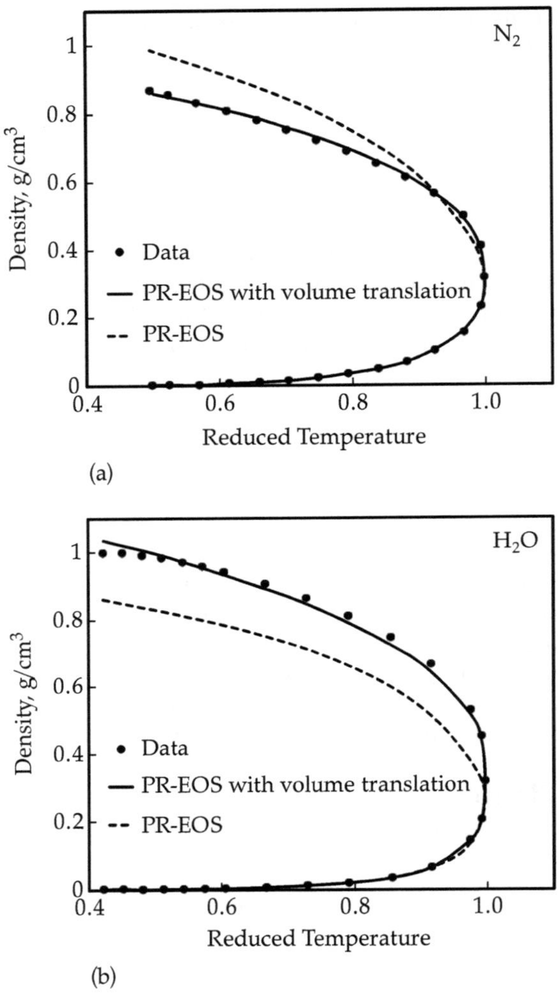

FIGURE 3.11 Comparison of data and predicted saturated densities of (a) N_2 and (b) H_2O (adapted from Mathias et al., 1989).

the phase-behavior description. Essentially, phase-behavior calculations rely on the use of the expression for the fugacity of component i in the mixture given by Eq. (3.32) for the PR-EOS.

Mixtures with well-defined components comprise of substances with a known boiling point. One may restrict the definition to mixtures with components of known critical properties.

The critical properties (T_c and P_c) and either the boiling point or acentric factor, ω, are needed to describe pure components by an EOS. For mixtures, the only additional parameter is the binary interaction parameter (often called interaction parameter). This coefficient is often empirically determined from one or more data points of binary

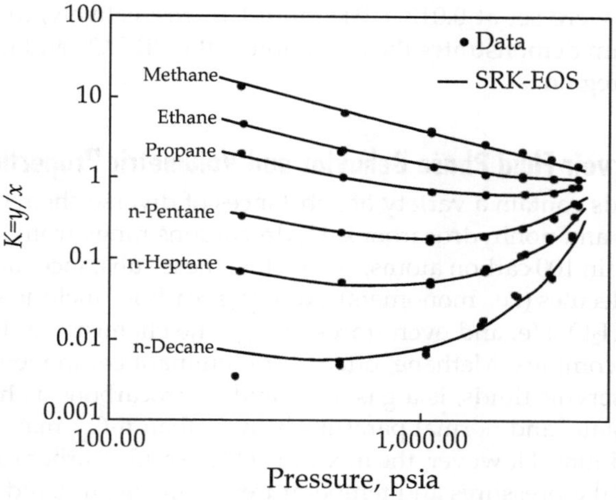

FIGURE 3.12 Predicted and measured data of equilibrium ratios of a well-defined hydrocarbon mixture (adapted from Yarborough, 1979). $T = 200°F$.

mixtures in the form of bubblepoint or *K*-values (equilibrium ratios). Once these coefficients are available, VLE for multicomponent systems can be predicted. Figure 3.12 shows a comparison of the measured and computed *K*-values of a well-defined mixture (Yarborough, 1979; Firoozabadi, 1988a). The SRK-EOS was used in the calculations. Note that the agreement between the measured and calculated values is good. For mixtures of different hydrocarbons where critical properties and normal boiling points or acentric factors are available, the VLE predictions from the PR-EOS and other similar equations are surprisingly good, except in the critical region. Figure 3.13 depicts the measured and predicted phase behavior (in terms of mol fractions) of the C_1/n-C_4/n-C_{10} system at 3,000 psia and 160°F. The interaction coefficients between C_1/n-C_4, C_1/n-C_{10},

FIGURE 3.13 Phase diagram showing the data and prediction of the C_1/n-C_4/n-C_{10} mixture at 160°F (adapted from Firoozabadi and Aziz, 1986).

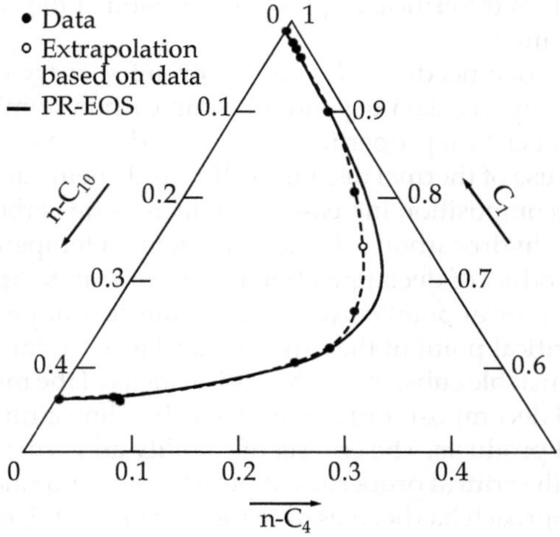

and C_4/n-C_{10} were set at 0.012, 0.044, and 0.01, respectively, in the calculations. This ternary diagram demonstrates the limitation of the PR-EOS and other similar equations in the critical region.

3.1.4 Reservoir Fluid Phase Behavior and Volumetric Properties

Reservoir fluids contain a variety of substances of diverse chemical nature that include hydrocarbons and nonhydrocarbons. Hydrocarbons range from methane to substances that may contain 100 carbon atoms, even when these substances are in the form of singly dispersed molecules (i.e., monomers). Nonhydrocarbons include substances such as N_2, CO_2, H_2S, S, H_2O, He, and even traces of Hg. The chemistry of hydrocarbon-reservoir fluids is very complex. Methane, often a predominant component of natural gases and petroleum-reservoir fluids, is a gas, n-C_5 and hydrocarbons as heavy as n-C_{15} may be in the liquid state, and normal paraffins heavier than n-C_{15} may be in the solid state at ambient conditions. However, the mixture of these hydrocarbons may be in a gaseous or liquid state at the pressures and temperatures often encountered in hydrocarbon reservoirs. The mixture may also be a solid as in waxes (Firoozabadi, 1999). The majority of reservoirs fall within the temperature range of 80 to 350°F, and the pressure range of 50 to 20,000 psia. When steam is injected into hydrocarbon reservoirs, the temperature may exceed 550°F and for in-situ combustion, the temperature may be even higher.

In spite of the complexity of hydrocarbon fluids found in the subsurface, simple cubic equations of state have shown surprising performance in the phase-behavior calculations, for both vapor-liquid and vapor-liquid-liquid equilibria of these complex fluids. For the phase behavior of water-hydrocarbon fluids, water-CO_2, water-H_2S and their mixtures at high pressure, the incorporation of association of water molecules may become necessary and could be added to the cubic equations. For the purpose of asphaltene-rich phase precipitation from petroleum fluids, the same concept of association for water can be used. The addition of association does not result in complexity; the simplicity of the calculations remains intact.

Reservoir fluids are a complex mixture of thousands of components that may exhibit very complex phase behavior. It is, however, surprising that a simple two-constant equation of state such as the PR-EOS can do an excellent job for vapor-liquid equilibria description away from the critical region. A discussion of the various parameters for the description is presented.

As stated earlier, one needs T_c, P_c, and T_b or ω for every component in a mixture. However, critical property data beyond a certain carbon number are not yet available. The measurement of critical properties of heavy hydrocarbons is difficult and is subject to uncertainty because of thermal decomposition and chemical reactions at high temperatures. Thermal decomposition increases with increasing carbon number and temperature. When a heavy hydrocarbon is heated, it reaches a temperature at which it starts to decompose. The product of decomposition increases with temperature and time. Therefore, the measured critical point corresponds to the critical point of a mixture. In order to determine the critical point of the original substance, certain assumptions have to be made. For mildly unstable substances, the critical locus of the mixture (original substance and the products of decomposition) is assumed to be a linear function of the mol fractions of the decomposed products. The criteria of stability and criticality from Chap. 4 can be used to determine the critical properties of pure heavy components (Anselme, 1988; Teja et al., 1990). This approach has been used to measure T_c and P_c of normal alkanes to n-C_{18}

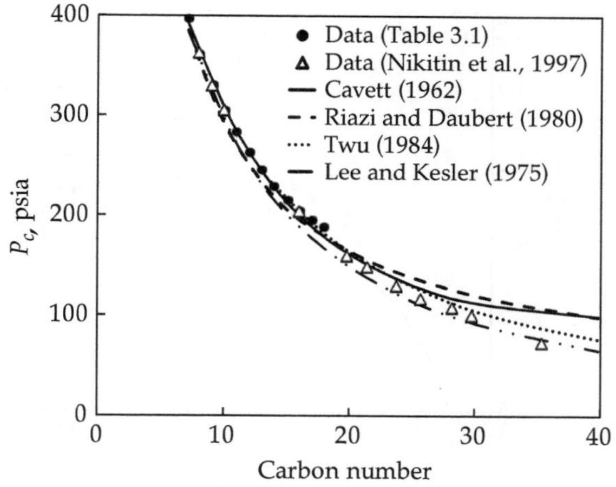

FIGURE 3.14 P_c of n-alkanes.

by Anselme (1988). The pulse-heating method has also been used to measure the critical pressure and temperature (Nikitin et al., 1993). Using this method, Nikitin et al., (1997) measured critical pressure and temperature of several normal alkanes including n-C_{36}.

For heavy n-alkanes, one could use various correlations. Figures 3.14 and 3.15 plot critical P_c and T_c from various correlations. These two figures show that the uncertainty in T_c estimation is less than that in P_c. Extrapolation of the correlation to n-C_{100} may result in unrealistic values of T_c and P_c. Also note that the trend of P_c data from Anselme (1988) and Nikitin et al. (1997) are not the same. Until the data of Nikitin et al. (1997) is verified, it should be used with caution.

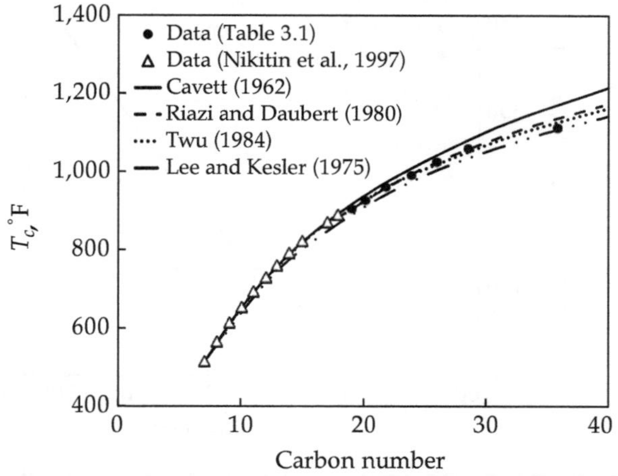

FIGURE 3.15 T_c of n-alkanes.

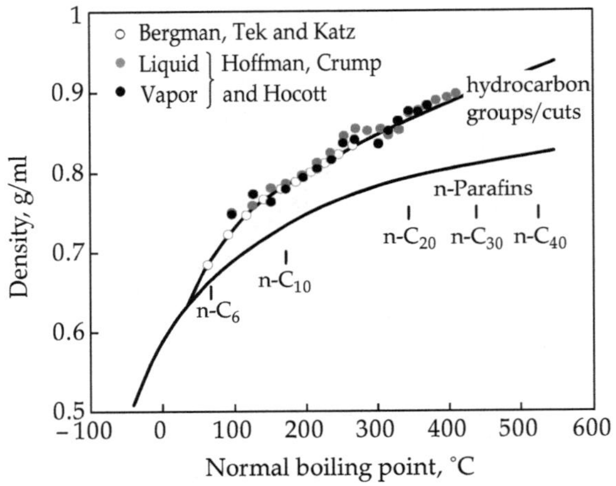

FIGURE 3.16 Density at 15.5°C vs. boiling points of hydrocarbon groups in crude-oil and condensate systems (adapted from Katz and Firoozabadi, 1978).

For crude oils, it is not practical to provide the analysis to thousands of components. Instead, for the C_{6+} residue, average boiling point, liquid density, and molecular weight of groups or cuts are measured. When only the amount and the density and the molecular weight of the C_{6+} or a heavier plus fraction is available, then it may become necessary for some calculations to estimate an extended analysis. The two-parameter gamma distribution function (Johnson and Kotz, 1970; Whitson, 1983) can be used for generating the molar distributions for the plus fractions. Figures 3.16 and 3.17 give average density and molecular weight for groups of compounds boiling between 0.5°C above the previous n-paraffin and 0.5°C above the n-paraffin carbon number used to identify the group. Table 3.2 shows the same data in tabular form. Crude oils, with few exceptions,

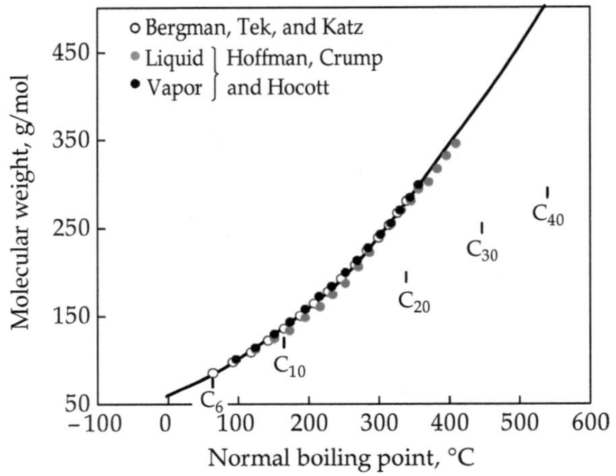

FIGURE 3.17 Molecular weight vs. normal boiling points of hydrocarbon groups in crude-oil and condensate systems (adapted from Katz and Firoozabadi, 1978).

Hydrocarbon Group	Boiling Range		Average Boiling Point		Density at 60°F	Molecular Weight
	°C	°F	°C	°F	g/cm³	g/gmol
C_6	36.5 to 69.2	97.9 to 156.7	63.9	147	0.685	84
C_7	69.2 to 98.9	156.7 to 210.1	91.9	197.5	0.722	96
C_8	98.9 to 126.1	210.1 to 259.1	116.7	242	0.745	107
C_9	126.1 to 151.3	259.1 to 304.4	142.2	288	0.764	121
C_{10}	151.3 to 174.6	304.4 to 346.4	165.8	330.5	0.778	134
C_{11}	174.6 to 196.4	346.3 to 385.5	187.2	369	0.789	147
C_{12}	196.4 to 216.8	385.5 to 422.2	208.3	407	0.800	161
C_{13}	216.8 to 235.9	422.2 to 456.7	227.2	441	0.811	175
C_{14}	235.9 to 253.9	456.7 to 489.2	246.4	475.5	0.822	190
C_{15}	253.9 to 271.1	489.2 to 520	266	511	0.832	206
C_{16}	271.1 to 287.3	520 to 547	283	542	0.839	222
C_{17}	287 to 303	547 to 577	300	572	0.847	237
C_{18}	303 to 317	577 to 603	313	595	0.852	251
C_{19}	317 to 331	603 to 628	325	617	0.857	263
C_{20}	331 to 344	628 to 652	338	640.5	0.862	275
C_{21}	344 to 357	652 to 675	351	664	0.867	291
C_{22}	357 to 369	675 to 696	363	686	0.872	305
C_{23}	369 to 381	696 to 717	375	707	0.877	318
C_{24}	381 to 392	717 to 737	386	727	0.881	331
C_{25}	392 to 402	737 to 756	397	747	0.885	345
C_{26}	402 to 413	756 to 775	408	766	0.889	359
C_{27}	413 to 423	775 to 793	419	784	0.893	374
C_{28}	423 to 432	793 to 810	429	802	0.896	388
C_{29}	432 to 441	810 to 825	438	817	0.899	402
C_{30}	441 to 450	826 to 842	446	834	0.902	416
C_{31}	450 to 459	842 to 857	455	850	0.906	430
C_{32}	459 to 468	857 to 874	463	866	0.909	444
C_{33}	468 to 476	874 to 888	471	881	0.912	458
C_{34}	476 to 483	888 to 901	478	895	0.914	472
C_{35}	483 to 491	901 to 915	486	908	0.917	486
C_{36}			493	922	0.919	500
C_{37}			500	934	0.922	514
C_{38}			508	947	0.924	528
C_{39}			515	959	0.926	542

TABLE 3.2 Generalized Properties of Petroleum Hexane-Plus Groups (from Katz and Firoozabadi, 1978)

Hydrocarbon Group	Boiling Range		Average Boiling Point		Density at 60°F	Molecular Weight
	°C	°F	°C	°F	g/cm³	g/gmol
C_{40}			522	972	0.928	556
C_{41}			528	982	0.930	570
C_{42}			534	993	0.931	584
C_{43}			540	1004	0.933	598
C_{44}			547	1017	0.935	612
C_{45}			553	1027	0.937	625

TABLE 3.2 Generalized Properties of Petroleum Hexane-Plus Groups (from Katz and Firoozabadi, 1978) (*Continued*)

follow these two graphs. The practice is to use (from measurements or estimates from Figs. 3.16 and 3.17) average normal boiling-point and liquid density instead of measured properties available for pure components. From the density and boiling point, correlations such as those of Cavett (1962) or Lee and Kesler (1975) are used to estimate critical properties. In addition to critical properties and normal boiling point, one needs binary interaction coefficients for phase-behavior calculation of petroleum fluids. It has been shown (Katz and Firoozabadi, 1978) that interaction coefficients for methane-heavy component binaries significantly improve the phase-behavior calculation of reservoir fluids. Measured phase behavior of binary mixtures provide interaction coefficients between hydrocarbon-hydrocarbon, hydrocarbon-nonhydrocarbon, and nonhydrocarbon-nonhydrocarbon pairs. In crude oils only one data point, such as a bubblepoint pressure, which is often available, can provide the interaction coefficient between methane and the heavy end. Various interaction coefficients given by Arbabi and Firoozabadi (1995) are summarized in Table 3.3. Figure 3.18 compares predicted and measured equilibrium ratios for the lean natural gas condensate of Hoffman et al. (1953). The data of Hoffman et al. (1953) are perhaps the most complete condensate measurement in the literature. The data and prediction results are in good agreement. Figure 3.19 compares the predicted and experimental values for a crude oil from Roland (1945). From two data points at 120°F, the interaction coefficient between C_1 and C_{28+} residue was estimated (Katz and Firoozabadi, 1978). The estimated interaction coefficient was used to predict the *K*-values at 200°F. The figure shows excellent results from the PR-EOS.

1	$\delta_{C_1-C_i} = 0.0289 + 1.633 \times 10^{-4} MW_{C_i}$ (Arbabi and Firoozabadi, 1995), where C_i is the hydrocarbon heavier than C_1
2	$\delta_{C_1-CO_2} = 0.15$
3	$\delta_{C_1-N_2} = 0.10$
4	$\delta_{C_1-H_2S} = 0.10$

TABLE 3.3 Binary Interaction Coefficients for the PR-EOS

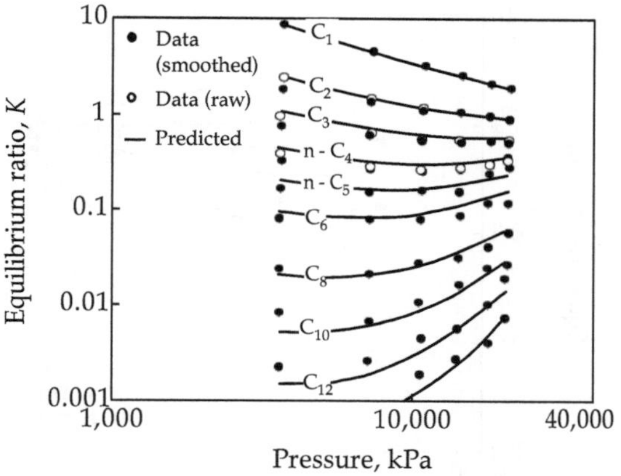

FIGURE 3.18 Equilibrium ratios from the PR-EOS predictions and data of Hoffman et al. (1953) condensate at 201°F (from Katz and Firoozabadi, 1978).

Figure 3.20 shows the computed results and data for a near-critical condensate. The figure shows that as the *K*-values approach one (i.e., critical point), the prediction results deviate from the measured data.

Both the usefulness and promise of the cubic equations to describe gas and liquid phases and the problems of the description in the critical region and density deviation have forced some users to adjust several parameters of the EOS to match measured and calculated phase and volumetric behavior. These adjusted parameters, in essence, include critical properties and interaction coefficients. The PR-EOS can readily describe the phase behavior of CO_2-crude systems. Due to interest in CO_2 improved oil recovery and unique features of CO_2 in petroleum fluids, we will briefly discuss CO_2 dissolution in petroleum fluids.

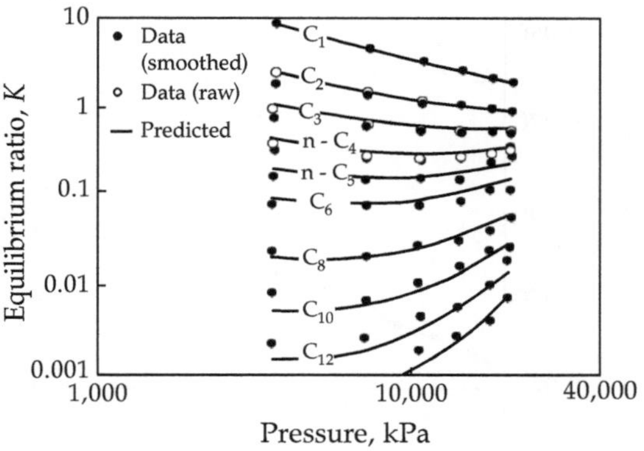

FIGURE 3.19 Predictions from the PR-EOS and data of equilibrium ratios of Roland crude oil (1945) at 200°F (adapted from Katz and Firoozabadi, 1978).

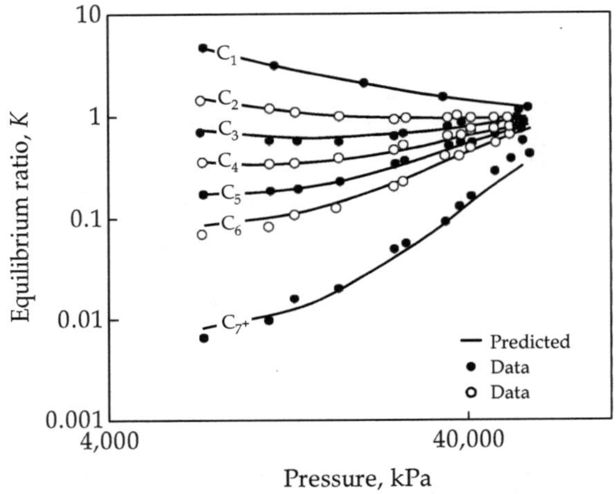

FIGURE 3.20 Predictions and data of equilibrium ratios of the GPA gas-condensate fluid at 100°F (from Firoozabadi, 1988a).

3.1.5 CO_2 Dissolution in Petroleum Fluids and Property Changes

Dissolution of CO_2 in petroleum fluids has three interesting features. One feature relates to increase in density in most crudes. The density increase can be as high as 5 to 7 percent (Ahmed et al., 2012). The density increase in water from CO_2 dissolution can be 1 to 2%. Figure 3.21 shows the density increase in two oils (A and B). Fluid A is a relatively heavy oil with density = 0.9433 g/cm^3 and viscosity of 256 cp at 75°F. Fluid B is a heavy oil with density = 0.9725 g/cm^3 and viscosity of 5392 cp at 75°F. The density increase from CO_2 dissolution in oils can be properly modeled by the Peng Robinson EOS using the volume translation. A volume translation parameter of $(c/b) = 0.25$ provides excellent match of CO_2 density prediction with the Peng-Robinson EOS to the measured data at

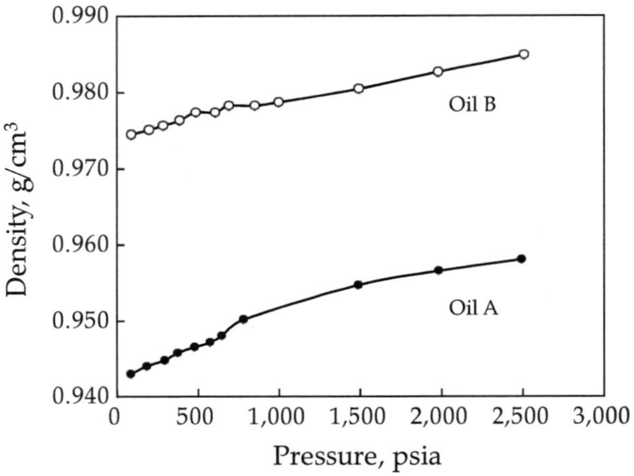

FIGURE 3.21 CO_2-saturated oil density at 75°F (adapted from DeRuiter et al., 1994).

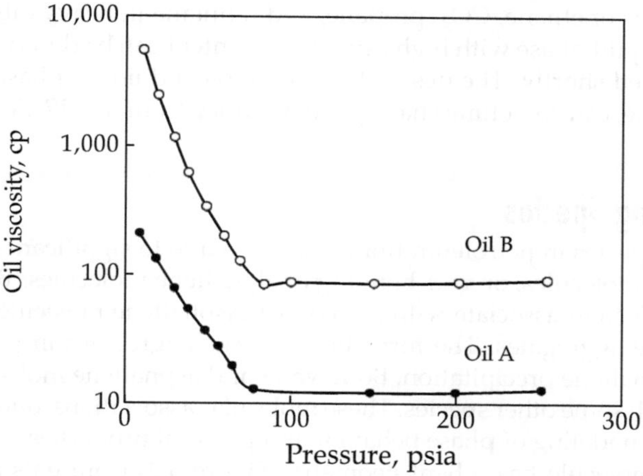

FIGURE 3.22 CO_2-saturated oil viscosity at 75°F (adapted from DeRuiter et al., 1994). The pronounced viscosity reduction is from CO_2 solubility.

46.7°C. The same shift parameter gives a good prediction of oil density with dissolved CO_2 (Ahmed et al., 2012). At other temperatures, a different value of the volume translation may be used. Another interesting feature in CO_2 dissolution in petroleum fluids is the swelling (increase in volume). The swelling in heavy oils can be as high as 40% or more. Figure 3.22 shows the pronounced viscosity reduction for petroleum fluids A and B discussed above. Note that there are more than two orders of magnitude of viscosity reduction in fluid B which is heavier than fluid A. For completeness, we have included viscosity reduction from solubility of methane, ethane, CO_2, propane, and n-butane in the same oils in Fig. 3.23. Generally, there is formation of a second liquid phase at high

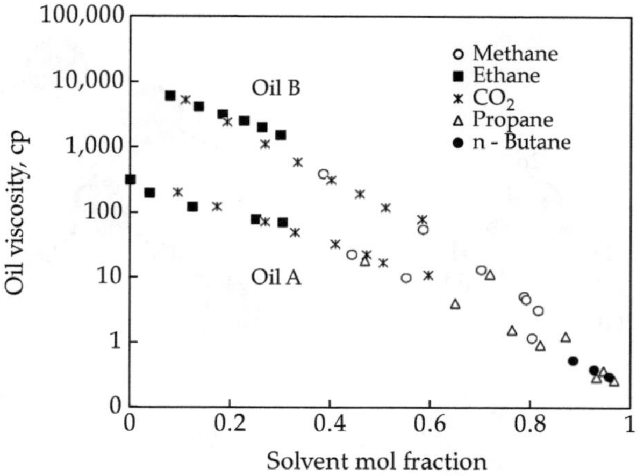

FIGURE 3.23 Oil saturated with various hydrocarbons and CO_2 at 75°F (adapted from DeRuiter et al., 1994).

concentrations of ethane, CO_2, propane, and n-butane in heavy oils. The description of the second liquid phase with high asphaltene content can be described by the CPA-EOS to be discussed shortly. The description of the second liquid phase from light alkanes, such as ethane, can sometimes have good accuracy from the PR-EOS.

3.2 Associating Species

Two of the species in petroleum fluids associate due to significant partial charge of the atoms in the molecules, or π–π bonding of alphaltene molecules (Breure et al., 2013).

Water molecules associate with each other. Asphaltene molecules may also associate to form large aggregates. The formation of large aggregates in petroleum fluids may result in asphaltene precipitation. Both water and asphaltene molecules may also cross-associate with some other species. These molecular associations, once accounted for, provide proper modeling of phase behavior and physical properties.

A water molecule has a bent geometry. Figure 3.24a presents a sketch of covalent bonds between the oxygen atom and the two hydrogen atoms. The electrons in the oxygen-hydrogen bonds are more strongly attracted to the oxygen atom than to the hydrogen atoms. As a result, the atoms in the water molecule are partially charged, despite the fact that the water molecule is electrically neutral. The partial charge of hydrogen atoms is assumed to be 0.59 e in the TIP4P-ice water model (Abascal et al., 2005). Note that −e is the charge of an electron. The partial charge of an oxygen atom in water is therefore −1.18 e. There are various charge models for water. Water forms tetrahedral bonding with neighboring water molecules as shown in Fig. 3.24b. We can account for water association and cross-association with other species in the cubic-plus-association approach.

Asphaltene molecules are the most complex species in the crude. First we describe the different types of hydrocarbons in petroleum fluids before showing a sketch of an asphaltene molecule. Figure 3.25 presents a sketch of three different types of

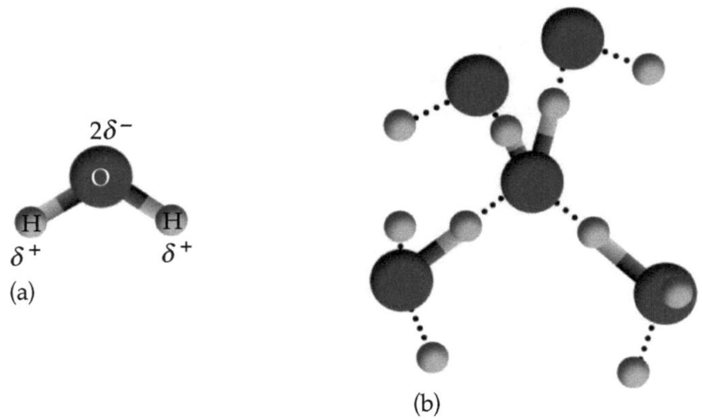

FIGURE 3.24 (a) Sketch of H_2O molecule with partial charge of atoms; (b) Association of water molecules (adapted from Dill and Bomberg, 2003).

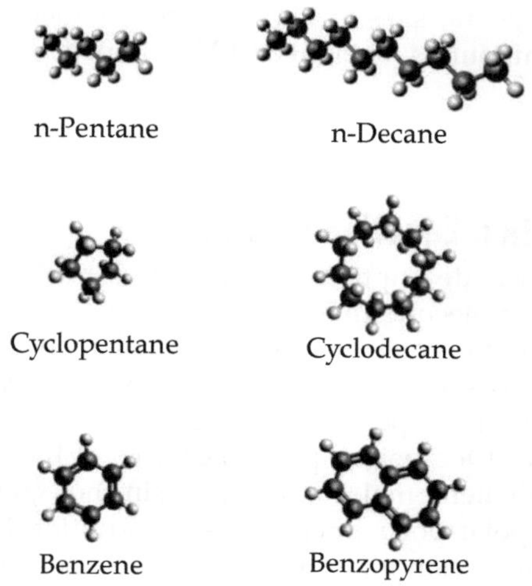

FIGURE 3.25 Hydrocarbon molecules. Top row, normal pentane and normal decane; middle row, cyclopentane and cyclodecane; bottom row, benzene and benzopyrene. Carbon atoms are shown in gray and hydrogen atoms are shown in light gray.

hydrocarbons. The first row shows linear molecules of normal alkanes, n-C_5 and n-C_{10}. The second row shows the naphthene molecules, cyclopentane and cyclodecane. The third row shows single- and two-ring aromatics, benzene and benzopyrene, respectively. Note the bonding differences in alkane/cyclo-alkanes, and aromatics. Asphaltenes are members of the aromatics group. They are the most polar fraction of a petroleum fluid (Goual and Firoozabadi, 2002). Asphaltenes consist of polyaromatic rings and aliphatic chains. They are made of carbon and hydrogen, in addition to heteroatoms such as nitrogen (occurring as amines or amides), oxygen (occurring as carbonylic, phenolic, or ketonic), and sulfur (occurring as thiobs and thiophenes) (Speight and Moschopedis, 1982). Asphaltenes often contain trace amounts of metals such as vanadium, nickel, zinc, and iron (Hashmi and Firoozabadi, 2013). The metals carry positive charge, and the heteroatoms may have negative charge. The asphaltenes, despite positive and negative charges in different parts may have low net charge (Hashmi and Firoozabadi, 2012). Figure 3.26 presents the sketch of an asphaltene molecule. Note that the heteroatoms

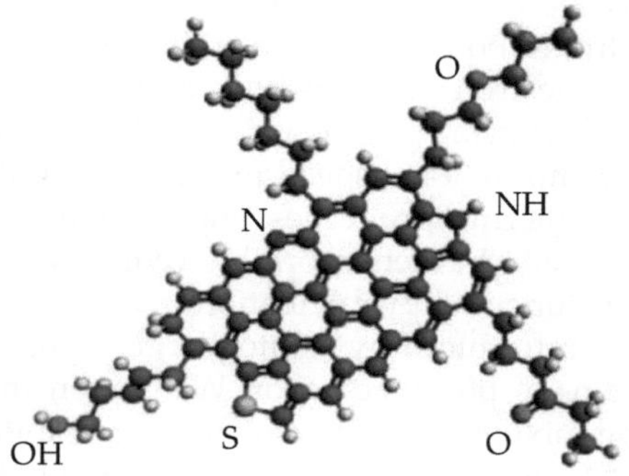

FIGURE 3.26 Molecular structure of asphaltenes which consist of a core of condensed polyaromatic rings with aliphatic chains. The heteroatoms (N, O, S) occur on peripheral aromatics, while the metals exist as chelates within porphyrin compounds (not shown).

(N, O, S) are at the peripheral sites in the asphaltene molecule. The metals exist as a chelate or porphyrin compounds (Speight and Moschopedis, 1982).

3.3 Cubic-Plus-Association Equation of State

We have given a very brief description of water and asphaltene molecule. In addition to self-association, these molecules may also cross-associate with other molecules. The association between water molecules and between asphaltene molecules is due to polar-polar interactions. The cross-association of water molecules with carbon dioxide, H_2S and other species is either from permanent polar moment of CO_2 and H_2S molecules, or from the water-induced temporary polar moments of hydrocarbon molecules. The cross-association of asphaltene molecules with resin molecules and heavy aromatics is believed to be from polar-polar interactions. In addition to asphaltene molecules, a petroleum fluid contains resin molecules which are also polar. The resin molecules are less polar than asphaltene molecules (Goual and Firoozabadi, 2002). They may not contain metals but may contain heteroatoms to a lesser degree than asphaltenes. Resins have more alkane tails than asphatenes which make them soluble in normal alkanes.

When there is a significant association between molecules, a cubic equation of state cannot describe accurately the phase behavior. Cubic equations of state such as the Peng-Robinson describe physical interactions between molecules. Physical interactions include London dispersion forces (instantaneous dipole-induced dipole interactions), dipole-dipole interactions (Keesom forces), and dipole-induced dipole interactions (Debye forces). The interactions from these forces are referred to as van der Waals interactions. There are many approaches to account for various interactions between molecules. The Peng-Robinson equation of state describes reservoir fluids in vapor-liquid equilibria. For liquid-liquid equilibria and vapor-liquid-liquid equilibria in reservoir fluids when one of the liquid phases is very rich in asphaltenes, we may need cubic-plus association EOS. This will allow association description in addition to physical interactions. Association interactions between molecules is due to hydrogen bonds between a molecule which has electronegative atoms such as oxygen, and nitrogen bond to hydrogen atoms in another molecule. Hydrogen bonds are stronger than physical interactions. In association, there is overlap of the electron clouds; in physical interactions there is no significant overlap of electron cloud. The cubic-plus-association (CPA) EOS may be the method of choice for accuracy and computational speed for associating fluids. The accuracy is in both phase behavior and in thermal properties. The thermal property description by the CPA-EOS will not be covered. A recent work by Myint, Hao, and Firoozabadi (2015) demonstrates good results for enthalpy calculations.

The CPA-EOS was first introduced by Kontogeorgis et al. (1996). It can be used for broad range of applications in phase behavior. We demonstrate powerful features of CPA for water-containing mixtures and for asphaltene precipitation.

We start our derivations with the expression for the Helmholtz free energy departure function by writing,

$$A^{departure} = A - A^{ideal\ gas} \tag{3.41a}$$

The ideal gas by definition consists of noninteracting molecules of zero volume. The departure contribution is a measure of the contribution from intermolecular forces and nonzero volume of the molecules. The contributions to $A^{departure}$ are divided into two categories,

$$A^{departure} = A^{departure}_{physical} + A^{departure}_{association} \tag{3.41b}$$

The physical Helmholtz free energy relates to short-range repulsion and dispersion attractions which are described by the PR-EOS. The association part describes hydrogen bonding between water molecules and between opposite partial charges in atoms of different molecules. When there are electrostatic interactions between species, additional terms are introduced in Eq. (3.41b). Based on the expression $P = -(\partial A/\partial V)_{T,\underline{n}}$, we obtain the EOS. The expression for $A^{departure}_{physical}$ for the Peng-Robinson EOS is given by (Wu and Prausnitz, 1998).

$$\frac{A^{departure}_{physical}}{nRT} = \ln\left(\frac{v-b}{v}\right) - \left(\frac{a}{2\sqrt{2}bRT}\right)\ln\left(\frac{v+(1+\sqrt{2})b}{v+(1-\sqrt{2})b}\right) \tag{3.42}$$

The association contribution to the Helmholtz free energy is based on the thermodynamic perturbation theory of Wetheim derived from statistical mechanics (Wertheim, 1984a, 1984b, 1986a, 1986b, 1987). In this approach, the molecules can form chains or rings through linking of the association of sites. The association strength between molecules is the key. Development of the statistical association fluid theory (SAFT) represents an important application of Wertheim's perturbation theory (Chapman et al., 1989, 1990; Huang et al., 1990, 1991; Kontogeoris and Flos, 2010). Based on Wertheim's perturbation theory, the association contribution to the Helmholtz free energy is given by

$$\frac{A^{departure}_{association}}{nRT} = \sum_{i=1}^{n_c} x_i \left\{ \sum_{j=\alpha,\beta,\ldots} \left[\ln \chi_{ij} - \frac{\chi_{ij}}{2}\right] + \frac{1}{2}M_i \right\} \tag{3.43}$$

where x_i is the mol fraction of molecule i which forms association (or cross-association), χ_{ij} is the mol fraction of molecule i not bonded at the site j, and M_i is the number of association sites of molecule i; n_c type of molecules in the mixture form association or cross-association. In Eq. (3.43), association sites are represented by $\alpha, \beta, \ldots$ The expression for $\chi_{i\alpha}$ is given by

$$\chi_{i\alpha} = \left[1 + \sum_{j=1}^{n_c} \sum_{k=\alpha,\beta,\ldots} c x_j \chi_{jk} \Delta^{jk}_{i\alpha}\right]^{-1} \tag{3.44}$$

where c is the molar density ($c = 1/v$) and $\Delta^{jk}_{i\alpha}$ is the association strength which is key in association or cross-association. It represents the association strength between site α in molecule i and site k in molecule j. For a single-component system such as water ($i = 1$), the Helmholtz free energy departure function due to association [from Eq. (3.43)] is given by

$$\frac{A^{departure}_{association}}{nRT} = \sum_{j=\alpha,\beta,\ldots} \left[\ln \chi_{1j} - \frac{\chi_{1j}}{2}\right] + \frac{1}{2}M \tag{3.45}$$

Suppose there are 4 sites of association (i.e., $M=4$; say α, β, δ, and γ sites). Then Eq. (3.45) gives

$$\frac{A^{departure}_{association}}{nRT} = \left(\ln\chi_{1\alpha} - \frac{\chi_{1\alpha}}{2}\right) + \left(\ln\chi_{1\beta} - \frac{\chi_{1\beta}}{2}\right) + \left(\ln\chi_{1\delta} - \frac{\chi_{1\delta}}{2}\right) + \left(\ln\chi_{1\gamma} - \frac{\chi_{1\gamma}}{2}\right) + 2 \tag{3.46}$$

The expression for mol fraction of molecules not bonded at site α is from Eq. (3.44)

$$\chi_{1\alpha} = \left[1 + \sum_{k=\alpha,\beta,\ldots} c\chi_{1k}\Delta^{1k}_{1\alpha}\right]^{-1} \tag{3.47}$$

Note that the extension of Eqs. (3.45) and (3.47) to mixtures is a simple geometrical averaging [see Eqs. (3.43) and (3.44)].

In the following derivations, we use the 4-site model. For water, the 4-site model produces good results. Figure 3.27 shows the 4-site model for water. There are two different association-site types (α and β). There are two sites of each type. The rationale for two types of sites (α and β) for water molecules is that two are electron donors and two are electron acceptors. The electron acceptors are designated by α and the electron donors are designated by β. In Fig. 3.27, the lone pairs on oxygen represent the electron acceptor and the hydrogens represent the electron donors. We will use the same 4-site model for the cross-associating species.

The following assumptions are made to derive expressions for the Helmholtz free energy of association for water (Li and Firoozabadi, 2009) and asphaltene association and cross-association (Li and Firoozabadi, 2010a and 2010b). We assume only one associating molecule. This molecule is designated by index "1." There are four sites in the associating molecule for association with like molecules and cross-association with other molecules.

1. All sites are symmetric for both associating species, and species that cross-associate. There is no difference in the probability of association in different sites.
2. The associating species can associate with like species and cross-associate with unlike species. Cross-association does not occur among unlike species. There is no cross-association of cross-associating species: $\Delta^{j\beta}_{i\alpha} = 0 \; i, j = 2, \ldots, n_c$.

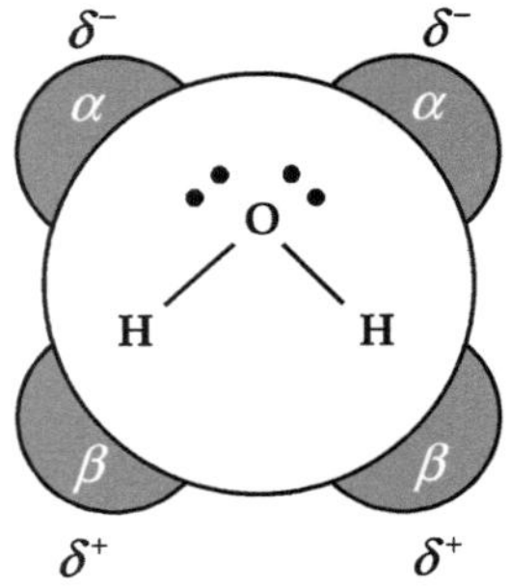

FIGURE 3.27 Sketch of a 4-site model of water.

3. There is neither association nor cross-association from the sites that have the same signs. This implies that

$$\Delta_{i\alpha}^{j\alpha} = \Delta_{i\beta}^{j\beta} = 0 \quad i, j = 1, \ldots n_c$$

4. The association and cross-association strength is symmetric between two different sites:

$$\Delta_{i\alpha}^{j\beta} = \Delta_{i\beta}^{j\alpha} \quad i = 1, \ldots, n_c$$

Based on the above criteria, we derive the following expressions using Eq. (3.44). For the associating species, we use index 1 as mentioned above. (There is only one associating species in the mixture.)

$$\chi_{1\alpha} = \chi_{1\beta} = \left[1 + 2c\left(x_1 \chi_{1\alpha} \Delta_{1\alpha}^{1\beta} + \sum_{j=2}^{n_c} x_j \left(\sum_{k=\alpha,\beta} \chi_{jk} \Delta_{1\alpha}^{j\beta}\right)\right)\right]^{-1} \tag{3.48}$$

For the cross-associating species

$$\chi_{i\alpha} = \chi_{i\beta} = \left[1 + 2c x_1 \chi_{i\alpha} \Delta_{i\beta}^{i\alpha}\right]^{-1} \quad i = 2, \ldots, n_c \tag{3.49}$$

The association Helmholtz free energy departure function is obtained from Eq. (3.43)

$$\frac{A_{association}^{departure}}{nRT} = \sum_{i=1}^{n_c} 4x_i \left(\ln \chi_{i\alpha} - \frac{\chi_{i\alpha}}{2} + \frac{1}{2}\right) \tag{3.50}$$

The self-association strength between site α and site β in the self-associating species is given by

$$\Delta_{1\alpha}^{1\beta} = g\kappa_{1\alpha}^{1\beta} b \left[\exp\left(\frac{\varepsilon_{1\alpha}^{1\beta}}{kT}\right) - 1\right] \tag{3.51}$$

where $\kappa_{1\alpha}^{1\beta}$ and $\varepsilon_{1\alpha}^{1\beta}$ are the self-association bonding volume and energy parameters, respectively. The association energy is related to the well depth, and the association volume is related to the well width in the square well potential. In Eq. (3.51) g is the radial distribution function, a probabilistic measure of finding a neighboring molecule at a distance from the reference molecule. The analytical expression for g based on the Carnahan-Starling (1969) hard sphere EOS reads

$$g \approx \frac{2 - \eta}{2(1 - \eta)^3} \tag{3.52}$$

where η is the reduced density defined by

$$\eta = \frac{b}{4v} = \frac{bc}{4} = \frac{\mathcal{B}}{4Z} \tag{3.53}$$

Equations (3.51), (3.52), and (3.53) provide the association strength of self-associating molecules. Li and Firoozabadi (2009) suggest the relation between self-association and cross-association strengths

$$\Delta_{1\alpha}^{j\beta} = S_{1j}\Delta_{1\alpha}^{1\beta} \quad j = 2, \ldots, n_c \tag{3.54}$$

where S_{1j} is a factor that relates the association strength of species 1 to the cross-association strength of species j. We are now set to present the CPA-EOS.

3.3.1 CPA-EOS

The pressure in the CPA comprises physical and association contributions

$$P = P^{physical} + P^{association} \tag{3.55}$$

The physical part is given by the Peng-Robinson EOS. The association contribution to pressure is given by $P^{association} = -(\partial A_{association}^{departure}/\partial V)_{T,\underline{n}}$. Based on Eq. (3.50), one can derive the following expression

$$P^{association} = -\frac{2RT}{v}\left(1 - v\frac{\partial \ln g}{\partial v}\right)\sum_{i=1}^{n_c} x_i(1 - \chi_{i\alpha}) \tag{3.56}$$

The derivation of the above equation as well as the derivation of fugacity coefficient expression requires the evaluation of the derivatives of site fractions. Michelsen and Hendricks (2001) outline a procedure to carry out the derivatives in the CPA–EOS. In the problems at the end of the chapter, we make remarks in relation to the derivations in the CPA–EOS.

The expression for the compressibility factor $Z = Z^{physical} + Z^{association}$ is obtained readily [$Z^{association}$ from Eq. (3.56) is used]

$$Z = \frac{Z}{Z - \mathcal{B}} - \frac{\mathcal{A}Z}{(Z^2 + 2\mathcal{B}Z - \mathcal{B}^2)} - 2\left(1 + c\frac{\partial \ln g}{\partial c}\right)\sum_{i=1}^{n_c} x_i(1 - \chi_{i\alpha}) \tag{3.57}$$

Equation (3.57) is not cubic. It can be solved by the bisection method to obtain all the roots. Alternately, we can provide the initial guess from the PR-EOS. The root to be selected is the one which has the lowest Gibbs free energy as we will discuss in Chap. 5. There are simple and efficient methods to solve Eq. (3.57).

We will use the following expressions to obtain fugacity coefficients

$$\ln \varphi_i = \frac{1}{RT}\left(\frac{\partial A^{departure}}{\partial n_i}\right)_{T,V,\boldsymbol{n}_j} - \ln Z \tag{3.58a}$$

$$\ln \varphi_i^{physical} = \frac{1}{RT}\left(\frac{\partial A_{physical}^{departure}}{\partial n_i}\right)_{T,V,\boldsymbol{n}_j} - \ln Z \tag{3.58b}$$

$$\ln \varphi_i^{association} = \frac{1}{RT}\left(\frac{\partial A_{association}^{departure}}{\partial n_i}\right)_{T,V,\boldsymbol{n}_j} \tag{3.58c}$$

The combination of physical and association contributions gives

$$\ln \varphi_i = \left\{ -\ln\left(\frac{Z-\mathcal{B}}{Z}\right) + \frac{b_i}{b}\left[\frac{\mathcal{B}}{Z-\mathcal{B}} + \frac{\mathcal{A}Z}{Z^2+2\mathcal{B}Z-\mathcal{B}^2}\right] - \frac{a}{2\sqrt{2}bRT}\left[\frac{2\sum_{j=1}^{n_c} x_j a_{ij}}{a} - \frac{b_i}{b}\right] \ln\left(\frac{Z+(1+\sqrt{2})\mathcal{B}}{Z+(1-\sqrt{2})\mathcal{B}}\right)\right\} + \left\{\sum_{j=\alpha,\beta} \ln \chi_{ij} - \frac{1}{2}\frac{\partial \ln g}{\partial n_i}\left[\sum_{i=1}^{n_c} n_i \sum_{j=\alpha,\beta}(1-\chi_{ij})\right]\right\} - \ln Z \tag{3.59}$$

In the following we present the results from the use of the CPA-EOS in water-hydrocarbons, water-CO_2, water-H_2S, and their mixtures. Then another use in relation to asphaltene precipitation will be presented. Parameters of the CPA-EOS will be also discussed.

3.3.2 Water-Containing Mixtures

Phase behavior of water-containing mixtures is of interest in a variety of disciplines in the field of energy and environmental sciences. Water is a predominant species in hydrocarbon reservoirs and may cover the rock surfaces. It is also injected in both light and in heavy oil reservoirs. In heavy oil reservoirs, steam is injected to reduce the oil viscosity from heating. Some of the species in reservoir fluids such as CO_2 and H_2S dissolve in water significantly. Sequestration of CO_2 in saline aquifers is affected by solubility of CO_2 in water (Firoozabadi and Cheng, 2010). Compared to other substances of similar molecular size and mass, water has a high boiling point. Water and hydrocarbon mixtures have strong asymmetric solubility. In liquid-liquid equilibria, the solubility of water in the hydrocarbon-rich phase is often several orders of magnitude greater than that of solubility of hydrocarbons in water-rich phase, provided the mixture is away from the critical region. The pronounced nonideality is because of the unique molecular structure of water, (i.e., hydrogen bonding that is prevalent in water, but not in hydrocarbons).

For water, the energy parameter a in the Peng-Robinson EOS is given by (Mathias et al., 1989)

$$a = a^o\left[1 + c_1(1-\sqrt{T_r}) + c_2(1-\sqrt{T_r})^2 + c_3(1-\sqrt{T_r})^3\right] \tag{3.60}$$

The parameters a^o, c_1, c_2, and c_3 are provided in Table 3.4. The above expression provides goods fit of the vapor pressure data of water and some other polar substance in the Peng-Robinson EOS.

The CPA-EOS for pure water has five parameters from physical part (a^o, c_1, c_2, c_3, and b) and two parameters from the association part ($\varepsilon_{\alpha\beta}$ and $\kappa_{\alpha\beta}$). Note that we have

a^o (J · m³/mol²)	0.09627
c_1	1.7557
c_2	0.003518
c_3	−0.2746
$b \times 10^5$ (m³/mol)	1.4584
$\varepsilon_{\alpha\beta}/k$ (K)	1738.4
$\kappa_{\alpha\beta} \times 10^6$ (m³/mol)	1.8015

TABLE 3.4 Parameters of Water in CPA-EOS

dropped the subscript 1 and have moved β to the subscript. In the association strength symbol we will do the opposite. These constants are estimated simultaneously from the saturation pressure and the liquid density of water in the temperature range of 273.16 ($T_r = 0.42$) to 633.15 K ($T_r = 0.98$). The seven constants are listed in Table 3.4. For water-containing mixtures, interaction coefficients between water and the other species are also required. Figure 3.28a depicts the interaction coefficients between water and C_1, C_2, CO_2, and H_2S. These interaction coefficients, unlike the ones for hydrocarbons, are a function of temperature because of the change in the association of water with temperature. The plots in Fig. 3.28 are versus the reduced temperatures of C_1, C_2, CO_2, and H_2S. Figure 3.28b shows the cross-association factors versus reduced temperature of C_1, C_2, CO_2, and H_2S. The association parameters of water ($\varepsilon_{\alpha\beta}$ and $\kappa_{\alpha\beta}$), k_{1j} and S_{1j} allow the prediction of the phase behavior of mixtures of water, C_1, C_2, CO_2, and H_2S. Note that H_2S has the

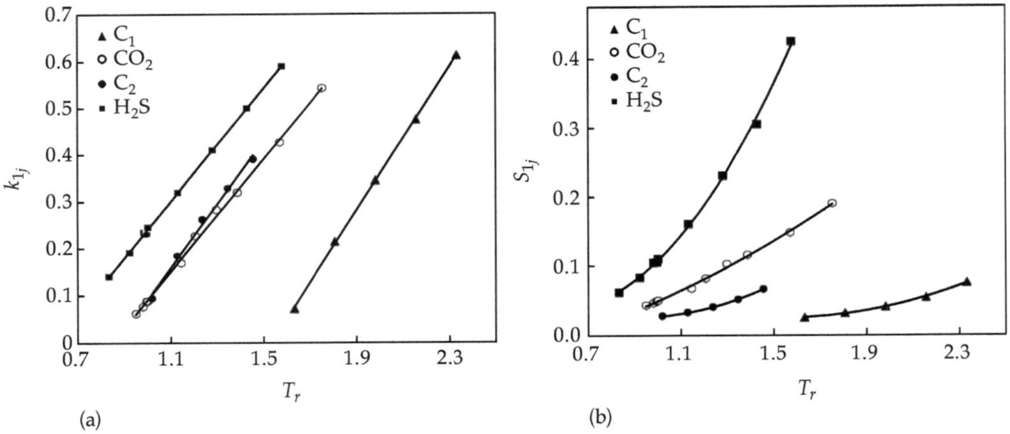

FIGURE 3.28 (a) Binary interaction coefficients and (b) cross-association factors between water (species 1) and C_1, C_2, CO_2, and H_2S (species j) as a function of reduced temperature of the latter (adapted from Li and Firoozabadi, 2009).

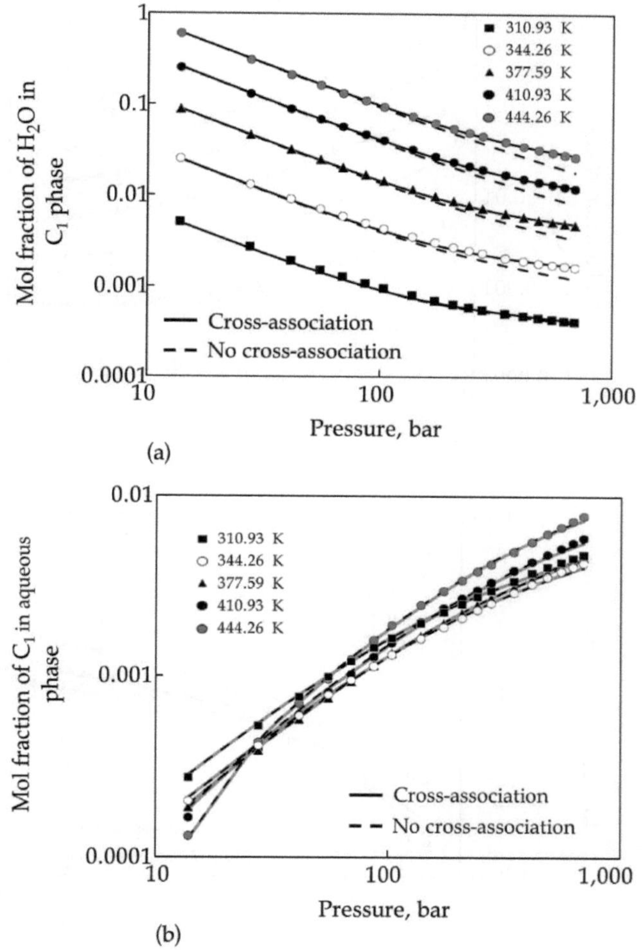

FIGURE 3.29 Composition of (a) C_1-rich phase and (b) aqueous phase of H_2O/C_1 mixture (adapted from Li and Firoozabadi, 2012).

largest cross-association parameter among the 4 species; it is the most polar among them. We will first start with binary mixtures.

Figure 3.29 presents the composition of the gas and liquid phases of binary system of C_1 and H_2O. As the figure shows, the solubility of C_1 in the aqueous phase is small. In the gas phase the concentration of H_2O increases as the temperature increases, but decreases with increase in pressure. Note that cross-association between C_1 and H_2O molecules increases the accuracy of predictions at high pressures in the nonaqueous phase. The distribution of species in the gas phase and aqueous phase for the H_2O/C_2 system is shown in Fig. 3.30. The trends are similar to those of the C_1/H_2O system. Figure 3.31 shows

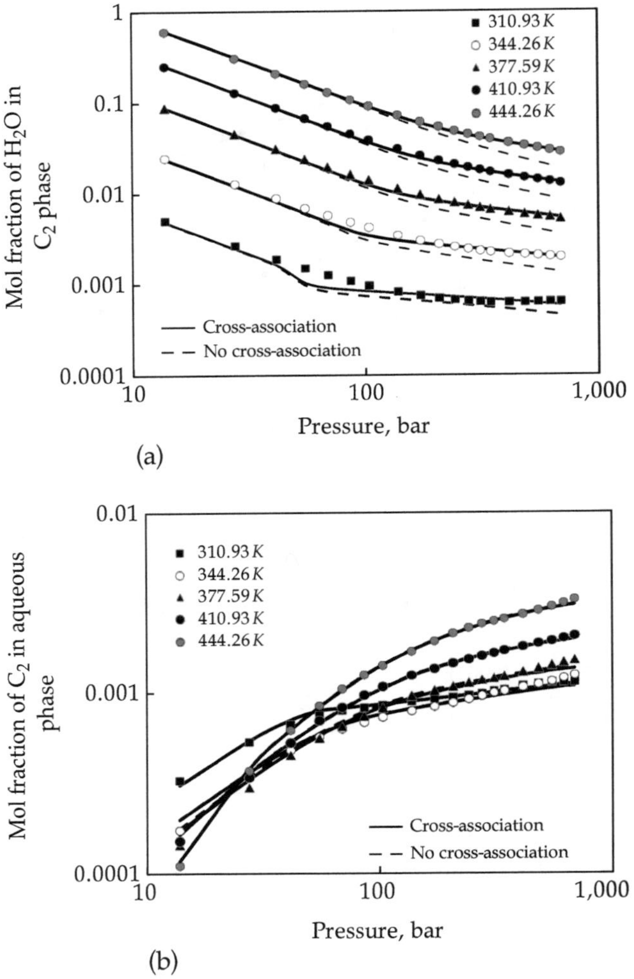

FIGURE 3.30 Composition of (a) H_2O in C_2 phase, and (b) composition of C_2 in the aqueous phase of H_2O/C_2 mixture.

the composition of the gas phase and the aqueous phase of the CO_2/H_2O system. There is a strong cross-association between CO_2 and H_2O, especially at low temperature and high pressure. With only association between water molecules and no cross-association between H_2O and CO_2 molecules, there will be substantial error in the prediction of the CPA-EOS. Figure 3.32 presents the composition of the phases in the H_2O/H_2S mixtures. The cross-association is stronger than the H_2O/CO_2 mixture. Figure 3.32b indicates that the solubility of H_2S in the aqueous phase is higher than CO_2 solubility at the same pressure and temperature. Note that cross-association does not make much difference in the description of the aqueous phase (see Figs. 3.29b, 3.30b, 3.31b, and 3.32b).

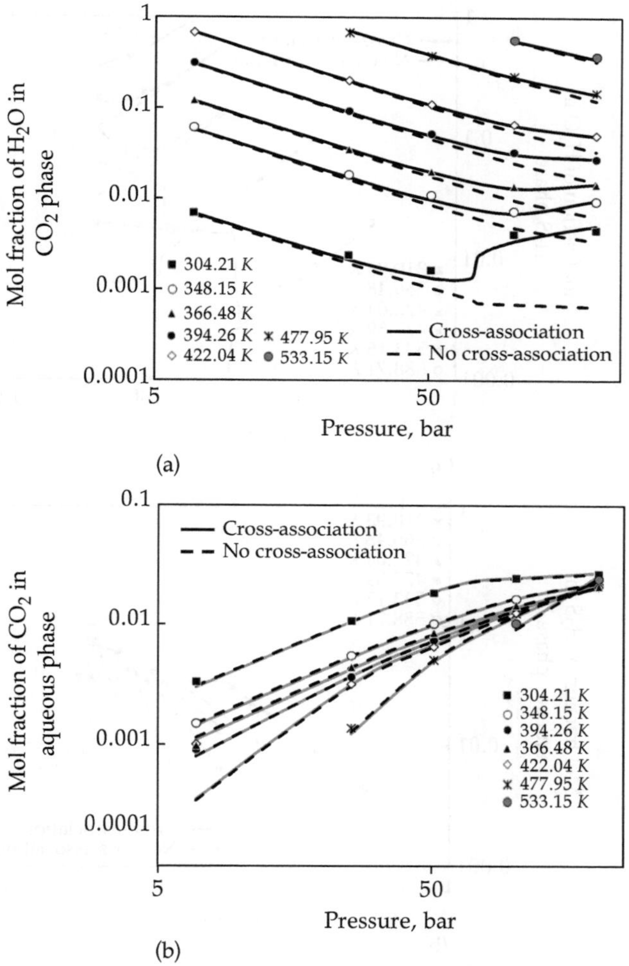

FIGURE 3.31 Composition of (a) H_2O in CO_2-rich phase, and (b) composition of CO_2 in the aqueous phase of H_2O/CO_2 mixture.

Distribution of the species in the $H_2S/CO_2/C_1/H_2O$ mixture in the H_2S-rich phase, the aqueous phase and the gas phase are listed in Table 3.5. There is good agreement between the predictions from the CPA-EOS and measured data. Note that for the conditions of temperature and pressure, the solubility of C_1, and to a lesser degree the performance of the CPA-EOS with parameters listed in Table 3.4 and the interaction coefficients of Fig. 3.28 has been examined to higher pressure than the pressure shown in Figs. 3.29 to 3.32. Table 3.6 displays the predicted compositions of CO_2 in the CO_2-rich phase of H_2O-CO_2 at very high pressures. Predictions are in agreement with data. The CPA-EOS for water-containing mixtures show remarkable accuracy for extreme conditions of temperature and pressure. The limit of solubility of methane in water will be

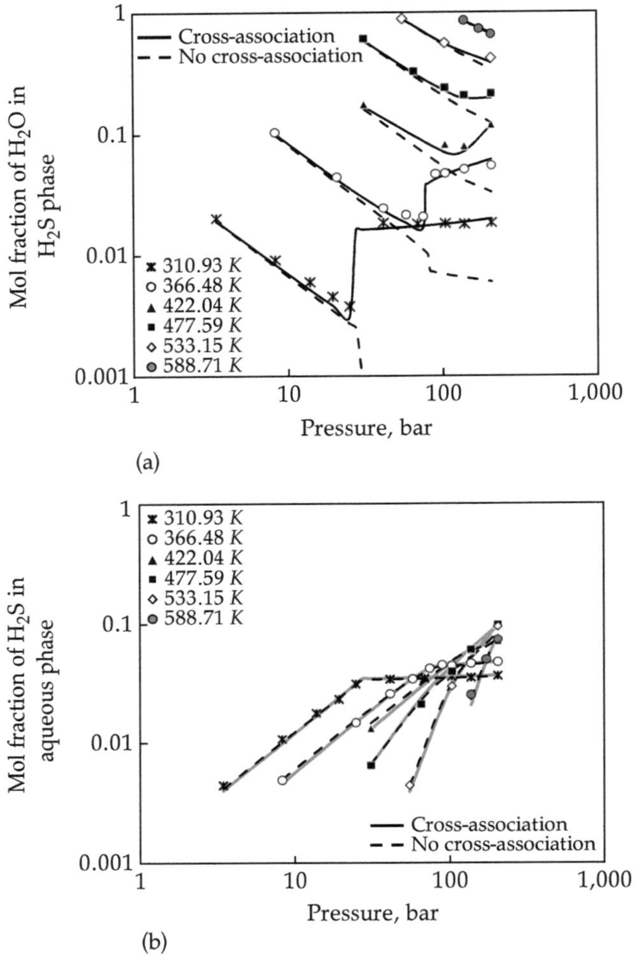

FIGURE 3.32 Composition of (a) H_2O in H_2S-rich phase, and (b) H_2S in the aqueous phase of the H_2O/H_2S mixture (adapted from Li and Firoozabadi, 2009).

discussed in Example 4.11. In that example, we compare the limit of solubility of methane in water at 285 K and 50 MPa. The predicted limit of solubility of methane in water is around 9 mol %, in agreement with molecular dynamics simulations (Jiménez-Angeles and Firoozabadi, 2014).

3.3.3 Asphaltene Precipitation Modeling

Petroleum fluids are comprised of a diverse group of species. For the purpose of asphaltene precipitation, the asphaltene species play a significant role. Asphaltenes are operationally defined as the fraction of a petroleum fluid which are insoluble in light alkanes such as heptane and soluble in aromatic solvents such as toluene. Under

Temperature K	Pressure bar	Component	Composition*					
			H_2S-rich liquid		Aqueous		Vapor	
			Exp.	Cal.	Exp.	Cal.	Exp.	Cal.
310.95	62.6	CH_4	0.0653	0.0616	4.9e-4	5.98e-4	0.3213	0.3596
		CO_2	0.1049	0.0971	3.5e-3	2.61e-3	0.1739	0.1219
		H_2S	0.8197	0.8292	0.0284	0.0310	0.5028	0.5169
		H_2O	0.0101	0.0122	0.9677	0.9659	2.14e-3	1.63e-3
338.75	84.3	CH_4	0.0580	0.0641	3.85e-4	4.15e-4	0.1872	0.2057
		CO_2	0.0904	0.0922	2.72e-3	2.47e-3	0.1484	0.1185
		H_2S	0.8287	0.8264	0.0321	0.0326	0.6557	0.6704
		H_2O	0.0212	0.0174	0.9684	0.9645	8.66e-3	5.44e-3

*compositions are in mol fractions
**from Li and Firoozabadi, 2009

TABLE 3.5 Vapor-Liquid-Liquid Equilibrium in C_1 (0.05)*, CO_2 (0.05)*, H_2S (0.4)*, and H_2O (0.5)* System**

certain conditions of pressure and composition, an asphaltene-rich phase separates from a petroleum fluid. As the name implies, the asphaltene-rich phase has a high concentration of asphaltene species. In addition to asphaltenes, two other groups of species in crudes, namely resins and heavy aromatics, have a strong influence on asphaltene precipitation. The CPA-EOS is used to model asphaltene precipitation from petroleum fluids (Li and Firoozabadi, 2010a, 2010b). Li and Firoozabadi (2010a and 2010b) assume that asphaltene molecules associate. The resin and heavy aromatic molecules may cross-associate with asphaltene molecules. In applying the concept of association and cross-association, the number of association sites are required. One may assume asphaltene molecules have four identical association sites. The resin molecules are also assumed to

P (MPa)	*T* (K)	Predicted (x_{CO_2})	Data (x_{CO_2})
169.2	500.95	0.765⁺	0.789*
264.5	534.95	0.625⁺	0.625*
280.0	538.15	0.608⁺	0.578**
300.0	538.15	0.607⁺	0.670**
311.1	546.45	0.562⁺	0.530*

⁺ Mutoru et al., 2011
* Mather and Franck, 1992
** Takenouchi and Kennedy, 1964

TABLE 3.6 Predicted composition of CO_2 (in CO_2-H_2O mixtures) from the CPA-EOS compared to experimental data at the given temperatures and pressure have four identical sites (heavy aromatics are counted as part of resins). As a result of these assumptions the Helmholtz free energy of association [from Eq. (3.43)] becomes

$$\frac{A^{departure}_{association}}{nRT} = 4x_a\left(\ln\chi_a + \frac{1-\chi_a}{2}\right) + 4x_r\left(\ln\chi_r + \frac{1-\chi_r}{2}\right) \tag{3.61}$$

In the above expression x_a and x_r represent the mol fraction of asphaltene and the resin/aromatic species, respectively. χ_a and χ_r are mol fractions of asphaltenes and resin/aromatic species not bonded at one of the association sites, respectively. The expressions for χ_a and χ_r [from Eqs. (3.48) and (3.49)] are given by

$$\chi_a = \frac{1}{1 + 4cx_a\chi_a\Delta^{aa} + 4cx_r\chi_r\Delta^{ar}} \tag{3.62a}$$

$$\chi_r = \frac{1}{1 + 4cx_a\chi_a\Delta^{ar}} \tag{3.62b}$$

where $\Delta^{ij} = g\kappa_{ij}b_{ij}[\exp(\varepsilon_{ij}/kT) - 1]$ $(i = a, j = a, r)$ is the association strength; $b_{ij} = (b_i + b_j)/2$.

The association sites of asphaltenes and cross-associated sites of resins are assumed each to be four (this is a simplification of the complex problem). The 4-site association is the same as that of water. The four association strength parameters are ε_{aa}, ε_{ar}, κ_{aa}, and κ_{ar}. The asphaltene association energy parameter ε_{aa} is based on the maximum precipitation data of, say, heptane dilution of the petroleum fluid in atmospheric pressure. A value of $\varepsilon_{aa}/k = 2800$ K is obtained for few petroleum fluids (Li and Firoozabadi, 2010a). The cross-association energy parameter ε_{ar}/k has been found to vary in the range of 1000 to 1500 K at room temperature. Figure 3.33 shows the fractional yield of asphaltenes versus the mass fraction of heptane for an Athabasca bitumen

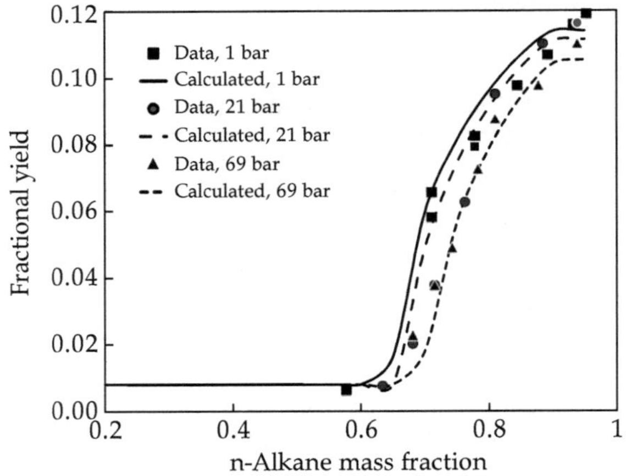

FIGURE 3.33 Effect of pressure on fractional asphaltene yield of Athabasca bitumen with n-C_7 at 296.15 K (adapted from Li and Firoozabadi, 2010a).

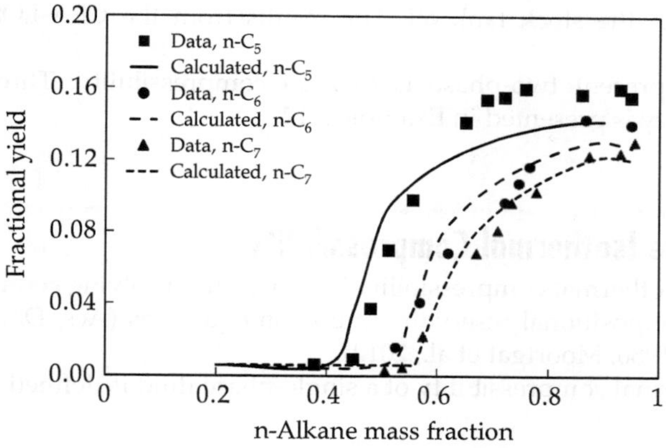

FIGURE 3.34 Effect of n-alkanes on the asphaltene fractional yield from Lloydminster heavy oil at 296.15K and 1 bar pressure (adapted from Li and Firoozabadi, 2010a).

at 296.15 K and three different pressures of 1, 21, 69 bar. With the increase in pressure, high amount of heptane is required for precipitation. The precipitation (i.e., fractional yield) is lower at high pressure. Figure 3.34 depicts the effect of the type of alkane on precipitation from Cold Lake bitumen at room conditions. There is higher precipitation with lighter alkanes. The association energy parameter increases with alkane molecular weight (Li and Firoozabadi, 2010a).

Figure 3.35 shows the upper asphaltene onset point, bubblepoint pressure, and the lower asphaltene onset point of a petroleum fluid that contains 1.4 weight percent

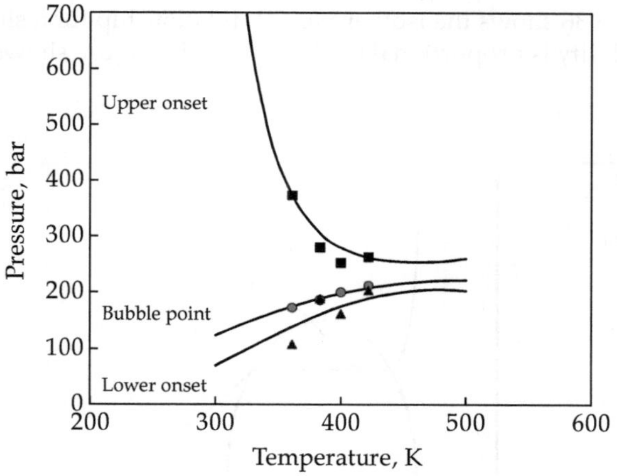

FIGURE 3.35 Bubblepoint pressure and asphaltene precipitation at upper and lower onset pressures as a function of temperature of reservoir fluid (adapted from Li and Firoozabadi, 2010b).

asphaltenes in the stock tank oil. The results from the CPA-EOS are in agreement with data.

Next, we present two-phase isothermal compressibility. Three-phase isothermal compressibility is presented in Example 3.12.

3.4 Two-Phase Isothermal Compressibility

Two-phase isothermal compressibility is important in solving certain well-test (Macias, 1985) and compositional reservoir simulation equations (Acs, DoleschDoll, and Farkas, 1985; Watts, 1986; Moortgat et al., 2012).

The isothermal compressibility of a single-phase fluid is defined in a straightforward fashion:

$$C_T = -\frac{1}{V}\left(\frac{\partial V}{\partial P}\right)_T \tag{3.63}$$

Similarly, the isothermal compressibility of a two-phase multicomponent system is defined by

$$C_T = -\frac{1}{V}\left(\frac{\partial V}{\partial P}\right)_{T,\underline{n}} \tag{3.64}$$

In the above equation, $\underline{n} = (n_1, n_2, \ldots, n_c)$ represents the total mols of the components in both phases, and the derivative term on the right side represents the change in the total fluid volume, V, caused by a small change in the pressure of the closed system at constant temperature.

Figures 3.36 and 3.37 highlight two-phase isothermal compressibility in the two-phase region. Figure 3.36 shows the isothermal P-V relationship for a single-component fluid; the compressibility is proportional to $(\partial V/\partial P)_T$. This figure shows that the $(\partial V/\partial P)_T$ of

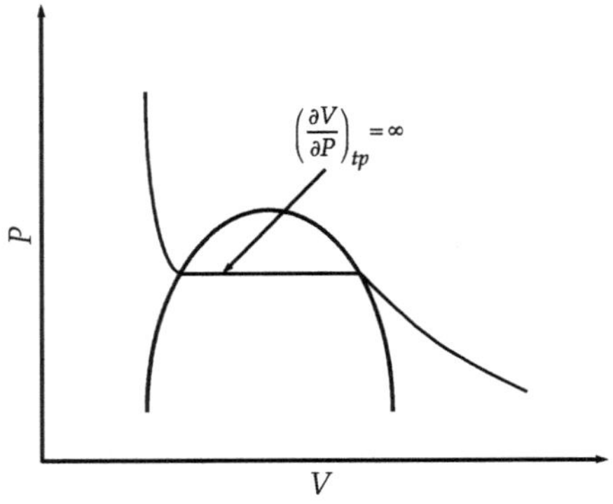

FIGURE 3.36 P-V plot of a single-component fluid implying infinite compressibility in two-phase.

FIGURE 3.37 *P-V* plot of a multicomponent mixture showing that compressibility in two-phase can be higher than the gas phase compressibility.

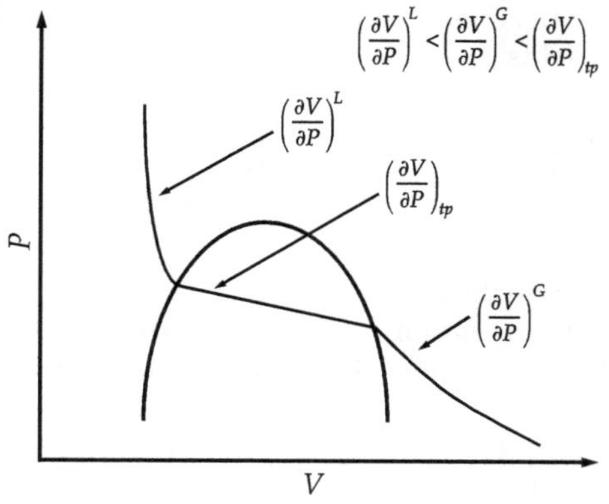

the gas phase is higher than that of the liquid phase, as expected. However, in the two-phase region, $(\partial V/\partial P)_T = \infty$, implying that the isothermal two-phase compressibility for a single-component fluid is infinite.

Figure 3.37 reveals that $(\partial V/\partial P)_{T,\underline{n}}$ for a mixture in the two-phase region might be higher than in the single-phase gas region. The implication is that the isothermal two-phase compressibility might be higher than the isothermal gas-phase compressibility. Practicing engineers often use the following relationship to obtain the two-phase compressibility:

$$C_T = C_T^G S^G + C_T^L S^L \tag{3.65}$$

where C_T is the two-phase isothermal compressibility and S^G and S^L are the gas and liquid saturations, respectively. This equation is invalid when there is mass transfer between the phases; at the bubblepoint for example, $S^G = 0$ and $C_T = C_T^L$ according to Eq. (3.65). As we will see shortly, when we get even one bubble of gas, C_T may become several times higher than the compressibility of the compressed liquid just before gas-phase appearance.

The equations for the calculation of two-phase isothermal compressibility (Firoozabadi et al., 1988b) follow. The key term in the evaluation of two-phase compressibility is the derivative of volume with respect to pressure at constant temperature and overall composition. To calculate this term, one writes

$$V = V_1 + V_2 \tag{3.66}$$

where V_1 and V_2 are the gas- and liquid-phase volumes, respectively. From the relation $PV_j = Z_j n_{j,t} RT$, where $n_{j,t}$ is the total number of mols in phase j,

$$V = \frac{RT}{P} \sum_{j=1}^{2} Z_j n_{j,t} \tag{3.67}$$

Taking the derivative of Eq. (3.67) with respect to pressure and holding temperature and overall mols constant yields

$$\left(\frac{\partial V}{\partial P}\right)_{T,\underline{n}} = -\frac{RT}{P^2}\sum_{j=1}^{2} Z_j n_{j,t} + \frac{RT}{P}\sum_{j=1}^{2}\left[n_{j,t}\left(\frac{\partial Z_j}{\partial P}\right)_{T,\underline{n}} + Z_j\left(\frac{\partial n_{j,t}}{\partial P}\right)_{T,\underline{n}}\right] \tag{3.68}$$

The unknown derivatives in the right side of Eq. (3.68) are $(\partial n_{j,t}/\partial P)_{T,\underline{n}}$ and $(\partial Z_j/\partial P)_{T,\underline{n}}$. The change in the total number of mols of phase j with pressure can be expressed as the sum of the changes in mols of each component in phase j with pressure:

$$\left(\frac{\partial n_{j,t}}{\partial P}\right)_{T,\underline{n}} = \sum_{i=1}^{n_c}\left(\frac{\partial n_{j,i}}{\partial P}\right)_{T,\underline{n}} \quad j = 1, 2 \tag{3.69}$$

where $n_{j,i}$ is the number of mols of component i in phase j. There are $2n_c$ unknown derivatives $(\partial n_{j,i}/\partial P)_{T,\underline{n}}$ in Eq. (3.69); $2n_c$ equations are needed. These equations come from n_c material balance equations,

$$n_i = n_{1,i} + n_{2,i} \quad i = 1, \ldots, n_c \tag{3.70}$$

which yield

$$\left(\frac{\partial n_{1,i}}{\partial P}\right)_{T,\underline{n}} + \left(\frac{\partial n_{2,i}}{\partial P}\right)_{T,\underline{n}} = 0 \quad i = 1, \ldots, n_c \tag{3.71}$$

and the remaining n_c equations result from the equilibrium conditions before and after the change in pressure of the system:

$$\left(\frac{\partial f_{1,i}}{\partial P}\right) - \left(\frac{\partial f_{2,i}}{\partial P}\right)_{T,\underline{n}} = \left[\sum_{k=1}^{n_c}\left(\frac{\partial f_{1,i}}{\partial n_{1,k}}\right)_{\boldsymbol{n}_{1,k}}\left(\frac{\partial n_{1,k}}{\partial P}\right)_{\underline{n}} - \sum_{k=1}^{n_c}\left(\frac{\partial f_{2,i}}{\partial n_{2,k}}\right)_{\boldsymbol{n}_{2,k}}\left(\frac{\partial n_{2,k}}{\partial P}\right)_{\underline{n}}\right]$$
$$+\left[\left(\frac{\partial f_{1,k}}{\partial P}\right)_{\underline{n}_1} - \left(\frac{\partial f_{2,i}}{\partial P}\right)_{\underline{n}_2}\right] = 0 \quad i = 1, \ldots, n_c \tag{3.72}$$

The subscripts $\boldsymbol{n}_{1,k} = (n_{1,1}, \ldots, n_{1,k-1}, n_{1,k+1}, \ldots n_{1,n_c})$, and $\underline{n}_1 = (n_{1,1}, \ldots, n_{1,n_c})$. A similar definition applies to $\boldsymbol{n}_{2,k}$ and $\underline{n}_2$. The subscript T has been dropped from the terms in the right side of the above equation. The derivatives of fugacity in the right side of Eq. (3.72) for the PR-EOS can be calculated readily by using software such as Mathematica. For the sake of brevity, we will drop some of the subscripts in the following equations.

Combining Eqs. (3.71) and (3.72) yields

$$\sum_{k=1}^{n_c}\left(\frac{\partial f_{1,i}}{\partial n_{1,k}} + \frac{\partial f_{2,i}}{\partial n_{2,k}}\right)\left(\frac{\partial n_{2,k}}{\partial P}\right) = \left[\left(\frac{\partial f_{1,i}}{\partial P}\right)_{T,\underline{n}_1} - \left(\frac{\partial f_{2,i}}{\partial P}\right)_{T,\underline{n}_2}\right] \quad i = 1, \ldots, n_c \tag{3.73}$$

After solving for $(\partial n_{j,k}/\partial P)_{T,\underline{n}}$, $j=1,2$, the following expression gives the various terms contributing to $(\partial Z_j/\partial P)_{T,\underline{n}}$:

$$\left(\frac{\partial Z_j}{\partial P}\right)_{T,\underline{n}} = \left(\frac{\partial Z_j}{\partial P}\right)_{T,\underline{n}_j} + \sum_{k=1}^{n_c}\left(\frac{\partial Z_j}{n_{j,k}}\right)\left(\frac{\partial n_{j,k}}{\partial P}\right) \quad j=1,2 \tag{3.74}$$

An EOS can be used to calculate the $(\partial Z_j/\partial P)$ and $(\partial Z_j/\partial n_{j,k})$ terms on the right side of Eq. (3.74).

The following example illustrates the use of the above equations in computing the two-phase isothermal compressibility of multicomponent systems. Consider a three-component system in the two-phase gas-liquid region. The calculation of the $(\partial n_{j,k}/\partial P)_{T,\underline{n}}$ terms is the major task of computing two-phase compressibility. The matrix representation of Eq. (3.73) for the three-component system is

$$\begin{pmatrix} \left(\frac{\partial f_{1,1}}{\partial n_{1,1}}+\frac{\partial f_{2,1}}{\partial n_{2,1}}\right) & \left(\frac{\partial f_{1,1}}{\partial n_{1,2}}+\frac{\partial f_{2,1}}{\partial n_{2,2}}\right) & \left(\frac{\partial f_{1,1}}{\partial n_{1,3}}+\frac{\partial f_{2,1}}{\partial n_{2,3}}\right) \\ \left(\frac{\partial f_{1,2}}{\partial n_{1,1}}+\frac{\partial f_{2,2}}{\partial n_{2,1}}\right) & \left(\frac{\partial f_{1,2}}{\partial n_{1,2}}+\frac{\partial f_{2,2}}{\partial n_{2,2}}\right) & \left(\frac{\partial f_{1,2}}{\partial n_{1,3}}+\frac{\partial f_{2,2}}{\partial n_{2,3}}\right) \\ \left(\frac{\partial f_{1,3}}{\partial n_{1,1}}+\frac{\partial f_{2,3}}{\partial n_{2,1}}\right) & \left(\frac{\partial f_{1,3}}{\partial n_{1,2}}+\frac{\partial f_{2,3}}{\partial n_{2,2}}\right) & \left(\frac{\partial f_{1,3}}{\partial n_{1,3}}+\frac{\partial f_{2,3}}{\partial n_{2,3}}\right) \end{pmatrix} \tag{3.75}$$

$$\times \begin{pmatrix} \frac{\partial n_{2,1}}{\partial P} \\ \frac{\partial n_{2,2}}{\partial P} \\ \frac{\partial n_{2,3}}{\partial P} \end{pmatrix} = \begin{pmatrix} \left(\frac{\partial f_{1,1}}{\partial P}-\frac{\partial f_{2,1}}{\partial P}\right) \\ \left(\frac{\partial f_{1,2}}{\partial P}-\frac{\partial f_{2,2}}{\partial P}\right) \\ \left(\frac{\partial f_{1,3}}{\partial P}-\frac{\partial f_{2,3}}{\partial P}\right) \end{pmatrix}$$

A similar procedure can be derived to compute two-phase thermal expansivity for use in thermal models. The defining equation for thermal expansivity is

$$e = \frac{1}{V}\left(\frac{\partial V}{\partial T}\right)_{P,\underline{n}} \tag{3.76}$$

The procedure is virtually identical to the calculation of isothermal compressibility.

Taking the derivative of Eq. (3.67) with respect to temperature, holding pressure and overall mol numbers constant, yields

$$\left(\frac{\partial V}{\partial T}\right)_{P,\underline{n}} = \frac{R}{P}\sum_{j=1}^{2} Z_j n_{j,t} + \frac{RT}{P}\sum_{j=1}^{2}\left[n_{j,t}\left(\frac{\partial Z_j}{\partial T}\right)_{P,\underline{n}} + Z_j\left(\frac{\partial n_{j,t}}{\partial T}\right)_{P,\underline{n}}\right] \tag{3.77}$$

The $(\partial n_{j,t}/\partial T)_{P,\underline{n}}$ term is expressed as

$$\left(\frac{\partial n_{j,t}}{\partial T}\right)_{P,\underline{n}} = \sum_{i=1}^{n_c} \left(\frac{\partial n_{j,i}}{\partial T}\right)_{P,\underline{n}} \qquad j = 1, 2 \tag{3.78}$$

The evaluation of $(\partial n_{j,t}/\partial T)_{P,\underline{n}}$ and $(\partial Z_j/\partial T)_{P,\underline{n}}$ terms are straightforward (see Example 3.8).

3.5 Two-Phase Isentropic Compressibility and Two-Phase Sonic Velocity

In the previous section, the two-phase isothermal compressibility of multicomponent systems was formulated using the equilibrium assumption that there is no gradient of chemical potentials in the systems. In this section, the two-phase isentropic compressibility and the two-phase sonic velocity for multicomponent systems will be formulated. We will make the assumption of equilibrium, which implies the gradients of chemical potentials are zero. The equilibrium assumption regarding the two-phase compressibilities depends on the problem and may or may not be justified. As long as there is no supersaturation, and adequate time is allowed to reach the state of equilibrium, then the equilibrium criterion can be invoked.

The isentropic compressibility and the thermodynamic sonic velocity are related to each other both in the single-phase and two-phase states. They are used in problems in the exploration and production of hydrocarbon reservoirs and in different disciplines. We should also make a comment on the thermodynamic sonic velocity, which is a purely thermodynamic property. The thermodynamic sonic velocity is equal to the true sonic velocity over a wide range of frequencies and amplitudes. However, at high frequencies where the thermal properties depend on the rate of heating, then the true sonic velocity and the thermodynamic sonic velocity are not the same (see Chaps. 2 and 3 of Rowlinson and Swinton, 1982; Voronel, 1976).

In the following, we will first derive the expressions that can be used to calculate the two-phase isentropic compressibility. The thermodynamic sonic velocity then can be readily calculated from the isentropic compressibility. We could have combined the derivations for the isothermal and isentropic compressibilities but have decided on separate derivations for the sake of simplicity.

Compressibility is defined on the basis of the thermodynamic path. For isothermal compressibility, the path is constant temperature; the isentropic compressibility C_S with a flat interface is defined by

$$C_S = -\frac{1}{V}\left(\frac{\partial V}{\partial P}\right)_{S,\underline{n}} \tag{3.79}$$

The isothermal and isentropic compressibilities are related by a simple expression in the single-phase state,

$$C_T = \frac{c_P}{c_V} C_S \tag{3.80}$$

where c_P and c_V are heat capacities at constant pressure and volume, respectively. The derivation of Eq. (3.80) is provided in Example 3.7. Since $c_P \geq c_V$ (see Example 3.7), then $C_T \geq C_S$. The difference between C_T and C_S depends on pressure, temperature, and composition and may vary from ten to several hundred percent in the single-phase state.

The two-phase isentropic compressibility C_S in a system with curved interface is defined as,

$$C_S = -\frac{1}{V}\left(\frac{\partial V}{\partial P_2}\right)_{S,\underline{n}} \tag{3.81}$$

Note that in Eq. (3.81), we define C_S using pressure of phase 2, say the liquid phase to take into account the effect of interface curvature. Figure 3.38 shows the schematics of the process for the estimation of C_S by a small increase of volume from V to $V + \Delta V$. The volume of each phase j is given by

$$V_j = RT\frac{Z_j n_{j,t}}{P_j} \tag{3.82}$$

where P_j is the pressure of phase j. The total volume V is then

$$V = RT\sum_{j=1}^{2}\left(\frac{Z_j n_{j,t}}{P_j}\right) \tag{3.83}$$

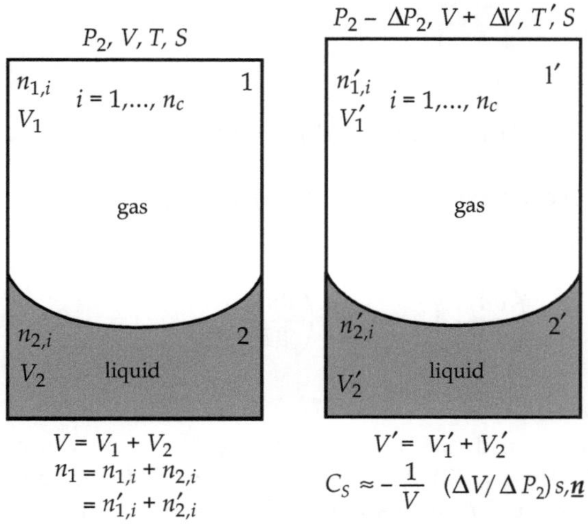

FIGURE 3.38 Schematic of changes associated with the calculation of two-phase compressibility C_s (adapted from Firoozabadi and Pan, 2000).

The variation of V with respect to P_2 at constant S and $\underline{n}$ can be obtained from Eq. (3.83)

$$\left(\frac{\partial V}{\partial P_2}\right)_{S,\underline{n}} = RT\sum_{j=1}^{2}\left[\left(\frac{Z_j}{P_j}\right)\left(\frac{\partial n_{j,t}}{\partial P_2}\right)_{S,\underline{n}} + \left(\frac{n_{j,t}}{P_j}\right)\left(\frac{\partial Z_j}{\partial P_2}\right)_{S,\underline{n}} - \left(\frac{Z_j n_{j,t}}{P_j^2}\right)\left(\frac{\partial P_j}{\partial P_2}\right)_{S,\underline{n}}\right] + \left[R\sum_{j=1}^{2}\left(\frac{Z_j n_{j.t}}{P_j}\right)\left(\frac{\partial T}{\partial P_2}\right)_{S,\underline{n}}\right] \tag{3.84}$$

Various derivative expressions on the right side of Eq. (3.84) are evaluated next. $(\partial Z_j/\partial P_2)_{S,\underline{n}}$ - One can write

$$dZ_j = \left(\frac{\partial Z_j}{\partial T}\right)_{P,\underline{n}_j} dT + \left(\frac{\partial Z_j}{\partial P_j}\right)_{T,\underline{n}_j} dP_j + \sum_{i=1}^{n_c}\left(\frac{\partial Z_j}{\partial n_{j,i}}\right)_{T,P_j,\boldsymbol{n}_{j,i}} dn_{j,i} \tag{3.85}$$

where $\underline{n}_j = (n_{j,1}, \ldots, n_{j,n_c})$ and $\boldsymbol{n}_{j,i} = (n_{j,1}, \ldots, n_{j,i-1}, n_{j,i+1}, \ldots, n_{j,n_c})$; $n_{j,i}$ is the number of mols of component i in phase j. By dividing Eq. (3.85) by ∂P_2 at constant $\underline{n}$ and S,

$$\left(\frac{\partial Z_j}{\partial P_2}\right)_{S,\underline{n}} = \left(\frac{\partial Z_j}{\partial T}\right)_{P_j,\underline{n}_j}\left(\frac{\partial T}{\partial P_2}\right)_{S,\underline{n}} + \left(\frac{\partial Z_j}{\partial P_j}\right)_{T,\underline{n}_j}\left(\frac{\partial P_j}{\partial P_2}\right)_{S,\underline{n}} + \sum_{i=1}^{n_c}\left(\frac{\partial Z_j}{\partial n_{j,i}}\right)_{T,P_j,\boldsymbol{n}_{j,i}}\left(\frac{\partial n_{j,i}}{\partial P_2}\right)_{S,\underline{n}} \quad j = 1, 2 \tag{3.86}$$

where coefficients $(\partial Z_j/\partial T)_{P_j,\underline{n}_j}$, $(\partial Z_j/\partial P_j)_{T,\underline{n}_j}$, and $(\partial Z_j/\partial n_{j,i})_{T,P_j,\boldsymbol{n}_{j,i}}$ can be calculated from an EOS.

$(\partial n_{j,t}/\partial P_2)_{S,\underline{n}}$ - The expression for $(\partial n_{j,t}/\partial P_2)$ is given by

$$\left(\frac{\partial n_{j,t}}{\partial P_2}\right)_{S,\underline{n}} = \sum_{i=1}^{n_c}\left(\frac{\partial n_{j,i}}{\partial P_2}\right)_{S,\underline{n}} \quad j = 1, 2 \tag{3.87}$$

Equations (3.84), (3.85), and (3.87) contain $2(n_c + 1)$ unknowns on the right side: (1) one unknown $(\partial P_1/\partial P_2)_{S,\underline{n}}$, (2) $2n_c$ unknowns $(\partial n_{j,i}/\partial P_2)_{S,\underline{n}}$, and (3) one unknown $(\partial T/\partial P_2)_{S,\underline{n}}$. Once these unknowns are available, the expression for $(\partial V/\partial P_2)_{S,\underline{n}}$ can be evaluated. We need, therefore, $2(n_c + 1)$ equations to solve for the same number of unknowns. The material balance, equilibrium criterion, entropy constraint, and the expression that relates P_1 and P_2 provide $2(n_c + 1)$ equations.

Material Balance

From Eq. (3.70), the variation in mol number of component i in phases 1 and 2 can be expressed by

$$\left(\frac{\partial n_i}{\partial P_2}\right)_{S,\underline{n}} = \left(\frac{\partial n_{1,i}}{\partial P_2}\right)_{S,\underline{n}} + \left(\frac{\partial n_{2,i}}{\partial P_2}\right)_{S,\underline{n}} = 0 \qquad i = 1, \ldots, n_c \tag{3.88}$$

Therefore, material balance provides n_c equations.

Equilibrium Criterion

The equilibrium criterion $f_{1,i} = f_{2,i}$ can be written in differential form,

$$\mathrm{d}f_{1,i} = \mathrm{d}f_{2,i} \qquad i = 1, \ldots, n_c \tag{3.89}$$

where $f_{j,i} = f_{j,i}(T, P_j, n_{j,1}, \ldots, n_{j,n_c})$. Therefore,

$$\begin{aligned}\mathrm{d}f_{j,i} = {} & \left(\frac{\partial f_{j,i}}{\partial T}\right)_{P_j,\underline{n}_j} \mathrm{d}T \\ & + \left(\frac{\partial f_{j,i}}{\partial P_j}\right)_{T,\underline{n}_j} \mathrm{d}P_j + \sum_{k=1}^{n_c} \left(\frac{\partial f_{j,i}}{\partial n_{j,k}}\right)_{T,P_j,\boldsymbol{n}_{j,k}} \mathrm{d}n_{j,k} \\ & j = 1, 2; \qquad i = 1, \ldots, n_c\end{aligned} \tag{3.90}$$

where $\boldsymbol{n}_{j,k} = (n_{j,1}, \ldots, n_{j,k-1}, n_{j,k+1}, \ldots, n_{j,n_c})$. Substituting Eq. (3.90) into Eq. (3.89) and then dividing by ∂P_2 while keeping S and $\underline{n}$ constant, one obtains

$$\begin{aligned}& \left[\left(\frac{\partial f_{1,i}}{\partial T}\right)_{P_1,\underline{n}_1} - \left(\frac{\partial f_{2,i}}{\partial T}\right)_{P_2,\underline{n}_2}\right] \left(\frac{\partial T}{\partial P_2}\right)_{S,\underline{n}} + \left[\left(\frac{\partial f_{1,i}}{\partial P_1}\right)_{T,\underline{n}_1} \right. \\ & \left. \cdot \left(\frac{\partial P_1}{\partial P_2}\right)_{S,\underline{n}} - \left(\frac{\partial f_{2,i}}{\partial P_2}\right)_{T,\underline{n}_2}\right] + \left[\sum_{k=1}^{n_c} \left(\frac{\partial f_{1,i}}{\partial n_{1,k}}\right)_{T,P_1,\boldsymbol{n}_{1,k}} \right. \\ & \left. \cdot \left(\frac{\partial n_{1,k}}{\partial P_2}\right)_{S,\underline{n}} - \sum_{k=1}^{n_c} \left(\frac{\partial f_{2,i}}{\partial n_{2,k}}\right)_{T,P_2,\boldsymbol{n}_{2,k}} \left(\frac{\partial n_{2,k}}{\partial P_2}\right)_{S,\underline{n}}\right] = 0 \qquad i = 1, \ldots, n_c\end{aligned} \tag{3.91}$$

The coefficients $(\partial f_{j,i}/\partial T)_{P_j,\underline{n}j}$, $(\partial f_{j,i}/\partial P_j)_{T,\underline{n}_j}$, and $(\partial f_{j,i}/\partial n_{j,k})_{T,P,\boldsymbol{n}_{j,k}}$ can be obtained from an EOS. We need two more equations to complete the $2(n_c + 1)$ equations.

Entropy Constraint

From $S = \sum_{j=1}^{2} S_j$,

$$\left(\frac{\partial S}{\partial P_2}\right)_{S,\underline{n}} = \sum_{j=1}^{2} \left(\frac{\partial S_j}{\partial P_2}\right)_{S,\underline{n}} = 0 \tag{3.92}$$

Since $S_j = S_j(T, P_j, n_{j,1}, \ldots, n_{j,n_c})$, one can write the differential form of S_j, divide it by ∂P_2 at constant $\underline{n}$ and S, and substitute the results into Eq. (3.92),

$$\sum_{j=1}^{2} \left[\left(\frac{\partial S_j}{\partial T}\right)_{P_j,\underline{n}j} \left(\frac{\partial T}{\partial P_2}\right)_{S,\underline{n}} + \left(\frac{\partial S_j}{\partial P_j}\right)_{T,\underline{n}_j} \left(\frac{\partial P_j}{\partial P_2}\right)_{S,\underline{n}} + \sum_{k=1}^{n_c} \left(\frac{\partial S_j}{\partial n_{j,k}}\right)_{T,P_j n_{j,k}} \left(\frac{\partial n_{j,k}}{\partial P_2}\right)_{S,\underline{n}} \right] = 0 \tag{3.93}$$

The coefficients $(\partial S_j/\partial T)_{P_j,\underline{n}_j}$, $(\partial S_j/\partial P_j)_{T,\underline{n}_j}$, and $(\partial S_j/\partial n_{j,k})_{T,P_j,n_{j,k}}$ can be estimated from the expression for entropy of phase j derived in Example 3.2. The entropy constraint provides $(\partial T/\partial P_2)_{S,\underline{n}}$.

Curvature Effect

For a curved interface, the gas and liquid phase pressures are related by the well-known Young-Laplace equation derived in Chap. 2,

$$P_c = P_1 - P_2 = \frac{2\sigma}{r} \tag{3.94}$$

In the process sketched in Fig. 3.38, when P_2 changes not only P_1 changes, but also σ and r may change. From these changes, we wish to establish the unknown $(\partial P_1/\partial P_2)_{S,\underline{n}}$, which depends on how σ and r change. In the special case that r stays constant (such as in a capillary tube with a fixed contact angle) then $(\partial P_1/\partial P_2)$ is related only to the variations of σ. The variation of σ is due to the variations in temperature, pressure, and composition and curvature. The Weinaugh-Katz model (1943) can be used to calculate composition, pressure, and temperature effect on σ:

$$\sigma^{1/4} = \sum_{i=1}^{n_c} P_i \left[x_{2,i} \left(\frac{d_2}{M_2}\right) - x_{1,i} \left(\frac{d_1}{M_1}\right) \right] \tag{3.95}$$

In Eq. (3.95), $x_{2,i}$ and $x_{1,i}$ are the mol fractions of component i in the liquid and gas phases, respectively, and d and M are mass density and molecular weight; P_i is the parachor of component i.

The curvature may vary with pressure P_2. In porous media, the relationship between P_1 and P_2 is given by

$$P_c = P_1 - P_2 = \sigma F(S_2) \tag{3.96}$$

where S_2 is the saturation of the liquid phase defined by $S_2 = V_2/(V_1 + V_2)$. Eq. (3.96) takes into account the variation of curvature with P_1 and P_2. Taking the derivative of Eq. (3.96) with respect to P_2 at constant S and $\underline{n}$,

$$\left(\frac{\partial P_1}{\partial P_2}\right)_{S,\underline{n}} = 1 + F(S_2)\left(\frac{\partial \sigma}{\partial P_2}\right)_{S,\underline{n}} + \sigma\left(\frac{\partial F}{\partial P_2}\right)_{S,\underline{n}} \tag{3.97}$$

The terms $(\partial\sigma/\partial P_2)_{S,\underline{n}}$ and $(\partial F/\partial P_2)_{S,\underline{n}}$ are evaluated next. For evaluation of the term $(\partial\sigma/\partial P_2)_{S,\underline{n}}$, a more useful form of Eq. (3.95) is employed,

$$\sigma^{1/4} = \sum_{i=1}^{n_c} P_i\left[\left(\frac{n_{2,i}}{V_2}\right) - \left(\frac{n_{1,i}}{V_1}\right)\right] \tag{3.98}$$

From the above equation,

$$\left(\frac{\partial \sigma}{\partial P_2}\right)_{S,\underline{n}} = 4\sigma^{3/4}\sum_{i=1}^{n_c} P_i \times\left[-\left(\frac{n_{2,i}}{V_2^2}\right)\left(\frac{\partial V_2}{\partial P_2}\right)_{S,\underline{n}} + \left(\frac{n_{1,i}}{V_1^2}\right)\left(\frac{\partial V_1}{\partial P_2}\right)_{S,\underline{n}} + \frac{\left(\frac{\partial n_{2,i}}{\partial P_2}\right)_{S,\underline{n}}}{V_2} - \frac{\left(\frac{\partial n_{1,i}}{\partial P_2}\right)_{S,\underline{n}}}{V_1}\right] \tag{3.99}$$

The second derivative term in Eq. (3.97) is estimated from

$$\left(\frac{\partial F}{\partial P_2}\right)_{S,\underline{n}} = \left(\frac{dF}{dS_2}\right)\left(\frac{\partial S_2}{\partial P_2}\right)_{S,\underline{n}} \tag{3.100}$$

where

$$\left(\frac{\partial S_2}{\partial P_2}\right)_{S,\underline{n}} = \frac{V_1(\partial V_2/\partial P_2)_{S,\underline{n}} - V_2(\partial V_1/\partial P_2)_{S,\underline{n}}}{(V_1 + V_2)^2} \tag{3.101}$$

We showed earlier how to calculate $(\partial V_1/\partial P_2)_{S,\underline{n}}$ and $(\partial V_2/\partial P_2)_{S,\underline{n}}$. Note that in Eq. (3.99) the effect of interface curvature on σ is neglected; Eq. (3.98) takes into account only pressure, temperature, and composition variation of σ.

The above system of equations may then be solved iteratively to calculate C_S. In the first iteration, we may assume that the interface is flat.

Two-Phase Sonic Velocity

The thermodynamic speed of sound, a, is given by the following expression:

$$a^2 = -v^2 \left(\frac{\partial P}{\partial v} \right)_s \tag{3.102}$$

which assumes that all irreversibilities, including heat conduction and diffusion, are excluded. We also neglect, in the thermodynamic definition, the frequency dependence of the sonic velocity. Note that in the above equation, v is molar volume. Equation (3.102) applies to both single-phase and two-phase systems. In the two-phase state, one of the phases should be dispersed in the other phase. For two-phase sonic velocity, Eq. (3.102) becomes

$$a^2 = \frac{v}{C_S} \tag{3.103}$$

where v is the two-phase molar volume. Since v and C_S are available once the two-phase isentropic compressibility is calculated, then the two-phase sonic velocity is calculated readily.

Using the equations presented in this section, one can calculate the two-phase isentropic compressibility and the two-phase sonic velocity. In the following, numerical results are presented from Firoozabadi and Pan (2000), who employed the PR-EOS for the calculation of coefficient derivatives.

Figure 3.39 shows the calculated compressibilties and sonic velocity for a mixture of C_1/C_3 (30 mol% C_1, 70 mol% C_3) at 130°F. In Fig. 3.39a, C_T and C_S are plotted vs. pressure. This figure indicates that there is a discontinuity in both isothermal and isentropic compressibilities, when the phase boundaries are crossed. From a pressure of 1,200 psia to a bubblepoint pressure of about 977 psia, there is a small increase in C_S of the undersaturated liquid; the C_T increase is, however, more noticeable. At the bubblepoint, there is a sudden increase in both C_T and C_S. Similar behavior is also observed at the dewpoint of about 453 psia. It is interesting to note that the compressibilities in the two-phase region approaching the dewpoint are higher than the corresponding gas-phase compressibilities. Figure 3.39a also reveals that the variation of C_S in the two-phase region is less than the variation of C_T. This figure also provides the experimental isothermal compressibility data of Sage et al. (1933). The results in Fig. 3.39a are for a flat interface between the phases. Figures 3.39b and 3.39c show the effect of a curved interface on C_T and C_S, respectively. For the preparation of these two figures, $F(S_2) = 25 - 5S_2$ is assumed with the units of (psia · cm)/dyne [see Eq. (3.96)]. As we have discussed in Chap. 2, a curved interface may increase or decrease the saturation pressure of mixtures. The calculated bubblepoint pressures of the flat and the curved interfaces are 977.2 and 966.4 psia, respectively. The dewpoint pressures of the flat and curved interfaces are 453.4 and 389.3 psia, respectively. The parachors [see Eq. (3.95)] used in the calculation of interfacial tension are $P_{C_1} = 77$ and $P_{C_3} = 150$ (Katz et al., 1959). The results in Figs. 3.39b and 3.39c show that while the interface curvature does not have a significant effect on C_T, its effect on C_S is pronounced. Figure 3.39d depicts the sonic velocity both in the single-phase and two-phase regions. In the single-phase liquid, the sonic velocity decreases as the pressure decreases. There is a sharp decrease in the sonic velocity at the bubblepoint, from 370 to about 100 m/s. There is much data on the sonic velocity of two-phase gas-liquid mixtures of water-air and water-steam (Kieffer, 1977). Those data reveal that: (1) the presence of

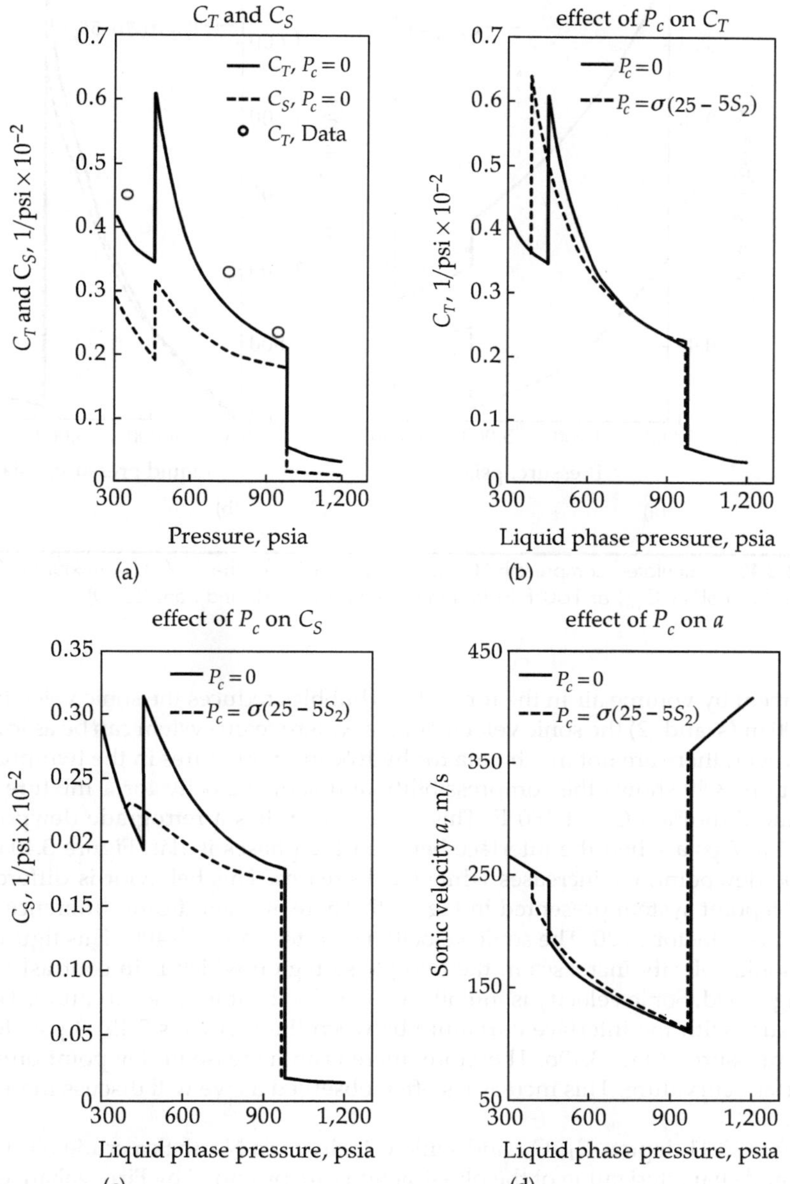

FIGURE 3.39 Compressibility and sonic velocity for the C_1/C_3 mixtures (30 mol % C_1, 70 mol % C_3) at 130°F (adapted from Firoozabadi and Pan, 2000).

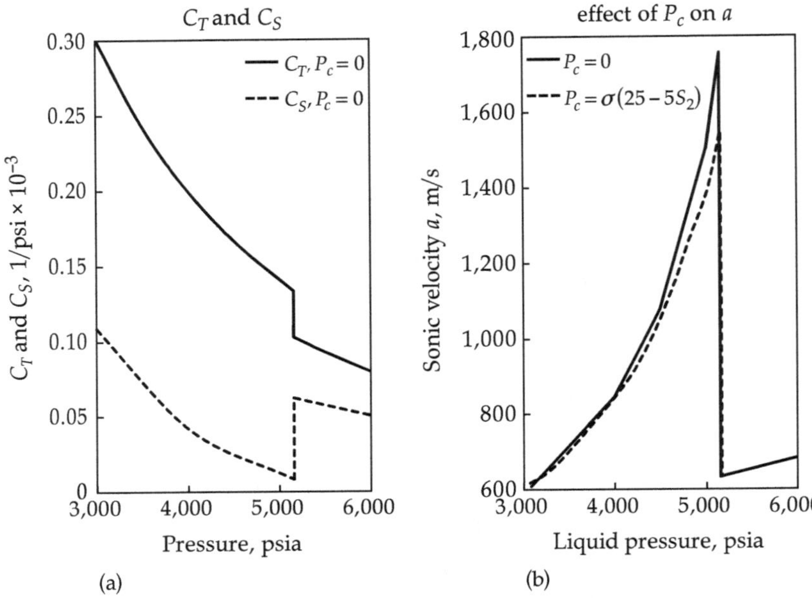

FIGURE 3.40 Calculated compressibility and sonic velocity for the C_1/n-C_{10} mixture (95 mol precent C_1 and 5 mol% n-C_{10}) at 160°F (adapted from Firoozabadi and Pan, 2000).

1 percent by volume air in the form of gas bubbles reduces the sonic velocity from 1,500 to 100 m/s, and (2) the sonic velocity in the water-steam system can be as low as 10 m/s. However, there are not much data for hydrocarbon mixtures in the two-phase region.

Figure 3.40 shows the compressibility and sonic velocity for a mixture of 95 mol% C_1 and 5 mol% n-C_{10} at 160°F. This fluid system has a retrograde dewpoint pressure of 5,146.7 psia when the interface between the phases is flat. Figure 3.40a shows that at the dewpoint, C_T increases while C_S decreases. This behavior is different from the bubblepoint system presented in Fig. 3.39. There is a significant difference between C_T and C_S—a factor of 20. The sonic velocity is plotted in Fig. 3.40b. This figure shows that the sonic velocity increases in the two-phase region which is in contrast to the results in Fig. 3.39d. Sonic velocity is initially affected by the interface curvature. The dewpoint pressure with the interface curvature between the phases is 5,151.3 psia for the capillary pressure in Fig. 3.40b. Therefore, there is an increase in dewpoint pressure due to interface curvature. This increase is often observed as we will discuss more in Example 3.14.

Figure 3.41 depicts C_T, C_S, and sonic velocity for a North Sea crude oil. The composition and characterization of the plus-fractions are provided by Firoozabadi et al. (1988b). In the single-phase liquid, C_T and C_S do not change appreciably with pressure; the difference between C_T and C_S is also not appreciable. However, this difference becomes appreciable in the two-phase region as is indicated in Fig. 3.41a. The measured isothermal compressibility is also plotted in Fig. 3.41a which shows good agreement with the calculated results. The effects of interface curvature on C_T and C_S are shown in Figs. 3.41b and 3.41c, respectively. Because of interface curvature, the bubblepoint pressure

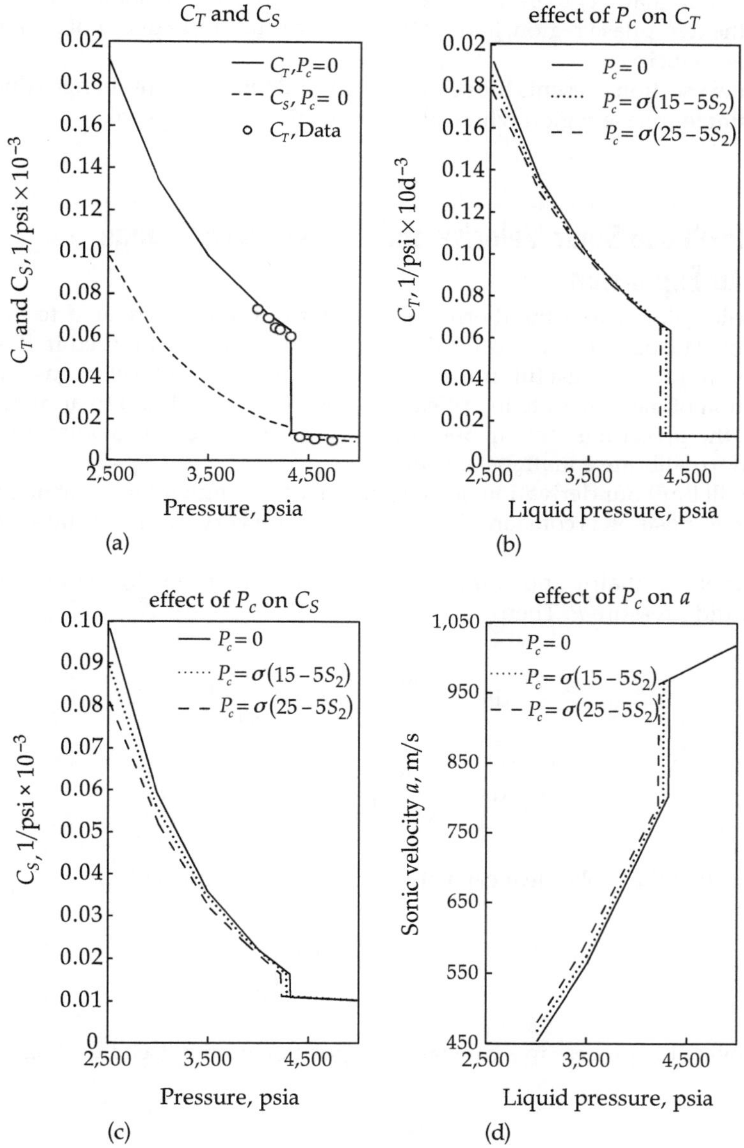

FIGURE 3.41 Compressibility and sonic velocity for the North Sea petroleum fluid at 224.6°F (adapted from Firoozabadi and Pan, 2000).

decreases, but the effect on C_T and C_S is not significant. The parachor for plus-fractions of the crude was adopted from Firoozabadi et al. (1988c). Figure 3.41d shows the calculated sonic velocity. The effect of interface curvature is also shown in the same figure. The results show that: (1) there is not a significant drop in the sonic velocity as the crude enters the two-phase region, and (2) the interface curvature does not appreciably change the sonic velocity.

The next section presents the sonic velocity and temperature changes due to expansion in the single–phase region where all relationships are analytical.

3.6 Single-Phase Sonic Velocity and Temperature Change Due to Expansion

The sonic velocity and the thermal properties of fluids are related to their volumetric behavior. The basic equations for the single-phase state are derived in this section.

Speed of sound is useful in the determination of liquid level in gas wells and in the estimation of gas flow rate in critical flow provers, in addition to applications enumerated in the preceding section. Speed of sound correlates with volumetric data and can be used to calibrate equations of state.

We will limit our derivation of sonic velocity to a single-phase system [see Eq. (3.102)] where composition is constant. The derivations are very similar to those of Thomas et al. (1970).

The molar enthalpy and entropy, h and s, can be assumed to be functions of temperature T, and pressure P. Then

$$dh = \left(\frac{\partial h}{\partial T}\right)_P dT + \left(\frac{\partial h}{\partial P}\right)_T dP \tag{3.104}$$

$$ds = \left(\frac{\partial s}{\partial T}\right)_P dT + \left(\frac{\partial s}{\partial P}\right)_T dP \tag{3.105}$$

Let us define the molar heat capacity at constant pressure as

$$c_P = \left(\frac{\partial h}{\partial T}\right)_P \tag{3.106}$$

For a reversible process in a constant-composition system [see Eq. (1.52)]

$$dh = T ds + v dP \tag{3.107}$$

Dividing Eq. (3.107) by dT and holding P constant results in

$$\left(\frac{\partial h}{\partial T}\right)_P = T\left(\frac{\partial s}{\partial T}\right)_P \tag{3.108}$$

Combining Eqs. (3.106) and (3.108) provides

$$c_P = T\left(\frac{\partial s}{\partial T}\right)_P \tag{3.109}$$

Now divide Eq. (3.107) by $\mathrm{d}P$, holding the temperature constant:

$$\left(\frac{\partial h}{\partial P}\right)_T = T\left(\frac{\partial s}{\partial P}\right)_T + v \tag{3.110}$$

Using the Maxwell relation given by Eq. (1.200),

$$\left(\frac{\partial h}{\partial P}\right)_T = -T\left(\frac{\partial v}{\partial T}\right)_P + v \tag{3.111}$$

Combining Eqs. (3.104), (3.106), and (3.111),

$$\boxed{\mathrm{d}h = c_P \mathrm{d}T + \left[v - T\left(\frac{\partial v}{\partial T}\right)_P\right]\mathrm{d}P} \tag{3.112}$$

An expression for $\mathrm{d}s$ can be obtained by combining Eqs. (3.105), (3.109), and Eq. (1.200):

$$\mathrm{d}s = \frac{c_P}{T}\mathrm{d}T - \left(\frac{\partial v}{\partial T}\right)_P \mathrm{d}P \tag{3.113}$$

Now express $u = u(T, v)$ and $s = s(T, v)$. Then

$$\mathrm{d}u = \left(\frac{\partial u}{\partial T}\right)_v \mathrm{d}T + \left(\frac{\partial u}{\partial v}\right)_T \mathrm{d}v \tag{3.114}$$

$$\mathrm{d}s = \left(\frac{\partial s}{\partial T}\right)_v \mathrm{d}T + \left(\frac{\partial s}{\partial v}\right)_T \mathrm{d}v \tag{3.115}$$

Define the molar heat capacity at constant v as

$$c_V = \left(\frac{\partial u}{\partial T}\right)_v \tag{3.116}$$

For a reversible process,

$$\mathrm{d}u = T\mathrm{d}s - P\mathrm{d}v \tag{3.117}$$

Dividing Eq. (3.117) by $\mathrm{d}T$ while holding v constant and using the definition of c_V from Eq. (3.116),

$$\left(\frac{\partial u}{\partial T}\right)_v = T\left(\frac{\partial s}{\partial T}\right)_v = c_V \tag{3.118}$$

Also dividing Eq. (3.117) by $\mathrm{d}v$ and holding T constant,

$$\left(\frac{\partial u}{\partial v}\right)_T = T\left(\frac{\partial s}{\partial v}\right)_T - P \tag{3.119}$$

Using the Maxwell relation given by Eq. (1.201),

$$\left(\frac{\partial u}{\partial v}\right)_T = T\left(\frac{\partial P}{\partial T}\right)_v - P \tag{3.120}$$

Combining Eqs. (3.114), (3.116), and (3.120), the following expression for $\mathrm{d}u$ is obtained:

$$\boxed{\mathrm{d}u = c_V\mathrm{d}T + \left[T\left(\frac{\partial P}{\partial T}\right)_v - P\right]\mathrm{d}v} \tag{3.121}$$

An expression for $\mathrm{d}s$ is also obtained by combining Eqs. (3.115) and (3.118) and the Maxwell relation given by Eq. (1.202):

$$\mathrm{d}s = \frac{c_V}{T}\mathrm{d}T + \left(\frac{\partial P}{\partial T}\right)_v \mathrm{d}v \tag{3.122}$$

By combining Eqs. (3.113) and (3.122), an expression for $\mathrm{d}T$ can be obtained:

$$\mathrm{d}T = \frac{T}{(c_P - c_V)}\left(\frac{\partial v}{\partial T}\right)_P \mathrm{d}P + \frac{T}{(c_P - c_V)}\left(\frac{\partial P}{\partial T}\right)_v \mathrm{d}v \tag{3.123}$$

Now express T as a function of P and v, $T = T(P, v)$; then

$$\mathrm{d}T = \left(\frac{\partial T}{\partial P}\right)_v \mathrm{d}P + \left(\frac{\partial T}{\partial v}\right)_P \mathrm{d}v \tag{3.124}$$

From Eqs. (3.123) and (3.124),

$$\boxed{c_P - c_V = T\left(\frac{\partial v}{\partial T}\right)_P \left(\frac{\partial P}{\partial T}\right)_v} \tag{3.125}$$

Note that in the above expression, the volumetric behavior fully describes the heat-capacity difference. Note also that for an ideal gas, $c_P - c_V = R$.

From Eqs. (3.109) and (3.118),

$$\left(\frac{C_P}{C_V}\right) = \frac{\left(\frac{\partial s}{\partial T}\right)_P}{\left(\frac{\partial s}{\partial T}\right)_v} \tag{3.126}$$

Using Eq. (1.79), Eq. (3.126) becomes

$$\left(\frac{\partial P}{\partial v}\right)_s = \left(\frac{c_P}{c_V}\right)\left(\frac{\partial P}{\partial v}\right)_T \tag{3.127}$$

The term $(\partial P/\partial v)_T$ can be evaluated from $Pv = ZRT$.

$$\left(\frac{\partial P}{\partial v}\right)_T = -\frac{ZRT}{v^2} + \frac{RT}{v}\left(\frac{\partial Z}{\partial P}\right)_T\left(\frac{\partial P}{\partial v}\right)_T \tag{3.128}$$

Rearranging the above equation and using $Pv = ZRT$ provides

$$\left(\frac{\partial P}{\partial v}\right)_T = \frac{-\frac{P}{v}}{\left[1 - \frac{P}{Z}\left(\frac{\partial Z}{\partial P}\right)_T\right]} \tag{3.129}$$

Combining Eqs. (3.127) and (3.129),

$$\left(\frac{\partial P}{\partial v}\right)_s = \left(\frac{c_P}{c_V}\right)\left[\frac{\frac{-P}{v}}{1 - \left(\frac{P}{Z}\right)\left(\frac{\partial Z}{\partial P}\right)_T}\right] \tag{3.130}$$

The isothermal compressibility expression $C_T = -(1/v)(\partial v/\partial P)_T$ and $Pv = ZRT$ can be combined to provide

$$C_T = \frac{1}{P} - \frac{1}{Z}\left(\frac{\partial Z}{\partial P}\right)_T \tag{3.131}$$

From Eqs. (3.102), (3.130), and (3.131), the expression for the sonic velocity becomes

$$a = \left[\frac{\left(\frac{c_P}{c_V}\right) v}{C_T} \right]^{\frac{1}{2}} \tag{3.132}$$

Equation (3.132) gives the sonic velocity of a single-phase fluid of constant composition in terms of the volumetric behavior and the c_P/c_V ratio. The next step is the calculation of the c_P/c_V ratio in terms of volumetric behavior. Let us derive the expression for c_P. At constant temperature, Eq. (3.112) reduces to

$$dh = \left[v - T\left(\frac{\partial v}{\partial T}\right)_P \right] dP \tag{3.133}$$

Integrating Eq. (3.133) from pressure 0 to pressure P,

$$h(T, P) - h^*(T, 0) = \int_0^P \left[v - T\left(\frac{\partial v}{\partial T}\right)_P \right] dP \tag{3.134}$$

or

$$h(T, P) - h^*(T, 0) = RT(Z - 1) + \int_\infty^v \left[T\left(\frac{\partial P}{\partial T}\right)_v - P \right] dv \tag{3.135}$$

where $h^*(T, 0)$ is the ideal gas enthalpy at zero pressure, which is a function of temperature only. Using the PR-EOS, Eq. (3.135) simplifies to

$$h(T, P) - h^*(T, 0) = RT(Z - 1) + \frac{T\left(\frac{da}{dT}\right) - a}{2\sqrt{2}b} \ln \frac{Z + 2.414\mathcal{B}}{Z - 0.414\mathcal{B}} \tag{3.136}$$

From the definition of c_P given by Eq. (3.106),

$$c_P = c_P^* + R(Z - 1) + RT\left(\frac{\partial Z}{\partial T}\right)_P + \frac{\partial}{\partial T}\left\{ \frac{T\left(\frac{da}{dT}\right) - a}{2\sqrt{2}b} \ln \frac{Z + 2.414\mathcal{B}}{Z - 0.414\mathcal{B}} \right\} \tag{3.137}$$

where c_P^* is the ideal-gas heat capacity. Passut and Danner (1972) provide correlations for ideal gas enthalpy, heat capacity, and entropy of some 90 substances–mostly hydrocarbons.

Coefficient	Value
A	3.7771
B	−0.00110
C	7.5281
D	0.65621
E	0.014609

TABLE 3.7 Coefficients of Eq. (3.138)

When c_P is available, then Eq. (3.125) can be used to calculate c_V; the ratio c_P/c_V becomes readily known. Thomas et al. (1970) provide the following equation for the low-pressure heat capacity of natural-gas mixtures as a function of temperature and gas gravity G ($G = M$ of gas/M of air):

$$c_P^* = A + BT + CG + DG^2 + E(TG) \tag{3.138}$$

Table 3.7 gives the coefficients A to E of the above equation. Equation (3.138) is valid for gas gravity G in the range of 0.55 to 2.0 and temperature in the 0 to 600°F interval; the units are c_P^* in Btu/lb °R and T in °F.

Figures 3.42 to 3.44 show the measured and computed sonic velocities for methane, nitrogen, and n-hexane, respectively. The computed enthalpy is compared

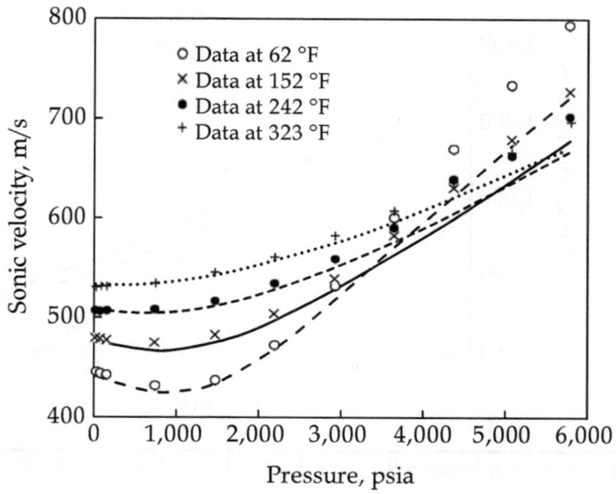

FIGURE 3.42 Sonic velocity in methane gas (adapted from Nutakki, 1989). The lines represent calculations from the PR-EOS.

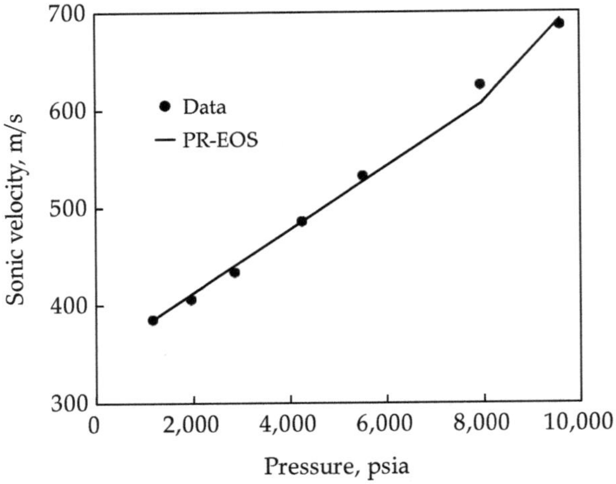

FIGURE 3.43 Sonic velocity in nitrogen gas at 122°F (adapted from Nutakki, 1989).

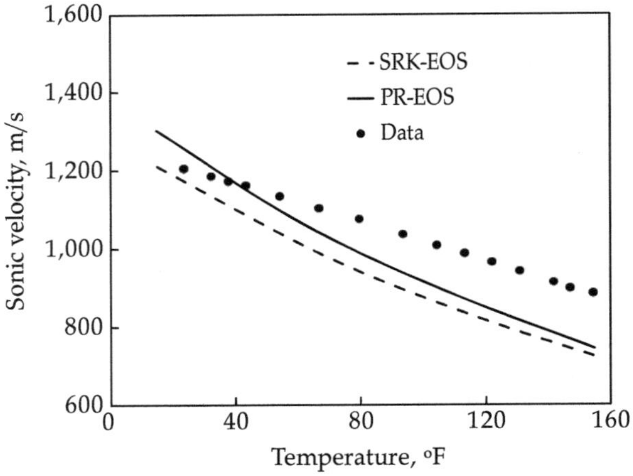

FIGURE 3.44 Sonic velocity in hexane liquid (adapted from Nutakki, 1989).

Temperature, °F	Pressure, psia	$(H^\circ - H)$, J/mol			
		Exp.	SRK-EOS	SW-EOS	PR-EOS
194	435	722	725	727	763
194	1015	1715	1739	1734	1807
194	1450	2489	2504	2485	2575
194	1886	3242	3223	3185	3283
140	435	841	865	864	897
140	1015	2113	2156	2138	2203
140	1450	3163	3188	3140	3211
140	1886	4175	4146	4061	4127
104	435	967	983	981	1011
104	1015	2534	2565	2533	2586
104	1450	3924	3924	3836	3878
104	1886	5174	5095	4956	4979

TABLE 3.8 Enthalpy of Ternary Mixture of Methane, Ethane, and CO_2 (data from Ng and Mather, 1978)

with experimental data in Tables 3.8 and 3.9. In these two tables, the computed results are from the Peng-Robinson EOS (1976) and the Schmidt-Wenzel EOS (1980).

3.6.1 Heating and Cooling Due to Expansion

Hydrocarbon fluid systems undergo rapid expansion around the wellbore and in production facilities. As a result of expansion, the temperature may rise or fall. The following establishes the criteria for cooling and heating due to free expansion. The derivations are limited to single-phase systems where the phase composition does not change. In Example 3.4, we provide an expression that can be used in the free expansion of a two-phase system.

Consider the sketch, Fig. 3.45, in which a fluid is expanding from pressure P_1 to pressure P_2 at steady-state conditions. The system is rigid and insulated. Assume two imaginary pistons at points A and B at time t. During time interval Δt, piston 1 moves from point A to A′ and piston 2 moves from point B to B′. The fluid contained between pistons 1 and 2 is our thermodynamic system. The work done on the system in the time interval Δt is given by

$$W = -P_2 v_2 \Delta m + P_1 v_1 \Delta m \tag{3.139}$$

where v_1 and v_2 are volume per unit mass (specific volume) at the upstream and downstream of the system, respectively, and Δm is the mass within the volume between the piston in the old and new positions, which is the same in the upstream and downstream sides. If one further assumes that there is no heat flow in the upstream and downstream sides, then

$$\Delta U = W \tag{3.140}$$

Temperature, °F	Pressure, psia	Enthalpy of n-C_5, Btu/lbmol		
		Exp.	PR-EOS	SW-EOS
190.6	600	3827	3485	3524
190.6	800	3850	3499	3539
190.6	1000	3875	3516	3557
190.6	1250	3907	3541	3583
190.6	1500	3941	3570	3613
190.6	1750	3975	3601	3646
190.6	2000	4015	3635	3681
280.2	600	8107	7855	7847
280.2	800	8089	7812	7804
280.2	1000	8080	7782	7776
280.2	1250	8082	7760	7754
280.2	1500	8091	7750	7745
280.2	1750	8108	7748	7745
280.2	2000	8129	7754	7752
370.7	600	13070	13267	13163
370.7	800	12844	12925	12815
370.7	1000	12716	12727	12612
370.7	1250	12612	12568	12449
370.7	1500	12547	12462	12340
370.7	1750	12506	12389	12266
370.7	2000	12482	12339	12215

TABLE 3.9 Enthalpy of n-Pentane (data of Erbar et al., 1964)

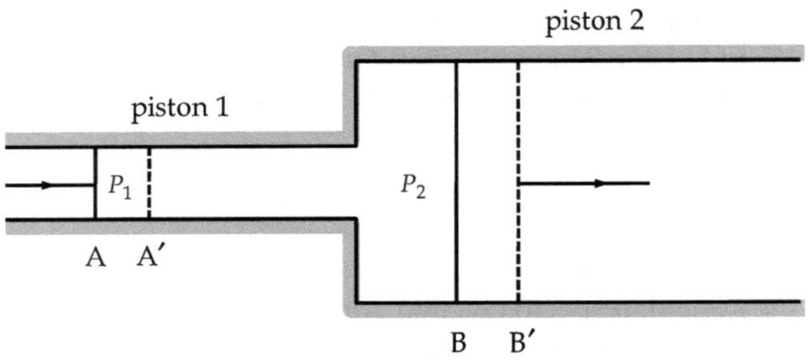

FIGURE 3.45 Sketch of free expansion process.

where

$$\Delta U = \Delta m(u_2 - u_1) \tag{3.141}$$

u_2 and u_1 are the internal energies per unit mass (specific internal energy) in the upstream and downstream sides, respectively. Combining Eqs. (3.139) to (3.141),

$$u_2 - u_1 = -P_2 v_2 + P_1 v_1 \tag{3.142}$$

Since $h = u + Pv$,

$$h_2 = h_1 \tag{3.143}$$

which implies that in free expansion, the enthalpy remains constant. Now, apply Eq. (3.112) to a free-expansion process:

$$c_P \mathrm{d}T = -\left[v - T\left(\frac{\partial v}{\partial T}\right)_P\right]\mathrm{d}P \tag{3.144}$$

Rearranging the above equation,

$$\left(\frac{\partial T}{\partial P}\right)_h = \frac{T\left(\frac{\partial v}{\partial T}\right)_P - v}{c_P} \tag{3.145}$$

From the EOS, $v = ZRT/P$, one obtains

$$\left(\frac{\partial v}{\partial T}\right)_P = \frac{RZ}{P} + \frac{RT}{P}\left(\frac{\partial Z}{\partial T}\right)_P \tag{3.146}$$

Combining Eqs. (3.145), (3.146), and the EOS, $Pv = ZRT$,

$$\left(\frac{\partial T}{\partial P}\right)_h = \frac{\left(\frac{RT^2}{P}\right)\left(\frac{\partial Z}{\partial T}\right)_P}{c_P} \tag{3.147}$$

The Joule-Thomson coefficient μ is defined as

$$\mu = \left(\frac{\partial T}{\partial P}\right)_h \tag{3.148}$$

Equation (3.148) implies that if μ is positive, the free expansion results in cooling. For a negative μ, the free expansion leads to heating.

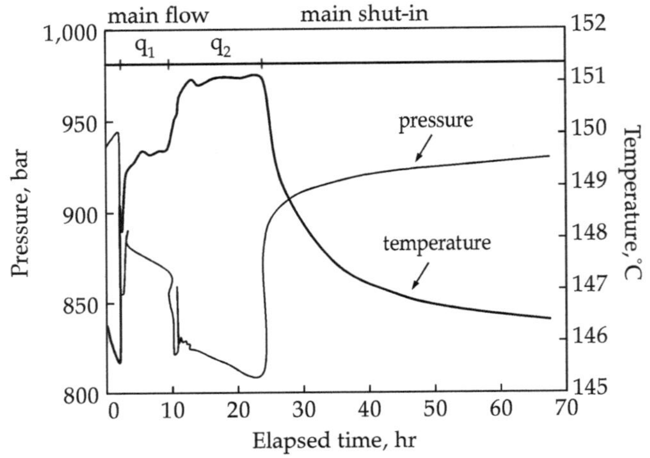

FIGURE 3.46 Downhole pressure and temperature profile during well testing (adapted from Jones, 1988).

Combining Eqs. (3.147)and (3.148),

$$\mu = \frac{\left(\frac{RT^2}{P}\right)\left(\frac{\partial Z}{\partial T}\right)_P}{c_P} \tag{3.149}$$

Equation (3.149) indicates that if the term $(\partial Z/\partial T)_P$ is positive, the free expansion results in cooling and if it is negative, the expansion leads to an increase in temperature.

For natural gases, the term $(\partial Z/\partial T)_P$ may become negative at pressures above 4,000 psia. An EOS conveniently can be used to estimate this term. The method of corresponding states can also be used to compute $(\partial Z/\partial T)_P$.

Jones (1988) reports temperature data during production testing. The data show an increase in bottomhole temperature during flow. The temperature rise from several fields was of the order of 2 to 5°C. All these reservoirs had an undersaturated oil. Figure 3.46 shows a typical downhole pressure and temperature profile during a production test. Jones suggests that the interpretation of temperature fall-off may exhibit the same features as seen on the pressure build-up. Therefore, temperature data may be a useful source of information that can supplement the pressure data.

3.7 Examples and Theory Extension

3.1: Gas, Oil, Near-Critical Oil, and Gas and Volatile Oil Sometimes when a hydrocarbon reservoir is discovered, it is not an easy task to describe the state of the fluid in the reservoir. How would one describe a gas, an oil, and a volatile oil? Is the density of the liquid of the fluid flashed to ambient conditions a good measure?

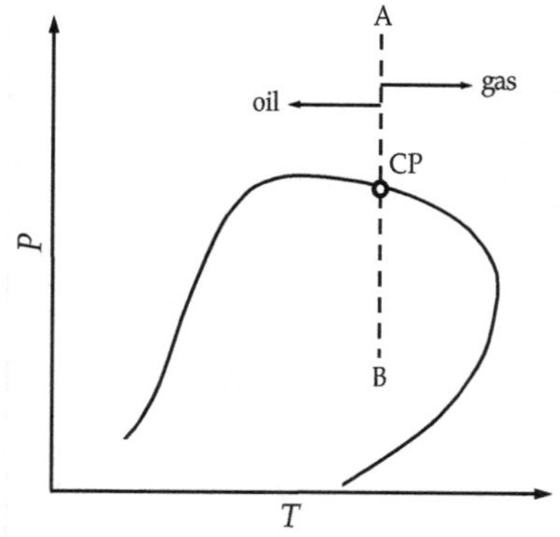

FIGURE 3.47 Definition of oil and gas fluids.

Solution Perhaps the best way to describe a complex high-pressure hydrocarbon mixture is through a pressure-temperature diagram. Figure 3.47 is a plot of saturation pressure of a given reservoir fluid as a function of temperature. In this figure, a point on the right side of line AB is called a "gas" and on the left side is called an "oil" (line AB passes through the critical point CP). Neither the API gravity nor the color of the stock tank liquid could be used as a yardstick to designate a reservoir fluid as a gas or an oil. As an example, a gas-condensate fluid could produce a liquid with an API gravity of 29° (Kilgrin, 1966).

Figure 3.48 shows a further subdivision of a reservoir fluid. The area in the vicinity of line AB, both on the right and left, is the near-critical region. One simple indication of a fluid being in the near-critical region is the magnitude of methane *K*-values. Whenever methane *K*-value is less than 1.1 and the residual (least volatile fraction) has a *K*-value greater than, say 0.3, near-criticality is assured. On the left side of line AB, the region next to near-critical is the volatile-oil region. The production characteristics of the near-critical and volatile-oil fluids can be examined by inspecting Fig. 3.49. This figure represents the constant-volume depletion of a fluid. The dashed curves represent an oil fluid and the

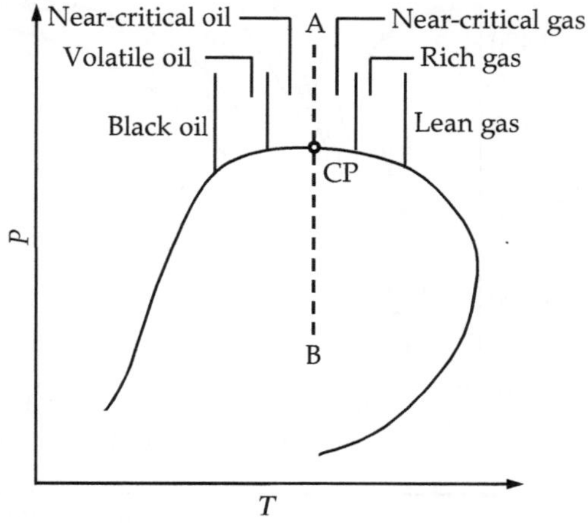

FIGURE 3.48 Definition of various reservoir fluid systems.

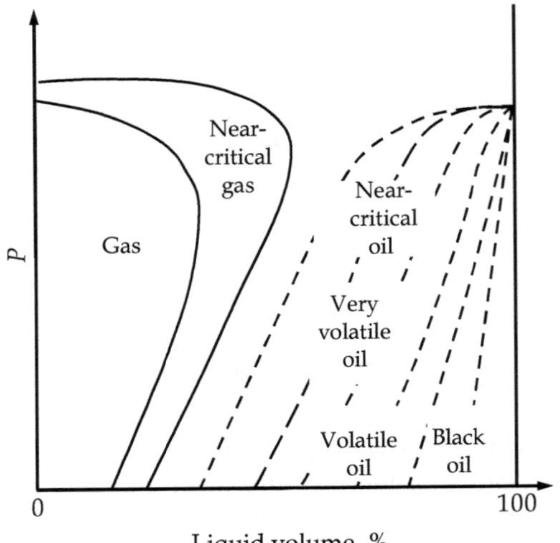

FIGURE 3.49 Constant volume depletion or constant mass expansion of various reservoir fluids.

solid curves a gas-condensate system. On the oil fluid side, this figure shows that as the degree of volatility increases, the producing GOR below the bubblepoint increases. This means that the recovery factor for a very volatile oil, and especially for a near-critical oil, could be low (provided gravity drainage and rock compressibility effects are negligible). Therefore, natural depletion (in the absence of water drive) could be very inefficient. The reason for high GOR production and low recovery is a sudden increase in gas saturation. At high gas saturation, relative permeability to gas is high and therefore, gas is produced at high rates. As a result, oil recovery from pressure depletion is low. Fluid injection, either gas or water, could become viable options.

The diagram in Fig. 3.49 can be generated by starting with a black-oil fluid and then a series of successively increasing temperatures to change the fluid to near-critical oil and then a gas condensate. Figure 3.50, taken from Katz et al. (1940), shows the behavior

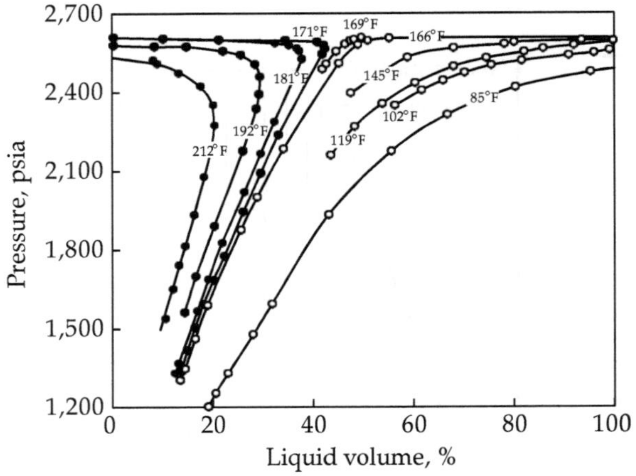

FIGURE 3.50 Liquid isotherms of a reservoir fluid at various temperatures (adapted from Katz et al., 1940).

of a hydrocarbon mixture as temperature changes. It should be mentioned that not all reservoir fluids could undergo such a behavior by changing the temperature. An oil with asphaltene materials may not have a measurable dewpoint and, therefore, does not follow the trends of Figs. 3.49 and 3.50. Figure 3.50 shows the effect of temperature on the volatility of a reservoir oil and the degree of near-criticality for a rich gas condensate. For an oil fluid, the degree of volatility increases as the temperature increases. For a gas-condensate fluid, the trend is the reverse. As the temperature increases, the gas moves away from critical behavior. For a gas-condensate reservoir, the liquid recovery also decreases as the gas becomes more near-critical. This is mainly due to high gas mobility compared to condensate liquids, even at high liquid saturations. Recovery of condensate liquids could be around 15 percent or less for some near-critical gas-condensate reservoirs. Recycling is an attractive option, provided liquid saturation is above 15 to 20 percent.

3.2: Molar Entropy from the PR-EOS Derive the following entropy expression for mixtures using the PR-EOS:

$$s = R\ln\frac{(Z-B)}{Z} + \left\{\frac{\frac{da(T)}{dT}}{2\sqrt{2}b}\right\}\ln\left\{\frac{Z+(1+\sqrt{2})B}{Z+(1-\sqrt{2})B}\right\} + R\sum_{i=1}^{n_c} x_i \ln\left(\frac{Z}{x_i P}\right) + \sum_{i=1}^{n_c} x s_i^o(T)$$

In the above equation s is the molar entropy of the mixture at T, P, and $\mathbf{x}$.

Solution From Example 1.5,

$$S = \int_V^\infty \left[\left(\frac{nR}{V}\right) - \left(\frac{\partial P}{\partial T}\right)_{V,\underline{n}}\right] dV + R\sum_{i=1}^{n_c} n_i \ln\frac{V}{n_i RT} + \sum_{i=1}^{n_c} n_i s_i^0(T)$$

The above equation in terms of molar entropy is

$$s = \int_V^\infty \left[\left(\frac{R}{v}\right) - \left(\frac{\partial P}{\partial T}\right)_{v,x}\right] dv + R\sum_{i=1}^{n_c} x_i \ln\frac{v}{x_i RT} + \sum_{i=1}^{n_c} x_i s_i^0(T)$$

From the PR-EOS,

$$\left(\frac{\partial P}{\partial T}\right)_{v,x} = \frac{R}{(v-b)} - \frac{\frac{da(T)}{dT}}{[v(v+b)+b(v-b)]}$$

Combining the above equations and using

$$\int_v^{\infty}\left[\frac{R}{v}-\frac{R}{(v-b)}\right]dv = R\ln\frac{(v-b)}{v}$$

and

$$\int_v^{\infty}\frac{dv}{v(v+b)+b(v-b)} = \frac{1}{2\sqrt{2}b}\ln\left\{\frac{v+(1+\sqrt{2})b}{v+(1-\sqrt{2})b}\right\}$$

the expression sought will be obtained.

3.3: Effect of Volume Translation on Phase Behavior Show that the simple volume-translation expression, $v^{true} = v^{EOS} + c$ where $c = \sum_{i=1}^{n_c} x_i c_i$, does not affect the prediction of phase compositions. It affects only the phase densities.

Solution Consider the expression for fugacity coefficient given by Eq. (1.100). Let us denote the fugacity coefficient using V^{true} as φ_i^{true} and the one from V^{EOS} by φ_i^{EOS}. Then

$$V^{true} = V^{EOS} + \sum_{j=1}^{n_c} n_j c_j$$

$$\bar{V}_i^{true} = \frac{\partial}{\partial n_i}\left[V^{EOS} + \sum_{j=1}^{n_c} n_j c_j\right]_{T,P,n_i} = \bar{V}_i^{EOS} + c_i$$

Substituting $\bar{V}_i^{true}$ and $\bar{V}_i^{EOS}$ into Eq. (1.100):

$$\frac{\varphi_i^{true}}{\varphi_i^{EOS}} = \exp\left[\frac{-c_i P}{RT}\right]$$

The equilibrium ratios which provide the phase compositions are the same whether one uses an EOS with or without volume translation,

$$K_i = \left[\frac{\varphi_i^L}{\varphi_i^V}\right]^{true} = \left[\frac{\varphi_i^L}{\varphi_i^V}\right]^{EOS}$$

The term $\exp[-c_i P/RT]$ drops out from both the numerator and denominator.

3.4: Cooling and Heating Due to Expansion Consider the following three gases at the given temperature and pressure. If these gases undergo a free expansion around the wellbore, would you expect heating or cooling? Compute the final temperature for each case (use the EOS approach to perform all of the computations).

Component	Gas A, mol%	Gas B, mol%	Gas D, mol%
CO_2	2.37	2.44	100
N_2	0.31	0.08	
C_1	73.19	82.10	
C_2	7.8	5.78	
C_3	3.55	2.87	
i-C_4	0.71	0.56	
n-C_4	1.45	1.23	
i-C_5	0.64	0.52	
n-C_5	0.68	0.60	
C_6	1.09	0.72	
C_{7+}	8.21	3.10	
$M_{C_{7+}}$	184.0	132.0	
Initial pressure, psia	8,500	6,500	12,000 to 7,500
Final pressure, psia	7,500	5,500	6,000 to 3,000
Initial temperature, °F	280	180	180

Solution Equation (3.143) can be used to solve this problem. In a free expansion,

$$h_{initial}(T, P) = h_{final}(T', P')$$

Initial pressure and temperature and final pressure, as well as the composition, are provided. Solving the above equation requires knowledge of $\sum_{i=1}^{n_c} x_i h_i^0(T)$ and $\sum_{i=1}^{n_c} x_i h_i^0(T')$. Passut and Danner (1972) provide the ideal-gas enthalpy data. The results of free expansion for gases A, B, and C are shown in Fig. 3.51. The PR-EOS was used in the calculations. Note that the above equation is not limited to the single-phase state; it also applies to the two-phase state. In the two-phase state at T' and P',

$$h(T', P') = Vh^G(T', P') + (1 - V)h^L(T', P')$$

where h^G and h^L are the molar enthalpies of the gas and liquid phases, respectively, and V is the mol fraction of the gas phase. To obtain V and the composition of the phase, one needs to include phase equilibrium expressions (to be discussed in Chap. 5).

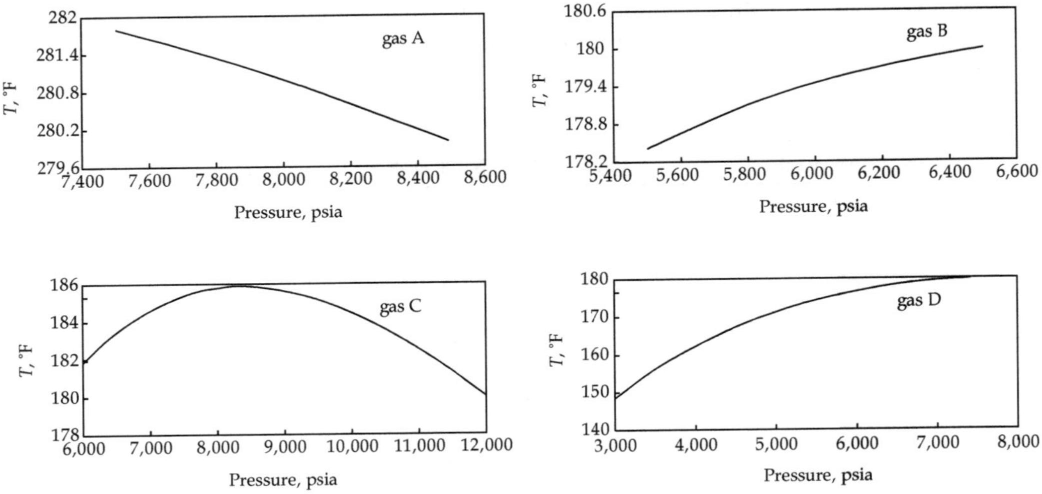

FIGURE 3.51 Cooling and heating due to expansion.

3.5: Effect of Composition Change from Curvature on the Interfacial Tension of Binary Mixture Consider a mixture of 95 mol% C_1 and 5 mol% n-C_{10} at 327.5 K and 276 bar. Use the PR-EOS and the expression for interfacial tension given by Eq. (3.95) to calculate the influence of compositional changes due to curvature on the interfacial tension for $r=\infty$, 10^{-4}, 10^{-5}, and 10^{-6} cm. Keep the liquid pressure constant at 276 bar.

Solution The following expressions define the problem:

$$x_1 + x_2 = 1$$

$$y_1 + y_2 = 1$$

$$z_1 = x_1 L + (1 - L) y_1$$

$$f_i^L(T, P^L, x_1) = f_i^G(T, P^G, y_1) \quad i = 1, 2$$

$$P^G = P^L + \frac{2}{r}\left[\sum_{i=1}^{2} P_i\left(x_i \frac{d^L}{M^L} - y_i \frac{d^G}{M^G}\right)\right]^{\frac{1}{4}}$$

where P_i is the parachor of component i ($P_{C_1}=70$; $P_{n\text{-}C_{10}}=500$) and L is the liquid mol fraction (see Chap. 5, section Two-Phase Split Calculations). The unknowns in the above set of 6 equations are x_1, x_2, y_1, y_2, L, and P^G. Once these unknowns are calculated, at given r, T, P^L, z_1, and z_2, then σ (the multiplier of $2/r$ in the last equation above) can be estimated. We use the PR-EOS for the fugacity and phase-density calculations. The predicted densities are not adjusted by volume translation. The results are summarized below.

r, cm	σ, dyne/cm	L, liquid mol fraction	$P^G - P^L$, bar
∞	1.06	0.1071	0
10^{-3}	1.06	0.1071	0.002
10^{-4}	1.05	0.1072	0.02
10^{-5}	1.05	0.1078	0.21
10^{-6}	0.97	0.1135	1.95

Note that as r decreases, σ also decreases. In the above calculation we have neglected the effect of curvature itself on σ, which will be discussed in Chap. 7.

3.6: Partial Molar Volumes from the PR-EOS Use the PR-EOS to calculate the partial molar volume of C_2 and n-C_7 at 80°C and 57 atm for $x_{\text{n-C}_7}$ in the range of 0 to 0.2. Compare the results with the data of Wu and Ehrlich (1973).

Solution The PR-EOS is pressure-explicit. Therefore,

$$\bar{V}_i = \left(\frac{\partial V}{\partial n_i}\right)_{T,P,n_j} = -\frac{\left[\left(\dfrac{\partial P}{\partial n_i}\right)_{T,V,n_j}\right]}{\left[\left(\dfrac{\partial P}{\partial V}\right)_{T,\underline{n}}\right]}$$

$(\partial P/\partial n_i)_{T,V,n_j}$ and $(\partial P/\partial V)_{T,\underline{n}}$ can be easily calculated from Eq. (3.6). The results are shown in Fig. 3.52. There is a good agreement between data and predicted results.

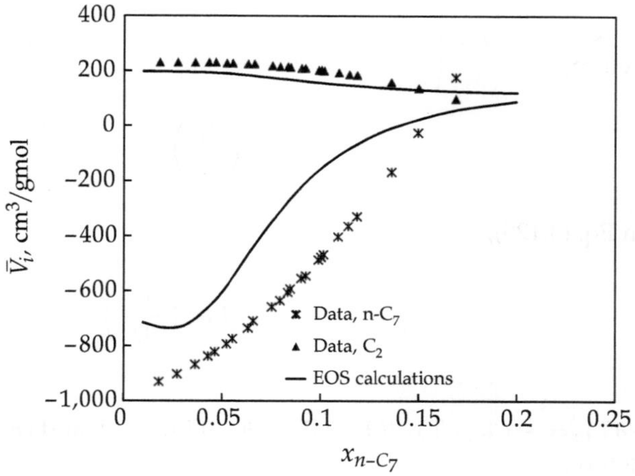

FIGURE 3.52 Partial molar volumes of C_2 and n-C_7 at 80°C and 57 atm (data from Wu and Ehrlich, 1973).

3.7: Relations between C_T, c_P, c_V, C_S Derive the following relationships between:

(a) The isothermal and isentropic compressibilities: $C_T = (c_P/c_V)C_S$, and

(b) c_P and c_V: $c_P = c_V + Tve^2/C_T$.

Solution

a. We can start from Eq. (3.131):

$$C_T = \frac{1}{P} - \frac{1}{Z}\left(\frac{\partial Z}{\partial P}\right)_T$$

From Eq. (3.130),

$$\left(\frac{\partial P}{\partial v}\right)_s = -\frac{\left(\dfrac{c_P}{c_V}\right)}{v}\left[\frac{1}{\dfrac{1}{P} - \left(\dfrac{1}{Z}\right)\left(\dfrac{\partial Z}{\partial P}\right)_T}\right]$$

Combining the two equations above results in

$$\left(\frac{\partial P}{\partial v}\right)_s = -\frac{\left(\dfrac{c_P}{c_V}\right)}{vC_T}$$

The definition of isentropic compressibility is

$$C_S = -\frac{1}{V}\left(\frac{\partial V}{\partial P}\right)_S$$

Therefore,

$$C_T = \left(\frac{c_P}{c_V}\right)C_S$$

b. From Eq. (3.125),

$$c_P - c_V = Tve\left(\frac{\partial P}{\partial T}\right)_v$$

but $(\partial P/\partial T)_v = -(\partial v/\partial T)_P/(\partial v/\partial P)_T$, $(\partial v/\partial T)_P = ve$, and $(\partial v/\partial P)_T = -vC_T$. Therefore,

$$c_P = c_V + \frac{Tve^2}{C_T}$$

3.8: Derivatives $(\partial n_{j,t}/\partial T)_{P,\underline{n}}$ and $(\partial Z_j/\partial T)_{P,\underline{n}}$ Calculate $(\partial n_{j,t}/\partial T)_{P,\underline{n}}$ and $(\partial Z_j/\partial T)_{P,\underline{n}}$ terms in Eq. (3.77) which are necessary for the calculation of thermal expansivity.

Solution $(\partial n_{j,t}/\partial T)_{P,\underline{n}}$ can be expressed in terms of $(\partial n_{j,i}/\partial T)_{P,\underline{n}}$ [see Eq. (3.78)], and the latter is obtained from the material balance and equilibrium criterion:

$$\left(\frac{\partial n_{1,i}}{\partial T}\right)_{P,\underline{n}} + \left(\frac{\partial n_{2,i}}{\partial T}\right)_{P,\underline{n}} = 0 \quad i = 1, \ldots, n_c$$

$$\left[\sum_{k=1}^{n_c}\left(\frac{\partial f_{1,i}}{\partial n_{1,k}}\right)_{n_{1,k}}\left(\frac{\partial n_{1,k}}{\partial T}\right)_{P,\underline{n}} - \sum_{k=1}^{n_c}\left(\frac{\partial f_{2,i}}{\partial n_{2,k}}\right)_{n_{2,k}}\left(\frac{\partial n_{2,k}}{\partial T}\right)_{P,\underline{n}}\right] + \left[\left(\frac{\partial f_{1,i}}{\partial T}\right)_{P,\underline{n}_1} - \left(\frac{\partial f_{2,i}}{\partial T}\right)_{P,\underline{n}_2}\right] = 0 \quad i = 1, \ldots, n_c$$

Note that we have dropped some of the subscripts on the fugacity derivatives. The above $2n_c$ equations provide the $2n_c$ composition variations. $(\partial Z_j/\partial T)_{P,\underline{n}}$ can be calculated from

$$\left(\frac{\partial Z_j}{\partial T}\right)_{P,\underline{n}} = \left(\frac{\partial Z_j}{\partial T}\right)_{P,\underline{n}_j} + \sum_{k=1}^{n_c}\left(\frac{\partial Z_j}{\partial n_{j,k}}\right)\left(\frac{\partial n_{j,k}}{\partial T}\right) \quad j = 1, 2$$

An EOS can be used to calculate $(\partial Z_j/\partial T)_{P,\underline{n}_j}$, $(\partial Z_j/\partial n_{j,k})_{P,T,n_{j,k}}$, as well as the fugacity derivatives of phase j. T he calculation of $(\partial Z_j/\partial T)_{P,\underline{n}_j}$ can be simplified if one uses

$$\frac{\partial Z_j}{\partial T} = \frac{\partial Z_j}{\partial \mathcal{A}}\frac{\partial \mathcal{A}}{\partial T} + \frac{\partial Z_j}{\partial \mathcal{B}}\frac{\partial \mathcal{B}}{\partial T}$$

where the expressions for $\mathcal{A}$ and $\mathcal{B}$ are provided in Eqs. (3.15) and (3.16) for the PR-EOS. In Prob. 3.6 we comment on how to calculate the derivative of Z with respect to P, which is required for the estimation of isothermal compressibility.

3.9: Liquid Heating from Expansion Consider pure n-C_{10} at 5,000 psia and 100°F. What would be the temperature after free expansion to 1,000 psia?

Solution A procedure similar to Example 3.4 can be used. The results are plotted in Fig. 3.53. There is heating due to expansion, which is often the case for liquids. The temperature of liquids often increases upon expansion.

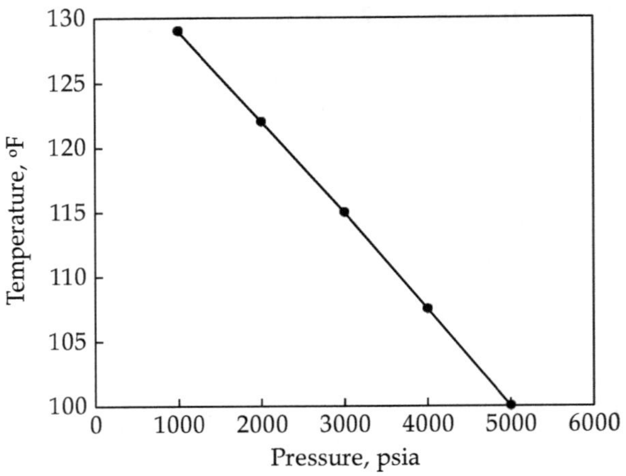

FIGURE 3.53 Temperature rise of n-C_{10} due to expansion.

3.10: Mol Fraction of Water Molecules Not Bonded at Saturation The mol fractions of H_2O molecules that do not associate are depicted in Fig. 3.54 along the vapor and the liquid equilibrium locus. The corresponding P-T plot of water is also shown in the same figure. In the liquid and vapor phases the trends are opposite. Explain the reason for the different trends.

Solution The expression for mol fraction of water molecules not bonded at sites α and β is obtained from Eq. (3.47)

$$\chi = \frac{-1 + \sqrt{1 + 8c\Delta^{\alpha\beta}}}{4c\Delta^{\alpha\beta}} \tag{3.151}$$

where the association strength, $\Delta^{\alpha\beta}$, is given in Eq. (3.51). Solving for v and χ when P and T are the saturation pressure and temperature of water, we can calculate v from Eq. (3.57) and then χ from Eq. (3.151). As pressure increases at high temperature, water molecules in the vapor become more associated due to stronger interaction. In the liquid, as temperature decreases, due to low kinetic effect, there is more association.

3.11: Maximum Superheating of Water at 1 atm from the CPA-EOS Compute the maximum superheating of water from the CPA-EOS at atmospheric pressure.

Solution The limit of stability is given by:

$$\left(\frac{\partial P}{\partial v}\right)_T = 0 \tag{3.152}$$

From Eq. (3.6) we obtain

$$\left(\frac{\partial P^{physical}}{\partial v}\right)_T = -\frac{RT}{(v-b)^2} + \frac{2a(v+b)}{(v^2+2bv-b^2)^2} \tag{3.153}$$

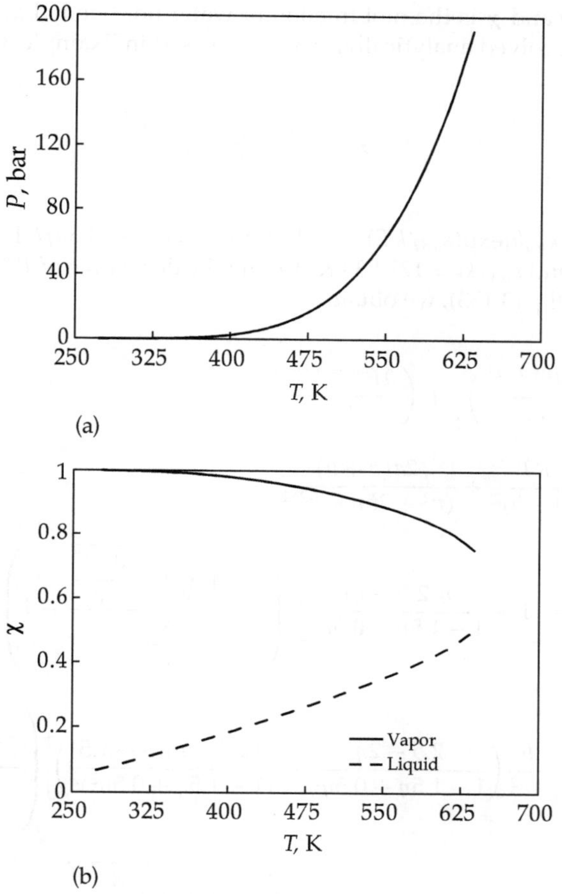

FIGURE 3.54 (a) Saturation pressure, and (b) χ of the vapor and liquid phases of H_2O.

Combining Eqs. (3.6), (3.152), and (3.153), we obtain at 1 atm pressure, $T = 288.36\ K$ ($\sim 15°C$) and molar volume $v = 3.525 \times 10^{-5}\ m^3/mol$. This temperature is far from the experimental value of 270°C. Now we use the CPA-EOS

$$P = P^{physical} + P^{association} \tag{3.154}$$

The association pressure for water, can be obtained from Eq. (3.56),

$$P^{association} = (2RT/v)\left[1 + \frac{\eta(2.5 - \eta)}{1 - 1.5\eta + 0.5\eta^2}\right](\chi - 1) \tag{3.155}$$

where $\eta = b/4v$ and χ is the mol fraction of water not bonded at sites α or β. For bulk water, χ can be solved analytically, as we discussed in Example 3.10.

$$\chi = \frac{-1 + \sqrt{1 + 8c\Delta^{\alpha\beta}}}{4c\Delta^{\alpha\beta}} \tag{3.156}$$

where $\Delta^{\alpha\beta} = g\kappa_{\alpha\beta} b[\exp(\varepsilon_{\alpha\beta}/kT) - 1]$ with $g = (1 - 0.5\eta)/(1 - \eta)^3, \kappa_{\alpha\beta} = 1.8015 \times 10^{-6}$ m^3/mol, and $\varepsilon_{\alpha\beta}/k_b = 1738.39$ K. Taking the derivative of $P^{association}$ from Eq. (3.155) and adding to Eq. (3.153), we obtain

$$\begin{aligned}
\left(\frac{\partial P}{\partial v}\right)_T &= \left(\frac{\partial P^{physical}}{\partial v}\right)_T + \left(\frac{\partial P^{association}}{\partial v}\right)_T \\
&= -\frac{RT}{(v-b)^2} + \frac{2a(v+b)}{(v^2+2bv-b^2)^2} \\
&\quad -\frac{2RT}{v^2}\left[1 + \frac{\eta(2.5-\eta)}{1-1.5\eta+0.5\eta^2}\right]\left(\frac{-1+\sqrt{1+\frac{8\Delta^{\alpha\beta}}{v}}}{\frac{4\Delta^{\alpha\beta}}{v}} - 1\right) + \frac{2RT}{v} \\
&\quad \times \left[1 - \frac{b}{4v^2}\left(\frac{2.5-2\eta}{1-1.5\eta+0.5\eta^2} - \frac{\eta(2.5-\eta)(\eta-1.5)}{(1-1.5\eta+0.5\eta^2)^2}\right)\right]\left(\frac{-1+\sqrt{1+\frac{8\Delta^{\alpha\beta}}{v}}}{\frac{4\Delta^{\alpha\beta}}{v}} - 1\right) \\
&\quad + \frac{2RT}{v}\left[1 + \frac{\eta(2.5-\eta)}{1-1.5\eta+0.5\eta^2}\right]\left(\frac{-1+\sqrt{1+\frac{8\Delta^{\alpha\beta}}{v}}}{\frac{4\Delta^{\alpha\beta}}{v}} - \frac{1}{v\sqrt{1+\frac{8\Delta^{\alpha\beta}}{v}}}\right) + \frac{2RT}{v} \\
&\quad \times \left[1 + \frac{\eta(2.5-\eta)}{1-1.5\eta+0.5\eta^2}\right]\left(\frac{\partial\Delta^{\alpha\beta}}{v}\right)_T \left[\frac{-1+\sqrt{\frac{1+8\Delta^{\alpha\beta}}{v}}}{\frac{4(\Delta^{\alpha\beta})^2}{v}} + \frac{1}{v\sqrt{1+\frac{8\Delta^{\alpha\beta}}{v}}}\right] = 0
\end{aligned} \tag{3.157}$$

where

$$\left(\frac{\partial\Delta^{\alpha\beta}}{\partial v}\right)_T = -\frac{b}{4v^2}\kappa_{\alpha\beta} b\left[\exp\left(\frac{\varepsilon_{\alpha\beta}}{kT}\right) - 1\right]\left[-\frac{1}{2(1-\eta)^3} + \frac{3(1-0.5\eta)}{(1-\eta)^4}\right]$$

Equation (3.157) has two unknowns, v and T, when P is given. We obtain $v = 3.636 \times 10^{-5}$ m^3/mol, $T = 606.1$ K $\sim 333°$C from Eqs. (3.57) and (3.157). At these conditions $\chi = 0.4882$.

Note that the computed temperature is higher than the measured value of 270°C but much closer to the measured value than the prediction from the PR-EOS.

3.12: Total Compressibility and Total Partial Molar Volume in Three-Phase Conditions In the text we presented working equations to compute two-phase compressibility. We neglected volume shift in our derivations. Here we will derive the expression for the computation of total isothermal compressibility in three-phase systems. We will also include the derivation of total partial molar volume in three-phase. The total partial molar volume in three-phase is given by

$$\bar{V}_i = \left(\frac{\partial V}{\partial n_i}\right)_{T,P,\boldsymbol{n}_i} \tag{3.158}$$

where $V = V_1 + V_2 + V_3$ and indeces, 1, 2, and 3 stand for phase identity. The total partial molar volume is part of the formulation in three-phase compositional modeling (Moortgart et al., 2012). In three-phase we write

$$n_{j,t} = \sum_{i=1}^{n_c} n_{j,i} \quad j = 1,\ 2,\ \text{and } 3$$

$$V = \sum_{j=1}^{3}\left[\frac{Z_j n_{j,t} RT}{P} + \sum_{i=1}^{n_c} n_{j,i} c_i\right]$$

From the definition of C_T and total partial molar volume $\bar{V}_i$, one can derive the expressions given below.

When the mixture is in three-phase state, the total compressibility and total partial molar volumes are given by

$$C_T = \frac{V^{EOS}}{PV} - \frac{1}{V}\sum_{j=1}^{3}\left\{\frac{n_{j,t}RT}{P}\left(\frac{\partial Z_j}{\partial P}\right)_{T,\underline{n}j} + \sum_{k=1}^{n_c}\left[\frac{n_{j,t}RT}{P}\left(\frac{\partial Z_j}{\partial n_{j,k}}\right)_{T,P,\boldsymbol{n}j,k} + v_j^{EOS} + c_k\right]\left(\frac{\partial n_{j,k}}{\partial P}\right)_{T,\underline{n}}\right\} \tag{3.159}$$

$$\bar{V}_i = \sum_{j=1}^{3}\sum_{k=1}^{n_c}\left[\frac{n_{j,t}RT}{P}\left(\frac{\partial Z_j}{\partial n_{j,k}}\right)_{T,P,\underline{n}_{j,k}} + v_j^{EOS} + c_k\right] \times \left(\frac{\partial n_{i,k}}{\partial n_i}\right)_{T,P,\underline{n_i}} \quad i = 1, \ldots, n_c \tag{3.160}$$

Note that v_j^{EOS} is the molar volume of phase j from the EOS. The unknowns in Eq. (3.159) and Eq. (3.160) are $(\partial n_{j,k}/\partial P)_{T,\underline{n}}$ and $(\partial n_{j,k}/\partial n_i)_{T,P,\boldsymbol{n}_i}$.

From $n_i = \sum_{j=1}^{3} n_{j,i} \quad i=1,\ldots,n_c$

$$\sum_{j=1}^{3}\left(\frac{\partial n_{j,i}}{\partial n_i}\right)_{T,P,\boldsymbol{n}_i} = \delta_{i,j} \qquad i=1,\ldots,n_c \tag{3.161}$$

$$\sum_{j=1}^{3}\left(\frac{\partial n_{j,i}}{\partial P}\right)_{T,\underline{\boldsymbol{n}}} = 0 \qquad i=1,\ldots,n_c \tag{3.162}$$

From $Z_j = Z_j(T, P, n_{j,1}, \ldots, n_{j,n_c})$, one can write

$$\left(\frac{\partial Z_j}{\partial P}\right)_{T,\underline{\boldsymbol{n}}} = \left(\frac{\partial Z_j}{\partial P}\right)_{T,\underline{\boldsymbol{n}}_j} + \sum_{k=1}^{n_c}\left(\frac{\partial Z_j}{\partial n_{j,k}}\right)\left(\frac{\partial n_{j,k}}{\partial P}\right) \qquad j=1,\ 2,\ 3 \tag{3.163}$$

Based on $\mathrm{d}f_{1,i} = \mathrm{d}f_{2,i} = \mathrm{d}f_{3,i}$ (i.e., equilibrium criterion), one can write the following additional expressions:

$$\sum_{k=1}^{n_c}\left\{\left[\left(\frac{\partial f_{1,i}}{\partial n_{1,k}}\right)_{T,P,\boldsymbol{n}_{1,k}} + \left(\frac{\partial f_{2,i}}{\partial n_{2,k}}\right)_{T,P,\boldsymbol{n}_{2,k}}\right]\left(\frac{\partial n_{2,k}}{\partial P}\right)_{T,\underline{\boldsymbol{n}}} + \left(\frac{\partial f_{1,i}}{\partial n_{1,k}}\right)_{T,P,\boldsymbol{n}_{1,k}}\left(\frac{\partial n_{3,k}}{\partial P}\right)_{T,\underline{\boldsymbol{n}}}\right\}$$

$$= \left(\frac{\partial f_{1,i}}{\partial P}\right)_{T,\underline{\boldsymbol{n}}_1} - \left(\frac{\partial f_{2,i}}{\partial P}\right)_{T,\underline{\boldsymbol{n}}_2} \qquad i=1,\ldots,n_c \tag{3.164a}$$

$$\sum_{k=1}^{n_c}\left\{\left(\frac{\partial f_{1,i}}{\partial n_{1,k}}\right)_{T,P,\underline{\boldsymbol{n}}_{1,k}}\left(\frac{\partial n_{2,k}}{\partial P}\right)_{T,\boldsymbol{n}} + \left[\left(\frac{\partial f_{1,i}}{\partial n_{1,k}}\right)_{T,P,\boldsymbol{n}_{1,k}} + \left(\frac{\partial f_{3,i}}{\partial n_{3,k}}\right)_{T,P,\boldsymbol{n}_{3,k}}\right]\left(\frac{\partial n_{3,k}}{\partial P}\right)_{T,\underline{\boldsymbol{n}}}\right\}$$

$$= \left(\frac{\partial f_{1,i}}{\partial P}\right)_{T,\underline{\boldsymbol{n}}_1} - \left(\frac{\partial f_{3,i}}{\partial P}\right)_{T,\underline{\boldsymbol{n}}_3} \qquad i=1,\ldots,n_c \tag{3.164b}$$

and

$$\sum_{k=1}^{n_c}\left\{\left[\left(\frac{\partial f_{1,i}}{\partial n_{1,k}}\right)_{T,P,\boldsymbol{n}_{1,k}} + \left(\frac{\partial f_{2,i}}{\partial n_{2,k}}\right)_{T,P,\boldsymbol{n}_{2,k}}\right]\left(\frac{\partial n_{2,k}}{\partial n_i}\right)_{T,P,\boldsymbol{n}_i} + \left(\frac{\partial f_{1,i}}{\partial n_{1,k}}\right)_{T,P,\boldsymbol{n}_{1,k}}\left(\frac{\partial n_{3,k}}{\partial n_i}\right)_{T,P,\boldsymbol{n}_i}\right\}$$

$$= \sum_{k=1}^{n_c}\delta_{k,i}\left(\frac{\partial f_{1,i}}{\partial n_{1,k}}\right)_{T,P,\boldsymbol{n}_{1,k}} \qquad i=1,\ldots,n_c \tag{3.165a}$$

$$\sum_{k=1}^{n_c}\left\{\left(\frac{\partial f_{1,j}}{\partial n_{1,k}}\right)_{T,P,\boldsymbol{n}_{1,k}}\left(\frac{\partial n_{2,k}}{\partial n_i}\right)_{T,P,\boldsymbol{n}_i} + \left[\left(\frac{\partial f_{1,j}}{\partial n_{1,k}}\right)_{T,P,\boldsymbol{n}_{1,k}} + \left(\frac{\partial f_{3,j}}{\partial n_{3,k}}\right)_{T,P,\boldsymbol{n}_{3,k}}\right]\left(\frac{\partial n_{3,k}}{\partial n_i}\right)_{T,P,\boldsymbol{n}_i}\right\}$$

$$= \sum_{k=1}^{n_c}\delta_{ki}\left(\frac{\partial f_{1,i}}{\partial n_{1,k}}\right)_{T,P,\boldsymbol{n}_{1,k}} \qquad i=1,\ldots,n_c \tag{3.165b}$$

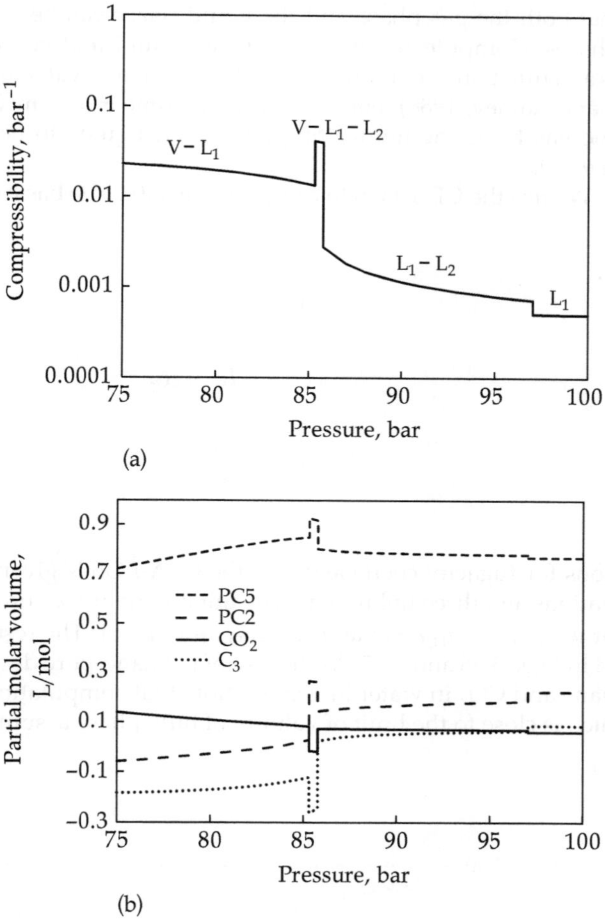

FIGURE 3.55 Total compressibility and total partial molar volumes of CO_2, C_3 and fractions PC2 and PC5 in three-phase (adapted from Li and Firoozabadi, 2012). V-L_1, V-L_1-L_2, L_1-L_2, and L_1 represent vapor-liquid phase 1, vapor-liquid phase 1-liquid phase 2, liquid phase 1-liquid phase 2, and liquid phase 1, respectively. L_1 represents the lighter liquid phase.

The above equations can be used to obtain C_T and $\bar{V}_i$. Li and Firoozabadi (2012) present C_T, and total partial molar volume of some select components of a reservoir fluid using the PR-EOS. Figure 3.55 shows computed C_T and $\bar{V}_i$ in a reservoir fluid with CO_2 in three-phase. Note the scale in Fig. 3.55a is logarithmic.

3.13: Solubility of C_1 and CO_2 in Water in Nanopores Compute the solubility of methane (C_1) and CO_2 in water in capillary tubes as a function of tube radius at the following conditions:

1. Methane: $T = 350$ K, external pressure $P^G = 500$ bar
2. CO_2: $T = 350$ K, external pressure $P^G = 500$ bar

Assume that both the gas phase and the liquid phase can be described as bulk homogenous phases. Compute the pressure in the liquid and composition of gas and liquid phases as a function of curvature ($1/r$). For methane/water assume $\sigma = 44$ mN/m (Firoozabadi and Ramey, 1988). For CO_2/water assume $\sigma = 24$ mN/m (Espinoza et al., 2010). Assume gas to be the nonwetting phase and liquids to be the wetting phase; contact angle $\theta = 0$.

Solution We use the CPA-EOS to compute fugacity. The basic expressions are

$$f^L_{H_2O} = f^G_{H_2O} \tag{3.166}$$

$$f^L_{C_1 \text{ or } CO_2} = f^G_{C_1 \text{ or } CO_2} \tag{3.167}$$

$$P^G - P^L = \frac{2\sigma}{r} \tag{3.168}$$

The expressions for fugacity coefficients for the CPA-EOS is given in Eq. (3.59). There are three equations and three unknowns. The unknowns are y_{C_1} or y_{CO_2}, x_{C_1} or x_{CO_2}, and P^L. Note that $y_{C_1 \text{ or } CO_2} + y_{H_2O} = 1$ and $x_{C_1 \text{ or } CO_2} + x_{H_2O} = 1$. The results of computations are presented in Figs. 3.56 and 3.57. As the size of the nano-pore decreases, the solubility of both methane and CO_2 in water increases. Note that computations are carried out to $r = 5$ nm, which is close to the limit of validity of bulk phase assumption.

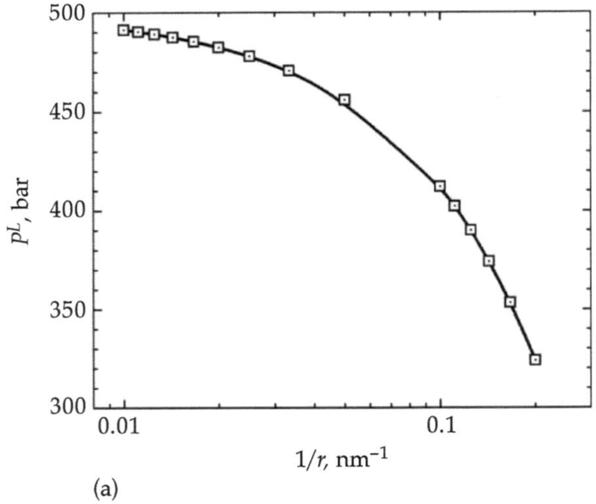

FIGURE 3.56 Effect of curvature on (a) liquid pressure, composition of (b) liquid and (c) gas phases of C_1-H_2O system; $P^G = 500$ bar, $T = 350$ K (Fig. 3.56b adapted from Jin and Firoozabadi, 2015).

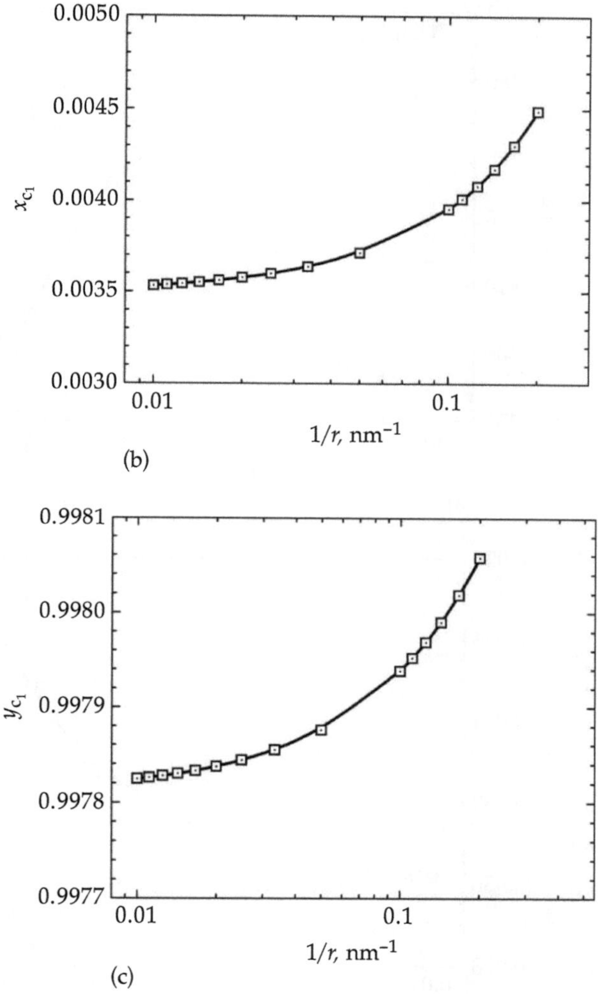

FIGURE 3.56 *(Continued)*

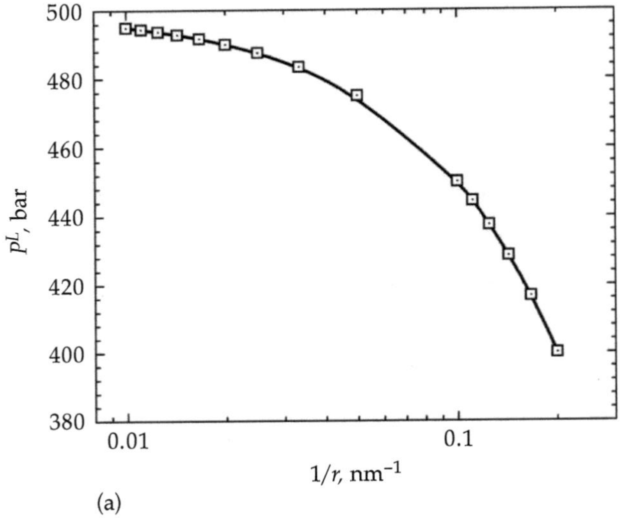

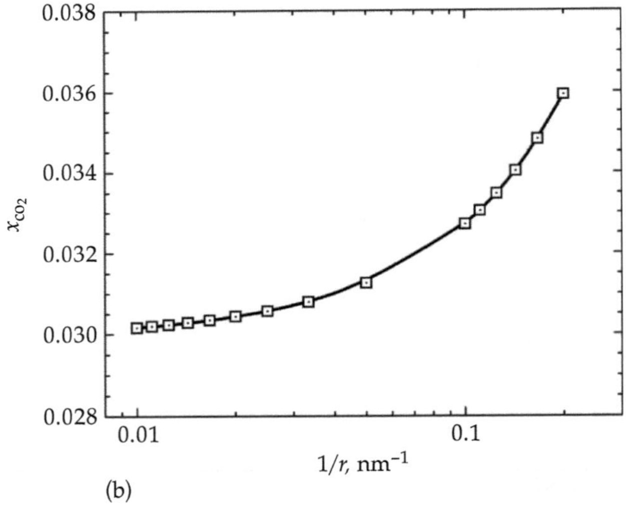

FIGURE 3.57 Effect of curvature on (a) liquid pressure, and composition of (b) liquid and (c) gas phases of CO_2-H_2O mixture; $P^G = 500$ bar, $T = 350$ K (Fig. 3.57b adapted from Jin and Firoozabadi, 2015).

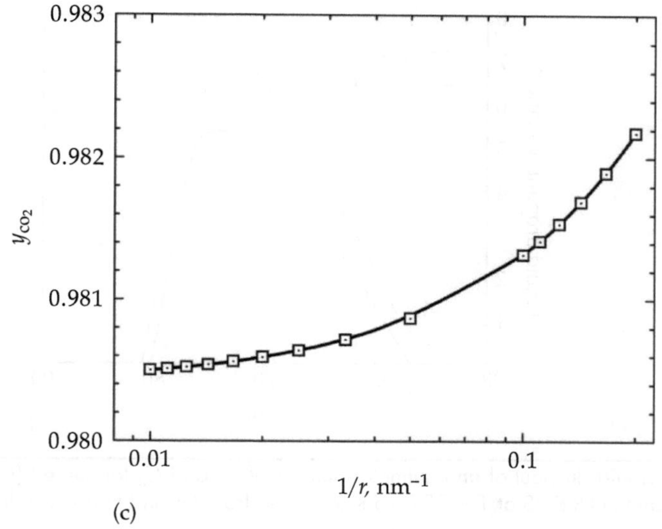

FIGURE 3.57 *(Continued)*

3.14: Retrograde Condensation in Nanopores Consider a capillary tube with inside diameter of 10 nm ($r = 5$ nm). We are interested in studying the effect of interface curvature on retrograde condensation of an equimolar mixture of C_1 and C_3 at 54°C. Plot the liquid dropout versus pressure in the gas phase. We would like to point out that when pores are less than some few nm, there is no two-phase in such nanopores (Jin and Firoozabadi, 2015).

Solution We use the PR-EOS to describe the gas and liquid phases. The basic expressions are:

$$f^L_{C_1} = f^G_{C_1}$$

$$f^L_{C_3} = f^G_{C_3}$$

$$P^G - P^L = \frac{2\sigma}{r}$$

$$\sigma^{\frac{1}{4}} = \sum_{i=1}^{2} P_i \left[y_i \left(\frac{d^G}{M^G} \right) - x_i \left(\frac{d^L}{M^L} \right) \right]$$

$$z_i = V^L x_i + (1 - V^L) y_i \quad i = 1 \text{ (methane)}, 2 \text{ (propane)}$$

New symbols in the above equations are overall molar composition, $z_{C_1} = 0.5$, $z_{C_3} = 0.5$, V^L = molar fraction of the liquid phase. The parachors are:. $P_{C_1} = 77$ and $P_{C_3} = 150.3$.

The unknowns in the above equations are P^L, x_i, y_i and V^L. The Newton method can be used to solve the unknowns. The results for constant volume depletion are shown in Fig. 3.58. Note that, as expected, there is an increase in the upper dewpoint pressure, a decrease in the lower dewpoint pressure, and there is an increase in retrograde

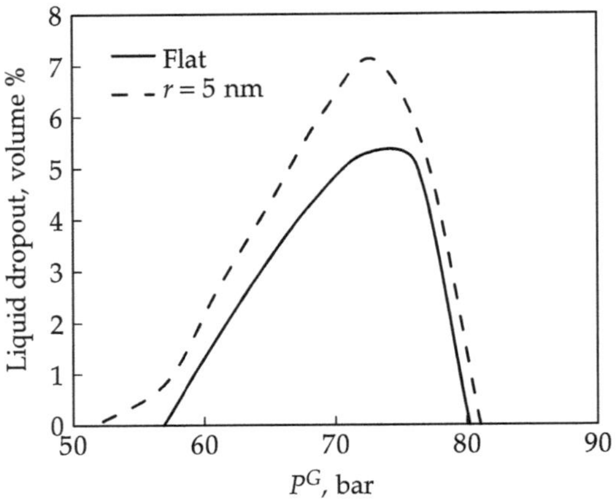

FIGURE 3.58 Liquid dropout of an equimolar mixture of C_1 and C_3 for curved ($r = 5$ nm) and flat interface from the PR-EOS at $T = 327.15$ K (adapted from Jin and Firoozabadi, 2015).

condensation. The implication is that more liquids will dropout in the formation with nanopores than in a flat interface system.

3.8 Problems

3.1 From the PR-EOS, derive $Z_c = 0.307$.

3.2 Derive the van der Waals equation of state in the reduced form given by Eq. (3.4).

3.3 Use the van der Waals equation of state to derive the following expression for the compressibility factor in terms of T_r and P_r:

$$Z^3 - \left[1 + \frac{P_r}{8T_r}\right] Z^2 + \left[\frac{27P_r}{64T_r}\right] Z - \frac{27P_r}{512T_r} = 0$$

3.4 Use the PR-EOS to prove that at the critical point of a pure substance $(\partial^3 P/\partial v^3) < 0$. Hint: You may use software such as Mathematica for this purpose.

3.5 Derive the expression for the partial molar volume of component i in a mixture using the PR-EOS. Hint: You may use software such as Mathematica for this purpose.

3.6 Derive the following expression for the derivative of Z with respect to pressure at constant T and composition from the PR-EOS

$$\frac{\partial Z}{\partial P} = \frac{a}{R^2T^2}\left[\frac{\mathcal{B} - Z}{3Z^2 - 2(1 - \mathcal{B}) + 2\mathcal{A} - 3\mathcal{B}^2 - 2\mathcal{B}}\right]$$

$$+ \frac{b}{RT}\left[\frac{-Z^2 + 2(3\mathcal{B} + 1)Z + (\mathcal{A} - 2\mathcal{B} - 3\mathcal{B}^2)}{3Z^2 - 2(1 - \mathcal{B})Z + (\mathcal{A} - 3\mathcal{B}^2 - 2\mathcal{B})}\right]$$

Note: To facilitate the above derivation, use

$$\frac{\partial Z}{\partial P} = \frac{\partial Z}{\partial \mathcal{A}}\frac{\partial \mathcal{A}}{\partial P} + \frac{\partial Z}{\partial \mathcal{B}}\frac{\partial \mathcal{B}}{\partial P}$$

3.7 Derive Eq. (3.32) of the text.

3.8 Prove that at the critical point of a pure component, $(\partial Z/\partial v) = P_c/RT_c$ and $(\partial^2 Z/\partial v^2) = 0$. Use the criterion of the second derivative to calculate the critical compressibility factor for the SRK equation of state from $Z^3 - Z^2 + Z(\mathcal{A} - \mathcal{B}^2 - \mathcal{B}) - \mathcal{AB} = 0$, where $\mathcal{A} = aP/(RT)^2$ and $\mathcal{B} = (bP/RT)$.

3.9: Relationship between Infinite Dilution Activity Coefficient and Interaction Coefficient in Binary Mixtures at Infinite Dilution The activity coefficient is solely a measure of solvent-solute interactions. The activity coefficient of component i at infinite dilution is shown by γ_i^∞, that is, $x_i \to 0$. Derive the following expression for γ_1^∞ from the PR-EOS for a binary mixture and show that $\ln \gamma_1^\infty$ is linear in δ_{12}.

$$\ln \gamma_1^\infty = \frac{b_1}{b_2}(Z_2 - 1) - (Z_1 - 1) + \ln\left(\frac{Z_1 - \mathcal{B}_1}{Z_2 - \mathcal{B}_2}\right) + \frac{\mathcal{A}_1}{2\sqrt{2}\mathcal{B}_1}\ln\left[\frac{Z_1 + 2.414v\mathcal{B}_1}{Z_1 - 0.414\mathcal{B}_1}\right]$$
$$- \left(\frac{a_{12}P}{R^2T^2}\right)\frac{1}{\sqrt{2}\mathcal{B}_2}\ln\left[\frac{Z_2 + 2.414\mathcal{B}_2}{Z_2 - 0.414\mathcal{B}_2}\right] + \frac{b_1\mathcal{A}_2}{2\sqrt{2}\mathcal{B}_2 b_2}\ln\left[\frac{Z_2 + 2.414\mathcal{B}_2}{Z_2 - 0.414\mathcal{B}_2}\right]$$

In the above equation subscripts 1 and 2 denote component indices; Z_1, Z_2, $\mathcal{B}_1$, $\mathcal{B}_2$, $\mathcal{A}_1$, and $\mathcal{A}_2$ are pure liquid parameters (at T and P), and $a_{12} = a_1^{1/2} a_2^{1/2}(1 - \delta_{12})$.

3.10 Use the definition of the solubility parameter from Eqs. (1.155) and (1.156), and the PR-EOS to derive the following expression for the solubility parameter,

$$\delta = \left[\frac{1}{2\sqrt{2}bv^L}\left(a - T\frac{da}{dT}\right)\ln\frac{v^L + (\sqrt{2}+1)b}{v^L - (\sqrt{2}-1)b}\right]^{1/2}$$

3.11: Effect of Temperature on Pressure in Gas and Liquid Systems Two students were arguing in a laboratory on the effect of temperature increase on two cylinders, one containing methane gas and the other cylinder containing n-decane liquid. The pressures of the two cylinders at the laboratory temperature of 25°C were the same–both were 100 atm.

They were planning to move the two cylinders outside the laboratory on a hot summer day to a room at 40°C temperature and leave them there for the whole day. Student A was concerned that the pressure of liquid decane may increase to 195 atm while he thought the methane cylinder pressure would increase to only 110 atm.

Use the PR-EOS to calculate the pressure of methane and decane cylinders at 40°C and to test the skill of the student A in EOS calculations. If student A is correct, then there is a basis that airlines would accept only fluids in the two-phase state for the transportation of liquids in a container.

3.12 Consider the P-T diagram of a hydrocarbon mixture when the interface between the phases is flat (see Fig. 3.59). Now suppose the interface between the phases is curved with the gas on the convex side. The dewpoint curve is sketched below for the curved interface system. Sketch the bubblepoint curve of the same system with a curved interface.

FIGURE 3.59

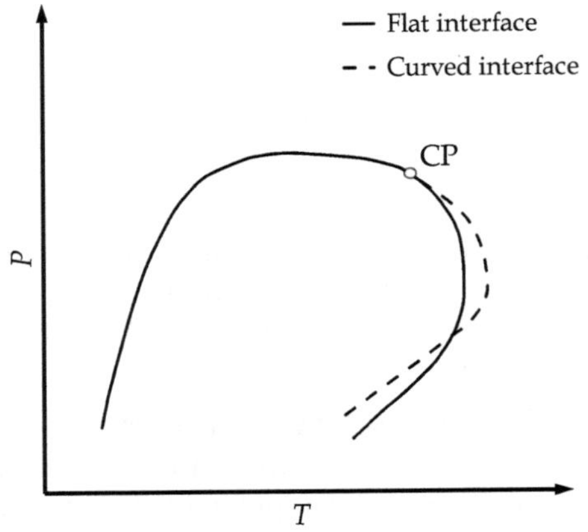

3.13 Figure 3.60 shows a plot of volume vs. pressure for a reservoir crude at constant temperature both in the single-phase and two-phase states. The change in compressibility provides the bubblepoint pressure, P_b.

(a) When a fluid is in the near-critical region, then the V vs. P plot at constant temperature is sketched in the Fig. 3.61.

FIGURE 3.60

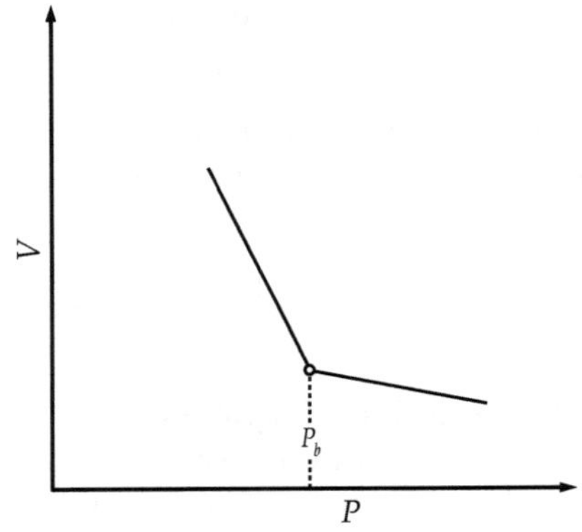

FIGURE 3.61

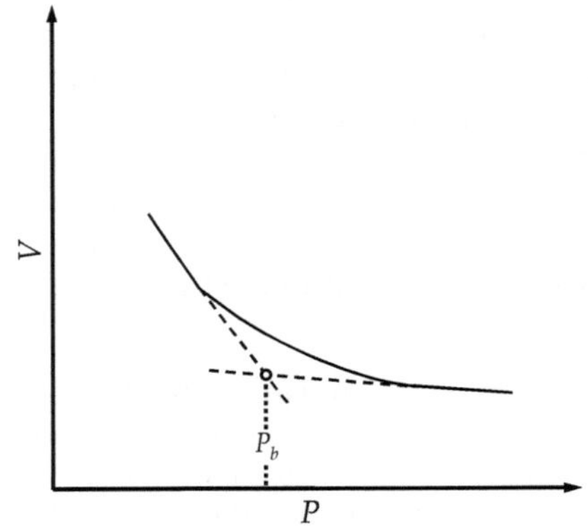

Explain the reason for the gradual increase in the compressibility from single-phase to two-phase states.

(b) Why is that the V vs. P plot cannot be used to estimate the retrograde dewpoint pressure?

3.14: Working Expression for Z in the CPA-EOS Assuming for asphaltene association and resin cross-association, derive the following expression for Z based on the CPA-EOS.

$$Z = \frac{Z}{Z - \mathcal{B}} - \frac{\mathcal{A}Z}{Z^2 - 2\mathcal{B}Z - \mathcal{B}^2} + \frac{4 + 4\eta - 2\eta^2}{2 - 3\eta + \eta^2}[x_a(\chi_a - 1) + x_r(\chi_r - 1)]$$

where x_a is the asphaltene mol fraction and x_r is the resin mol fraction. χ_a and χ_r are the mol fraction of asphaltene and resin molecules no bonded.

3.15 Repeat calculations in Example 3.13 when the contact angle is $\theta = 60°$ and $140°$. Your results will indicate a decrease in solubility as the contact angle increases.

3.16 Derive Eqs. (3.58a), (3.58b), and (3.58c)

3.17 Evaluate Eq. (3.58c). This evaluation is complicated because the fraction of molecules $1, 2, \ldots, n_c$ not associated or cross-associated are implicit functions of each other. See the remark in the next problem.

3.18 Evaluate $P^{association} = (\partial A^{departure}_{association}/\partial V)_{T,\underline{n}}$. The same complexity as in problem 3.17 is encountered here. In both problems you can use the method outlined by Michelsen and Hendricks (2001) to perform the derivatives.

3.9 References

Abascal, J. L. F., E. Sanz, R. Garcia Fernández, and C. Vega, A potential model for the study of ices and amorphous water: TIP4P/Ice. *Journal of Chemical Physics*, 2005. 122(23): p. 234511.

Acs, G., S. Dolesch〈oll, and E. Farkas, General purpose compositional model. *SPE Journal*, 1985. 25(4): pp. 543–553.

Ahmed, T., H. Nasrabadi, and A. Firoozabadi, Complex flow and composition path in CO_2 injection schemes from density effects. *Energy Fuels*, 2012. 26: pp. 4590–4598.

Anselme, M. J., The Critical Properties of Thermally Stable and Unstable Fluids and Dilute Fluid Mixtures. Ph.D. dissertation, 1988, Georgia Institute of Technology.

Arbabi, S., and A. Firoozabadi, Near-critical phase behavior of reservoir fluids using equations of state SPE advanced technology series. *SPE Advanced Technology Series*, 1995. 3(1): pp. 139–145.

Benedict, M., G. B. Webb, and L. C. Rubin, An empirical equation for thermodynamic properties of light hydrocarbons and their mixtures II. Mixtures of methane, ethane, propane, and n-butane. *Journal of Chemical Physics*, 1942. 10(12): pp. 747–758.

Bergman, D. F., M. R. Tek, and D. L. Katz, *Retrograde Condensation in Natural Gas Pipelines*. Monograph Series, American Gas Association, New York City, 1975.

Bjorlykke, O., and A. Firoozabadi, Measurement and computation of near-critical phase behavior of a C1/nC24 binary mixture. *SPE Reservoir Engineering*, 1992. 7(2): pp. 271–277.

Breure, B., D. Subramanian, J. Leys, C. J. Peters, and M. A. Anisimov, Modeling asphaltene aggregation with a single compoud. *Energy Fuels*, 2013. pp. 27–37.

Briggs, L. J., Limiting negative pressure of water. *The Journal of Applied Physics*, 1950. 21: pp. 721–722.

Carnahan, N. F., and K. E. Starling, Equation of state for nonattracting rigid spheres. *Journal of Chemical Physics*, 1969. 51: pp. 635–636.

Cavett, R. H., *Physical Data for Distillation Calculations, Vapor-Liquid Equilibria*, in 27th Mid-year Meeting of the API Division of Refining. 1962: San Francisco, CA. pp. 351–366.

Chapman, W. G., K. E. Gubbins, G. Jackson, and M. Radosz, New reference equation of state for associating liquids. *Industrial & Engineering Chemistry Research*, 1990. 29(8): pp. 1709–1721.

Chapman, W. G., K. E. Gubbins, G. Jackson, and M. Radosz, SAFT: Equation-of-state solution model for associating fluids. *Fluid Phase Equilibria*, 1989. 52(0): pp. 31–38.

Chou, G. F. and J. M. Prausnitz, A phenomenological correction to an equation of state for the critical region. *AIChE Journal*, 1989. 35(9): pp. 1487–1496.

Debenedetti, P. G., *Metastable Liquids, Concepts and Principles*. 1996, Princeton, NJ: Princeton University Press.

DeRuiter, R. A., L. J. Nash, and M. S. Singletary, Solubility and displacement behavior of a viscous crude with CO_2 and hydrocarbon gases. *SPE Reservoir Engineering*, 1994. 9(2): pp. 101–106.

Dill, K., and S. Bromberg, *Molecular Driving Forces: Statistical Thermodynamics in Chemistry and Biology*, Garland Science, 2003.

Dodson, C. R., and M. B. Standing, *Pressure-Volume-Temperature and Solubility Relations for Natural-Gas-Water Mixtures*. Drill & Prod. Prac. API, 1944. pp. 173–180.

Erbar, J. H., C. L. Persyn, and W. C. Edmister, Enthalpies and K-ratios for Hydrocarbon Mixtures by New Improved Computer Program, in Proceedings of the 1964 GPA Annual Convention, 1964. pp. 26–34.

Espinoza, D. N., and J. C. Santamarina, Water- CO_2-mineral systems: Interfacial tension, contact angle, and diffusion–Implications to CO_2 geological storage. *Water Resources Research*, 2010. 46(7): pp. 1944–1973.

Firoozabadi, A., *Thermodynamics of Hydrocarbon Reservoirs*. 1999, McGraw-Hill.

Firoozabadi, A., Reservoir-fluid phase behavior and volumetric prediction with equations of state. *Journal of Petroleum Technology*, 1988a. 40(4): pp. 397–406.

Firoozabadi, A., and H. Pan, Two-Phase Isentropic compressibility and two-phase sonic velocity for multicomponent-hydrocarbon mixtures. *SPE Reservoir Evaluation & Engineering*, 2000. 3(4): pp. 335–341.

Firoozabadi, A., and K. Aziz, Analysis and correlation of nitrogen and lean-gas miscibility pressure. *SPE Reservoir Engineering*, 1986. 1(6): pp. 575–582.

Firoozabadi, A., R. Nutakki, T. W. Wong, and K. Aziz, EOS Predictions of compressibility and phase behavior in systems containing water, hydrocarbons, and CO_2. *SPE Reservoir Engineering*, 1988b. 3(2): pp. 673–684.

Firoozabadi, A., D. Katz, H. Soroosh, and V. A. Sajjadian, Surface tension of reservoir crude oil/gas systems recognizing the asphalt in the heavy fraction. *SPE Reservoir Engineering*, 1988c. 3(1): pp. 265–272.

Firoozabadi, A., and H. J. Ramey, Jr., Surface tension of water-hydrocarbon systems at reservoir conditions. *Journal of Canadian Petroleum Technology*, (May–June 1988d): pp. 41–48.

Firoozabadi, A., and P. Cheng, Prospects for subsurface CO_2 sequestration. *AIChE Journal*, 2010. 56(6): pp. 1398–1405.

Gas-Processors-Association. *Engineering Data Book*. 9th ed. 1972, Tulsa.

Goual, L., and A. Firoozabadi, Measuring asphaltenes and resins, and dipole moment in petroleum fluids. *AIChE Journal*, 2002. 48(11): pp. 2646–2663.

Hashmi, S., and A. Firoozabadi, Self-assembly of resins and asphaltenes facilitiates asphaltene dissolution by an organic acid. *Journal of Colloid and Interface Science*, 2013. (394): p. 39.

Hashmi, S., K. X. Zhong, and A. Firoozabadi, Acid-base chemistry enables reversible colloid-to-solution transition of asphaltenes in non-polar systems. *Soft Matter*, 2012. 8: pp. 8778–8785.

Hoffman, A. E., J. S. Crump, and C. R. Cocott, Equilibrium constants for a gas-condensate system. *Journal of Petroleum Technology*, 1953. 5(1): pp. 1–10.

Holbrook, N. M., M. J. Burns, and C. B. Field, Negative xylem pressures in plants: a test of the balancing pressure technique. *Science*, 1995. 270(5239): pp. 1193–1194.

Huang, S. H., and M. Radosz, Equation of state for small, large, polydisperse, and associating molecules. *Industrial & Engineering Chemistry Research*, 1990. 29(11): pp. 2284–2294.

Huang, S. H., and M. Radosz, Equation of state for small, large, polydisperse, and associating molecules: Extension to fluid mixtures. *Industrial & Engineering Chemistry Research*, 1991. 30(8): pp. 1994–2005.

Jhaveri, B. S., and G. K. Youngren, Three-parameter modification of the peng-robinson equation of state to improve volumetric predictions. *SPE Reservoir Engineering*, 1988. 3(3): pp. 1033–1040.

Jiménez-Ángeles, F., and A. Firoozabad, Nucleation of methane hydrates at moderate sub-cooling by molecular dynamics simulations, *Journal of Chemical Physics C*, 2014. 118: pp. 11310–11318.

Jin, Z., and A. Firoozabadi, Thermodynamic modeling of phase behavior in shale media. *SPE Journal*, (To appear in 2015).

Johnson, N. L., and S. Kotz, *Continuous Univariate Distributions*. 1970, Boston, MA: Houghton Mifflin Corporation.

Jones, C., *The Use of Bottomhole Temperature Variations in Production Testings*, in European Petroleum Conference. 1988: London, United Kingdom.

Katz, D. L., D. J. Vink, and R. A. David, Phase diagram of a mixture of natural gas and natural gasoline near the critical conditions transactions of the AIME, 1940. 136(1): pp. 106–118.

Katz, D. L., and A. Firoozabadi, Predicting phase behavior of condensate/crude-oil systems using methane interaction coefficients. *Journal of Petroleum Technology*, 1978. 30(11): pp. 1649–1655.

Katz, D. L., D. Cornell, R. Kobayashi, F. H. Poettmann, J. A. Vary, J. R. Elenbaas, and C. F. Weinaug, *Handbook of Natural Gas Engineering*. 1959, New York: McGraw-Hill Book Company.

Kenrick, F. B., C. S. Gilbert, and K. L. Wismer, The superheating of liquids. *Journal of Physical Chemistry*, 1923. 28(12): pp. 1297–1307.

Kieffer, S. W., Sound speed in liquid-gas mixtures: water-air and water-steam. *Journal of Geophysical Research*, 1977. 82(20): pp. 2895–2904.

Kilgren, K. H., Phase behavior of a high-pressure condensate reservoir fluid. *Journal of Petroleum Technology*, 1966. 18(8): pp. 1001–1005.

Kontogeorgis, G. M., and G. K. Folas, *Thermodynamic Models for Industrial Applications: from Classical and Advanced Mixing Rules to Association Theories*. 2010, John Wiley and Sons, Ltd.

Kontogeorgis, G. M., E. C. Voutsas, I. V. Yakoumis, and D. P. Tassios, An equation of state for associating fluids. *Industrial & Engineering Chemistry Research*, 1996. 35(11): pp. 4310–4318.

Kontogeorgis, G. M., I. V. Yakoumis, H. Meijer, E. Hendriks, and T. Moorwood, multicomponent phase equilibrium calculations for water-methanol-alkane mixtures. *Fluid Phase Equilibria*, 1999. 158–160(0): pp. 201–209.

Lee, B. I., and M. G. Kesler, A generalized thermodynamic correlation based on three-parameter corresponding states. *AIChE Journal*, 1975. 21(3): pp. 510–527.

Li, Z., and A. Firoozabadi, Cubic-plus-association (CPA) equation of state for water-containing mixtures: is 'cross association' necessary? *AIChE Journal*, 2009. 55(7): pp. 1803–1813.

Li, Z., and A. Firoozabadi, Cubic-plus-association equation of state for asphaltene precipitation in live oils. *Energy & Fuels*, 2010b. 24(5): pp. 2956–2963.

Li, Z., and A. Firoozabadi, General strategy for stability testing and phase-split calculation in two and three phases. *SPE Journal*, 2012. pp. 1096–1107.

Li, Z., and A. Firoozabadi, Modeling asphaltene precipitation by n-alkanes from heavy oils and bitumens using cubic-plus-association equation of state. *Energy & Fuels*, 2010a. 24(2): pp. 1106–1113.

Li, Z., Z. Jin, and A. Firoozabadi, Phase behavior and adsorption of pure substances and mixtures and 4 characterization in nanopore structures by density functional theory. *SPE Journal*, 2014. 19 (6).

Macias, L., *Multiphase, Multicomponent Compressibility in Petroleum Engineering*. Ph.D. Thesis, 1985, Stanford University: Stanford, CA.

Mather, A. E., and E. U. Franck, Phase equilibria in the system carbon dioxide-water at elevated pressures. *Journal of Physical Chemistry*, 1992. 96(1): pp. 6–8.

Mathias, P. M., and T. W. Copeman, Extension of the Peng-Robinson equation of state to complex mixtures: Evaluation of the various forms of the local composition concept. *Fluid Phase Equilibria*, 1983. 13(0): pp. 91–108.

Mathias, P. M., T. Naheiri, and E. M. Oh, A density correction for the peng-robinson equation of state. *Fluid Phase Equilibria*, 1989. 47(1): pp. 77–87.

McKetta, J. J., and D. L. Katz, Methane-n-butane-water system in two-and three-phase regions. *Industrial & Engineering Chemistry*, 1948. 40(5): pp. 853–863.

Michelsen, M. L., Hendricks E, M., Physical properties from association models, *Fluid Phase Equilibria*, 2001. 180: pp. 165–174.

Moortgat, J., Z. Li, and A. Firoozabadi, Three-phase compositional modeling of CO_2 injection by higher-order finite element methods with PR and CPA equations of state. *Water Resour Res*, 2012. 48: W12511.

Mutoru, J. W., A. Leahy-Dios, and A. Firoozabadi, Modeling infinite dilution and fickian diffusion coefficients of carbon dioxide in water. *AIChE Journal*, 2011. 57(6): p. 111.

Myint, P. C., Hao, Y., and A. Firoozabadi, The CPA Equation of State and an Activity Coefficient Model for Accurate Molar Enthalpy Calculations of Mixtures with Carbon Dioxide and Water/Brine. Lawrence Livermore National Laboratory Technical Report (2015, LLNL-TR-672077).

Ng, H. -J., and A. E. Mather, Enthalpy of gaseous mixtures of methane + ethane + carbon dioxide under pressure. *Journal of Chemical & Engineering Data*, 1978. 23(3): pp. 224–227.

Ng, H. J., C. J. Chen, and D. B. Robinson, Vapor Liquid Equilibrium and Condensing Curves in the Vicinity of the Critical Point for a Typical Gas Condensate. 1986, Gas Processors Association: Tulsa.

Nikitin, E. D., P. A. Pavlov, and A. P. Popov, Vapour-liquid critical temperatures and pressures of normal alkanes with from 19 to 36 carbon atoms, naphthalene and m-terphenyl determined by the pulse-heating technique. *Fluid Phase Equilibria*, 1997. 141(1–2): pp. 155–164.

Nikitin, E. D., P. A. Pavlov, and P. V. Skripov, Measurement of the critical properties of thermally unstable substances and mixtures by the pulse-heating method. *Journal of Chemical Thermodynamics*, 1993. 25(7): pp. 869–880.

Nutakki, R., *Phase Behavior Calculations for Systems with Hydrocarbons, Water, and* CO_2. Ph.D. Dissertation. 1989, Stanford University: Stanford, CA.

Passut, C. A., and R. P. Danner, Correlation of ideal gas enthalpy, heat capacity and entropy. *Industrial & Engineering Chemistry Process Design and Development*, 1972. 11(4): pp. 543–546.

Péneloux, A., E. Rauzy, and R. Fréze, A consistent correction for redlich-kwong-soave volumes. *Fluid Phase Equilibria*, 1982. 8(1): pp. 7–23.

Peng, D. -Y., and D. B. Robinson, A new two-constant equation of state. *Industrial & Engineering Chemistry Fundamentals*, 1976. 15(1): pp. 59–64.

Pitzer, K. S., Corresponding states for perfect liquids. *Journal of Chemical Physics*, 1939. 7(8): pp. 583–590.

Pitzer, K. S., D. Z. Lippmann, R. F. Curl, Jr., C. M. Huggins, and D. E. Petersen, The volumetric and thermodynamic properties of fluids. II. Compressibility factor, vapor pressure and entropy of vaporization1. *Journal of the American Chemical Society*, 1955. 77(13): pp. 3433–3440.

Prausnitz, J. M., R. N. Lichtenhaler, and E. G. de Azevedo, Molecular Thermodynamics of Fluid-Phase Equilibria 1986, Englewood Cliffs, NJ: Prentice-Hall Inc.

Riazi, M. R., and T. E. Daubert, Simplify property predictions. *Hydrocarbon Processing*, 1980. 59(3): pp. 115–116.

Robinson, D. B., D. -Y. Peng, and S. Y. K. Chung, The development of the Peng-Robinson equation and its application to phase equilibrium in a system containing methanol. *Fluid Phase Equilibria*, 1985. 24(1–2): pp. 25–41.

Roland, C. H., Vapor-liquid equilibria for natural gas-crude oil mixtures. *Industrial & Engineering Chemistry*, 1945. 37(10): pp. 930–936.

Rowlinson, J. S., and F. L. Swinton. *Liquid and Liquid Mixtures*. 3rd ed. 1982, London: Butterworth Scientific.

Sage, B. H., W. N. Lacey, and J. G. Schaafsma, Report on API research project 37-covers behavior of hydrocarbon mixtures. *Oil and Gas Journal*, 1933: pp. 12–14.

Schmidt, G., and H. Wenzel, A modified van der Waals type equation of state. *Chemical Engineering Science*, 1980. 35(7): pp. 1503–1512.

Skuse, B., A. Firoozabadi, and H. J. Ramey, Computation and interpretation of capillary pressure from a centrifuge. *SPE Formation Evaluation*, 1992. 7(1): pp. 17–24.

Soave, G., Equilibrium constants from a modified redlich-kwong equation of state. *Chemical Engineering Science*, 1972. 27(6): pp. 1197–1203.

Speight J. G., and E. Moschopedis Speros, On the molecular nature of petroleum asphaltenes, in chemistry of asphaltenes. *American Chemical Society*. 1982. pp. 1–15.

Standing, M. B., *Volumetric and Phase Behavior of Oil Field Hydrocarbon Systems*. 1977, Dallas, TX: SPE of AIME Monograph.

Starling, K. E., *Fluid Thermodynamic Properties for Light Petroleum Systems*. 1973, Gulf Publishing Company.

Steinhardt, P. J., A quintessential introduction to dark energy. *Philosophical Transactions of the Royal Society of London. Series A: Mathematical, Physical and Engineering Sciences*, 2003. 361(1812): pp. 2497–2513.

Takenouchi, S., and G. C. Kennedy, The binary system H_2O-CO_2 at high temperatures and pressures. *American Journal of Science*, 1964. 262(9): pp. 1055–1074.

Teja, A. S., R. J. Lee, D. Rosenthal, and M. Anselme, correlation of the critical properties of alkanes and alkanols. *Fluid Phase Equilibria*, 1990. 56(0): pp. 153–169.

Thomas, L. K., R. W. Hankinson, and K. A. Phillips, Determination of acoustic velocities for natural gas. *Journal of Petroleum Technology*, 1970. 22(7): pp. 889–895.

Twu, C. H., An internally consistent correlation for predicting the critical properties and molecular weights of petroleum and coal-tar liquids. *Fluid Phase Equilibria*, 1984. 16(2): pp. 137–150.

Voronel, A. V., *In Phase Transitions and Critical Phenomena*, ed. C. Domb and M. S. Green. Vol. 5. 1976, London: Academic Press.

Watts, J. W., A compositional formulation of the pressure and saturation equations. *SPE Reservoir Engineering*, 1986. 1(3): pp. 243–252.

Weinaug, C. F., and D. L. Katz, Surface tensions of methane-propane mixtures. *Industrial & Engineering Chemistry*, 1943. 35(2): pp. 239–246.

Wertheim, M. S., Fluids with highly directional attractive forces. I. Statistical thermodynamics. *Journal of Statistical Physics*, 1984a. 35(1–2): pp. 19–34.

Wertheim, M. S., Fluids with highly directional attractive forces. II. Thermodynamic perturbation theory and integral equations. *Journal of Statistical Physics*, 1984b. 35(1–2): pp. 35–47.

Wertheim, M. S., Fluids with highly directional attractive forces. III. Multiple attraction sites. *Journal of Statistical Physics*, 1986a. 42(3–4): pp. 459–476.

Wertheim, M. S., Fluids with highly directional attractive forces. IV. Equilibrium polymerization. *Journal of Statistical Physics*, 1986b. 42(3–4): pp. 477–492.

Wertheim, M. S., Thermodynamic perturbation theory of polymerization. *Journal of Chemical Physics*, 1987. 87(12): pp. 7323–7331.

Whitson, C. H., Characterizing hydrocarbon plus fractions. *SPE Journal*, 1983. 23(4): pp. 683–694.

Wu, J., and J. M. Prausnitz, Phase equilibria for systems containing hydrocarbons, water, and salt:? An extended peng-robinson equation of state. *Industrial & Engineering Chemistry Research*, 1998. 37(5): pp. 1634–1643.

Wu, P. C. and P. Ehrlich, Volumetric properties of supercritical ethane-n-heptane mixtures: Molar volumes and partial molar volumes. *AIChE Journal*, 1973. 19(3): pp. 533–540.

Yarborough, L., Application of a generalized equation of state to petroleum reservoir fluids, in equations of state in engineering and research. *American Chemical Society*, 1979. pp. 385–439.

Zudkevitch, D., and J. Joffe, Correlation and prediction of vapor-liquid equilibria with the redlich-kwong equation of state. *AIChE Journal*, 1970. 16(1): pp. 112–119.

CHAPTER 4

Stability and Criticality

Thermodynamic stability is one of the most important concepts in physical sciences and has broad applications in a wide range of problems. We have already seen that sedimentation in multicomponent mixtures is related to the determinant of chemical potential derivatives in Chap. 2. This determinant is connected to the thermodynamic stability. In Chap. 6, we will derive basic expressions for diffusion flux and diffusion coefficient. The expression for thermodynamic stability is linked to the diffusion coefficients and diffusion fluxes. In Chap. 7, we will show that adsorption in multicomponent mixtures is also linked to thermodynamic stability.

In Chap. 3, stability and limit of stability was linked to $(\partial P/\partial v)_T$ in single-component systems. We stated (without proof) that for a single-component fluid to be stable, $(\partial P/\partial v)_T < 0$. In this chapter we will derive the stability condition for both single- and multicomponent mixtures. The general criterion for thermodynamic stability is $y^{(n_c)}_{n_c+1,n_c+1} \geq 0$, where $y^{(n_c)}$is the Legendre Transformation of any of our thermodynamic functions, and the index of $n_c + 1$ represents derivative with respect to variable $n_c + 1$. For a single-component system, $y^{(n_c)}_{n_c+1,n_c+1}$ will become $-(\partial P/\partial v)_T$ in certain arrangement of variables. In a different arrangement of variables of a thermodynamic function, we will have $(\partial T/\partial s)_P > 0$ as the condition for thermodynamic stability in a single-component system. Closely connected to thermodynamic stability is the criticality. The criteria for criticality are based on $y^{(n_c)}_{n_c+1,n_c+1} = 0$ and $y^{(n_c)}_{n_c+1,n_c+1,n_c+1} = 0$. In one of the examples, based on criteria of criticality, we will derive the condition for complete miscibility of a solvent and a polymer, $\chi \leq (1 + \sqrt{r})^2/2r$ where χ is the Flory interaction parameter, and r is the number of segments in a polymer molecule. In another example, we show that the criteria of stability can be used to increase the speed of molecular dynamic simulations by three orders of magnitude in hydrate nucleation.

In the following, we will first present the mathematical derivations for stability analysis and then discuss criticality. In the *Examples and Theory Extension* section various related mathematical and analytical expressions will be derived and discussed. In the end, seventeen problems are outlined.

4.1 Stability Analysis and Stability Limit

Considerable effort will be made in this chapter to derive the criteria of stability and criticality from basic principles. In order to appreciate these concepts, we first derive the criteria of stability for a pure component and then proceed briefly to binary mixtures. The stability criteria for multicomponent systems will be derived in detail. The derivation for multicomponent systems is very general and relies on the Legendre and Jacobian transformations discussed in Chap. 1 and the mathematical principles for definitiveness that will be discussed in this chapter.

4.1.1 Stability Analysis for a Single-Component

The basic derivations for the stability analysis of a single-component system are similar to Haase's presentation (1956) with certain changes.

Stability Criteria Based on Entropy

Let us consider N mols of a single-component single-phase system with molar internal energy u and molar volume v. At this condition, if the single-phase system is not stable, it may split into two phases; N' mols of the primed phase and N'' mols of the double-primed phase. The molar internal energy and the molar volume of the primed phase are represented by $u+\Delta u$ and $v+\Delta v$, and those of the double-primed phase are represented by $u + \Delta^\circ u$ and $v + \Delta^\circ v$. Let us represent the molar entropy of the original single phase by s, the primed phase by $s + \Delta s$, and the double-primed phase by $s + \Delta^\circ s$. The criteria of stability of the original single phase in comparison to the two-phase state can be developed in terms of the entropy difference between the single-phase and two-phase states. The problem is sketched in Fig. 4.1: which one of the two states is more stable, a or b?

The expression for the entropy change, ΔS, between the two states can be written as

$$\Delta S = S_{two\ phase} - S_{single\ phase} \tag{4.1}$$

and the entropy of each system is given by

$$S_{two\ phase} = N'(s + \Delta s) + N''(s + \Delta^\circ s) \tag{4.2}$$

$$S_{single\ phase} = Ns = (N' + N'')s \tag{4.3}$$

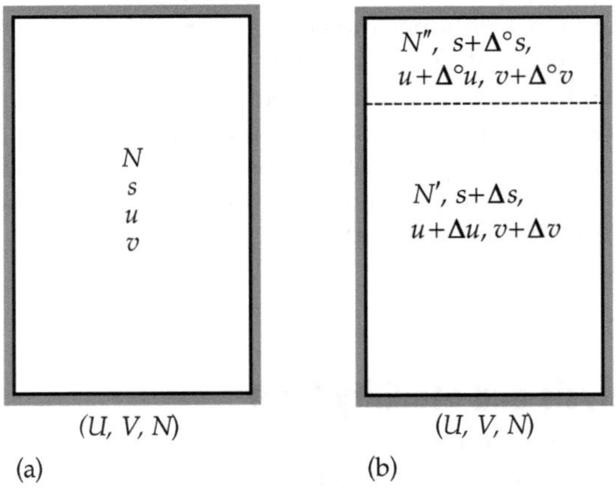

FIGURE 4.1 Stability of a single-component at constant U, V and $N(N = N' + N'')$.

Therefore,

$$\Delta S = N'\Delta s + N''\Delta^{\circ}s \tag{4.4}$$

Based on the second law, the single-phase state is stable if $\Delta S < 0$; the entropy of the original single-phase state is greater than the entropy of the two-phase state for the isolated path. The constraints of the isolated path are

$$\Delta U = U_{two\ phase} - U_{single\ phase} = N'\Delta u + N''\Delta^{\circ}u = 0 \tag{4.5}$$

$$\Delta V = V_{two\ phase} - V_{single\ phase} = N'\Delta v + N''\Delta^{\circ}v = 0 \tag{4.6}$$

$$N = N' + N'' = constant \tag{4.7}$$

The ΔU and ΔV expressions in Eqs. (4.5) and (4.6) are derived in the same way as Eq. (4.4) was derived. Next, we perform the Taylor's series expansion of $\Delta s(u, v)$, and $\Delta^{\circ}s(u, v)$ in terms of independent variables u and v.

$$\Delta s = \left(\frac{\partial s}{\partial u}\right)\Delta u + \left(\frac{\partial s}{\partial v}\right)\Delta v + \left(\frac{1}{2}\right)\left[\left(\frac{\partial^2 s}{\partial u^2}\right)(\Delta u)^2 + 2\left(\frac{\partial^2 s}{\partial u \partial v}\right)(\Delta u \Delta v) + \left(\frac{\partial^2 s}{\partial v^2}\right)(\Delta v)^2\right] + \cdots \tag{4.8}$$

and

$$\Delta^{\circ}s = \left(\frac{\partial s}{\partial u}\right)\Delta^{\circ}u + \left(\frac{\partial s}{\partial v}\right)\Delta^{\circ}v + \left(\frac{1}{2}\right)\left[\left(\frac{\partial^2 s}{\partial u^2}\right)(\Delta^{\circ}u)^2 + 2\left(\frac{\partial^2 s}{\partial u \partial v}\right)(\Delta^{\circ}u \Delta^{\circ}v) + \left(\frac{\partial^2 s}{\partial v^2}\right)(\Delta^{\circ}v)^2\right] + \cdots \tag{4.9}$$

The Taylor's series expansions are based on the assumptions that Δu, Δv, $\Delta^{\circ}u$, and $\Delta^{\circ}v$ are small. Note that in Eqs. (4.8) and (4.9), the partial derivatives are evaluated at the original single-phase state.

From Eqs. (4.5) and (4.6),

$$\Delta u = -\left(\frac{N''}{N'}\right)\Delta^{\circ}u \tag{4.10}$$

and

$$\Delta v = -\left(\frac{N''}{N'}\right)\Delta^{\circ}v \tag{4.11}$$

The substitution of Δu and Δv from Eqs. (4.10) and (4.11) into Eq. (4.8) results in

$$\Delta s = -\left(\frac{N''}{N'}\right)\left[\left(\frac{\partial s}{\partial u}\right)\Delta^{\circ}u + \left(\frac{\partial s}{\partial v}\right)\Delta^{\circ}v\right] + \left(\frac{1}{2}\right)\left(\frac{N''}{N'}\right)^2 \times \left[\left(\frac{\partial^2 s}{\partial u^2}\right)(\Delta^{\circ}u)^2 + 2\left(\frac{\partial^2 s}{\partial u \partial v}\right)(\Delta^{\circ}u \Delta^{\circ}v) + \left(\frac{\partial^2 s}{\partial v^2}\right)(\Delta^{\circ}v)^2\right] + \cdots \tag{4.12}$$

Combining Eqs. (4.4), (4.7), (4.8), and (4.12),

$$\Delta S = \frac{1}{2}\frac{N''N}{N'}\left[\left(\frac{\partial^2 s}{\partial u^2}\right)(\Delta^\circ u)^2 + 2\left(\frac{\partial^2 s}{\partial u \partial v}\right)(\Delta^\circ u \Delta^\circ v) + \left(\frac{\partial^2 s}{\partial v^2}\right)(\Delta^\circ v)^2\right] + \cdots \quad (4.13)$$

Since $\frac{1}{2}(N''N/N')$ is positive, then the condition for the stability of the original single phase is after dropping the "∘" superscript,

$$\boxed{\left(\frac{\partial^2 s}{\partial u^2}\right)(\Delta u)^2 + 2\left(\frac{\partial^2 s}{\partial u \partial v}\right)(\Delta u \Delta v) + \left(\frac{\partial^2 s}{\partial v^2}\right)(\Delta v)^2 < 0} \quad (4.14)$$

Therefore, the stability of the original phase is defined with respect to the second-order variations of s. If the second-order variations are zero, third-order variations must also be zero, and then fourth-order variations should be negative.

Let us review some simple concepts on the negativeness and positiveness of a function $f(u, v)$ of two independent variables u and v to make use of the above expression in terms of simple criteria. The function $f(u, v)$ is said to be positive definite if it is strictly positive at all points other than $u = 0$, $v = 0$, (i.e., the origin). The same function f, if it is strictly negative at all points except at the origin $u = 0$, $v = 0$ is called negative definite. If it can take either sign, both $f > 0$ and $f < 0$, it is called indefinite. Semidefiniteness will be defined shortly.

Consider the quadratic form for f:

$$f(u, v) = au^2 + 2buv + cv^2 \quad (4.15)$$

Equation (4.15) can be written as

$$f(u, v) = a\left(u + \frac{b}{a}v\right)^2 + \left(c - \frac{b^2}{a}\right)v^2 \quad (4.16)$$

The first term on the right side of Eq. (4.16) is never negative provided $a > 0$. When $a < 0$, this term is never positive. For the second term to be always positive, $(c - b^2/a) > 0$. From the above simple discussion, the criteria for positive definiteness and negative definiteness of the quadratic $f = au^2 + 2buv + cv^2$ are then established: (1) f is positive definite if and only if $a > 0$ and $(ac - b^2) > 0$, and (2) f is negative definite if and only if $a < 0$ and $(ac - b^2) > 0$. For positive definiteness, the above conditions demand $c > 0$ and for negative definiteness $c < 0$. However, the sign of c is necessary but not sufficient for either positiveness or negativeness. When $ac = b^2$, the second term on the right side of Eq. (4.16) disappears and only the first term remains. Then f is said to be positive semidefinite when $a > 0$ and negative semidefinite when $a < 0$. The quadratic f is indefinite if $ac - b^2 < 0$; f can take either sign, both $f > 0$ and $f < 0$. Later we will discuss the definitiveness of functions of more than two variables. Strang (1986, 1988) presents a detailed analysis of definiteness in a simple manner.

With the above background on definitiveness, we now examine the quadratic given by Eq. (4.14) for negative definiteness:

$$\left(\frac{\partial^2 s}{\partial u^2}\right) < 0 \tag{4.17}$$

$$\left(\frac{\partial^2 s}{\partial u^2}\right)\left(\frac{\partial^2 s}{\partial v^2}\right) - \left(\frac{\partial^2 s}{\partial u \partial v}\right)^2 > 0 \tag{4.18}$$

Note that $(\partial^2 s/\partial v^2) < 0$ is a necessary but not sufficient condition of negative definiteness. To derive a simple criterion for stability from Eqs. (4.17) and (4.18), it is necessary to derive the expressions for $(\partial^2 s/\partial u^2)$, $(\partial^2 s/\partial v^2)$, and $(\partial^2 s/\partial u \partial v)$ in terms of pressure, volume, temperature, and heat capacity.

$(\partial^2 s/\partial u^2)_v$ Expression The first derivative of s with respect to u at constant volume is given by $(\partial s/\partial u)_v = 1/T$ (from $du = T\,ds - P\,dv$). The second derivative of s with respect to u at constant v is simply

$$\left(\frac{\partial^2 s}{\partial u^2}\right)_v = -\frac{1}{T^2}\left(\frac{\partial T}{\partial u}\right)_v = -\frac{1}{(T^2 c_V)} \tag{4.19}$$

In the derivation of the above equation, the definition of c_V from Eq. (3.116) is used.

$(\partial^2 s/\partial v^2)_u$ Expression From Eq. (1.22), $(\partial s/\partial v)_u = P/T$. The second derivative is

$$\left(\frac{\partial^2 s}{\partial v^2}\right)_u = \left(\frac{1}{T}\right)\left(\frac{\partial P}{\partial v}\right)_u - \left(\frac{P}{T^2}\right)\left(\frac{\partial T}{\partial v}\right)_u \tag{4.20}$$

but

$$\left(\frac{\partial T}{\partial v}\right)_u = -\frac{\left(\frac{\partial u}{\partial v}\right)_T}{\left(\frac{\partial u}{\partial T}\right)_v} = -\left(\frac{1}{c_V}\right)\left(\frac{\partial u}{\partial v}\right)_T \tag{4.21}$$

From Eq. (1.21),

$$\left(\frac{\partial u}{\partial v}\right)_T = T\left(\frac{\partial s}{\partial v}\right)_T - P \tag{4.22}$$

Using the Maxwell relation given by Eq. (1.202), the above equation simplifies to

$$\left(\frac{\partial u}{\partial v}\right)_T = T\left(\frac{\partial P}{\partial T}\right)_v - P \tag{4.23}$$

Combining Eqs. (4.21) and (4.23),

$$\left(\frac{\partial T}{\partial v}\right)_u = \left(\frac{1}{c_V}\right)\left[P - T\left(\frac{\partial P}{\partial T}\right)_v\right] \tag{4.24}$$

We also need to evaluate $(\partial P/\partial v)_u$ in Eq. (4.20) using $P = P(v, T)$

$$dP = \left(\frac{\partial P}{\partial v}\right)_T dv + \left(\frac{\partial P}{\partial T}\right)_v dT \tag{4.25}$$

and, dividing by ∂v at constant u,

$$\left(\frac{\partial P}{\partial v}\right)_u = \left(\frac{\partial P}{\partial v}\right)_T + \left(\frac{\partial P}{\partial T}\right)_v \left(\frac{\partial T}{\partial v}\right)_u \tag{4.26}$$

Substituting $(\partial T/\partial v)_u$ from Eq. (4.24) into the above equation,

$$\left(\frac{\partial P}{\partial v}\right)_u = \left(\frac{\partial P}{\partial v}\right)_T + \left(\frac{\partial P}{\partial T}\right)_v \left(\frac{1}{c_V}\right) \left[P - T\left(\frac{\partial P}{\partial T}\right)_v\right] \tag{4.27}$$

Combining Eqs. (4.20), (4.24), and (4.27) gives the sought expression,

$$\left(\frac{\partial^2 s}{\partial v^2}\right)_u = \left(\frac{1}{T}\right)\left(\frac{\partial P}{\partial v}\right)_T - \left(\frac{1}{T^2}\right)\left(\frac{1}{c_V}\right)\left[P - T\left(\frac{\partial P}{\partial T}\right)_v\right]^2 \tag{4.28}$$

$(\partial^2 s/\partial v \partial u)$ Expression From $(\partial s/\partial u)_v = 1/T$, one obtains

$$\left(\frac{\partial^2 s}{\partial u \partial v}\right) = -\left(\frac{1}{T^2}\right)\left(\frac{\partial T}{\partial v}\right)_u \tag{4.29}$$

Substituting Eq. (4.24) into Eq. (4.29),

$$\left(\frac{\partial^2 s}{\partial u \partial v}\right) = \left[\frac{1}{(T^2 c_V)}\right]\left[T\left(\frac{\partial P}{\partial T}\right)_v - P\right] \tag{4.30}$$

From Eqs. (4.17) and (4.19), $-[1/(T^2 c_V)] < 0$ and since T is positive, then

$$\boxed{c_V > 0} \tag{4.31}$$

The above relationship is referred to as thermal stability. The substitution of Eqs. (4.19), (4.28), and (4.30) into Eq. (4.18) gives $-[1/c_V T^3][\partial v/\partial P]_T > 0$ and since $c_V > 0$, then

$$\boxed{\left(\frac{\partial P}{\partial v}\right)_T < 0} \tag{4.32}$$

which is referred to as mechanical stability. From the definition of isothermal compressibility $C_T = -(1/v)(\partial v/\partial P)_T$, an alternative statement of mechanical stability is

$$\boxed{C_T > 0} \tag{4.33}$$

Therefore, $c_V > 0$ and $C_T > 0$ guarantee that the original single phase shown in Fig. 4.1 is stable. Note that for a single-component fluid to be at stable-equilibrium state, both of the relations given by Eqs. (4.31) and (4.32) should be satisfied simultaneously. In Eq. (4.31), the stable equilibrium is based on a thermal quantity and in Eq. (4.32), stable equilibrium is based on mechanical quantities, therefore, they are referred to as thermal and mechanical stabilities, respectively.

Figure 4.2 shows a sketch of the *P-v* plot for a pure substance at three different temperatures; T_c is the critical temperature, which will be discussed in the presentation on criticality. For $T_2 < T_c$, the ABC branch has a negative slope that is, $-(\partial P/\partial v)_{T_2} > 0$, and according to Eq. (4.32) is stable. Similarly, the EFG branch also has a negative slope and

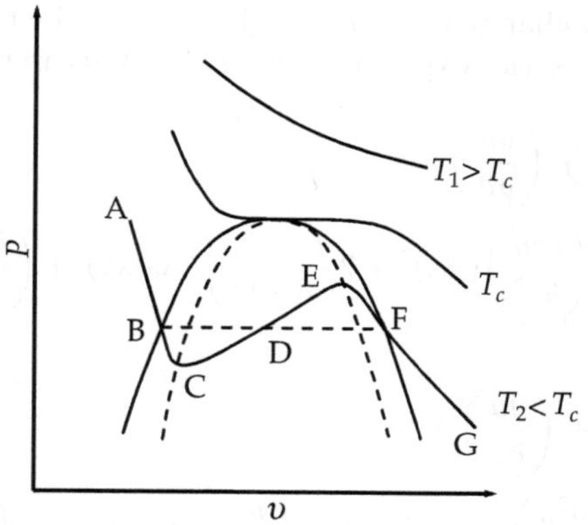

FIGURE 4.2 P-v plot showing stable and unstable regions. Note that the sketch is idealized. Real fluids behave as in Fig. 3.6.

is stable. At C and E, $(\partial P/\partial v)_T = 0$, which defines the limit of stability. At $T_1 > T_c$, the whole curve has a negative slope and therefore is stable. Note that the segment of the $P - v$ curve between C and E has a positive slope, $(\partial P/\partial v) > 0$, and according to Eq. (4.32) is unstable. The criteria can also be derived from stability analysis of the internal energy.

Stability Criteria from Internal Energy

The stability of the original single phase in comparison to the two phases demands that the internal energy change be positive (see Fig. 4.3):

$$\Delta U = N'(u + \Delta u) + N''(u + \Delta^\circ u) - (N' + N'')u = N'\Delta u - N''\Delta^\circ u > 0 \quad (4.34)$$

subject to the constraints

$$\Delta S = N'\Delta s + N''\Delta^\circ s = 0 \quad (4.35)$$

$$\Delta V = N'\Delta v + N''\Delta^\circ v = 0 \quad (4.36)$$

$$N = N' + N'' = constant \quad (4.37)$$

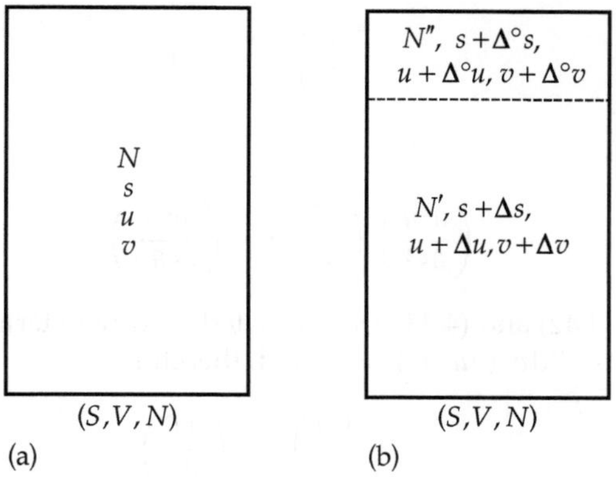

FIGURE 4.3 Stability of a single-component at constant S, V and $N(N = N' + N'')$.

where ΔS is the entropy change from the single-phase to the two-phase state. Since $u = u(s, v)$, then the Taylor's series expansion of Δu and $\Delta^\circ u$ can be expressed by

$$\Delta u = \left(\frac{\partial u}{\partial s}\right)\Delta s + \left(\frac{\partial u}{\partial v}\right)\Delta v + \left(\frac{1}{2}\right)\left[\left(\frac{\partial^2 u}{\partial s^2}\right)(\Delta s)^2 + 2\left(\frac{\partial^2 u}{\partial s \partial v}\right)(\Delta s \Delta v) + \left(\frac{\partial^2 u}{\partial v^2}\right)(\Delta v)^2\right] + \cdots \tag{4.38}$$

$$\Delta^\circ u = \left(\frac{\partial u}{\partial s}\right)\Delta^\circ s + \left(\frac{\partial u}{\partial v}\right)\Delta^\circ v + \left(\frac{1}{2}\right)\left[\left(\frac{\partial^2 u}{\partial s^2}\right)(\Delta^\circ s)^2 + 2\left(\frac{\partial^2 u}{\partial s \partial v}\right)(\Delta^\circ s \Delta^\circ v) + \left(\frac{\partial^2 u}{\partial v^2}\right)(\Delta^\circ v)^2\right] + \cdots \tag{4.39}$$

Similarly to the entropy derivation, Δs, Δv, $\Delta^\circ s$, and $\Delta^\circ v$ should be small. From the constraints given by Eqs. (4.35) and (4.36), $\Delta s = -(N''/N')\Delta^\circ s$ and $\Delta v = -(N''/N')\Delta^\circ v$; substitution of these equations into Eq. (4.38) and combining the results with Eqs. (4.34) and (4.39) gives (after dropping the "$\circ$" superscript)

$$\Delta U = \left(\frac{1}{2}\frac{N''N}{N'}\right)\left[\left(\frac{\partial^2 u}{\partial s^2}\right)(\Delta s)^2 + \left(\frac{\partial^2 u}{\partial v^2}\right)(\Delta v)^2 + 2\left(\frac{\partial^2 u}{\partial v \partial s}\right)(\Delta s \Delta v)\right] + \cdots > 0 \tag{4.40}$$

For ΔU to be positive, since $\frac{1}{2}(N'N/N'')$ is positive, the second-order variations with respect to s and v have to be positive. In case the second-order variations are zero, higher-order even variations should be positive. In other words, the first nonvanishing terms of order two- or even higher-order should be positive. Therefore, stability of the single-phase state demands that

$$\boxed{\left(\frac{\partial^2 u}{\partial s^2}\right)(\Delta s)^2 + \left(\frac{\partial^2 u}{\partial v^2}\right)(\Delta v)^2 + \left(\frac{\partial^2 u}{\partial v \partial s}\right)(\Delta v \Delta s) > 0} \tag{4.41}$$

The positive definiteness of Eq. (4.41) is based on

$$\left(\frac{\partial^2 u}{\partial s^2}\right)_v > 0 \tag{4.42}$$

and

$$\left(\frac{\partial^2 u}{\partial s^2}\right)\left(\frac{\partial^2 u}{\partial v^2}\right) - \left(\frac{\partial^2 u}{\partial s \partial v}\right)^2 > 0 \tag{4.43}$$

The terms in Eqs. (4.42) and (4.43) are evaluated readily in terms of T, P, and c_V.

From $du = T ds - P dv$, $(\partial u/\partial s)_v = T$ and, therefore,

$$\left(\frac{\partial^2 u}{\partial s^2}\right)_v = \left(\frac{\partial T}{\partial s}\right)_v \tag{4.44}$$

From Eq. (3.122), we write

$$\left(\frac{\partial^2 u}{\partial s^2}\right)_v = \frac{T}{c_V} \tag{4.45}$$

From $du = Tds - Pdv$, $(\partial u/\partial v)_s = -P$, and $(\partial^2 u/\partial v^2)_s = -(\partial P/\partial v)_s$ which when combined with Eq. (3.127) gives

$$\left(\frac{\partial^2 u}{\partial v^2}\right)_s = -\left(\frac{c_P}{c_V}\right)\left(\frac{\partial P}{\partial v}\right)_T \tag{4.46}$$

From the expression for du, $(\partial u/\partial s)_v = T$ and $(\partial^2 u/\partial s \partial v)_s = (\partial T/\partial v)_s$. The expression for $(\partial T/\partial v)_s$ is given by

$$\left(\frac{\partial T}{\partial v}\right)_s = -\frac{\left(\frac{\partial s}{\partial v}\right)_T}{\left(\frac{\partial s}{\partial T}\right)_v} = -\frac{\left(\frac{\partial s}{\partial v}\right)_T}{\frac{c_V}{T}} \tag{4.47}$$

Combining Eqs. (1.202), and (4.47),

$$\left(\frac{\partial^2 u}{\partial s \partial v}\right) = -T\frac{\left(\frac{\partial P}{\partial T}\right)_v}{c_V} \tag{4.48}$$

We also can write

$$\left(\frac{\partial P}{\partial T}\right)_v = -\frac{\left(\frac{\partial v}{\partial T}\right)_P}{\left(\frac{\partial v}{\partial P}\right)_T} = \frac{e}{C_T} \tag{4.49}$$

where e is the thermal expansivity and C_T is the isothermal compressibility. Combining Eqs. (4.48) and (4.49),

$$\left(\frac{\partial^2 u}{\partial s \partial v}\right) = -\frac{Te}{c_V C_T} \tag{4.50}$$

From Eqs. (4.42) and (4.45), $T/c_V > 0$ which gives $c_V > 0$; therefore, the thermal stability condition is established. This is the same relationship as that developed earlier from the entropy approach. Substituting Eqs. (4.45), (4.46), and (4.50) into Eq. (4.43) and using $(\partial P/\partial v)_T = -vC_T$ and $c_P = c_V + Tve^2/C_T$ [see Example 3.7 and Eq. (3.125)],

$$\frac{T}{(c_V C_T v)} > 0 \tag{4.51}$$

Therefore, $C_T > 0$. Note that since c_V and C_T are positive for a stable system, the implication is that $c_P > 0$ when a system is stable.

4.1.2 Stability Analysis in a Two-Component System

Consider N mols of a binary mixture of composition x_1 and x_2 (mol fractions) at temperature T and pressure P. Let us compare the stability of the single-phase system with that of the two-phase system at the same temperature and pressure. Similarly to the single-component system in the preceding section, we assume N' mols of primed phase and N'' mols of double-primed phase, in the two-phase state. The composition of primed and double-primed phases are $x_1 + \Delta x_1$ and $x_1 + \Delta^\circ x_1$, respectively, for component 1 (see Fig. 4.4). This time we will work with the Gibbs free energy function. The Gibbs free energy difference between the two states shown in Fig. 4.4 is

$$\Delta G = G_{two\ phase} - G_{single\ phase} \tag{4.52}$$

where $G_{two\ phase} = N'(g + \Delta g) + N''(g + \Delta^\circ g)$ and $G_{single\,phase} = (N' + N'')g$. Therefore,

$$\Delta G = N'\Delta g + N''\Delta^\circ g \tag{4.53}$$

In Eq. (4.53), Δg is the molar Gibbs free energy difference between the primed phase and the original single phase, and $\Delta^\circ g$ is the molar Gibbs free energy difference between the double-primed phase and the original phase. The constraints (see Fig. 4.4) are constant temperature and pressure and mol numbers of components 1 and 2; $N_i = N_i' + N_i'', i = 1, 2$ $(N = N_1 + N_2,\ N' = N_1' + N_2',$ and $N'' = N_1'' + N_2'')$. From the mol constraints,

$$N'\Delta x_1 + N''\Delta^\circ x_1 = 0 \tag{4.54}$$

The molar Gibbs free energy $g = g(T, P, x_1)$ and since T and P are held constant, then

$$\Delta g = \left(\frac{\partial g}{\partial x_1}\right)\Delta x_1 + \left(\frac{1}{2}\right)\left(\frac{\partial^2 g}{\partial x_1^2}\right)(\Delta x_1)^2 + \cdots \tag{4.55}$$

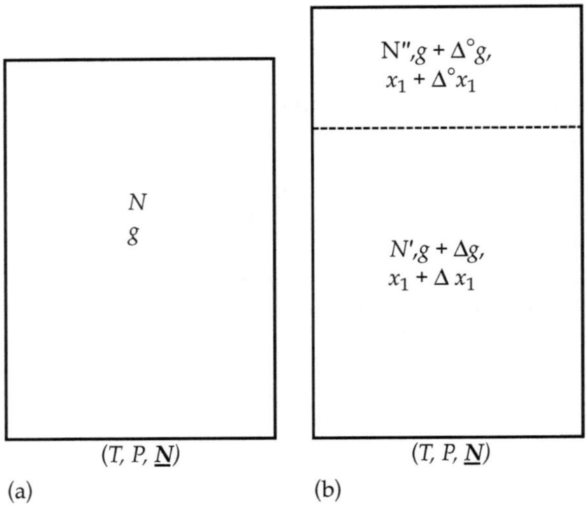

FIGURE 4.4 Stability of a two-component mixture at constant T, P, and $\underline{N}$.

and

$$\Delta^\circ g = \left(\frac{\partial g}{\partial x_1}\right)\Delta^\circ x_1 + \left(\frac{1}{2}\right)\left(\frac{\partial^2 g}{\partial x_1^2}\right)(\Delta^\circ x_1)^2 + \cdots \tag{4.56}$$

where Δx_1 and $\Delta^\circ x_1$ are assumed to be small. Combining Eqs. (4.52) to (4.56), and dropping the superscript "$\circ$",

$$\Delta G = \left(\frac{N''N}{2N'}\right)\left(\frac{\partial^2 g}{\partial x_1^2}\right)(\Delta x_1)^2 + \cdots \tag{4.57}$$

The stability of the single-phase state in comparison to that of the two-phase state requires $\Delta G > 0$. Since the multiplier of $(\partial^2 g/\partial x_1^2)$ is positive, then

$$\left(\frac{\partial^2 g}{\partial x_1^2}\right)_{T,P} > 0 \tag{4.58}$$

is the stability condition of the original single phase. Note that when the second-order variation is zero, then

$$\left(\frac{\partial^4 g}{\partial x_1^4}\right)_{T,P} > 0 \tag{4.59}$$

If fourth-order variations are zero, then the higher-order even variations should be positive.

Equation (4.58) can be transformed into the chemical potential form. Since $(\partial g/\partial x_1)_{T,P} = \mu_1 - \mu_2$, then

$$\left(\frac{\partial^2 g}{\partial x_1^2}\right)_{T,P} = \left(\frac{\partial \mu_1}{\partial x_1}\right)_{T,P} - \left(\frac{\partial \mu_2}{\partial x_1}\right)_{T,P} \tag{4.60}$$

From the Gibbs-Duhem relation at constant T and P, $\sum_{i=1}^{2} x_i(\partial \mu_i/\partial x_1)_{T,P} = 0$. Therefore, Eq. (4.58) transforms to $(\partial \mu_1/\partial x_1)_{T,P}/x_2$ and, since x_2 is positive,

$$\left(\frac{\partial \mu_1}{\partial x_1}\right)_{T,P} > 0 \tag{4.61}$$

Equation (4.61) is an alternative form of Eq. (4.58) for the stability criteria. Figure 4.5 shows a plot of μ_1 vs. x_1 at constant temperature and pressure in a binary mixture. For a system to be stable, $(\partial \mu_1/\partial x_1)_{T,P}$ should be positive. At T_1 and P_1, the system is in the stable single-phase state. At T_2 and P_2 only the points from A to B and from C to D are stable. At B and C, the limits of stability are defined by

$$\boxed{\left(\frac{\partial \mu_1}{\partial x_1}\right)_{T,P} = 0} \tag{4.62}$$

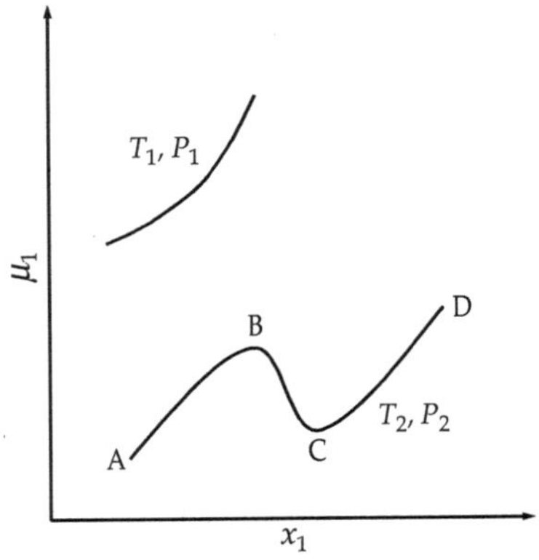

FIGURE 4.5 Plot of μ_1 vs. x_1 in a binary system at constant T and P. Note that the sketch is idealized.

We will show in Example 4.5 that the following inequalities hold for a stable binary system. At constant pressure and composition,

$$\left(\frac{\partial^2 g}{\partial T^2}\right)_{P,x_1} < 0 \tag{4.63}$$

and at constant temperature and composition,

$$\left(\frac{\partial^2 g}{\partial P^2}\right)_{T,x_1} < 0 \tag{4.64}$$

4.1.3 Stability Analysis for Multicomponent Mixtures

The stability criteria for single-component and binary systems are given above; those criteria are general, but restrictive assumptions were made in their derivations. These assumptions can be removed and the same results can be obtained. The generalized derivation relies on the use of the sum-of-the-squares expression, and Legendre and Jacobian transformations. The derivations of stability criteria presented above provide a physical sense of the problem using only the Taylor's series expansion. In the following, a more elaborate derivation for the general problem of multicomponent systems will be presented. An alternative derivation presented by Tester and Modell (1997) relies more on Legendre transforms.

Consider N mols of a multicomponent system consisting of n_c components. We wish to study the stability of the single-phase state in relation to the perturbed two-phase state. The perturbed two-phase state consists of N' mols of the primed phase and N'' mols of the double-primed phase. We assume that $N'' << N'$ solely for the purpose of determining whether the internal energy of a fixed mass could be reduced by introducing a new phase (i.e., the double-primed phase) of infinitesimal amount while keeping the volume and entropy fixed. We can use the criterion of the U minimum (as was just mentioned) or G minimum or the criterion of the maximum of S to study stability. Let us use the U-minimum principle. The Legendre transform can be used to transform the results from U to any other basic thermodynamic function, such as H, A, or G.

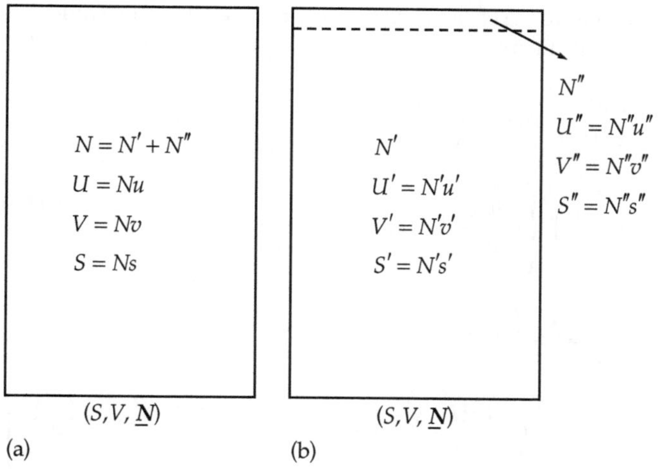

FIGURE 4.6 Stability of a multicomponent mixture at constant S, V, and $\underline{N}$.

The internal energy of the original homogeneous phase is a function of $(S, V, \underline{N})$. Figure 4.6 shows the state of the system before and after perturbation. The change in the internal energy from a single phase to a two-phase state is

$$\Delta U = \Delta U' + \Delta U'' \tag{4.65}$$

where $\Delta U'$ denotes the internal energy change of N' mols from the original single-phase state to the primed phase, and $\Delta U''$ represents the internal energy change of N'' mols from the original single-phase state to the double-primed phase. The restrictions of the internal energy change are

$$\Delta S = \Delta S' + \Delta S'' = 0 \tag{4.66}$$

$$\Delta V = \Delta V' + \Delta V'' = 0 \tag{4.67}$$

$$\Delta N_i = \Delta N_i' + \Delta N_i'' = 0 \quad i = 1, \ldots, n_c \tag{4.68}$$

In Eq. (4.66), $\Delta S'$ represents the entropy change of N' mols from the single-phase state to the primed phase, and $\Delta V'$ is the volume change of N' mols from the single-phase state to the primed phase. $\Delta S''$ and $\Delta V''$ have similar definitions. The expressions for various variations are given by

$$\Delta U' = N'(u' - u), \quad \Delta V' = N'(v' - v), \quad \Delta S' = N'(s' - s)$$
$$\Delta U'' = N''(u'' - u), \quad \Delta V'' = N''(v'' - v), \quad \Delta S'' = N''(s'' - s) \tag{4.69}$$

where u, u', and u'' are the molar internal energies of the original single-phase, primed, and double-primed phases, respectively; s, s', and s'' and v, v', and v'' have similar definitions. Note that in the above equations, there is no restriction on the magnitude of the molar property changes.

The expressions for $\Delta U'$ and $\Delta U''$ are provided by the Taylor's series expansions, which assume that the amount of the double-primed phase is small.

$$\begin{aligned}\Delta U' = {} & \left(\frac{\partial U}{\partial S}\right)' \Delta S' + \left(\frac{\partial U}{\partial V}\right)' \Delta V' + \sum_{i=1}^{n_c} \left(\frac{\partial U}{\partial N_i}\right)' \Delta N_i' \\ & + \left(\frac{1}{2}\right)\left\{\left(\frac{\partial^2 U}{\partial S^2}\right)' (\Delta S')^2 + \left(\frac{\partial^2 U}{\partial V^2}\right)' (\Delta V')^2 + 2\left(\frac{\partial^2 U}{\partial S \partial V}\right)' (\Delta S' \Delta V')\right. \\ & + \sum_{i=1}^{n_c}\sum_{j=1}^{n_c}\left(\frac{\partial^2 U}{\partial N_i \partial N_j}\right)' (\Delta N_i' \Delta N_j') \\ & \left. + 2\sum_{i=1}^{n_c}\left[\left(\frac{\partial^2 U}{\partial N_i \partial V}\right)' (\Delta N_i' \Delta V') + \left(\frac{\partial^2 U}{\partial N_i \partial S}\right)' (\Delta N_i' \Delta S')\right]\right\} \\ & + \textit{higher-order terms}\end{aligned} \tag{4.70}$$

and

$$\begin{aligned}\Delta U'' = {} & \left(\frac{\partial U}{\partial S}\right)'' \Delta S'' + \left(\frac{\partial U}{\partial V}\right)'' \Delta V'' + \sum_{i=1}^{n_c} \left(\frac{\partial U}{\partial N_i}\right)'' \Delta N_i'' \\ & + \left(\frac{1}{2}\right)\left\{\left(\frac{\partial^2 U}{\partial S^2}\right)'' (\Delta S'')^2 + \left(\frac{\partial^2 U}{\partial V^2}\right)'' (\Delta V'')^2 + 2\left(\frac{\partial^2 U}{\partial S \partial V}\right)'' (\Delta S'' \Delta V'')\right. \\ & + \sum_{i=1}^{n_c}\sum_{j=1}^{n_c}\left(\frac{\partial U^2}{\partial N_i \partial N_j}\right)'' (\Delta N_i'' \Delta N_j'') + 2\sum_{i=1}^{n_c}\left[\left(\frac{\partial^2 U}{\partial N_i \partial V}\right)'' (\Delta N_i'' \Delta V'')\right. \\ & \left.\left. + \left(\frac{\partial^2 U}{\partial N_i \partial S}\right)'' (\Delta N_i'' \Delta S'')\right]\right\} + \textit{higher-order terms}\end{aligned} \tag{4.71}$$

In Eq. (4.70), $(\partial U/\partial S)'$ represents the variation of the internal energy of N' mols with respect to the variation of the entropy of N' mols at the original state of the single phase. Similar representations for $(\partial U/\partial S)''$, $(\partial^2 U/\partial S^2)'$, etc. follow. From $dU = TdS - PdV + \sum_{i=1}^{n_c} \mu_i dN_i$, at the original state,

$$\begin{aligned}\left(\frac{\partial U}{\partial S}\right)' &= \left(\frac{\partial U}{\partial S}\right)'' = T \\ \left(\frac{\partial U}{\partial V}\right)' &= \left(\frac{\partial U}{\partial V}\right)'' = -P \\ \left(\frac{\partial U}{\partial N_i}\right)' &= \left(\frac{\partial U}{\partial N_i}\right)'' = \mu_i \quad i = 1, \ldots, n_c\end{aligned} \tag{4.72}$$

The first derivatives are intensive variables. However, the second and higher-order derivatives at the original state depend on mass. The second derivatives are related by

$$
\begin{aligned}
N'\left(\frac{\partial^2 U}{\partial S^2}\right)' &= N''\left(\frac{\partial^2 U}{\partial S^2}\right)'' \\
N'\left(\frac{\partial^2 U}{\partial V^2}\right)' &= N''\left(\frac{\partial^2 U}{\partial V^2}\right)'' \\
N'\left(\frac{\partial^2 U}{\partial S \partial V}\right)' &= N''\left(\frac{\partial^2 U}{\partial S \partial V}\right)'' \\
N'\left(\frac{\partial^2 U}{\partial N_i \partial S}\right)' &= N''\left(\frac{\partial^2 U}{\partial N_i \partial S}\right)'' \quad i = 1, \ldots, n_c \\
N'\left(\frac{\partial^2 U}{\partial N_i \partial V}\right)' &= N''\left(\frac{\partial^2 U}{\partial N_i \partial V}\right)'' \quad i = 1, \ldots, n_c \\
N'\left(\frac{\partial^2 U}{\partial N_i \partial N_j}\right)' &= N''\left(\frac{\partial^2 U}{\partial N_i \partial N_j}\right)'' \quad i = 1, \ldots, n_c, \; j = 1, \ldots, n_c
\end{aligned}
\tag{4.73}
$$

To prove the first relationship above, we write

$$
\begin{aligned}
\left(\frac{\partial^2 U}{\partial S^2}\right)' &= \frac{\partial}{\partial S'}\left(\frac{\partial U}{\partial S}\right)' = \left(\frac{\partial T}{\partial S}\right)' \\
\left(\frac{\partial^2 U}{\partial S^2}\right)'' &= \frac{\partial}{\partial S''}\left(\frac{\partial U}{\partial S}\right)'' = \left(\frac{\partial T}{\partial S}\right)''
\end{aligned}
\tag{4.74}
$$

But $S' = N's$ and $S'' = N''s$ at the original state, and, therefore,

$$
\begin{aligned}
\left(\frac{\partial^2 U}{\partial S^2}\right)' &= \left(\frac{1}{N'}\right)\frac{\partial T}{\partial s} \\
\left(\frac{\partial^2 U}{\partial S^2}\right)'' &= \left(\frac{1}{N''}\right)\frac{\partial T}{\partial s}
\end{aligned}
\tag{4.75}
$$

From the above two equations, the first relationship in Eq. (4.73) is established. (The last expression in Eq. (4.73) and some higher-order derivatives are derived in Example 4.3). Combining Eqs (4.65) to (4.73), we obtain,

$$
\begin{aligned}
\Delta U = \left(\frac{1}{2}\right)\left(\frac{N}{N'}\right)\Bigg\{&\left(\frac{\partial^2 U}{\partial S^2}\right)''(\Delta S')^2 + \left(\frac{\partial^2 U}{\partial V^2}\right)''(\Delta V')^2 \\
&+2\left(\frac{\partial^2 U}{\partial S \partial V}\right)''(\Delta S'\Delta V') + \sum_{i=1}^{n_c}\sum_{j=1}^{n_c}\left(\frac{\partial^2 U}{\partial N_i \partial N_j}\right)''(\Delta N_i'\Delta N_j') \\
&+2\sum_{i=1}^{n_c}\left(\frac{\partial^2 U}{\partial N_i \partial V}\right)''(\Delta N_i'\Delta V') + 2\sum_{i=1}^{n_c}\left(\frac{\partial^2 U}{\partial N_i \partial S}\right)''(\Delta N_i'\Delta S')\Bigg\} \\
&+ \textit{higher-order terms}
\end{aligned}
\tag{4.76}
$$

Equation (4.76) can be written as

$$\Delta U = \frac{1}{2!}\left[1+\frac{N''}{N'}\right]\mathrm{d}^2U + \frac{1}{3!}\left[1-\left(\frac{N''}{N'}\right)^2\right]\mathrm{d}^3U + \frac{1}{4!}\left[1+\left(\frac{N''}{N'}\right)^3\right]\mathrm{d}^4U + \cdots \tag{4.77}$$

where d^2U is the term in {} in Eq. (4.76); it represents the second-order variations of U. Note that there is no term to represent first-order variations of U in the above equations. The first-order variations drop out because of the equilibrium condition. ΔU in Eq. (4.77) is positive if the quadratic term representing second-order variations is positive for every small variation of variables ΔN_i, ΔS, and ΔV. (Note that the coefficient of d^2U is positive.) When the quadratic term is positive, the original homogeneous phase is stable and there is no need to examine higher-order terms. If the quadratic term is zero, then the cubic form must also be zero, and the fourth-order term must be positive for ΔU to be positive. In other words, the first nonvanishing even-order term must be positive for ΔU to be positive and odd-order terms below that must be zero. The expression for d^2U in a compact form can be written as (dropping the superscript),

$$\mathrm{d}^2U = \sum_{j=1}^{n_c+2}\sum_{k=1}^{n_c+2}\left(\frac{\partial^2 U}{\partial X_j \partial X_k}\right)\Delta X_j \Delta X_k \tag{4.78a}$$

Equation (4.78a) can also be written as

$$\mathrm{d}^2U = \sum_{j=1}^{n_c+2}\sum_{k=1}^{n_c+2} U_{jk}\Delta X_j \Delta X_k \tag{4.78b}$$

where $U_{jk} \equiv (\partial^2 U/\partial X_j \partial X_k)$.

Linear algebra provides a perfect shorthand for writing Eq. (4.78b) because of the symmetry of U_{jk}; $U_{jk} = U_{kj}$. This equation can be written as the inner product of the perturbation vector $\Delta\underline{\mathbf{X}}^T$ and the vector that results from the Hessian matrix of U on the perturbation vector $\Delta\underline{\mathbf{X}}$.

The Hessian matrix is the second-order-derivative matrix; the entries of the Hessian matrix of U are U_{jk}. Therefore,

$$\mathrm{d}^2U = \Delta\underline{\mathbf{X}}^T H_U \Delta\underline{\mathbf{X}} \tag{4.79}$$

where the Hessian matrix, H_U, is given by

$$H_U = \begin{pmatrix} U_{1,1} & U_{1,2} & \cdots & U_{1,n_c+2} \\ U_{2,1} & U_{2,2} & \cdots & U_{2,n_c+2} \\ \vdots & \vdots & & \vdots \\ U_{n_c+2,1} & U_{n_c+2,2} & \cdots & U_{n_c+2,n_c+2} \end{pmatrix} \tag{4.80}$$

and $\Delta\underline{\mathbf{X}}^T = (\Delta X_1\ \Delta X_2\ \ldots\ \Delta X_{n_c+2})$; $\Delta\underline{\mathbf{X}}^T$ is the transpose of the column vector $\Delta\underline{\mathbf{X}} = (\Delta X_1, \Delta X_2, \ldots, \Delta X_{n_c+2})$. Note that we have commas between the entries of the column vector $\Delta\underline{\mathbf{X}}$ to write it on a horizontal line! There is no comma between the entries of the row vector $\Delta\underline{\mathbf{X}}^T$. The Hessian matrix, which has a pure quadratic form, is used to study the problem of maxima and minima for functions of many variables. From the Taylor's series expansion of $U(\underline{X} + \Delta\underline{X})$ (see Example 4.1),

$$U(\underline{X} + \Delta\underline{X}) = U(\underline{X}) + \Delta\underline{X}^T(\nabla \boldsymbol{U}) + \left(\frac{1}{2}\right) \Delta\underline{X}^T H_U \Delta\underline{X} + \textit{higher-order terms} \tag{4.81}$$

where $\nabla \boldsymbol{U}$ represents the gradient of U. The gradient of U when set to zero provides the stationary point of $U(\underline{X})$; at a stationary point, $\nabla \boldsymbol{U} = (\partial U/\partial X_1, \partial U/\partial X_2, \ldots, \partial U/\partial X_{n_c+2})$ is a vector of zeros: $\partial U/\partial X_1 = 0$, $\partial U/\partial X_2 = 0 \ldots$, $\partial U/\partial X_{n_c+2} = 0$ (see Strang, 1988). However, a stationary point can be a maximum, a minimum, or a saddle point. The maxima and minima are, therefore, decided by the sign of $\Delta\underline{X}^T H_U \Delta\underline{X}$. U has a minimum when $\Delta\underline{X}^T H_U \Delta\underline{X}$ is positive. The Hessian matrix H_U should be positive definite in order to have $\Delta\underline{X}^T H_U \Delta\underline{X} > 0$. There are various tests for positive definiteness of the symmetric matrix H_U. One such test is that all the pivots of H_U should be positive (see Strang, 1988). Another test is that a positive-definite matrix has positive eigenvalues (also see Strang, 1988). For a single-component system, the derivations for positive and negative definiteness were straightforward because of dependence of S or U on two variables. When the number of variables exceeds two, a knowledge of the positive and negative definiteness of the Hessian matrix becomes helpful.

Let us return to Eq. (4.78b) and write it in the form of the sum-of-the-squares,

$$d^2U = \sum_{j=1}^{n_c+2} y_{jj}^{(j-1)} (\Delta Z_j)^2 \tag{4.82}$$

The expression for ΔZ_j and $y_{jj}^{(j-1)}$ will be derived next. Equation (4.82) is often written without full proof. The criteria of stability and criticality rely on the use of intermediate steps in the expressions for the sum-of-the-squares in our derivations.

The double summation can be expanded as

$$\begin{aligned} d^2U &= \sum_{j=1}^{n_c+2} \sum_{k=1}^{n_c+2} U_{jk} \Delta X_j \Delta X_k \\ &= \sum_{j=1}^{n_c+2} [U_{j1} \Delta X_j \Delta X_1 + U_{j2} \Delta X_j \Delta X_2 + U_{j3} \Delta X_j \Delta X_3 + \cdots] \end{aligned} \tag{4.83}$$

which can be rearranged to

$$\begin{aligned} d^2U &= \sum_{j=1}^{n_c+2} [U_{j1} \Delta X_j \Delta X_1] + \sum_{j=1}^{n_c+2} [U_{j2} \Delta X_j \Delta X_2 + U_{j3} \Delta X_j \Delta X_3] + \cdots \\ &= U_{1,1}(\Delta X_1)^2 + \sum_{j=2}^{n_c+2} (U_{j1} \Delta X_j \Delta X_1) + [U_{12} \Delta X_1 \Delta X_2 + U_{13} \Delta X_1 \Delta X_3] + \cdots \\ &\quad + \left[\sum_{j=2}^{n_c+2} U_{j2} \Delta X_j \Delta X_2 + U_{j3} \Delta X_j \Delta X_3 \right] + \cdots \end{aligned} \tag{4.84}$$

Use of $U_{j1} = U_{1j}$ and further rearrangement gives

$$d^2U = U_{1,1}(\Delta X_1)^2 + 2\sum_{j=2}^{n_c+2} U_{1j}\Delta X_1 \Delta X_j + \sum_{j=2}^{n_c+2}\sum_{k=2}^{n_c+2} U_{jk}\Delta X_j \Delta X_k \tag{4.85}$$

The above equation can be expressed as

$$\begin{aligned} d^2U = U_{1,1}&\left[\Delta X_1 + \left(\frac{1}{U_{1,1}}\right)\left(\sum_{j=2}^{n_c+2} U_{1j}\Delta X_j\right)\right]^2 \\ &- \left(\frac{1}{U_{1,1}}\right)\left(\sum_{j=2}^{n_c+2} U_{1j}\Delta X_j\right)^2 + \sum_{j=2}^{n_c+2}\sum_{k=2}^{n_c+2} U_{jk}\Delta X_j \Delta X_k \end{aligned} \tag{4.86}$$

provided $U_{1,1} \neq 0$. We have already shown that $U_{1,1} > 0$ for a single-component system to be stable; that also applies when there is no change in composition or when composition is held constant for multicomponent systems [see Eq. (4.42)]. The squared-single summation $(\sum_{j=2}^{n_c+2} U_{1j}\Delta X_j)^2$ in Eq. (4.86) in terms of double summation is

$$\left(\sum_{j=2}^{n_c+2} U_{1j}\Delta X_j\right)^2 = \sum_{j=2}^{n_c+2}\sum_{k=2}^{n_c+2} U_{1j}U_{1k}\Delta X_j \Delta X_k \tag{4.87}$$

Equation (4.87) is derived by opening the double summation:

$$\begin{aligned} \sum_{j=2}^{n_c+2}\sum_{k=2}^{n_c+2} U_{1j}U_{1k}\Delta X_j \Delta X_k &= \sum_{k=2}^{n_c+2}[U_{1,2}U_{1k}\Delta X_2 \Delta X_k + U_{1,3}U_{1k}\Delta X_3 \Delta X_k + \cdots] \\ &= \{[U_{1,2}^2(\Delta X_2)^2 + U_{1,2}U_{1,3}\Delta X_2\Delta X_3 + \cdots] \\ &\quad + [U_{1,2}U_{1,3}\Delta X_2 \Delta X_3 + U_{1,3}^2(\Delta X_3)^2 + \cdots] + \cdots\} \\ &= \left[\sum_{j=2}^{n_c+2}(U_{1j}\Delta X_j)\right]^2 \end{aligned} \tag{4.88}$$

Defining ΔZ_1 and $U_{jk}^{(1)}$ as

$$\Delta Z_1 = \Delta X_1 + \left(\frac{1}{U_{1,1}}\right)\sum_{j=2}^{n_c+2} \Delta U_{1j}\Delta X_j \tag{4.89}$$

$$U_{jk}^{(1)} = U_{jk} - \left(\frac{1}{U_{1,1}}\right) U_{1j}U_{1k} \qquad j, k \geq 2 \tag{4.90}$$

Combining Eqs. (4.86), (4.87), and the above definitions,

$$d^2U = U_{1,1}(\Delta Z_1)^2 + \sum_{j=2}^{n_c+2} \sum_{k=2}^{n_c+2} U_{jk}^{(1)} \Delta X_j \Delta X_k \tag{4.91}$$

The double summation in the above equation can be written as [similarly to Eq. (4.86)]

$$\begin{aligned}
&\sum_{j=2}^{n_c+2} \sum_{k=2}^{n_c+2} U_{jk}^{(1)} \Delta X_j \Delta X_k \\
&= U_{2,2}^{(1)} \left[\Delta X_2 + \left(\frac{1}{U_{2,2}^{(1)}} \right) \left(\sum_{j=3}^{n_c+2} U_{j2}^{(1)} \Delta X_j^2 \right) \right]^2 - \left(\frac{1}{U_{2,2}^{(1)}} \right) \left[\sum_{j=3}^{n_c+2} U_{j2} \Delta X_j \right]^2 \\
&+ \sum_{j=3}^{n_c+2} \sum_{k=3}^{n_c+2} U_{jk}^{(1)} \Delta X_j \Delta X_k
\end{aligned} \tag{4.92}$$

provided $U_{2,2}^{(1)} \neq 0$. The squared single summation can be written as [similarly to Eq. (4.87)]

$$\left(\sum_{j=3}^{n_c+2} U_{j2} \Delta X_j \right)^2 = \sum_{j=3}^{n_c+2} \sum_{k=3}^{n_c+2} (U_{j2} U_{k2})(\Delta X_j \Delta X_k) \tag{4.93}$$

Define ΔZ_2 and $U_{jk}^{(2)}$ as

$$\Delta Z_2 = \Delta X_2 + \left(\frac{1}{U_{2,2}^{(1)}} \right) \sum_{j=3}^{n_c+2} U_{j2}^{(1)} \Delta Z_j \tag{4.94}$$

$$U_{jk}^{(2)} = \left[U_{jk}^{(1)} - \left(\frac{1}{U_{2,2}^{(1)}} \right) (U_{j2} U_{k2}) \right], \quad j, k \geq 3 \tag{4.95}$$

From Eqs. (4.91) and (4.95),

$$d^2U = U_{1,1}(\Delta Z_1)^2 + U_{2,2}^{(1)}(\Delta Z_2)^2 + \sum_{j=3}^{n_c+2} \sum_{k=3}^{n_c+2} U_{jk}^{(2)} \Delta X_j \Delta X_k \tag{4.96}$$

The above procedure can be continued to obtain the final expressions:

$$
\begin{array}{|ll|}
\hline
U_{jk}^{(0)} = U_{jk} & j,k = 1,\ldots,n_c+2 \\
U_{jk}^{(l)} = U_{jk}^{(l-1)} - \left(\dfrac{1}{U_{ll}^{(l-1)}}\right)\left[U_{lj}^{(l-1)}U_{lk}^{(l-1)}\right] & j,k \geq l \\
 & j,k = 2\ldots n_c+2 \\
 & l = 1,\ldots,n_c+2 \\
\Delta Z_j = \Delta X_j + \left(\dfrac{1}{U_{jj}^{(j-1)}}\right)\displaystyle\sum_{k=j+1}^{n_c+2} U_{kj}^{(j-1)}\Delta X_k & j = 1\ldots n_c+2 \\
\mathrm{d}^2U = U_{1,1}(\Delta Z_1)^2 + U_{2,2}^{(1)}(\Delta Z_2)^2 + \cdots U_{n_c+2,n_c+2}^{(n_c+1)}(\Delta Z_{n_c+2})^2 & \\
\hline
\end{array}
\tag{4.97}
$$

Let us define

$$
\begin{aligned}
y_{1,1}^{(0)} &= U_{1,1} \\
y_{jj}^{(j-1)} &= U_{jj}^{(j-1)} \quad j = 2,\ldots,n_c+2
\end{aligned}
\tag{4.98}
$$

The $y_{1,1}^{(0)}$, $y_{jj}^{(j-1)}$, and ΔZ_j in Eqs. (4.97) and (4.98) define the variables of Eq. (4.82). The purpose of introducing y in place of U is to emphasize that Eq. (4.82) represents not only U but other thermodynamic functions, H, A, and G.

Alternative Expression for d^2U

Equation (4.78b) can also be written as

$$
\mathrm{d}^2U = U_{1,1}(\Delta Z_1)^2 + \frac{\begin{vmatrix} U_{1,1} & U_{1,2} \\ U_{2,1} & U_{2,2} \end{vmatrix}}{U_{1,1}}(\Delta Z_2)^2 + \frac{\begin{vmatrix} U_{1,1} & U_{1,2} & U_{1,3} \\ U_{2,1} & U_{2,2} & U_{2,3} \\ U_{3,1} & U_{3,2} & U_{3,3} \end{vmatrix}}{\begin{vmatrix} U_{1,1} & U_{1,2} \\ U_{2,1} & U_{2,2} \end{vmatrix}}(\Delta Z_3)^2 + \cdots
\tag{4.99}
$$

This form of d^2U is very useful for the derivation of the criteria of stability. The key steps in the derivation of Eq. (4.99) are provided below. The coefficient of $(\Delta Z_2)^2$ is $U_{2,2}^{(1)}$. From Eq. (4.97)

$$
U_{2,2}^{(1)} = U_{2,2} - \left(\frac{1}{U_{1,1}}\right)(U_{1,2}U_{1,2}) = \frac{\begin{vmatrix} U_{1,1} & U_{1,2} \\ U_{2,1} & U_{2,2} \end{vmatrix}}{U_{1,1}}
\tag{4.100}
$$

Similarly, the coefficient of $(\Delta Z_3)^2$, $U_{3,3}^{(2)}$, is

$$
U_{3,3}^{(2)} = \frac{\left[U_{3,3}^{(1)}U_{2,2}^{(1)} - U_{2,3}^{(1)}U_{2,3}^{(1)}\right]}{U_{2,2}^{(1)}}
\tag{4.101}
$$

The expressions for $U_{3,3}^{(1)}$, $U_{2,2}^{(1)}$, and $U_{2,3}^{(1)}$ from Eq. (4.97) when substituted into Eq. (4.101) result in

$$U_{3,3}^{(2)} = \frac{\begin{vmatrix} U_{1,1} & U_{1,2} & U_{1,3} \\ U_{2,1} & U_{2,2} & U_{2,3} \\ U_{3,1} & U_{3,2} & U_{3,3} \end{vmatrix}}{U_{2,2}^{(1)}} \tag{4.102}$$

Other higher-order expressions are obtained in the same manner. The algebra is tedious, but the method is straightforward. Another compact form of Eq. (4.99) is

$$\boxed{d^2U = \sum_{j=1}^{n_c+2} \left(\frac{\lambda^{(j)}}{\lambda^{(j-1)}}\right)(\Delta Z_j)^2} \tag{4.103}$$

where $\lambda^{(0)} = 1$, $\lambda^{(1)} = U_{1,1}$, and

$$\lambda^{(j)} = \begin{vmatrix} U_{1,1} & U_{1,2} & \cdots & U_{1j} \\ U_{2,1} & U_{2,2} & \cdots & U_{2j} \\ \vdots & \vdots & & \vdots \\ U_{j1} & U_{j2} & \cdots & U_{jj} \end{vmatrix}, \quad j = 2, \ldots, n_c + 2 \tag{4.104}$$

Since $y_{n_c+2,n_c+2}^{(n_c+1)} = 0$ [see Eq. (1.186)], then Eqs. (4.82) and (4.103) can be written as,

$$\boxed{d^2U = \sum_{j=1}^{n_c+1} y_{jj}^{(j-1)}(\Delta Z_j)^2} \tag{4.105}$$

and

$$\boxed{d^2U = \sum_{j=1}^{n_c+1} \left(\frac{\lambda^{(j)}}{\lambda^{(j-1)}}\right)(\Delta Z_j)^2} \tag{4.106}$$

respectively. Now we are ready to establish the criteria of stability. From the set of expressions in Eqs. (4.97) and (4.98),

$$y_{jj}^{(j-1)} = y_{jj}^{(j-2)} - \frac{\left[y_{j,j-1}^{(j-2)}\right]^2}{y_{j-1,j-1}^{(j-2)}} \quad j = 2, \ldots, n_c + 1 \tag{4.107}$$

An alternative derivation of the above equation was presented in Chap. 1 using Legendre and Jacobian transformations [see Eq. (1.197)]. In Eq. (4.107), if $y_{j-1,j-1}^{(j-2)}$ and $y_{jj}^{(j-1)}$ approach zero, $y_{jj}^{(j-1)}$ approaches zero first and $y_{j-1,j-1}^{(j-2)}$ approaches zero later, provided ($y_{j,j-1}^{(j-2)} \neq 0$). Therefore, the positivity of d^2U which is related to the positivity of the coefficients of $(\Delta Z_j)^2$ simplifies considerably; in place of $y_{jj}^{(j-1)} > 0$ for $j = 1, \ldots, n_c + 1$, one can demand only that

$$\boxed{y_{n_c+1,n_c+1}^{(n_c)} > 0} \tag{4.108}$$

The above simple relationship provides the criteria of stability for multicomponent systems. When

$$\boxed{y^{(n_c)}_{n_c+1,n_c+1} = 0} \tag{4.109}$$

the multicomponent system is said to be at the limit of the stability; $y^{(n_c)}_{n_c+1,n_c+1}$ provides the locus of the limit of stability (or in fact the limit of metastability) discussed in Chap. 3. Examples are provided at the end of the chapter to give a clear picture of stability criteria and their significance. Let us transform Eqs. (4.108) and (4.109) in terms of more familiar variables. From Eq. (4.98),

$$y^{(0)}(X_1, X_2, \ldots, X_{n_c+2}) = U(S, V, N_1, N_2, \ldots, N_{n_c}) \tag{4.110}$$

where $y^{(n_c)}_{n_c+1,n_c+1}$ is simply

$$y^{(n_c)}_{n_c+1,n_c+1} = \left(\frac{\partial^2 y^{(n_c)}}{\partial X^2_{n_c+1}}\right)_{C_1,C_2,\ldots,C_{n_c},X_{n_c+2}} = \left(\frac{\partial C_{n_c+1}}{\partial X_{n_c+1}}\right)_{C_1,C_2,\ldots,C_{n_c},X_{n_c+2}} \tag{4.111}$$

From Eqs. (1.184), and (1.185), $C_1 = T$, $C_2 = -P$, $C_3 = \mu_1, \ldots, C_{n_c} = \mu_{n_c-2}$, $C_{n_c+1} = \mu_{n_c-1}$, and $X_{n_c+2} = N_{n_c} \ldots$ an expression of mole fractions for stable systems derived from Eq. (4.108) is (see Prob. 4.19)

$$\boxed{\left(\frac{\partial \mu_{n_c-1}}{\partial x_{n_c-1}}\right)_{T,P,\mu_1,\mu_2,\ldots,\mu_{n_c-2}} > 0} \tag{4.112}$$

The expression in Eq. (4.112) implies that for a n_c−component system to be stable, $(\partial \mu_{n_c-1}/\partial x_{n_c-1})$ should be positive at constant T, P, $\mu_1, \ldots, \mu_{n_c-2}$, which is inconvenient in terms of the variables to which we are accustomed. Figure 4.7 is a plot of μ_{n_c-1} vs. x_{n_c-1}. The curve I has a positive slope and, therefore, is stable. The curve II shows the system can go through a two-phase region; between B and C, $(\partial \mu_{n_c-1}/\partial x_{n_c-1})$ is negative and the system is unstable. We will transform Eq. (4.112) in terms of the variables that are convenient for a pressure-explicit EOS in the examples at the end of this chapter.

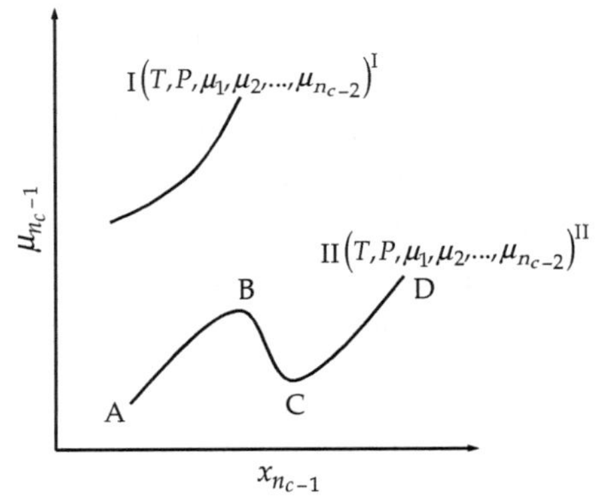

FIGURE 4.7 Plot of μ_{n_c-1} vs. x_{n_c-1}. Note that the sketch is idealized.

4.2 Criticality Analysis

In the derivation of the criteria of criticality, we will first use a simple approach that relies on geometrical representation. Then an alternative approach will be presented which is more suitable for multicomponent systems with a large number of components. Consider the vapor-liquid critical point of a single-component system by studying the P-v graph shown in Fig. 4.8. In this figure, the spinodal is represented by the dashed curve and the binodal by the solid curve. The P-v isotherms at four different temperatures T_1, T_2, T_c, and T_3 are shown. Points B and C represent the limits of stability at temperature T_1. Points B′ and C′ represent the same limits at temperature T_2. According to Eq. (4.32), the limits of stability at temperatures T_1 and T_2 are obtained from $(\partial P/\partial v)_{T_1} = 0$ and $(\partial P/\partial v)_{T_2} = 0$, respectively. Note that between points B and C, $(\partial P/\partial v)_{T_1} > 0$, and this is, therefore, the unstable segment of the isotherm. Also note that the curvature changes between B and C; therefore, an inflection point where $(\partial P^2/\partial v^2)_{T_1} = 0$ exists between these two points. There is another requirement for the inflection point; the third derivative (odd derivative) should be nonzero. At T_2 which is higher than T_1, points B′ and C′ are closer together than are points B and C, and the inflection point $(\partial^2 P/\partial v^2)_{T_2} = 0$ is an unstable point. As the temperature approaches T_c, the two limits of stability and the inflection point coincide, and since the inflection point is now located on the binodal curve, the inflection point $(\partial^2 P/\partial v^2)_{T_c} = 0$ is a stable point. Points A and D and A′ and D′ represent equilibrium phases at T_1 and T_2, respectively. Toward the critical point, the points corresponding to A and A′ and D and D′ also coincide with the limit of stability. At the critical point, the gas and liquid phases can be transformed into each other without going through the two-phase region, which implies the continuity of gas and liquid states. The criteria for the critical point of a single-component system are, therefore,

$$\left(\frac{\partial P}{\partial v}\right)_{T_c} = \left(\frac{\partial^2 P}{\partial v^2}\right)_{T_c} = 0, \quad \text{and} \quad \left(\frac{\partial^3 P}{\partial v^3}\right)_{T_c} < 0 \tag{4.113}$$

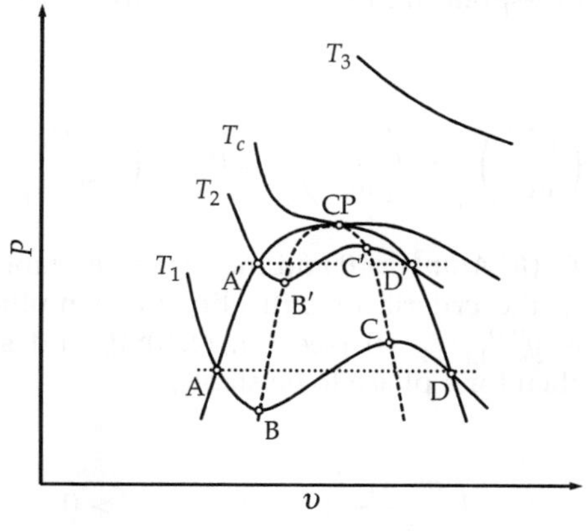

FIGURE 4.8 *P*-*v* plot for a single-component system. Note that this sketch is idealized. When phase change is undergone, a true sketch looks like the one in Fig. 3.6.

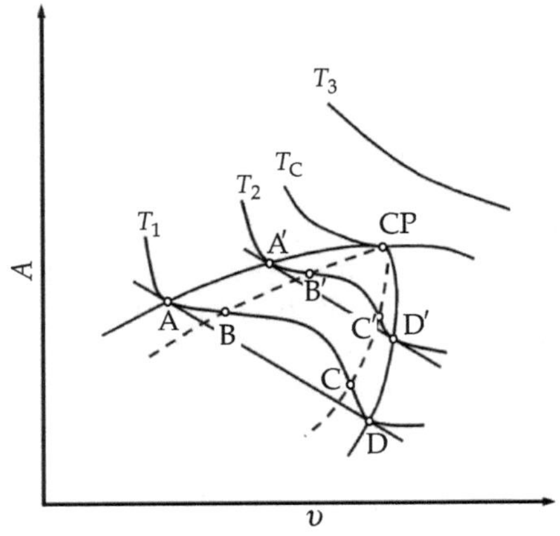

FIGURE 4.9 Plot of A vs. v for a single-component system.

The last relationship to the right in Eq. (4.113) implies that the critical point is neither a maximum nor a minimum. This inflection point from the algebraic standpoint has to have a nonvanishing odd-order derivative (see Example 4.2).

Now let us examine the plot of Helmholtz free energy, A vs. v, at constant temperature; $A(T, v)$ is a function of temperature and molar volume and a natural choice as the thermodynamic function for the criticality criteria. Figure 4.9 shows a schematic diagram of the A-v plot for a single-component system. From the general criteria of stability given by Eq. (4.108), $y_{2,2}^{(1)} > 0$ for a single-component system. If $y^{(0)} = U$, then $y^{(1)} = A$ and $y_{2,2}^{(1)} = (\partial^2 A/\partial V^2)_{T,N}$ from $dA = -SdT - PdV + \mu dN$. The stability is, therefore, given by $(\partial^2 A/\partial V^2)_{T,N} > 0$ or $(\partial^2 A/\partial v^2)_T > 0$, and the limit of stability by $(\partial^2 A/\partial v^2)_T = 0$. Let us now examine the isotherm at T_1 in Fig. 4.9. In the stable segment, $(\partial^2 A/\partial v^2)_{T_1} > 0$, and in the segment between B and C, $(\partial^2 A/\partial v^2)_{T_1} < 0$, implying instability. Note that when $(\partial^2 A/\partial v^2)_{T_1} < 0$, the Helmholtz free energy curve is convex (like the surface of a ball as seen from inside). Points A and D are on the concave side, $(\partial^2 A/\partial v^2)_{T_1} > 0$, and are stable. Since these two points have a common tangent, $(\partial A/\partial v)_{T_1}|_{@A} = (\partial A/\partial v)_{T_1}|_{@D}$, they are at the same equilibrium pressure. Between B and C because of convexity, $(\partial^3 A/\partial v^3)_{T_1} > 0$. As the temperature rises, points B′ and C′ get closer. Finally, when $T = T_c$, points corresponding to B′ and C′ as well as A′ and D′ coincide. At the critical point,

$$\left(\frac{\partial^2 A}{\partial v^2}\right)_{T_c} = \left(\frac{\partial^3 A}{\partial v^3}\right)_{T_c} = 0, \qquad \left(\frac{\partial^4 A}{\partial v^4}\right)_{T_c} > 0 \tag{4.114}$$

At temperature, $T_3 > T_c$, $(\partial^2 A/\partial v^2) > 0$ and the system is stable.

Now let us examine the criteria of criticality for a multicomponent system. The criterion of stability is $y_{n_c+1,n_c+1}^{(n_c)} > 0$ [see Eq. (4.108)], and since $y_{n_c+1,n_c+1}^{(n_c)} = (\partial C_{n_c+1}/\partial X_{n_c+1})_{C_1,C_2,\ldots,C_{n_c},X_{n_c+2}}$, then for a phase to be stable,

$$\left(\frac{\partial C_{n_c+1}}{\partial X_{n_c+1}}\right)_{C_1,C_2,\ldots,C_{n_c},X_{n_c+2}} > 0 \tag{4.115}$$

Let us examine the plot of C_{n_c+1} vs. X_{n_c+1} at fixed $C_1, \ldots, C_{n_c}, X_{n_c+2}$. Figure 4.10 shows three curves; for each curve, the $C_1, \ldots, C_{n_c}, X_{n_c+2}$ variables are fixed. The fixed variables for each curve are selected in such a way that curve I is in the two-phase region, curve II passes through the critical point, and curve III stays in the singe-phase region. For curve I, B and C are the stability limits; these two points are obtained from $(\partial C_{n_c+1}/\partial X_{n_c+1})_{C_1,C_2,\ldots,C_{nc},X_{nc+2}} = 0$. At the inflection point, E, between B and C, $(\partial^2 C_{n_c+1}/\partial X^2_{n_c+1})_{C_1,\ldots,C_{nc},X_{nc+2}} = 0$. Between B and C, $(C_{n_c+1}/\partial X_{n_c+1})_{C_1,\ldots,C_{nc},X_{nc+2}} < 0$, which implies instability. As X_{n_c+2} is varied, curve I may move upward and points B and C come closer. When points B and C and inflection point E coincide,

$$\left(\frac{\partial C_{n_c+1}}{\partial X_{n_c+1}}\right)_{C_1,\ldots,C_{nc},X_{nc+2}} = \left(\frac{\partial^2 C_{n_c+1}}{\partial X^2_{n_c+1}}\right)_{C_1,\ldots,C_{nc},X_{nc+2}} = 0,$$
$$\left(\frac{\partial^3 C_{n_c+1}}{\partial X^3_{n_c+1}}\right)_{C_1,\ldots,C_{nc},X_{nc+2}} > 0 \tag{4.116}$$

In terms of $y^{(n_c)}$, the criteria of criticality are

$$y^{(n_c)}_{n_c+1,n_c+1} = 0, \quad y^{(n_c)}_{n_c+1,n_c+1,n_c+1} = 0, \quad y^{(n_c)}_{n_c+1,n_c+1,n_c+1,n_c+1} > 0 \tag{4.117}$$

The inequality provides the criterion that the critical point is an inflection point. If the third derivative [in Eq. (4.116)] vanishes, the next higher-order odd derivative should be positive. We will use the above criteria to establish the critical point for a pure component and for binary and multicomponent systems.

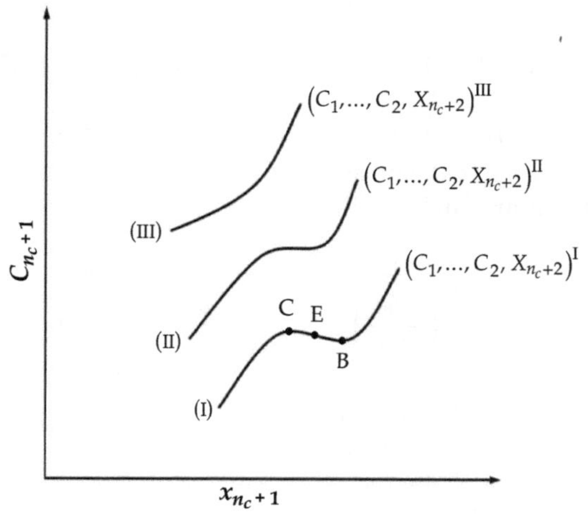

FIGURE 4.10 Plot of C_{n_c+1} vs. X_{n_c+1} for an n_c-component system.

4.2.1 Single-Component Fluid

For a single-component, $y^{(1)}_{2,2} = y^{(1)}_{2,2,2} = 0$ and $y^{(1)}_{2,2,2,2} > 0$. From $y^{(0)} = U$, $\mathrm{d}y^{(1)} = -S\mathrm{d}T - P\mathrm{d}V + \mu\mathrm{d}N$ [see Eq. (1.176)]. Therefore,

$$y^{(1)}_{2} = (\partial y^{(1)}/\partial V)_{T,N} = -P, y^{(1)}_{2,2} = -(\partial P/\partial V)_{T,N} = 0, \quad y^{(1)}_{2,2,2} = (-\partial^2 P/\partial V^2)_{T,N} = 0$$

and

$$y^{(1)}_{2,2,2,2} = (-\partial^3 P/\partial V^3)_{T,N} > 0$$

Since

$$V = Nv$$

then

$$(\partial P/\partial v)_T = (\partial^2 P/\partial v^2)_T = 0$$

and

$$-(\partial^3 P/\partial v^3)_T > 0$$

where

$$T = T_c$$

These are the same expressions as in Eq. (4.113).

4.2.2 Two-Component Fluid

From Eq. (4.117), $y^{(2)}_{3,3} = y^{(2)}_{3,3,3} = 0$ and $y^{(2)}_{3,3,3,3} > 0$. Rewriting Eq. (1.177),

$$\mathrm{d}y^{(2)} = \mathrm{d}G = -S\mathrm{d}T + V\mathrm{d}P + \mu_1\mathrm{d}N_1 + \mu_2\mathrm{d}N_2 \tag{4.118}$$

where $y^{(2)}_{3} = (\partial y^{(2)}/\partial N_1)_{T,P,N_2} = \mu_1, y^{(2)}_{3,3} = (\partial \mu_1/\partial N_1)_{T,P,N_2} = 0$
$y^{(2)}_{3,3,3} = (\partial \mu_1^2/\partial N_1^2)_{T,P,N_2} = 0$, and $y^{(2)}_{3,3,3,3} = (\partial \mu_1^3/\partial N_1^3)_{T,P,N_2} > 0$. The above expressions in terms of mol fractions can be written as

$$\boxed{\begin{aligned} &\left(\frac{\partial \mu_1}{\partial x_1}\right)_{T_c,P_c} = \left(\frac{\partial^2 \mu_1}{\partial x_1^2}\right)_{T_c,P_c} = 0 \\ &\left(\frac{\partial^3 \mu_1}{\partial x_1^3}\right)_{T_c,P_c} > 0 \end{aligned}} \tag{4.119}$$

where $x_1 = N_1/(N_1 + N_2)$, $(\partial \mu_1/\partial N_1)_{T,P,N_2} = (\partial \mu_1/\partial x_1)_{T,P}(\partial x_1/\partial N_1)_{T,P,N_2}$, and $(\partial x_1/\partial N_1)_{T,P,N_2} = (1 - x_1)/(N_1 + N_2)$.

In terms of fugacity, $(d\mu_1 = RTd\ln f_1)_T$, the criteria of criticality are, therefore,

$$\left(\frac{\partial f_1}{\partial x_1}\right)_{T_c,P_c} = 0, \quad \left(\frac{\partial^2 f_1}{\partial x_1^2}\right)_{T_c,P_c} = 0, \quad \text{and} \quad \left(\frac{\partial^3 f_1}{\partial x_1^3}\right)_{T_c,P_c} > 0 \tag{4.120}$$

4.2.3 Multicomponent Fluid

Equations (4.113) and (4.120) derived from the criteria of criticality are in a form that can readily be used to calculate the critical point of a single-component and a binary mixture. For a three-component system, the criteria are of the form $y_{4,4}^{(3)} = 0$ and $y_{4,4,4}^{(3)} = 0$. Let us use $y^{(0)} = U(S, V, N_1, N_2, N_3)$. Then $dy^{(3)} = -SdT + VdP - N_1 d\mu_1 + \sum_{i=2}^{3} \mu_i dN_i$ [see Eqs. (1.21) and (1.182)] and $y_4^{(3)} = \mu_2$. Therefore, at the critical point for a ternary mixture,

$$\left(\frac{\partial \mu_2}{\partial N_2}\right)_{T,P,\mu_1,N_3} = 0 \tag{4.121}$$

$$\left(\frac{\partial^2 \mu_2}{\partial N_2^2}\right)_{T,P,\mu_1,N_3} = 0 \tag{4.122}$$

The inequality constraint, $y_{4,4,4,4}^{(3)} > 0$, results in $(\partial^3 \mu_2/\partial N_2^3)_{T,P,\mu_1,N_3} > 0$ which is generally not tested. One may apply a stability analysis check on the critical point.

The above equations are not very practical to use! How can we keep μ_1 constant when a system is undergoing a change? An alternative procedure may be more convenient for the calculation of the critical point of mixtures with three or more components. Since we often use a pressure-explicit equation of state, $P = P(V, T, \underline{N})$, the following derivations are aimed toward the use of such equations. The first equation for criticality can be written as

$$y_{n_c+1,n_c+1}^{(n_c)} = \frac{\lambda^{(n_c+1)}}{\lambda^{(n_c)}} \tag{4.123}$$

where λ's are defined in Eq. (4.104). Since $y_{n_c+1,n_c+1}^{(n_c)} = 0$ at the critical point, therefore,

$$\lambda^{(n_c+1)} = 0 \tag{4.124}$$

The second equation for criticality is $y_{n_c+1,n_c+1,n_c+1}^{(n_c)} = 0$, which is $(\partial y_{n_c+1,n_c+1}^{(n_c)}/\partial X_{n_c+1}) = 0$. Taking the derivative of Eq. (4.123) with respect to X_{n_c+1},

$$y_{n_c+1,n_c+1,n_c+1}^{(n_c)} = \frac{\left[\lambda_{n_c+1}^{(n_c+1)}\lambda^{(n_c)} - \lambda_{n_c+1}^{(n_c)}\lambda^{(n_c+1)}\right]}{[\lambda^{(n_c)}]^2} = 0 \tag{4.125}$$

Since $\lambda^{(n_c+1)}=0$ at the critical point, therefore, $\lambda^{(n_c+1)}_{n_c+1}=0$, where the subscript n_c+1 is used to denote the derivative of $\lambda^{(n_c+1)}$ with respect to variable X_{n_c+1}.

$$\lambda^{(n_c+1)}_{n_c+1} = \left[\frac{\partial \lambda^{(n_c+1)}}{\partial X_{n_c+1}}\right]_{C_1,\ldots,C_{n_c},X_{n_c+2}} = 0 \tag{4.126}$$

Using the Jacobian transformations from Chap. 1,

$$\lambda^{(n_c+1)}_{n_c+1} = \frac{\partial(C_1,\ldots,C_{n_c},\lambda^{(n_c+1)},X_{n_c+2})}{\partial(C_1,\ldots,C_{n_c},X_{n_c+1},X_{n_c+2})} = 0 \tag{4.127}$$

The above equation can be reduced to

$$\left.\frac{\partial(C_2,\ldots,C_{n_c},\lambda^{(n_c+1)})}{\partial(C_2,\ldots,C_{n_c},X_{n_c+1})}\right|_{C_1,X_{n_c+2}} = 0 \tag{4.128}$$

The reduction is intended for the use of the Helmholtz free energy $A=A(T,V,\underline{N})$. Dividing the numerator and denominator by $\partial(X_2,\ldots,X_{n_c+1})$ [see Eq. (1.188)],

$$\left.\frac{\partial(C_2,\ldots,C_{n_c},\lambda^{(n_c+1)})/\partial(X_2,\ldots,X_{n_c},X_{n_c+1})}{\partial(C_2,\ldots,C_{n_c},X_{n_c+1})/\partial(X_2,\ldots,X_{n_c},X_{n_c+1})}\right|_{C_1,X_{n_c+2}} = 0 \tag{4.129}$$

In determinant form,

$$\frac{\begin{vmatrix} \left(\frac{\partial C_2}{\partial X_2}\right) & \left(\frac{\partial C_2}{\partial X_3}\right) & \cdots & \left(\frac{\partial C_2}{\partial X_{n_c+1}}\right) \\ \vdots & \vdots & & \vdots \\ \vdots & \vdots & & \vdots \\ \left(\frac{\partial C_{n_c}}{\partial X_2}\right) & \left(\frac{\partial C_{n_c}}{\partial X_3}\right) & & \left(\frac{\partial C_{n_c}}{\partial X_{n_c+1}}\right) \\ \left(\frac{\partial \lambda^{(n_c+1)}}{\partial X_2}\right) & \left(\frac{\partial \lambda^{(n_c+1)}}{\partial X_3}\right) & \cdots & \left(\frac{\partial \lambda^{(n_c+1)}}{\partial X_{n_c+1}}\right) \end{vmatrix}}{\begin{vmatrix} \left(\frac{\partial C_2}{\partial X_2}\right) & \left(\frac{\partial C_2}{\partial X_3}\right) & \cdots & \left(\frac{\partial C_2}{\partial X_{n_c+1}}\right) \\ \vdots & \vdots & & \vdots \\ \vdots & \vdots & & \vdots \\ \left(\frac{\partial C_{n_c}}{\partial X_2}\right) & \left(\frac{\partial C_{n_c}}{\partial X_3}\right) & & \left(\frac{\partial C_{n_c}}{\partial X_{n_c+1}}\right) \\ \left(\frac{\partial X_{n_c+1}}{\partial X_2}\right) & \left(\frac{\partial X_{n_c+1}}{\partial X_3}\right) & \cdots & \left(\frac{\partial X_{n_c+1}}{\partial X_{n_c+1}}\right) \end{vmatrix}} = 0 \tag{4.130}$$

Since $(\partial X_{n_c+1}/\partial X_i)=0$ for $i=2,\ldots,n_c$ and $=1$ for $i=n_c+1$, assuming that the resulting determinant in the denominator is not zero, then the numerator is zero. The $(\partial C_i/\partial X_j)$ elements are in the form $y^{(1)}_{ij}$, since for all of them C_1 and the X_i's are held constant, that is, $(\partial C_2/\partial X_2)_{C_1,X_3,\ldots,X_{n_c+2}} = y_{2,2}^{(1)}$ and $(\partial C_2/\partial X_3)_{C_1,X_2,X_4,\ldots,X_{n_c+2}} = y_{2,3}^{(1)}$ [see Eq. (1.183)]. From Eqs. (4.124) and (4.130), we can write the criticality conditions, respectively:

$$\lambda^{(n_c+1)} = \begin{vmatrix} y_{2,2}^{(1)} & y_{2,3}^{(1)} & \cdots & y_{2,n_c+1}^{(1)} \\ \vdots & \vdots & \vdots & \vdots \\ y_{n_c,2}^{(1)} & y_{n_c,3}^{(1)} & \cdots & y_{n_c,n_c+1}^{(1)} \\ y_{n_c+1,2}^{(1)} & y_{n_c+1,3}^{(1)} & \cdots & y_{n_c+1,n_c+1}^{(1)} \end{vmatrix} = 0 \tag{4.131}$$

and

$$\begin{vmatrix} y_{2,2}^{(1)} & y_{2,3}^{(1)} & \cdots & y_{2,n_c+1}^{(1)} \\ \vdots & \vdots & \vdots & \vdots \\ y_{n_c,2}^{(1)} & y_{n_c,3}^{(1)} & \cdots & y_{n_c,n_c+1}^{(1)} \\ \lambda_2^{(n_c+1)} & \lambda_3^{(n_c+1)} & \cdots & \lambda_{n_c+1}^{(n_c+1)} \end{vmatrix} = 0 \tag{4.132}$$

Since $y^{(1)} = A(T, V, N_1, \ldots, N_{n_c})$, the elements of the above equations can be expressed as the derivatives of A with respect to V and $(N_1, N_2, \ldots, N_{n_c-1})$. The above set of equations was used by Baker and Luks (1980) to calculate the critical points of multicomponent systems.

4.3 Alternative Approach for Critical-Point Calculation

As we have already seen, the critical point can be calculated in a variety of related ways. However, the basic expressions that define the critical point involve setting two determinants equal to zero [i.e., Eqs. (4.131) and (4.132)]. Of the two determinants, one [i.e., Eq. (4.132)] requires much effort, especially for multicomponent systems; it is necessary to evaluate the derivatives of the determinant in Eq. (4.132).

In 1980, Heidemann and Khalil introduced an alternative approach for the calculation of the critical point that is mathematically different from the expressions in Eqs. (4.131) and (4.132), but the concept is not. As we have already seen, the critical point is a stable point at the limit of stability. Let us expand on this concept. Consider the P-v diagram of a multicomponent system of fixed composition sketched in Fig. 4.11. The thick solid curve on the left represents the bubblepoints and the thin solid curve on the right represents the dewpoints. The bubblepoint and dewpoint are in stable equilibrium states; a perturbation in P, for example, results in a stable state. Critical point CP is the point at which bubblepoint and dewpoint converge and in addition to being a stable state, it is at the limit of stability. These two features were used by Gibbs in 1876 to derive the expressions for the critical point. We will repeat the use of stability and stability limit concepts in a more straightforward approach in this section. Let us consider the Helmhotz free energy of the closed n_c−component system at constant T and V sketched in Fig. 4.12. The perturbed two-phase state consists of N' mols of the primed phase and N'' mols

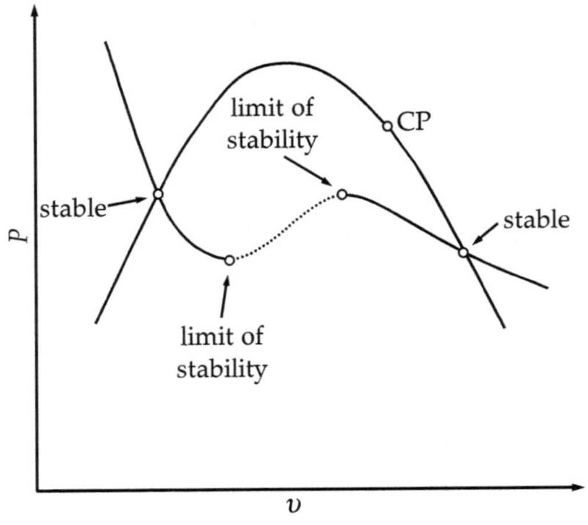

FIGURE 4.11 Critical point representation in a multicomponent system. Note that the sketch is idealized.

of the double-primed phase. We assume $N'' << N'$. Similar to the U representation, the Helmhotz free energy change from single-phase to two-phase state is given by,

$$\Delta A = \Delta A' + \Delta A'' \tag{4.133}$$

where $\Delta A'$ and $\Delta A''$ are the free energy changes of N' mols from the original single-phase state to the primed phase, and the free energy change of N'' mols from the original single-phase state to the double-primed phase, respectively. The restrictions of the Helmhotz free energy change are

$$\Delta V = \Delta V' + \Delta V'' = 0 \tag{4.134}$$

$$\Delta N_i = \Delta N_i' + \Delta N_i'' = 0 \quad i = 1, \ldots, n_c \tag{4.135}$$

$$T = constant \tag{4.136}$$

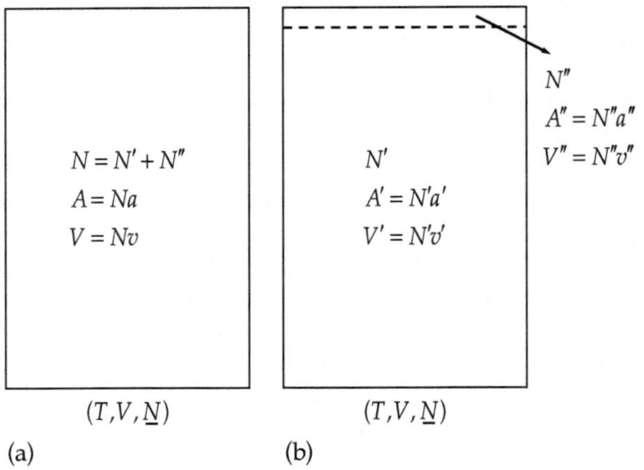

FIGURE 4.12 Stability of a multicomponent mixture at constant T, V and $\underline{N}$.

Variables $\Delta V'$, $\Delta V''$, $\Delta N_i'$, $\Delta N_i''$ are defined in Eqs. (4.67) and (4.68). Manipulations similar to those for Eqs. (4.69) to (4.75) result in the following expression for ΔA:

$$\Delta A = \frac{1}{2!}\left[1+\left(\frac{N''}{N'}\right)\right]\left\{\sum_{i=1}^{n_c}\sum_{j=1}^{n_c}\left(\frac{\partial^2 A}{\partial N_i \Delta N_j}\right)(\Delta N_i \Delta N_j) + \left(\frac{\partial^2 A}{\partial V^2}\right)(\Delta V)^2\right.$$
$$\left. + 2\sum_{i=1}^{n_c}\left(\frac{\partial^2 A}{\partial N_i \partial V}\right)(\Delta N_i \Delta V)\right\} + \frac{1}{3!}\left[1-\left(\frac{N''}{N'}\right)^2\right]$$
$$\left\{\sum_{i=1}^{n_c}\sum_{j=1}^{n_c}\sum_{k=1}^{n_c}\left(\frac{\partial^3 A}{\partial N_i \partial N_j \partial N_k}\right)(\Delta N_i \Delta N_j \Delta N_k)\right.$$
$$\left. + \left(\frac{\partial^3 A}{\partial V^3}\right)(\Delta V)^3 + 2\sum_{i=1}^{n_c}\sum_{j=1}^{n_c}\left(\frac{\partial^3 A}{\partial V \partial N_i \partial N_j}\right)(\Delta N_i \Delta N_j \Delta V)\right\}$$
$$+ \textit{higher-order terms} \tag{4.137}$$

Let us represent the second- and third-order variations by $d^2 A$ and $d^3 A$ without the coefficients $[1 + (N''/N')]$ and $[1 - (N''/N')^2]$, respectively. For the original state to be stable, $\Delta A > 0$, which requires $d^2 A > 0$. If $d^2 A = 0$, then $d^3 A$ should be zero and $d^4 A > 0$. Note that the coefficient of $d^4 A$, $[1 + (N''/N')^3]$, is positive. Therefore, at the critical point

$$d^2 A = 0 \tag{4.138}$$

$$d^3 A = 0 \tag{4.139}$$

$$d^4 A > 0 \tag{4.140}$$

If $d^4 A = 0$, $d^5 A$ should be zero, and the next higher-order even term, that is, $d^6 A$ should be positive, and so on. $d^2 A = \Delta \underline{\mathbf{X}}^T A_H \Delta \underline{\mathbf{X}}$ where $\Delta \underline{\mathbf{X}}^T$ is the row vector $\Delta \underline{\mathbf{X}}^T = (\Delta V \Delta N_1\ \Delta N_2 \ldots \Delta N_{nc})$, $\Delta \underline{\mathbf{X}}$ is the column vector $\Delta \underline{\mathbf{X}} = (\Delta V, \Delta N_1, \Delta N_2, \ldots, \Delta N_{nc})$, and A_H is the Hessian matrix of A given by

$$A_H = \begin{pmatrix} A_{V,V} & A_{V,N_1} & \cdots & A_{V,N_{nc}} \\ A_{N_1 V} & A_{N_1 N_1} & \cdots & A_{N_1,N_{nc}} \\ \vdots & & & \\ \vdots & & & \\ A_{N_{nc},V} & A_{N_{nc},N_1} & \cdots & A_{N_{nc},N_{nc}} \end{pmatrix} \tag{4.141}$$

This matrix is positive semidefinite (see Prob. 4.18). For $d^2 A > 0$, the $(n_c + 1)$ $(n_c + 1)$ matrix may be converted to an $(n_c)(n_c)$ A_H submatrix for positive definiteness. The principal submatrix B_H is obtained by removing any ith row and ith column of A_H, $i = 1, \ldots, (n_c + 1)$. Therefore, B_H is an $n_c \times n_c$ matrix, whereas A_H is an $(n_c + 1)(n_c + 1)$ matrix. Heidemann and Khalil (1980) based on physical grounds argued that when $\Delta V/V = (\Delta N_i/N_i) = k$ for all $i = 1, \ldots, n_c$ (k is a constant), such variation does not qualify as a variation in the phase; the Helmhotz free energy A is simply multiplied by the constant k. Therefore, the quadratic, cubic, and higher-order terms can be contracted to eliminate the case of $\Delta V/V = \Delta N_i/N_i = k, i = 1, \ldots, n_c$. The criticality criteria in the contracted (by removing volume variable) form can be written as

$$\sum_{i=1}^{n_c} \sum_{j=1}^{n_c} \left(\frac{\partial^2 A}{\partial N_i \partial N_j} \right) (\Delta N_i \Delta N_j)$$

$$= (\Delta N_1 \Delta N_2 \ldots \Delta N_{n_c}) \begin{pmatrix} A_{N_1,N_1} & \ldots & A_{N_1,N_{nc}} \\ \vdots & & \vdots \\ \vdots & & \vdots \\ A_{N_{nc},N_1} & \ldots & A_{N_{nc},N_{nc}} \end{pmatrix} \begin{pmatrix} \Delta N_1 \\ \vdots \\ \vdots \\ \Delta N_{n_c} \end{pmatrix} = \begin{pmatrix} 0 \\ \vdots \\ \vdots \\ 0 \end{pmatrix} \tag{4.142}$$

and

$$\boxed{\sum_{i=1}^{n_c} \sum_{j=1}^{n_c} \sum_{k=1}^{n_c} \left(\frac{\partial^3 A}{\partial N_i \partial N_j \partial N_k} \right) (\Delta N_i \Delta N_j \Delta N_k) = 0} \tag{4.143}$$

The inequality constraint given by Eq. (4.140) is generally not tested. Heidemann and Khalil (1980) proposed the use of Eqs. (4.142) and (4.143) to calculate the critical point. One distinct feature of the criticality criteria in terms of Eqs. (4.142) and (4.143) in comparison with Eqs. (4.131) and (4.132) is that the vector $\Delta \underline{N}$ should be evaluated. Equations (4.131) and (4.132) bypass this step. At the limit of stability, the quadratic term represented by Eq. (4.142) is positive semidefinite; there will be some variations in the phase for which the quadratic is positive, but for some other variations it will be exactly zero. These variations will make the cubic form in Eq. (4.143) zero.

From Eq. (4.142), it follows that the determinant

$$\begin{vmatrix} A_{N_1,N_1} & \ldots & A_{N_1,N_{nc}} \\ \vdots & & \vdots \\ A_{N_{nc},N_1} & \ldots & A_{N_{nc},N_{nc}} \end{vmatrix} = 0 \tag{4.144}$$

When the above determinant is zero, then from Eq. (4.142),

$$\begin{pmatrix} A_{N_1,N_1} & \ldots & A_{N_1,N_{nc}} \\ \vdots & & \vdots \\ A_{N_{nc},N_1} & \ldots & A_{N_{nc},N_{nc}} \end{pmatrix} \begin{pmatrix} \Delta N_1 \\ \vdots \\ \Delta N_{n_c} \end{pmatrix} = \begin{pmatrix} 0 \\ \vdots \\ 0 \end{pmatrix} \tag{4.145}$$

which can have a nonzero solution. By fixing one of the ΔN_i's ($i = 1, \ldots, n_c$), other ΔN_i's can be found from Eq. (4.145). With the column vector $\Delta \underline{N} = (\Delta N_1, \Delta N_2, \ldots, \Delta N_{n_c})$, the cubic Eq. (4.143) can be solved. In other words, Eqs. (4.142) and (4.143) can provide two unknowns such as critical pressure and critical temperature or any other two unknowns. Note that Eq. (4.144), which provides the limit of stability, is the same as the result from Eq. (4.131), when instead of the vector $(V, \Delta N_1, \ldots, \Delta N_{n_c-1})$, the vector $(\Delta N_1, \ldots, \Delta N_{n_c})$ is selected. Also note that in Eqs. (4.142) to (4.145), we could have selected the vector $(\Delta V, \Delta N_1, \ldots, \Delta N_{n_c-1})$. There are certain advantages in the vector of $\Delta \underline{N} = (\Delta N_1, \ldots, \Delta N_{n_c})$ (see Prob. 4.10).

In the following, we will further discuss the use of Eqs. (4.142), (4.144), and (4.145) for the calculation of the critical point of a single-component system and binary and ternary mixtures. We will also use Eq. (4.132) for the binary mixture to show that more computational effort is required in the calculation of the critical point when the cubic Eq. (4.143) is not used.

4.3.1 Single-Component Fluids

For a pure substance from Eq. (4.142), $d^2A = A_{N,N}(\Delta N)^2 = 0$, and from Eq. (4.143) $d^3A = A_{N,N,N}(\Delta N)^3 = 0$ are established at the critical point. Assign a nonzero value to ΔN, then $A_{N,N} = 0$, and $A_{N,N,N} = 0$. From $dA = -SdT - PdV + \mu dN$, one obtains $(\partial\mu/\partial N)_{T,V} = 0$ and $(\partial^2\mu/\partial N^2)_{T,V} = 0$. If ΔV is selected instead of ΔN, one obtains $(\partial P/\partial V)_{T,N} = 0$ and $(\partial^2 P/\partial V^2)_{T,N} = 0$; these are the same expressions as in Eq. (4.113).

4.3.2 Two-Component Fluids

For a binary mixture from Eq. (4.144),

$$A_{N_1,N_1}A_{N_2,N_2} - A^2_{N_1,N_2} = 0 \tag{4.146}$$

From Eq. (4.145),

$$A_{N_1,N_1}\Delta N_1 + A_{N_1,N_2}\Delta N_2 = 0 \tag{4.147}$$

$$A_{N_1,N_2}\Delta N_1 + A_{N_2,N_2}\Delta N_2 = 0 \tag{4.148}$$

Using Eq. (4.146), both Eqs. (4.147) and (4.148) collapse into the following equation:

$$A_{N_1,N_2}\Delta N_1 + A_{N_2,N_2}\Delta N_2 = 0 \tag{4.149}$$

We can set $\Delta N_1 = A_{N_1,N_2}$, then $\Delta N_2 = -A_{N_1,N_1}$. From Eq. (4.143),

$$A_{N_1,N_1,N_1}(\Delta N_1)^3 + A_{N_2,N_2,N_2}(\Delta N_2)^3 + 3A_{N_1,N_1,N_2}(\Delta N_1)^2(\Delta N_2) + 3A_{N_1,N_2,N_2}(\Delta N_1)(\Delta N_2)^2 = 0 \tag{4.150}$$

Substituting for ΔN_1 and ΔN_2 in Eq. (4.150), we obtain

$$A_{N_1,N_1,N_1}\left(A_{N_1,N_2}\right)^3 - A_{N_2,N_2,N_2}\left(A_{N_1,N_1}\right)^3 - 3A_{N_1,N_1,N_2}\left(A_{N_1,N_2}\right)^2\left(A_{N_1,N_1}\right) + 3A_{N_1,N_2,N_2}\left(A_{N_1,N_2}\right)\left(A_{N_1,N_1}\right)^2 = 0 \tag{4.151}$$

Substituting for $A_{N_1,N_1,N_2}A_{N_2,N_2} + A_{N_1,N_1}A_{N_2,N_2,N_2} - 2A_{N_1,N_2}A_{N_1,N_2,N_2}$ from Eq. (4.146) in the first term of the above equation results in

$$A_{N_1,N_1,N_1}\left(A_{N_1,N_2}A_{N_2,N_2}\right)^3 - A_{N_2,N_2,N_2}\left(A_{N_1,N_1}\right)^2 - 3A_{N_1,N_1,N_2}\left(A_{N_1,N_2}\right)^2 + 3A_{N_1,N_2,N_2}\left(A_{N_1,N_2}A_{N_1,N_1}\right) = 0 \tag{4.152}$$

Equation (4.152) provides the same expression as Eq. (4.131) when ΔN_1 and ΔN_2 are selected in place of ΔV and ΔN_1. However, Eq. (4.131) requires much additional work. We need to find the derivative of the determinant in Eq. (4.131) with respect to N_1; $A_{N_1,N_1,N_2}A_{N_2,N_2} + A_{N_1,N_2,N_2}A_{N_1,N_1} - 2A_{N_1,N_2}A_{N_1,N_1,N_2}$ and N_2; $A_{N_1,N_1,N_2}A_{N_2,N_2} + A_{N_1,N_1}A_{N_2,N_2,N_2} - 2A_{N_1,N_2}A_{N_1,N_2,N_2}$. Then these derivatives should be substituted in appropriate entries of the determinant in Eq. (4.132). We also need to combine the result with Eq. (4.146) to obtain Eq. (4.152).

4.3.3 Three-Component Fluids

For a three-component system, Eq. (4.144) results in

$$A_{N_1,N_1}\left(A_{N_2,N_2}A_{N_3,N_3} - A^2_{N_3,N_2}\right) - A_{N_1,N_2}\left(A_{N_1,N_2}A_{N_3,N_3} - A_{N_2,N_3}A_{N_1,N_3}\right) + A_{N_1,N_3}\left(A_{N_1,N_2}A_{N_2,N_3} - A_{N_1,N_3}A_{N_2,N_2}\right) = 0 \tag{4.153}$$

From Eq. (4.145),

$$\begin{pmatrix} A_{N_1,N_1} & A_{N_1,N_2} & A_{N_1,N_3} \\ A_{N_2,N_1} & A_{N_2,N_2} & A_{N_2,N_3} \\ A_{N_3,N_1} & A_{N_3,N_2} & A_{N_3,N_3} \end{pmatrix}\begin{pmatrix} \Delta N_1 \\ \Delta N_2 \\ \Delta N_3 \end{pmatrix} = \begin{pmatrix} 0 \\ 0 \\ 0 \end{pmatrix} \tag{4.154}$$

and from Eq. (4.143)

$$\begin{aligned} &A_{N_1,N_1,N_1}(\Delta N_1)^3 + A_{N_2,N_2,N_2}(\Delta N_2)^3 + A_{N_3,N_3,N_3}(\Delta N_3)^3 \\ &\quad + 3A_{N_1,N_1,N_2}(\Delta N_1)^2(\Delta N_2) + 3A_{N_1,N_1,N_2}(\Delta N_1)^2(\Delta N_3) \\ &\quad + 3A_{N_1,N_2,N_2}(\Delta N_1)(\Delta N_2)^2 + 3A_{N_2,N_3,N_3}(\Delta N_2)(\Delta N_3)^2 \\ &\quad + 3A_{N_1,N_3,N_3}(\Delta N_1)(\Delta N_3)^2 + 3A_{N_2,N_2,N_3}(\Delta N_2)^2(\Delta N_3) \\ &\quad + 6A_{N_1,N_2,N_3}(\Delta N_1 \Delta N_2 \Delta N_3) = 0 \end{aligned} \tag{4.155}$$

There are different ways to solve the above system of nonlinear equations [Eqs. (4.153) to (4.155)] to obtain two unknowns at the critical point. Heidemann and Khalil used Eq. (4.153) to solve for the critical temperature at a given critical volume, and then used the result to obtain ΔN_2 and ΔN_3 in Eq. (4.155). Note that one ΔN_i, for example, ΔN_1, can be fixed. With values of $\Delta \underline{N}$ known, Eq. (4.155) is used to estimate critical volume. With known critical volume and critical temperature, the EOS is used to calculate critical pressure. Use of the nested calculations has been successful for every mixture that had a vapor-liquid critical point, including mixtures with more than 40 components (Heidemann and Khalil, 1980). We will cover computational algorithm in Chap. 5.

4.4 Examples and Theory Extension

4.1: Vector Representation of Taylor Series Expansion of *U* Write the expression for the Taylor's series expansion of the internal energy $U = U(\underline{X}) = U(X_1, X_2, \ldots, X_{n_c+2})$ around point $\underline{X}$ in vector representation.

Solution The Taylor's series expansion of U around the perturbed point $\underline{\boldsymbol{X}}$ can be written as

$$U(\underline{\boldsymbol{X}} + \Delta\underline{\boldsymbol{X}}) = U(\underline{\boldsymbol{X}}) + \sum_{i=1}^{n_c+2} U_i \Delta X_i + \frac{1}{2!}\sum_{i=1}^{n_c+2}\sum_{j=1}^{n_c+2} U_{ij}\Delta X_i \Delta X_j$$
$$+ \textit{ higher-order terms}$$

where $U_i = (\partial U/\partial X_i)$ and $U_{ij} = (\partial^2 U/\partial X_i \partial X_j)$. The second term on the right side of the above equation can be written as the inner product or the dot product of the transpose vector $\Delta\underline{\boldsymbol{X}}^T = (\Delta X_1\ \Delta X_2 \ldots \Delta X_{n_c+2})$ and the gradient vector
$\nabla \boldsymbol{U} = (\partial U/\partial X_1, \partial U/\partial X_2, \ldots, \partial U/\partial X_{n_c+2})$

$$\Delta\underline{\boldsymbol{X}}^T \nabla U = (\Delta X_1\ \Delta X_2 \ldots \Delta X_{n_c+2}) \begin{pmatrix} \dfrac{\partial U}{\partial X_1} \\ \dfrac{\partial U}{\partial X_2} \\ \vdots \\ \dfrac{\partial U}{\partial X_{n_c+2}} \end{pmatrix} = \sum_{i=1}^{n_c+2} U_i \Delta X_i$$

We use matrix multiplication to show that

$$\Delta\underline{\boldsymbol{X}}^T H_U \Delta\underline{\boldsymbol{X}} = (\Delta X_1\ \Delta X_2 \ldots \Delta X_{n_c+2}) \begin{pmatrix} U_{1,1} & U_{1,2} & \ldots & U_{1,n_c+2} \\ U_{1,1} & U_{1,2} & \ldots & U_{1,n_c+2} \\ \vdots & \vdots & & \vdots \\ U_{n_c+2,1} & U_{n_c+2,2} & \ldots & U_{n_c+2,n_c+2} \end{pmatrix} \begin{pmatrix} \Delta X_1 \\ \Delta X_2 \\ \vdots \\ \Delta X_{n_c+2} \end{pmatrix}$$

$$= \left(\sum_{i=1}^{n_c+2} U_{i1}\Delta X_i\ \sum_{i=1}^{n_c+2} U_{i2}\Delta X_i \ldots \sum_{i=1}^{n_c+2} U_{i,n_c+2}\Delta X_i \right) \begin{pmatrix} \Delta X_1 \\ \Delta X_2 \\ \vdots \\ \Delta X_{n_c+2} \end{pmatrix}$$

$$= \sum_{i=1}^{n_c+2}\sum_{j=1}^{n_c+2} U_{ij}\Delta X_i \Delta X_j$$

4.2: Extremum of a Single-Variable Function Derive the criteria for the maximum and minimum of a single-variable function.

Solution Let us write the Taylor's series expansion of $f(x)$ around $x = 0$,

$$f(x) = \sum_{n=0}^{\infty} \frac{f^{(n)}(0)}{n!} x^n = f(0) + \sum_{n=1}^{\infty} \frac{f^{(n)}(0)}{n!} x^n$$

where $f^{(0)}(0) = f(0)$, and $f^{(1)}, f^{(2)}, \ldots,$ are the first-, second-, and higher-order derivatives of f, respectively. At the stationary point, $f^{(1)}(0) = 0$; to develop the criteria for a maximum or minimum, one needs to examine the higher-order derivatives at $x = 0$.

Close to the point $x = 0$,

$$f(x) - f(0) \approx \frac{f^{(m)}(0)x^m}{m!}$$

where m is the first nonvanishing term of the Taylor's series expansion. If m is odd, the function $f^{(m)}(0)x^m/m!$ may look like the graph in Fig. 4.13a. The function $f(x)$ has an inflection point at $x = 0$. However, if m is even, depending on the sign of $f^{(m)}(0)x^m/m!$, one may have either a maximum or a minimum. If $f^{(m)}(0)$ is positive, the graph of $f(x) - f(0)$ vs. x may look like the sketch in Fig. 4.13b and if $f^{(m)}(0)$ is negative, $f^{(m)}(0)x^m/m!$ vs. x may look like the sketch in Fig. 4.13c. Therefore, for $f(x)$ or $f(x) - f(0)$ to have a maximum at $x = 0$, the first nonvanishing higher-order odd derivative of $f(x)$ should be negative (see Fig. 4.13b). Similarly, for $f(x)$ or $f(x) - f(0)$ to have a minimum at $x = 0$, the first nonvanishing higher-order even derivative of $f(x)$ should be positive (see Fig. 4.13c).

4.3: Relation between Higher-Order Derivatives of *U* in Primed and Double-Primed Phases Derive the following expressions:

$$N'\left(\frac{\partial^2 U}{\partial N_i \partial N_j}\right)' = N''\left(\frac{\partial^2 U}{\partial N_i \partial N_j}\right)''$$

$$(N')^2\left(\frac{\partial^3 U}{\partial S^3}\right)' = (N'')^2\left(\frac{\partial^3 U}{\partial S^3}\right)''$$

$$(N')^2\left(\frac{\partial^3 U}{\partial N_i \partial N_j \partial N_k}\right)'' = (N'')^2\left(\frac{\partial^3 U}{\partial N_i \partial N_j \partial N_k}\right)''$$

Solution To derive the first expression, we write

$$\left(\frac{\partial^2 U}{\partial N_i \partial N_j}\right)' = \frac{\partial}{\partial N_i'}\left(\frac{\partial U}{\partial N_j}\right)'$$

where $(\partial U/\partial N_j)'$ is at the original state and is equal to μ_j. Similarly, $(\partial U/\partial N_j)''$ is at the original state and is equal to μ_j. Therefore,

$$\left(\frac{\partial^2 U}{\partial N_i \partial N_j}\right)'' = \left(\frac{\partial \mu_j}{\partial N_i'}\right)$$

Next we can write $\partial/\partial N_i' = (\partial/\partial x_i)(\partial x_i/\partial N_i')$, where x_i is the mol fraction of component i in the original single-phase state. Since $x_i = N_i'/\sum_{i=1}^{n_c} N_i' = N_i'/N'$, then

$$\frac{\partial}{\partial x_i}\left[\frac{\partial x_i}{\partial N_i'}\right] = \frac{\partial}{\partial x_i}\left[\frac{\partial}{\partial N_i'}\left(\frac{N_i'}{N'}\right)\right] = \frac{1}{N'}\frac{\partial}{\partial x_i}(1 - x_i)$$

and, therefore,

$$\frac{\partial}{\partial N_i'} = \frac{1}{N'}\frac{\partial}{\partial x_i}(1 - x_i)$$

FIGURE 4.13 Plots of $f(x) - f(0) \approx \frac{f^{(m)}(0)x^m}{m!}$ vs. x.

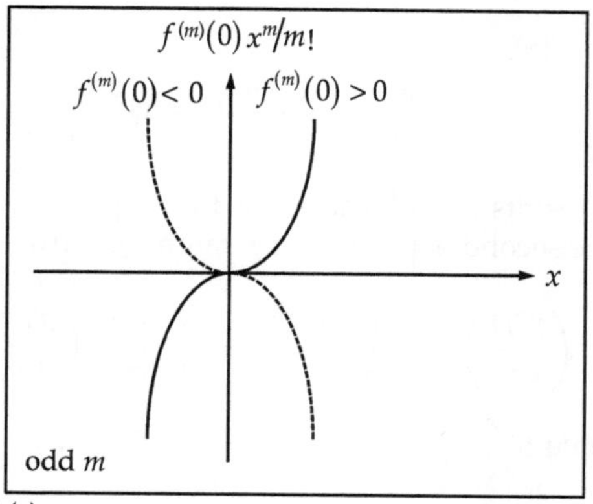

(a)

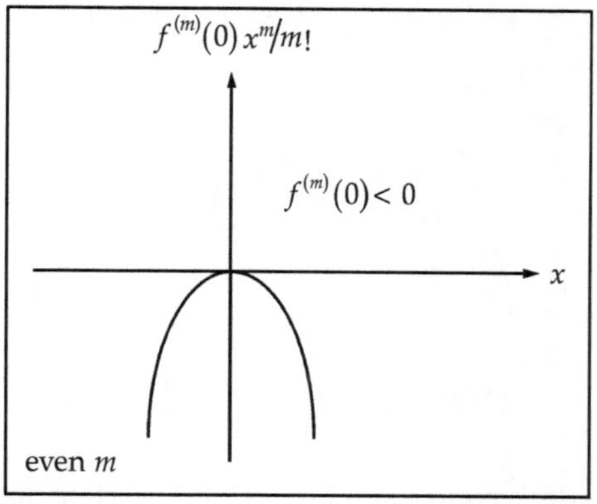

(b)

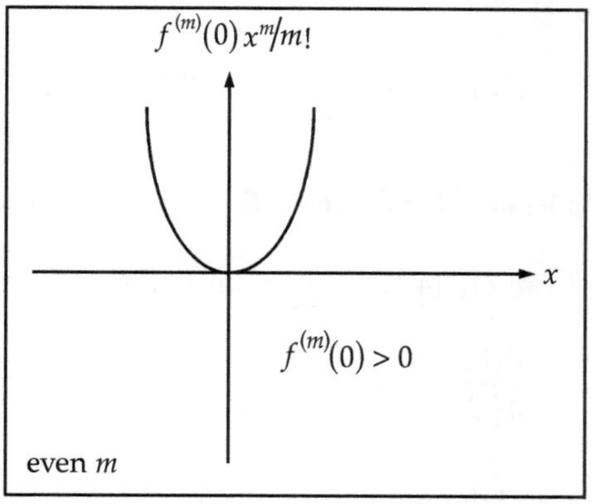

(c)

Similarly,

$$\frac{\partial}{\partial N_i''} = \frac{1}{N''}\frac{\partial}{\partial x_i}(1 - x_i)$$

Combining the above results would lead to the first expression.

The derivation of the second expression is straightforward:

$$\left(\frac{\partial^3 U}{\partial S^3}\right)' = \frac{\partial}{\partial S'}\left[\frac{\partial^2 U}{\partial S^2}\right]' = \left(\frac{\partial}{\partial S}\right)'\left(\frac{\partial T}{\partial S}\right)'$$

Since at the original state $S' = N's$, then

$$\left(\frac{\partial^3 U}{\partial S^3}\right)' = \left(\frac{1}{N'}\right)^2\left(\frac{\partial^2 T}{\partial s^2}\right)$$

Similarly,

$$\left(\frac{\partial^3 U}{\partial S^3}\right)'' = \left(\frac{1}{N''}\right)^2\left(\frac{\partial^2 T}{\partial s^2}\right)$$

Therefore,

$$(N')^2\left(\frac{\partial^3 U}{\partial S^3}\right)' = (N'')^2\left(\frac{\partial^3 U}{\partial S^3}\right)''$$

The third expression is derived in a similar manner. Higher-order derivatives follow the same trend:

$$(N')^{l-1}\left(\frac{\partial^l U}{\partial S^l}\right)' = (N'')^{l-1}\left(\frac{\partial^l U}{\partial S^l}\right)''$$

4.4: Criteria of Criticality in Terms of Derivatives of *G* Derive the criteria of criticality in terms of the derivatives of G.

Solution We start from Eq. (4.127) and reduce it to

$$\left.\frac{\partial\left(C_3, \ldots, C_{n_c}, \lambda^{n_c+1}\right)}{\partial\left(C_3, \ldots, C_{n_c}, X_{n_c+1}\right)}\right|_{C_1, C_2, X_{n_c+2}} = 0$$

Dividing the numerator and denominator by $\partial(X_3, \ldots, X_{n_c+1})$,

$$\left.\frac{\dfrac{\partial\left(C_3, \ldots, C_{n_c}, \lambda^{(n_c+1)}\right)}{\partial\left(X_3, \ldots, X_{n_c}, X_{n_c+1}\right)}}{\dfrac{\partial\left(C_3, \ldots, C_{n_c}, X_{n_c+1}\right)}{\partial\left(X_3, \ldots, X_{n_c}, X_{n_c+1}\right)}}\right|_{C_1, C_2, X_{n_c+2}} = 0$$

Since C_1, C_2, and the X_i's are held constant, $(\partial C_3/\partial X_3)_{C_1,C_2,X_4,\ldots,X_{nc+2}} = y_{3,3}^{(2)}$, $(\partial C_3/\partial X_{n_c+1})_{C_1,C_2,\ldots,X_{n_c},X_{n_c+2}} = y_{3,n_c+1}^{(2)}$, etc. [see Eq. (1.183)]. $\lambda^{(n_c+1)} = 0$ can be expressed as

$$\lambda^{(n_c+1)} = \begin{vmatrix} y_{3,3}^{(2)} & y_{3,4}^{(2)} & \cdots & y_{3,n_c+1}^{(2)} \\ \vdots & \vdots & \vdots & \vdots \\ y_{n_c,3}^{(2)} & y_{n_c,4}^{(2)} & \cdots & y_{n_c,n_c+1}^{(2)} \\ y_{n_c+1,3}^{(2)} & y_{n_c+1,4}^{(2)} & \cdots & y_{n_c+1,n_c+1}^{(2)} \end{vmatrix} = 0 \tag{4.156}$$

and the second criterion becomes

$$\boxed{\begin{vmatrix} y_{3,3}^{(2)} & y_{3,4}^{(2)} & \cdots & y_{3,n_c+1}^{(2)} \\ \vdots & \vdots & \vdots & \vdots \\ y_{n_c,3}^{(2)} & y_{n_c,4}^{(2)} & \cdots & y_{n_c,n_c+1}^{(2)} \\ \lambda_3^{(n_c+1)} & \lambda_4^{(n_c+1)} & \cdots & \lambda_{n_c+1}^{(n_c+1)} \end{vmatrix} = 0}$$

From $y^{(2)} = G(T, P, \underline{N})$, one obtains $y_{3,3}^{(2)} = (\partial^2 G/\partial N_1)^2_{T,P,N_1}$ and all the other elements of the two determinants above. Therefore, only derivatives of G with respect to $\underline{N}$ are required. These are the set of equations that were used by Peng and Robinson (1977) to predict the critical points of multicomponent systems.

4.5: Alternative Criteria of Stability in Binary Mixtures Show that for a stable binary homogeneous phase, the following inequalities should hold:

(a) $\left(\dfrac{\partial^2 g}{\partial T^2}\right)_{P,x_1} < 0$

(b) $\left(\dfrac{\partial^2 g}{\partial P^2}\right)_{T,x_2} < 0$

Solution For a stable homogeneous phase, whether a pure component, a binary mixture, or a multicomponent system, the heat capacities c_P and c_V and the isothermal compressibility C_T should be positive. The derivations for a single-component system were presented in the text. Example 4.9 provides the derivation for C_T. The derivations for c_V and c_P are in Prob. 4.5.

Let us write the expression for dG of a multicomponent system,

$$dG = -SdT + VdP + \sum_{i=1}^{n_c} \mu_i dn_i$$

At constant composition,

$$[dG = -SdT + VdP]_{\underline{n}}$$

Then

$$\left(\frac{\partial G}{\partial T}\right)_{P,\underline{n}} = -S$$

and

$$\left(\frac{\partial^2 G}{\partial T^2}\right)_{P,\underline{n}} = -\left(\frac{\partial S}{\partial T}\right)_{P,\underline{n}}$$

But $(\partial S/\partial T)_{P,\underline{n}} = nc_P/T$ [see Eq. (3.109)].

Since $c_P > 0$, then

$$\left(\frac{\partial^2 G}{\partial T^2}\right)_{P,\underline{n}} < 0$$

and for a binary system to be stable

$$\left(\frac{\partial^2 g}{\partial T^2}\right)_{P,x_1} < 0.$$

The second relationship, $(\partial^2 g/\partial P^2)_{T,x_1} < 0$, can be obtained from

$$\left(\frac{\partial G}{\partial P}\right)_{T,\underline{n}} = V$$

and

$$\left(\frac{\partial^2 G}{\partial P^2}\right)_{T,\underline{n}} = \left(\frac{\partial V}{\partial P}\right)_{T,\underline{n}}$$

Since $(\partial V/\partial P)_{T,\underline{n}} = -VC_T$ and $C_T > 0$ for a stable system, then

$$\left(\frac{\partial^2 g}{\partial P^2}\right)_{T,x_1} < 0$$

4.6: Stability of Pure Substances and Binary Mixtures The purpose of this example is to have an appreciation of supersaturation in pure hydrocarbons and binary mixtures. Supersaturation can be defined as the difference between the equilibrium pressure at the saturation point and the pressure at which an infinitesimal amount of a new phase is formed. The temperature is held constant in the process.

a. Consider pure hydrocarbons n-C_5 and n-C_7. Calculate the limits of stability in the gas and liquid states. Plot the results on a P-T diagram and compare them with the vapor pressure curve. Skripov and Ermakov (1964) have measured the limit of stability of liquid n-C_5 and n-C_7. Compare the results of calculation with the measured data of these authors.

b. Compute the stability limits of a mixture of 70% C_1 and 30% n-C_5, a mixture of 70% C_1 and 30% n-C_7, and a mixture of 70% C_1 and 30% n-C_{10} both in the gas and liquid states. Plot the results on a P-T diagram. Show also the saturation pressures of the above systems on the P-T plots.

Solution

a. For a pure substance, the stability limit is given by the first expression of Eq. (4.113), $(\partial P/\partial v)_T = 0$. The PR-EOS can be used to evaluate $(\partial P/\partial v)_T = 0$. The results are presented in Fig. 4.14 for both the vapor and liquid phases. In the same figure, the experimental data of Skripov and Ermakov (1964) for the liquid-phase stability limit are also presented. Note that there is qualitative agreement between the experimental data and calculated results. The difference between the pressure at the limit of stability and the vapor pressure at a given temperature is the maximum supersaturation in pressure. Note that close to the critical point, supersaturation is small.

Figure 4.15 provides the calculated limit of stability for the gas and liquid phases, the vapor pressure, and the measured limit of stability for the liquid phase (Skripov and Ermakov, 1964) for n-C_7. The PR-EOS was used in the stability limit calculations. The results presented in Fig. 4.15 for n-C_7 are very similar to the results in Fig. 4.14 for n-C_5. Similar to the n-C_5, the supersaturation in pressure for n-C_7 (at a given temperature) is higher for the liquid phase than the gas phase, as expected (see Prob. 2.22). Note that because of supersaturation, the pressure in the liquid phase can be zero or even negative.

b. The limit of stability for a binary mixture is calculated from $(\partial f_1/\partial x_1)_{T,P} = 0$ [the first expression in Eq. (4.120)]. The results of the calculation for the C_1/n-C_5 mixture are presented in Fig. 4.16a. The same figure also shows the saturation pressure. Note that on the bubblepoint side over a large range of pressure in the neighborhood of the critical point, the supersaturation in pressure is negligible.

Figures. 4.16b and 4.16c show both the stability limits and saturation pressures for the C_1/n-C_7 and C_1/n-C_{10} mixtures. The results are very similar to those presented in

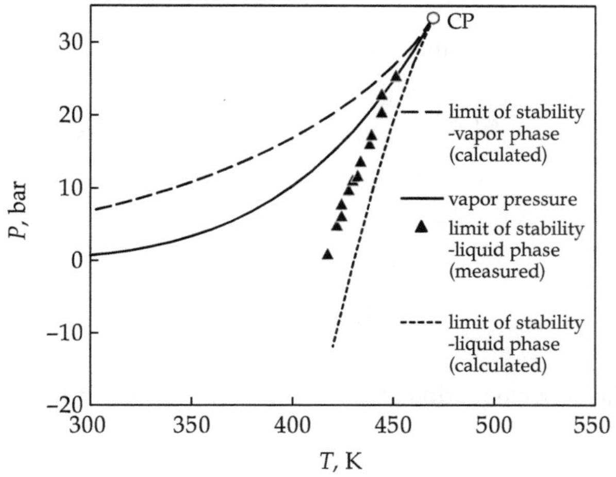

FIGURE 4.14 Limit of stability for n-C_5 in the gas and liquid states.

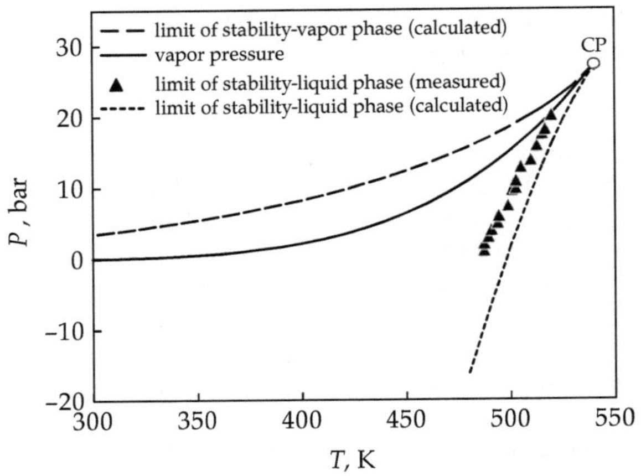

FIGURE 4.15 Limit of stability for n-C_7 in the gas and liquid states.

Fig. 4.16a for the C_1/n-C_5 mixture. The maximum supersaturation in pressure for the liquid state becomes more pronounced for the C_1/n-C_{10} mixture than for the C_1/n-C_7 and C_1/n-C_5 mixtures.

4.7: Various Derivatives of the Helmholtz Free Energy for the PR-EOS Derive the expressions for the following derivatives using the PR-EOS: $(\partial A/\partial V)$, $(\partial^2 A/\partial V^2)$, $(\partial^3 A/\partial V^3)$, $(\partial A/\partial N_i)$, $(\partial^2 A/\partial V \partial N_i)$, $(\partial^2 A/\partial N_i \partial N_j)$, $(\partial^3 A/\partial V \partial N_i \partial N_j)$, and $(\partial^3 A/\partial N_i \partial N_j \partial N_k)$.

Solution Software such as Mathematica have made life easy for deriving the expressions for derivatives of the equations of state. Many years ago, such software were not available and apparently Baker and Luks (1980) spent considerable effort to perform the above derivatives for the SRK-EOS, although the derivations for this EOS are simpler than those of the PR-EOS. In the following, the derivatives from Mathematica are presented.

Solution The PR-EOS can be written as [see Eq. (3.6)]

$$P = \frac{NRT}{V - \mathcal{B}} - \frac{\mathcal{A}}{V(V + \mathcal{B}) + \mathcal{B}(V - \mathcal{B})}$$

where

$$\mathcal{A} = N^2 a = \sum_{i=1}^{n_c} \sum_{j=1}^{n_c} N_i N_j a_{ij}$$

$$\mathcal{B} = Nb = \sum_{i=1}^{n_c} N_i b_i$$

Introducing the above equation into the expression given in Example 1.4 yields (using Mathematica)

$$A = -NRT \ln \frac{V - \mathcal{B}}{RT} - \frac{\mathcal{A}}{2\sqrt{2}\mathcal{B}} \ln \frac{V + (\sqrt{2} + 1)\mathcal{B}}{V - (\sqrt{2} - 1)\mathcal{B}} + RT \sum_{i=1}^{n_c} N_i \ln N_i + \sum_{i=1}^{n_c} N_i \left(u_i^0 - Ts_i^0\right)$$

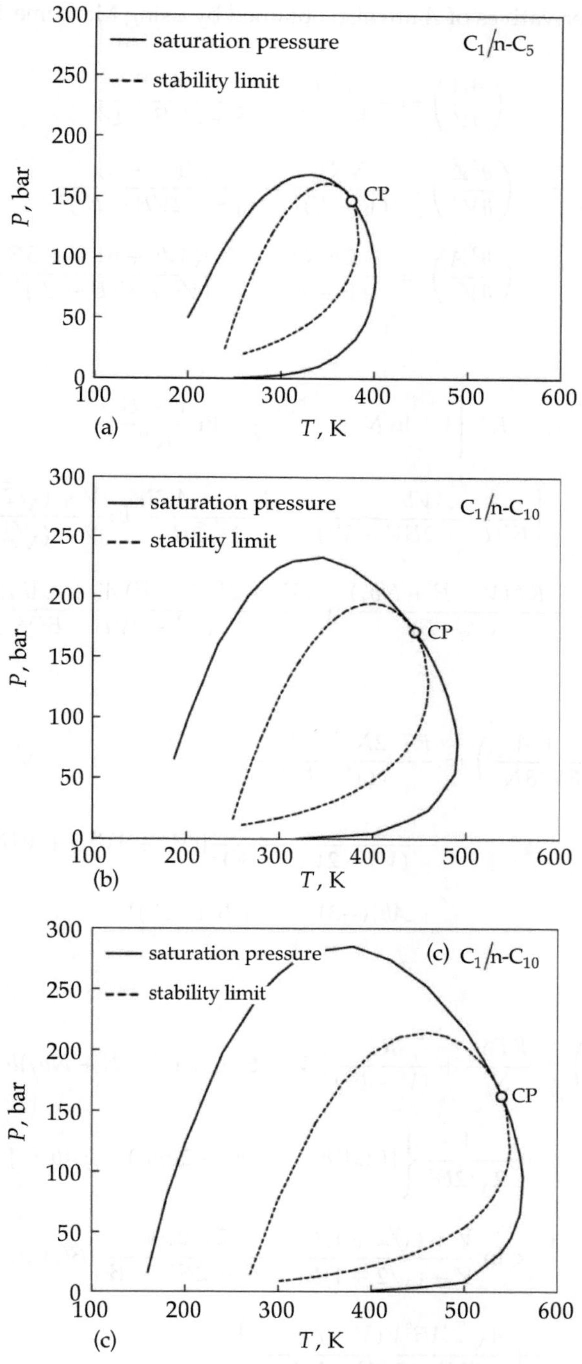

FIGURE 4.16 Calculated saturation pressure and stability limits for three binary hydrocarbon mixtures ($x_{c_1} = 70\%$).

The various derivatives of A are also obtained by using Mathematica:

$$\left(\frac{\partial A}{\partial V}\right) = -\frac{NRT}{V-\mathcal{B}} + \frac{\mathcal{A}}{V^2+2V\mathcal{B}-\mathcal{B}^2}$$

$$\left(\frac{\partial^2 A}{\partial V^2}\right) = \frac{NRT}{(V-\mathcal{B})^2} - \frac{2\mathcal{A}(V+\mathcal{B})}{(V^2+2V\mathcal{B}-\mathcal{B}^2)^2}$$

$$\left(\frac{\partial^3 A}{\partial V^3}\right) = -\frac{2NRT}{(V-\mathcal{B}^3)} + \frac{2\mathcal{A}(3V^2+6V\mathcal{B}+5\mathcal{B}^2)}{(V^2+2V\mathcal{B}-\mathcal{B}^2)^3}$$

$$\left(\frac{\partial A}{\partial N_i}\right) = \mu_i = RT\left[1+\ln N_i + \frac{Nb_i}{V-\mathcal{B}} - \ln\frac{V-\mathcal{B}}{RT}\right] + \left[\frac{\mathcal{A}Vb_i}{\mathcal{B}^2(\mathcal{B}^2-2\mathcal{B}V-V^2)} + \frac{(b_i\mathcal{A}-\mathcal{A}'_i\mathcal{B})}{2\sqrt{2}\mathcal{B}}\ln\frac{V+(\sqrt{2}+1)\mathcal{B}}{V-(\sqrt{2}-1)\mathcal{B}}\right] + (u_i^0 - Ts_i^0)$$

$$\left(\frac{\partial^2 A}{\partial V \partial N_i}\right) = -\frac{RT(V-\mathcal{B}+Nb_i)}{(V-\mathcal{B})^2} + \frac{(V^2+2V\mathcal{B}-\mathcal{B}^2)\mathcal{A}'_i + 2\mathcal{A}b_i(\mathcal{B}-V)}{(V^2+2V\mathcal{B}-\mathcal{B}^2)^2}$$

$$\left(\frac{\partial^3 A}{\partial V^2 \partial N_i}\right) = \frac{RT(2Nb_i+V-\mathcal{B})}{(V-\mathcal{B})^3} - \frac{2}{(V^2+2V\mathcal{B}-\mathcal{B}^2)^3}[(V^3+3V^2\mathcal{B}+V\mathcal{B}^2-\mathcal{B}^3)\mathcal{A}'_i + \mathcal{A}b_i(-3V^2+2V\mathcal{B}+3\mathcal{B}^2)]$$

$$\left(\frac{\partial^2 A}{\partial N_i \partial N_j}\right) = \frac{RT\delta_{ij}}{N_j} + \frac{RT}{(V-\mathcal{B}^2)}[(V-\mathcal{B})b_i + (V-\mathcal{B}+Nb_i)b_j] + \frac{1}{2\sqrt{2}\mathcal{B}^3}\left\{[\mathcal{B}(\mathcal{A}'_i b_j + \mathcal{A}'_j b_j - 2\mathcal{B}a_{ij}) - 2\mathcal{A}b_i b_j] \times \ln\frac{V+(\sqrt{2}+1)\mathcal{B}}{V-(\sqrt{2}-1)\mathcal{B}} - \frac{2\sqrt{2}\mathcal{B}V}{V^2+2\mathcal{B}V-\mathcal{B}^2}[\mathcal{B}(\mathcal{A}'_i b_j + \mathcal{A}'_j b_i) - 2\mathcal{A}b_i b_j] + \frac{4\sqrt{2}\mathcal{A}\mathcal{B}^2 V(V-b)b_i b_j}{(V^2+2\mathcal{B}V-\mathcal{B}^2)^2}\right\}$$

$$\left(\frac{\partial^3 A}{\partial V \partial N_i \partial N_j}\right) = \frac{-RT}{(V-\mathcal{B})^3}[(V-b)(b_i+b_j)+2Nb_ib_j]$$
$$+\frac{2}{(V^2+2\mathcal{B}V-\mathcal{B}^2)^3}\{(V^2+2\mathcal{B}V-\mathcal{B}^2)\left[a_{ij}(V^2+2\mathcal{B}V-\mathcal{B}^2)\right.$$
$$\left.+\mathcal{A}'_ib_j(\mathcal{B}-V)\right]-b_i\left[\mathcal{A}'_j(V^3+\mathcal{B}V^2-3\mathcal{B}^2V+\mathcal{B}^3)\right.$$
$$\left.+\mathcal{A}b_j(-5V^2+6\mathcal{B}V-3\mathcal{B}^2)\right]\}$$

$$\left(\frac{\partial^3 A}{\partial N_i \partial N_j \partial N_k}\right) = -\frac{RT\delta_{ij}\delta_{jk}}{N_k^2}$$
$$+\frac{RT}{(V-\mathcal{B})^3}[(V-\mathcal{B})(b_ib_j+b_ib_k+b_jb_k)+2Nb_ib_jb_k]+\frac{1}{2\sqrt{2}\mathcal{B}^4}$$
$$\times\left\{\left[\mathcal{B}^2(b_ia_{jk}+b_ja_{ik}+b_ka_{ij})-\mathcal{B}(\mathcal{A}'_ib_jb_k+\mathcal{A}'_jb_ib_k+\mathcal{A}'_kb_ib_j)\right.\right.$$
$$\left.+3\mathcal{A}b_ib_jb_k\right]\times\left[2\ln\frac{V+(\sqrt{2}+1)\mathcal{B}}{V-(\sqrt{2}-1)\mathcal{B}}-\frac{4\sqrt{2}\mathcal{B}V}{V^2+2\mathcal{B}V-\mathcal{B}^2}\right]$$
$$-\frac{4\sqrt{2}\mathcal{B}^2V}{(V^2+2\mathcal{B}V-\mathcal{B}^2)^2}\,[\mathcal{B}(\mathcal{B}-V)\times(\mathcal{A}'_ib_jb_k+\mathcal{A}'_jb_ib_k+\mathcal{A}'_kb_ib_j)$$
$$\left.+b_ib_jb_k\mathcal{A}(3V-2\mathcal{B})]-\frac{16\sqrt{2}\mathcal{A}\mathcal{B}^3(V-\mathcal{B})^2Vb_ib_jb_k}{(V^2+2\mathcal{B}V-\mathcal{B}^2)^3}\right\},$$

where $\mathcal{A}'_i=\partial\mathcal{A}/\partial N_i$, $\mathcal{A}'_j=\partial\mathcal{A}/\partial N_j$, $\mathcal{A}'_k=\partial\mathcal{A}/\partial N_k$, $\delta_{ij}=1$ for $i=j$, and $\delta_{ij}=0$ for $i\neq j$.

4.8: Heat Capacity at the Critical Point First show that at the critical point $(\partial T/\partial s)_P=0$ for a pure substance. Then estimate the isothermal compressibility and isobaric heat capacity at the limit of stability and at the critical point in pure components. Do you expect a different behavior for the isobaric heat capacity in mixtures around the critical point (with 2 species and more)? You may examine the heat capacity for D_2O in Fig 4.17 as we approach the critical point.

Solution Let

$$y^{(0)} = U(V,\ S,\ n) = -pV+TS+\mu n$$
$$dy^{(0)} = dU = -pdV+TdS+\mu dn$$
$$dy^{(1)} = dy^{(0)}-d(-PV) = TdS+VdP+\mu dn = dH$$

At the critical point, $y^{(1)}_{2,2}=0$

$$\left(\frac{\partial T}{\partial S}\right)_{P,n}=0$$

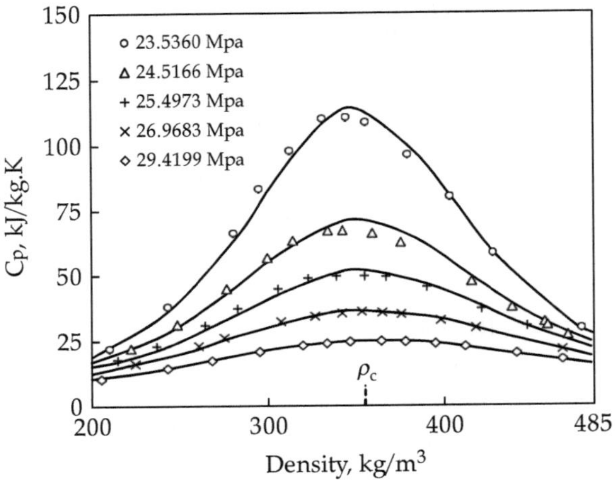

FIGURE 4.17 The specific heat at constant pressure of D_2O in the critical region at various pressures as a function of density. The data are those of Rivkin and Egorov (1962, 1963) and the curves represent the values calculated from the universal scaled fundamental equation. The critical pressure for D_2O is around 21.66 MPa (adapted from Kamgar-Parsi et al., 1983).

and since $S = ns$, then

$$\left(\frac{\partial T}{\partial s}\right)_P = 0$$

From the definitions of isobaric heat capacity and isothermal compressibility:

$$c_P = T\left(\frac{\partial s}{\partial T}\right)_P \quad \text{and} \quad C_T = -\frac{1}{v}\left(\frac{\partial v}{\partial P}\right)_T$$

At the critical point, $(\partial s/\partial T)_P = 0$ and $(\partial v/\partial P)_T = 0$

Therefore, $c_P \to \infty$ and $C_T \to \infty$.

Figure 4.17 shows data of c_P vs. density of D_2O in the critical region. There is a significant increase in c_P as the critical pressure is approached.

For multicomponent mixtures $(\partial T/\partial s)_p = 0$ and $(\partial P/\partial v)_T = 0$ are no longer valid for criticality. Therefore, the conclusion about isobaric heat capacity and isothermal compressibility above may not be valid.

4.9: Isothermal Compressibility in Multicomponent Mixtures Show that for a stable multicomponent homogeneous fluid similar to a single-component fluid, the isothermal compressibility should be positive.

Solution For a stable system the coefficient $y_{jj}^{(j-1)}$ of $(\Delta Z_j)^2$ in Eq. (4.82) should be positive, and since $y_{n_c+2,n_c+2}^{(n_c+1)} = 0$ [see Eq. (1.186)], $y_{jj}^{(j-1)} > 0, \quad j = 1, \ldots, n_c + 1$.

If $y^{(0)} = U(S, V, n_1, \ldots, n_c)$, then $y^{(1)} = A(T, V, n_1, \ldots, n_c)$, and $dy^{(1)} = -SdT - PdV + \sum_{i=1}^{n_c} \mu_i dn_i$ which gives $(-\partial P/\partial V)_{T,\underline{n}} > 0$ or $-(1/V)(\partial V/\partial P)_{T,\underline{n}} > 0$.

Therefore, the isothermal compressibility of a multicomponent stable system should be positive.

4.10: Miscibility in Polymer Solution Mixtures from Stability and Criticality Criteria In textbooks on polymer chemistry, the following expression provides the conditions for complete miscibility of a solvent and a polymer:

$$\chi \leq \frac{(1+\sqrt{r})^2}{2r}, \tag{4.157}$$

where $r \equiv$ number of segments in a polymer molecule, and $\chi \equiv$ Flory interaction parameter. $kT\chi$ represents the difference in energy of a solvent molecule immersed in the pure polymer and the energy of a solvent molecule in pure solvent. Polymer molecules are assumed to behave like flexible chains, each segment having the same size as the solvent molecules; the segments are mobile. To derive the above expression, we first require the expression for the chemical potentials of the polymer and solvent species in the mixture of solvent and polymer. The following steps are taken in our derivations:

When solvent and polymer molecules are mixed at constant temperature and pressure, the expression for Gibbs free energy of mixing is obtained from

$$\Delta G_{mix} = \Delta H_{mix} + T\Delta S_{mix} \tag{4.158}$$

ΔG_{mix} represents the difference between unmixed and mixed Gibbs free energy. The expression for ΔH_{mix}, the enthalpy of mixing is derived readily. The entropy of mixing can be derived according to liquid lattice theory. Chap. 12 of Flory (1953) presents the derivations for both ΔH_{mix} and ΔS_{mix}. Based on enthalpy and entropy of mixing, one can obtain the expression for Gibbs free energy of mixing of solvent and polymer molecules (Prausnitz, Lichtenthaler, & de Azevedo, 1998),

$$\Delta G_{mix} = RT[N_1 \ln \Phi_1 + N_2 \ln \Phi_2 + \chi\Phi_1\Phi_2(N_1 + rN_2)] \tag{4.159}$$

N_1 is the number of solvent molecules, and N_2 is the number of polymer molecules. In the above expression, Φ_1 and Φ_2 are volume fractions of solvent (component 1) and polymer molecule (component 2) given by

$$\Phi_1 = \frac{N_1}{N_1 + rN_2} \tag{4.160}$$

$$\Phi_2 = \frac{rN_2}{N_1 + rN_2} \tag{4.161}$$

Let us now obtain the expressions for chemical potentials.

We can write the expressions for ΔG_{mix} using the expressions for Φ_1 and Φ_2 in Eq. (4.159) above.

$$\Delta G_{mix} = RT\left(N_1 \ln\frac{N_1}{N_1 + rN_2} + N_2 \ln\frac{rN_2}{N_1 + rN_2} + \frac{\chi N_1 rN_2}{N_1 + rN_2}\right) \tag{4.162}$$

The definition for $\Delta\mu_1 = (\partial\Delta G_{mix}/\partial N_1)_{T,P,N_2}$ and $\Delta\mu_2 = (\partial\Delta G_{mix}/\partial N_2)_{T,P,N_1}$ can be used for derivation of the expressions for $\Delta\mu_1 = \mu_1 - \mu_1^0$ and $\Delta\mu_2 = \mu_2 - \mu_2^0$, where the superscript "0" represents the pure species. Using the $\Delta\mu_1$ defined above and

Eq. (4.162), one can obtain

$$\Delta\mu_1 = RT\left[\ln\frac{N_1}{N_1 + rN_2} + 1 - \frac{N_2}{N_1 + rN_2} - \frac{\chi N_1}{r(N_1 + rN_2)} + \chi\left(\frac{rN_1N_2 + r^2N_2^2 - rN_1N_2}{(N_1 + rN_2)^2}\right)\right] \tag{4.163}$$

Combining Eqs. (4.160), (4.161), and (4.163) and the fact that $\Phi_1 + \Phi_2 = 1$, we obtain

$$\Delta\mu_1 = RT\left[\ln\Phi_1 + \left(1 - \frac{1}{r}\right)\Phi_2 + \chi\Phi_2^2\right] \tag{4.164}$$

or

$$\boxed{\Delta\mu_1 = \mu_1 - \mu_1^0 = RT\left[\ln(1 - \Phi_2) + \left(1 - \frac{1}{r}\right)\Phi_2 + \chi\Phi_2^2\right]} \tag{4.165}$$

Similarly,

$$\boxed{\Delta\mu_2 = \mu_2 - \mu_2^0 = RT\left[\ln\Phi_2 + (1 - r)\Phi_1 + \chi r\Phi_1^2\right]} \tag{4.166}$$

The rest of the problem is divided into two parts.

(a) Show that at constant T and P, the limit of stability is given by

$$\left(\frac{\partial\Delta\mu_2}{\partial\Phi_2}\right)_{T,P} = 0 \tag{4.167}$$

Based on the above expression, derive

$$\chi = \frac{(1/\Phi_2) + (r - 1)}{2r(1 - \Phi_2)} \tag{4.168}$$

Solution For a two-component mixture the stability limit is given by $y_{3,3}^{(2)} = 0$. From $y^{(0)} = U$,

$$dy^{(2)} = dU - d(TS) - d(-PV) = -S\,dT + V\,dP + \mu_2\,dr N_2 + \mu_1 N_1 \tag{4.169}$$

$$y_{3,3}^{(2)} = \left(\frac{\partial\mu_2}{\partial rN_2}\right)_{T,P,N_1} = \left(\frac{\partial\Delta\mu_2}{\partial rN_2}\right)_{T,P,N_1} \tag{4.170}$$

At constant T and P, from Eq. (4.166)

$$\Delta\mu_2 = \Delta\mu_2(\Phi_2) \tag{4.171}$$

Another dependency of $\Delta\mu_2$ is expressed by

$$\Delta\mu_2 = \Delta\mu_2(N_1, rN_2) \tag{4.172}$$

We can write

$$\left(\frac{\partial\Delta\mu_2}{\partial rN_2}\right)_{T,P,N_1} = \left(\frac{\partial\Delta\mu_2}{\partial\Phi_2}\right)_{T,P}\left(\frac{\partial\Phi_2}{\partial rN_2}\right)_{T,P,N_1} = 0 \tag{4.173}$$

From Eq. (4.161),

$$\left(\frac{\partial\Phi_2}{\partial rN_2}\right) = \frac{1}{N_1 + rN_2} - \frac{rN_2}{(N_1 + rN_2)^2} = \frac{N_1 + rN_2 - rN_2}{(N_1 + rN_2)^2} = \frac{N_1}{(N_1 + rN_2)^2} \neq 0 \tag{4.174}$$

From Eqs. (4.173) and (4.174), it follows that

$$\boxed{\left(\frac{\partial\Delta\mu_2}{\partial\Phi_2}\right)_{T,P} = 0} \tag{4.175}$$

Similarly, we can obtain

$$\left(\frac{\partial\Delta\mu_1}{\partial\Phi_2}\right)_{T,P} = 0 \tag{4.176}$$

Next, we derive Eq. (4.168). Based on Eq. (4.165),

$$\left(\frac{\partial\Delta\mu_1}{\partial\Phi_2}\right)_{T,P} = RT\frac{\partial}{\partial\Phi_2}\left[\ln(1-\Phi_2) + \left(1 - \frac{1}{r}\right)\Phi_2 + \chi\Phi_2^2\right]$$

which gives

$$-\frac{1}{1-\Phi_2} + \left(1 - \frac{1}{r}\right) + 2\chi\Phi_2 = 0 \tag{4.177}$$

Therefore,

$$\boxed{\chi = \frac{{}^{1}\!/_{\Phi_2} + (r-1)}{2r(1-\Phi_2)}} \tag{4.178}$$

(b) The condition for complete miscibility is based on the inflection point in the $\Delta\mu_1$ vs. Φ_2 plot from criticality conditions: $y_{3,3}^{(2)} = 0$, and $y_{3,3,3}^{(2)} = 0$.

From the second criterion

$$\left(\frac{\partial^2\Delta\mu_1}{\partial\Phi_2^2}\right)_{T,P} = 0 \tag{4.179}$$

using the expression for $\Delta\mu_1$, Eq. (4.179) gives

$$-\frac{1}{(1-\Phi_2)^2} + 2\chi = 0 \tag{4.180}$$

Combining Eqs. (4.178) and (4.180)

$$\boxed{\chi_{critical} = \frac{(1+\sqrt{r})^2}{2r}} \tag{4.181}$$

Equation (4.181) can be approximated by

$$\chi_{critical} \approx \frac{1}{2} + \frac{1}{\sqrt{r}}$$

The above expression implies that for a large chain length, $\chi_{critical} = 0.50$. Now we have to establish that for complete miscibility $\chi \leq (1+\sqrt{r})^2/2r$. We have plotted in Fig. 4.18,

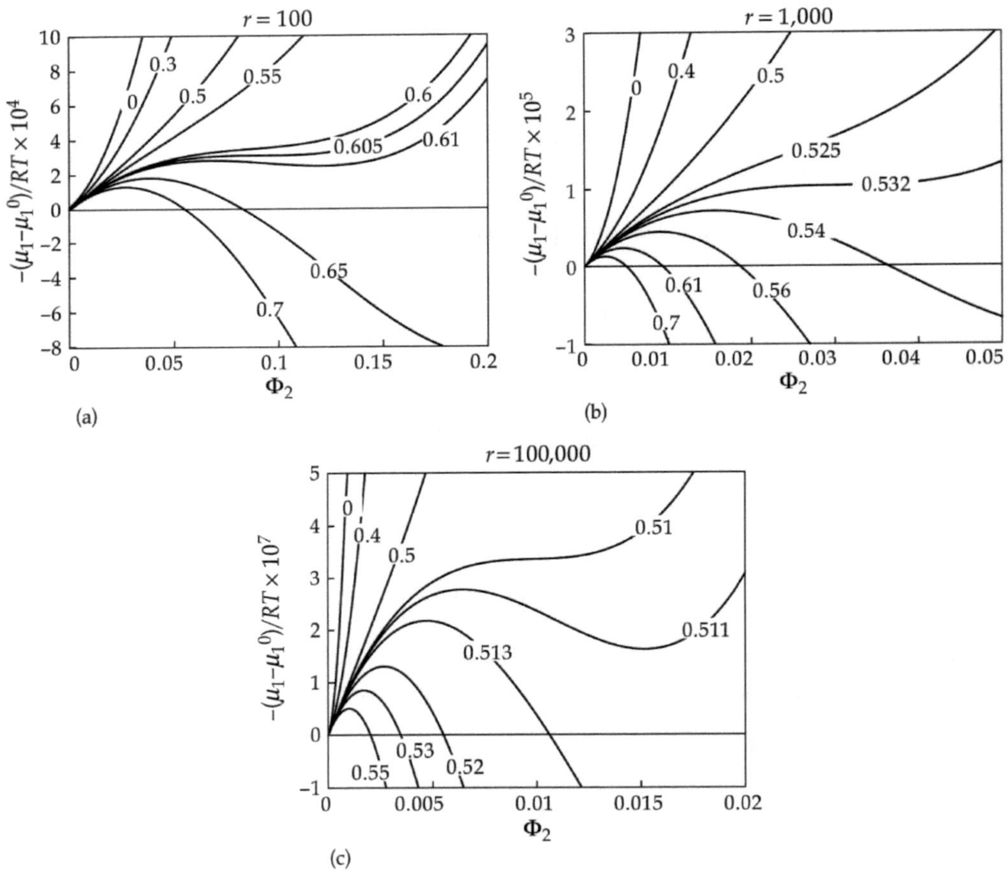

FIGURE 4.18 $\Delta\mu_1/RT$ vs. Φ_2 for (a) $r = 100$, (b) $r = 1000$ and (c) $r = 100,000$. In the above plots the value of χ is fixed; it is varied from 0 to 0.7.

$\Delta\mu_1/RT$ from Eq. (4.165) vs. Φ_2 for various values of χ with $r = 100$, 1000, and 10,000. Eq. (4.165) was used in our computations. Note that for a mixture to be stable, $(\partial\mu_2/\partial\Phi_2)_{T,P} \geq 0$. Figure 4.18 is in line with stability and limit of complete miscibility given by Eq. (4.181).

4.11: Solubility of Methane in Water at the Limit of Stability Molecular dynamics simulations of hydrates can be facilitated from thermodynamic stability analysis (Jimenez and Firoozabadi, 2014). By supersaturating water with methane at the limit of stability, the speed of molecular dynamics simulations can be increased by two to three orders of magnitude. Interestingly, in molecular dynamics simulations, slightly above the limit of stability a gas phase forms in a nanosecond. Compute the concentration of methane at the limit of stability at pressure of $P = 500$ bar.

Solution The stability of a supersaturated mixture of methane in water is obtained by solving numerically the following equations:

$$\frac{\partial \ln(\varphi_1 x_1)}{\partial x_1} = 0$$

where φ_1 is the fugacity coefficient of methane. The derivative in the above expression is at constant T and P. We use Eq. 3.59 to compute the derivative.

For a given T and P, we calculate composition at the limit of stability. Figure 4.19 provides both the equilibrium composition and composition at the limit of stability.

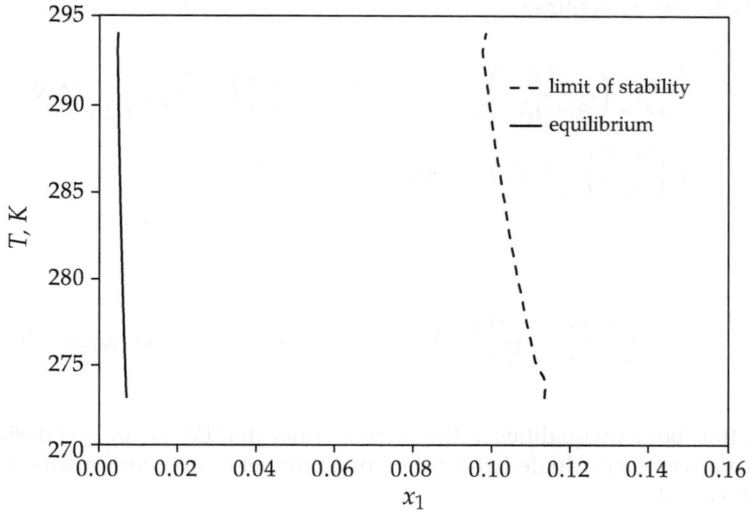

FIGURE 4.19 Concentration of methane (in mol fraction) at equilibrium and at the limit of stability in water at 500 bar.

4.5 Problems

4.1 Show that at the critical point of a pure substance,

$$\left(\frac{\partial^3 \mu}{\partial N^3}\right)_{T,V} \geq 0$$

Note: The above expression is equivalent to the criterion expressed by $(\partial^3 P/\partial V^3)_{T,N} \leq 0$.

4.2 Gibbs presented one of the alternative sets of the criteria for criticality in the following form:

$$\left(\frac{\partial \mu_{n_c}}{\partial N_{n_c}}\right)_{T,V,\mu_1,\ldots,\mu_{nc-1}} = 0$$

$$\left(\frac{\partial^2 \mu_{n_c}}{\partial N_{n_c}^2}\right)_{T,V,\mu_1,\ldots,\mu_{nc-1}} = 0$$

$$\left(\frac{\partial^3 \mu_{n_c}}{\partial N_{n_c}^3}\right)_{T,V,\mu_1,\ldots,\mu_{nc-1}} \geq 0$$

Derive the above set of equations.

4.3 The stability test for a homogeneous phase can be expressed in terms of A and G for the following inequalities to hold for every possible variation around the initial state of the homogeneous phase:

$$\sum_{i=1}^{n_c}\sum_{j=1}^{n_c}\left(\frac{\partial^2 A}{\partial N_i \partial N_j}\right)_{T,V}(\Delta N_i \Delta N_j) + 2\sum_{i=1}^{n_c}\left(\frac{\partial^2 A}{\partial N_i \partial V}\right)_{T,V}(\Delta N_i \Delta V) + \left(\frac{\partial^2 A}{\partial V^2}\right)_{T,V}(\Delta V)^2 + \textit{higher-order terms} > 0$$

and

$$\sum_{i=1}^{n_c}\sum_{j=1}^{n_c}\left(\frac{\partial^2 G}{\partial N_i \partial N_j}\right)_{T,P}\Delta N_i \Delta N_j + \textit{higher-order terms} > 0$$

Derive the above inequalities in the same manner that Eq. (4.76) was derived. Note that the derivatives are evaluated at constant mol number of the components other than those that are varied.

4.4 Show that Maxwell's equal-area rule for multicomponent systems takes the following form: a plot of $-\mu_{n_c-1}$ vs. N_{n_c-1} at constant T, P, $\mu_1, \ldots, \mu_{n_c-2}$, N_{n_c} in the two-phase region gives two equal areas shown by the dotted region of the Fig. 4.20. This sketch is idealized.

FIGURE 4.20

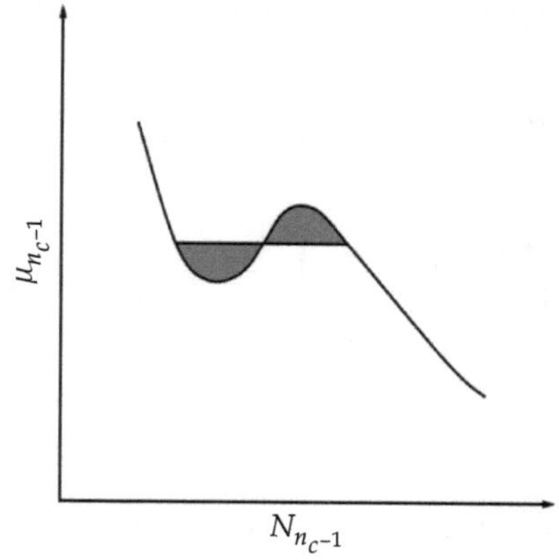

4.5 Show that for a stable multicomponent homogeneous mixture, similar to a single-component fluid, c_P and c_V are positive. Can these parameters be positive at the limit of stability? If the answer is "yes ", why?

4.6 The evaluation of the cubic in Eq. (4.143) can be performed using terms involving at most double sums (Heidemann and Khalil, 1980; Heidemann, 1994). The triple sums in Eq. (4.143) can be reduced as follows:

$$\sum_{i=1}^{n_c}\sum_{j=1}^{n_c}\sum_{k=1}^{n_c}\Delta N_i \Delta N_j \Delta N_k = \left(\sum_{i=1}^{n_c}\Delta N_i\right)^3$$

$$\sum_{i=1}^{n_c}\sum_{j=1}^{n_c}\sum_{k=1}^{n_c}\beta_i \Delta N_i \Delta N_j \Delta N_k = \left(\sum_{i=1}^{n_c}\beta_i \Delta N_i\right)\left(\sum_{i=1}^{n_c}\Delta N_i\right)^2$$

$$\sum_{i=1}^{n_c}\sum_{j=1}^{n_c}\sum_{k=1}^{n_c}\gamma_{ij} \Delta N_i \Delta N_j \Delta N_k = \left(\sum_{i=1}^{n_c}\sum_{j=1}^{n_c}\gamma_{ij}\Delta N_i \Delta N_j\right)\left(\sum_{i=1}^{n_c}\Delta N_i\right)$$

Derive the above relationships and relate β_i and γ_{ij} to the coefficients in Eq. (4.143).

4.7 Show that n_c different matrices of the form given by Eq. (4.142) can be composed. Why do permutations of the component index not produce a new matrix?

4.8 Use Gibbs free energy representation of the criticality criteria from Eqs. (4.142) and (4.143) to show that for a binary mixture $(\partial\mu_1/\partial N_1)_{T,P,N_2} = 0$ and $(\partial\mu_1^2/\partial N_1^2)_{T,P,N_2} = 0$. Note that these two expressions are the same expressions as those of Eq. (4.119).

4.9 Show that at the critical point of a n_c-component mixture, the following determinant is zero

$$\begin{vmatrix} \left(\dfrac{\partial \mu_1}{\partial y_1}\right) & \left(\dfrac{\partial \mu_1}{\partial y_2}\right) & \cdots & \left(\dfrac{\partial \mu_1}{\partial y_{n_c-1}}\right) \\ \left(\dfrac{\partial \mu_2}{\partial y_1}\right) & \left(\dfrac{\partial \mu_2}{\partial y_2}\right) & \cdots & \left(\dfrac{\partial \mu_2}{\partial y_{n_c-1}}\right) \\ \vdots & \vdots & \cdots & \vdots \\ \left(\dfrac{\partial \mu_{n_c-1}}{\partial y_1}\right) & \left(\dfrac{\partial \mu_{n_c-1}}{\partial y_2}\right) & \cdots & \left(\dfrac{\partial \mu_{n_c-1}}{\partial y_{n_c-1}}\right) \end{vmatrix} = 0$$

Note that the above is the determinant of the matrix in Eq. (2.19b).
Hint: You may use the results from Example 4.4 in your derivation.

4.10 Use the second criterion of criticality expressed by Eq. (4.132) for a three-component system with $y^{(1)} = A(T, V, N_1, N_2, N_3)$, with index 2 for V, index 3 for N_1, and index 4 for N_2. Show that the derivative with respect to V for the determinant given by Eq. (4.131) is

$$\begin{aligned} \lambda_2^{(4)} = {} & A_{2,2,2}\left(A_{3,3}A_{4,4} - A_{4,3}^2\right) + A_{3,3,2}\left(A_{2,2}A_{4,4} - A_{4,2}^2\right) + A_{4,4,2}\left(A_{2,2}A_{3,3} - A_{3,2}^2\right) \\ & + 2A_{2,2,3}(A_{4,2}A_{4,3} - A_{3,2}A_{4,4}) + 2A_{4,2,2}(A_{3,2}A_{4,3} - A_{3,3}A_{4,2}) \\ & + 2A_{4,3,2}(A_{3,2}A_{4,2} - A_{2,2}A_{4,3}) \end{aligned}$$

and the derivative of the same determinant with respect to N_1 is given by

$$\begin{aligned} \lambda_3^{(4)} = {} & A_{2,2,3}\left(A_{3,3}A_{4,4} - A_{4,3}^2\right) + A_{3,3,3}\left(A_{2,2}A_{4,4} - A_{4,2}^2\right) + A_{4,4,3}\left(A_{2,2}A_{3,3} - A_{3,2}^2\right) \\ & + 2A_{3,2,3}(A_{4,2}A_{4,3} - A_{3,2}A_{4,4}) + 2A_{4,2,3}(A_{3,2}A_{4,3} - A_{3,3}A_{4,2}) \\ & + 2A_{4,3,3}(A_{3,2}A_{4,2} - A_{2,2}A_{4,3}) \end{aligned}$$

Then derive the final form of the expression from Eq. (4.132) given below:

$$\begin{aligned} & A_{2,2,2}\ \left(A_{3,3}A_{4,4}A_{2,3}A_{3,4} - A_{3,3}A_{4,4}A_{2,4}A_{3,3} - A_{2,3}A_{3,4}^3 + A_{2,4}A_{3,3}A_{3,4}^2\right) \\ & \quad + A_{2,3,3}\left(3A_{2,2}A_{2,3}A_{3,4}A_{4,4} - 2A_{2,2}A_{3,4}^2A_{2,4} - A_{2,2}A_{2,4}A_{3,3}A_{4,4} + A_{2,3}A_{3,4}A_{2,4}^2\right. \\ & \quad \left. - 2A_{2,3}^2A_{2,4}A_{4,4} + A_{2,4}^3A_{3,3}\right) \\ & \quad + A_{2,2,3}\left(- A_{2,2}A_{3,3}A_{3,4}A_{4,4} + A_{2,2}A_{3,4}^3 + A_{2,3}A_{2,4}A_{3,3}A_{4,4}\right. \\ & \quad \left. - 2A_{2,4}^2A_{3,3}A_{3,4} + A_{2,3}A_{3,4}^2A_{2,4}\right) \\ & \quad + A_{3,3,3}\left(- A_{2,2}^2A_{3,4}A_{4,4} + A_{2,2}A_{2,3}A_{2,4}A_{4,4} + A_{2,2}A_{3,4}A_{2,4}^2 - A_{2,3}A_{2,4}^3\right) \\ & \quad + A_{2,4,4}\left(3A_{2,2}A_{2,3}A_{3,3}A_{3,4} - 3A_{2,2}A_{2,4}A_{3,3}^2 - 3A_{2,3}^3A_{3,4} + 3A_{2,3}^2A_{2,4}A_{3,3}\right) \\ & \quad + A_{2,2,4}\left(A_{2,2}A_{3,3}^2A_{4,4} - A_{2,2}A_{3,3}A_{3,4}^2 + 3A_{2,3}^2A_{3,4}^2 - A_{2,3}^2A_{3,3}A_{4,4}\right. \\ & \quad \left. - 4A_{2,3}A_{3,3}A_{2,4}A_{3,4} + 2A_{2,4}^2A_{3,3}^2\right) \end{aligned}$$

$$+A_{2,3,4}\big(-4A_{2,2}A_{2,3}A_{3,4}^2+6A_{2,2}A_{2,4}A_{3,3}A_{3,4}-2A_{2,2}A_{2,3}A_{3,3}A_{4,4}+2A_{2,3}^3A_{4,4}$$
$$+2A_{2,3}^2A_{2,4}A_{3,4}-4A_{2,3}A_{2,4}^2A_{3,3}\big)$$
$$+A_{3,4,4}\big(-3A_{2,2}^2A_{3,3}A_{3,4}+3A_{2,2}A_{2,3}^2A_{3,4}+3A_{2,2}A_{2,3}A_{2,4}A_{3,3}-3A_{2,3}^3A_{2,4}\big)$$
$$+A_{3,3,4}\big(A_{2,2}^2A_{3,3}A_{4,4}+2A_{2,2}^2A_{3,4}^2$$
$$-A_{2,2}A_{2,3}^2A_{4,4}-4A_{2,2}A_{2,3}A_{2,4}A_{3,4}-A_{2,2}A_{3,3}A_{2,4}^2+3A_{2,3}^2A_{2,4}^2\big)$$
$$+A_{4,4,4}\big(A_{2,2}^2A_{3,3}^2-2A_{2,2}A_{3,3}A_{3,2}^2+A_{2,3}^4\big)=0.$$

Hint: In order to obtain the derivative of determinant $\lambda^{(4)}$ with respect to V or any other variable, the following steps may facilitate the algebra: (1) take the derivative of each element of determinant $\lambda^{(4)}$ with respect to V or any other variable; call the matrix with derivative elements $\lambda*^{(4)}$, (2) form the cofactor or adjugate matrix (see Strang, 1988) of the determinant $\lambda*^{(4)}$, and (3) the sum of the element by element products of the co-factor matrix and the matrix $\lambda*^{(4)}$ is the determinant of $\lambda^{(4)}$ with respect to V or any other variable.

4.11 Show that the isothermal compressibility of a stable two-phase multicomponent mixture is positive.

4.12 Write five expressions for the criteria of stability of a binary mixture using $y_{3,3}^{(2)}=0$

4.13 Consider Eq. (4.105). Provide a detailed analysis that $y_{jj}^{(j-1)}$ approaches zero before $y_{j-1,j-1}^{(j-2)}$approaches zero. Is this analysis based on solid grounds?

4.14 Consider the general expression $y_{n_c+1,n_c+1}^{(n_c)}\geq 0$. Explain whether $c_V>0$ and $C_T>0$ can be extracted for multicomponent mixtures from this criterion.

4.15 Show that the critical point (T_c, P_c) in a binary mixture is given by:

$$\left(\frac{\partial P}{\partial v}\right)_{T,\mu_2}=0$$

$$\left(\frac{\partial^2 P}{\partial v^2}\right)_{T,\mu_2}=0$$

You may also demonstrate that the following relations are invalid at the critical point of a binary mixture:

$$\left(\frac{\partial P}{\partial v}\right)_{T,x_1}=0$$

$$\left(\frac{\partial^2 P}{\partial v^2}\right)_{T,x_1}=0$$

4.16 Consider the following plots for mixtures of methane and propane at $T=346$ K computed by the Peng-Robinson Equation of State (1976) from Shukla & Firoozabadi (1998). The computations are based on $\delta_{C_1-C_3}=0.01$. Compute $\ln f_1$ and

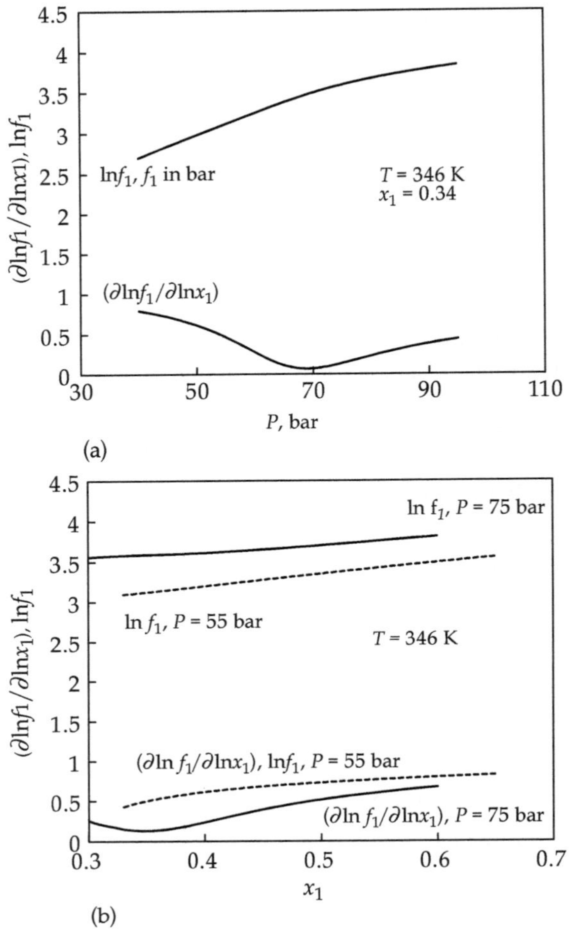

FIGURE 4.21

$(\partial \ln f_1/\partial \ln x_1)_{T=346\,\mathrm{K}}$ as a function of (1) pressure for $x_1 = 0.34$ and (2) composition x_1 at two pressures of 55 and 75 bar. Compare your computations with the plots shown in the Fig. 4.21(a) and (b). Note that as we approach the critical point, $(\partial \ln f_1/\partial \ln x_1)_T$ approaches zero. Also note that $(\partial \ln f_1/\partial \ln x_1)_T$ is always greater than zero as expected from thermodynamic stability.

4.17 Show that at the critical point of a polymer and solvent the volume fraction of the solvent is given by

$$\Phi_{1\,critical} = \frac{1}{1+\sqrt{r}}$$

The above expression indicates that for large r,

$$\Phi_{1\,critical} = \frac{1}{\sqrt{r}}$$

An implication of the above equation is that with increasing r, a polymer will not dissolve in a solvent.

4.18 The determinant of the Hessian matrix A_H in Eq. (4.141) can become zero because of the Gibbs-Duhem equation. Show how the Gibbs-Duhem equation can be incorporated in the determinant of A_H. As a result, A_H is said to be positive semidefinite.

4.19 Derive Eq. (4.112), using the definition of $x_{n_{c-1}} = N_{n_{c-1}} / \sum_{i=1}^{n_c} N_i$.

4.6 References

Baker, L. E., and K. D. Luks, Critical point and saturation pressure calculations for multipoint systems. *SPE Journal*, 1980. 20(1): pp. 15–24.

Haase, R., Thermodynamik der Mischphasen: mit einer Einfhrung in die Grundlagen der Thermodynamik. 1956, Berlin, Gttingen, Heidelberg: Springer-Verlag.

Heidemann, R. A., and A. M. Khalil, The calculation of critical points. *AIChE Journal*, 1980. 26(5): pp. 769–779.

Heidemann Robert, A., Critical points in reacting mixtures, in thermodynamics of aqueous systems with industrial applications. *American Chemical Society*, 1980. pp. 379–392.

Jiménez-Angeles, F., and A. Firoozabadi, Nucleation of methane hydrates at moderate subcooling by molecular dynamics simulations, *Journal of Physical Chemistry C*, 2014. 118 (21): pp. 11310–11318.

Kamgar-Parsi, B., J. M. H. Levelt Sengers, and J.V. Sengers, Thermodynamic properties of D_2O in the critical region. *Journal of Physical and Chemical Reference Data*, 1983. 12 (3): pp. 513–529.

Michelsen, M. L., and R. A. Heidemann, Calculation of critical points from cubic two-constant equations of state. *AIChE Journal*, 1981. 27(3): pp. 521–523.

Peng, D.Y., and D. B. Robinson, A rigorous method for predicting the critical properties of multicomponent systems from an equation of state. *AIChE Journal*, 1977. 23: pp. 137–144.

Rivkin, S. L., and B. N. Egorov, Experimental studies of the heat capacity of heavy water in the supercritical region of the parameters of state. *Teploenergetika*, 1962. 9(12): p. 60.

Rivkin, S. L., and B. N. Egorov, Experimental studies of heavy water thermal capacity at high pressures and temperatures. *Teploenergetika*, 1963. 10(7): p. 75.

Shukla, K., and A. Firoozabadi, A new model of thermal diffusion coefficients in binary hydrocarbon mixtures. *Industrial & Engineering Chemistry Research*, 1998. 37(8): pp. 3331–3342.

Strang, G., *Introduction to Applied Mathematics*. 1986, Wellesley-Cambridge Press.

Strang, G., *Linear Algebra and Its Applications*. 1988, New York: Harcourt Brace Jovanovich.

Tester, J. W., and M. Modell, *Thermodynamics and Its Applications*. 3rd ed. 1997, Prentice-Hall Inc.

CHAPTER 5

Phase Equilibrium Computations

The central ideas in this chapter relate to phase-split computations based on the equality of chemical potentials or fugacities of different species in different phases. Additionally, one needs to write the mass balance to complete the formulations. The mass balance requirement complicates the computations even for two species in two-phase. The equality of chemical potentials is necessary but not sufficient due to the shape of the Gibbs free energy function (for the case of constant temperature and constant pressure conditions in a closed container). The Gibbs free energy function may have more than one extremum. The phase-split computation should be based on global minimum of Gibbs free energy. In some systems, there may be very many minima. Another complexity is the solution of nonlinear algebraic equations and the need for very efficient computational methods. Newton method is the method of choice for solution of nonlinear algebraic equations because of quadratic convergence in many problems. However, there is a need for good enough initial guesses. In order to provide good enough initial guesses, we may start with an approach with linear convergence rate and then switch to Newton method where only two to three iterations will suffice to achieve convergence. We may start with good initial guesses based on stability testing.

In this chapter, we mainly cover phase-split computations for the conditions of constant temperature, constant pressure, and fixed overall composition. This case relates to the Gibbs free energy and its minimization. There are other types of phase-split calculations. Among them, phase-split calculations at constant volume and temperature have many applications. Minimization of Helmholtz free energy provides composition of the phases for the conditions of constant volume and temperature (Mikyška and Firoozabadi, 2011).

In this chapter, we start with the Gibbs free energy surface analysis to show the need for global minimum of Gibbs free energy with respect to appropriate variables. Then, the tangent-plane distance *TPD* analysis will follow. The *TPD* approach addresses the question whether a given system is stable or may split to more than one phase. In case the mixture is not stable, the tangent-plane distance analysis may provide a good guess for phase-split computations. Our presentation of phase-split computations covers step-by-step solution of nonlinear algebraic equations in two and three phases in different variables. The choice of variables can have significant effect on convergence rate and robustness. Chapter 5 contains twelve examples. The first example shows how to select the proper root of an equation of state based on the Gibbs free energy. Other examples cover critical point computation in multicomponent mixtures and root finding in the CPA-EOS. Twelve problems at the end of the chapter cover basic manipulations.

5.1 Gibbs Free Energy Surface Analysis

We will give a simple geometrical interpretation of the phase-stability analysis, mainly drawing from the work of Baker, Pierce, and Luks (1982) using the Gibbs free energy.

Because the equality of chemical potentials or fugacities imply stationary points of the Gibbs free energy, we should make sure the equilibrium corresponds to the global minimum of the Gibbs free energy. The phase-split computations should guarantee the lowest possible Gibbs free energy at system temperature T and pressure P. The global minimum of Gibbs free energy is the statement of the second law, which is equivalent to the statement that the entropy of an isolated system must be a maximum.

In three- and higher-phase split calculations, there are many instances where the equality of chemical potentials or fugacities does not guarantee a global Gibbs free energy minimum. As a result, the solution may be false. Examples of three and higher phases are CO_2-hydrocarbon systems and water-hydrocarbon systems, as well as mixtures of rich gases and crude oils. When there is solid precipitation, the occurrence of several phases is a norm rather than an exception (Firoozabadi, 1999). All these multiple phases occur in the temperature range of 50 to 300°F, commonly observed in the reservoir and in some petroleum production facilities.

Let us investigate a two-component mixture composed of a less volatile component 1 and a more volatile component 2 at temperature T. For this binary mixture, the pressure-composition diagram is sketched in Fig. 5.1; P_A and P_B are the vapor pressures at T of

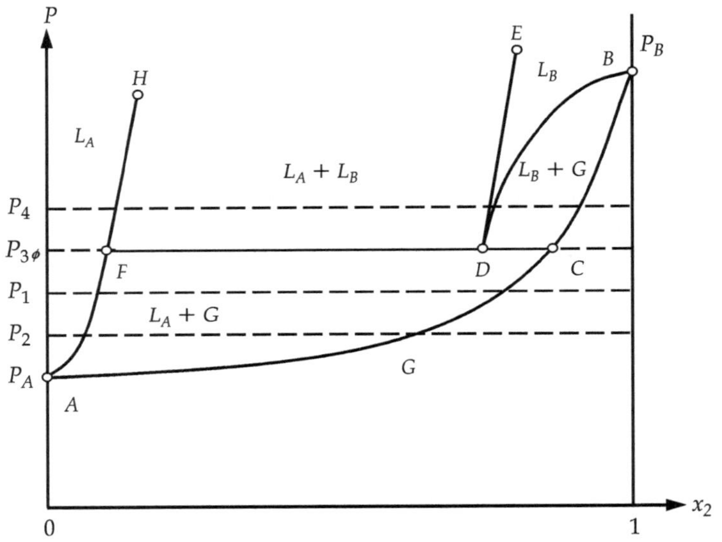

FIGURE 5.1 Pressure composition diagram of the binary mixture at temperature T (adapted from Baker et al., 1982).

pure components 1 and 2, respectively. Several different regions exist in this figure. Toward the right of curve AC, the state of the system is gas. Along the line FDC, the mixture exists as three phases, gas and two liquids; the pressure for this state is called the three-phase pressure, $P_{3\phi}$. In the region ACDF, gas phase and a liquid phase rich in the less volatile component 1 coexist. To the left of AFH, only a liquid phase rich in component 1 is present. Two liquid phases, L_A and L_B, exist in the region bounded by HFDE. Liquid L_B and gas exist in the CBDC region, and only liquid phase L_B exists in the region bounded by the BD and DE curves.

Now consider the mixing of components 1 and 2 at constant temperature and pressure. The molar Gibbs free energy of mixing, Δg_{mix}, can be expressed as:

$$\Delta g_{mix} = g - \sum_{i=1}^{2} x_i \mu_i^0(T, P) \tag{5.1}$$

where $g = \sum_{i=1}^{2} x_i \mu_i(T, P, x_i)$ is the molar Gibbs free energy of the mixture and the term $\sum_{i=1}^{2} x_i \mu_i^0(T, P)$ represents the Gibbs free energy of one mol of components 1 and 2 before mixing. Equation (5.1) can be written as,

$$\Delta g_{mix} = x_1\left(\mu_1 - \mu_1^0\right) + x_2\left(\mu_2 - \mu_2^0\right) \tag{5.2}$$

In Example 5.2, we will illustrate how to calculate Δg_{mix} over the whole range of composition (i.e., $0 \leq x_2 \leq 1$), making the assumption that over the whole range, the mixture stays in the hypothetical homogeneous single-phase state.

From Eq. (5.2), one may obtain the following expression:

$$\left(\frac{\partial \Delta g_{mix}}{\partial x_2}\right) = -\left(\mu_1 - \mu_1^0\right) + \left(\mu_2 - \mu_2^0\right) \tag{5.3}$$

The above relationship can be derived by taking the derivative of Eq. (5.2) with respect to x_2 and using the Gibbs-Duhem Equation $\sum_{i=1}^{2} x_i(\partial \mu_i/\partial x_2) = 0$ at constant temperature and pressure. Finally, by combining Eqs. (5.2) and (5.3),

$$\Delta g_{mix} = \left(\mu_1 - \mu_1^0\right) + x_2\left(\frac{\partial \Delta g_{mix}}{\partial x_2}\right) \tag{5.4}$$

Figure 5.2a shows a plot of Δg_{mix} vs. x_2 at pressure P_1 and temperature T for the binary mixture (see Fig. 5.1). At a given overall composition z_2, the composition and amount of the gas and liquid phases can be obtained from this figure. The compositions are obtained by drawing a tangent line to the Δg_{mix} curve. At the points of tangency, the chemical potentials of components 1 and 2 are equal; $\mu_1^{L_A} = \mu_1^V$ and $\mu_2^{L_A} = \mu_2^V$. The proof of the equality of chemical potentials is simple. From Fig. 5.2a,

$$\Delta g_{mix}^V = \Delta g_{mix}^{L_A} + \left(\frac{\partial \Delta g_{mix}}{\partial x_2}\right)\left(x_2^V - x_2^{L_A}\right) \tag{5.5}$$

where $(\partial \Delta g_{mix}/\partial x_2)$ is the slope of the common tangent, and superscripts L_A and V indicate the points of tangency. Writing Eq. (5.4) for the points of tangency and substituting the results in Eq. (5.5) give $\mu_1^{L_A} = \mu_1^V$ which is the sought relation. The molar Gibbs free energy of the hypothetical single-phase system is Δg_{mix1} (see Fig. 5.2a). The molar Gibbs free energy of the two-phase mixture is Δg_{mix2}. Then $z_2 = Lx_2^{L_A} + (1 - L)x_2^V$ (L is the mol

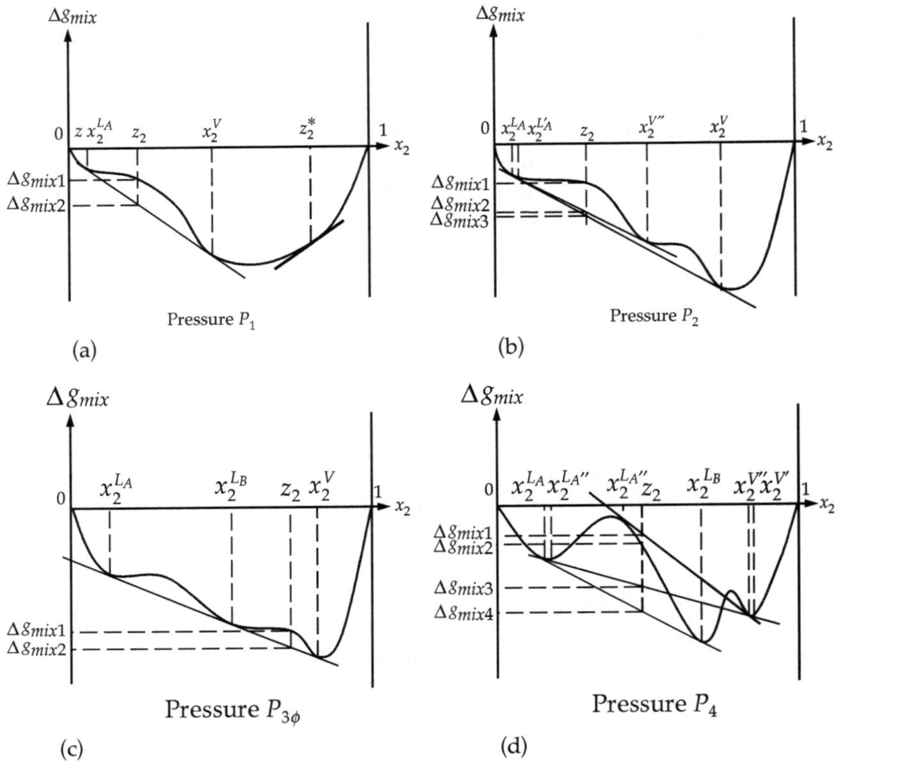

FIGURE 5.2 Δg_{mix} vs. x_2 of the binary mixture at temperature T (adapted from Baker et al., 1982).

fraction of liquid phase) can be computed. Note that x_2^{LA} and x_2^V are the compositions of the liquid and vapor phases. Since $\Delta g_{mix2} < \Delta g_{mix1}$, then the two-phase state for the given overall mol fraction, z_2, is more stable than the single phase. For $0 \leq z_2 < x_2^{LA}$ and $x_2^V < z_2 \leq 1$, there will be only one single phase. Outside the two-phase region, the tangent to Δg_{mix} curve lies below this curve, and only one phase exists. As an example at z_2^*, $x_2^V < z_2^* \leq 1$, only the gas phase exists.

At pressure P_2, the Δg_{mix} plot is sketched in Fig. 5.2b. At overall composition z_2, two different tangent lines can be drawn that satisfy the material balance and the equality of chemical potentials. One tangent line results in the vapor composition $x_2^{V'}$ and liquid composition $x_2^{L'A}$. The other tangent line provides vapor and liquid compositions x_2^V and x_2^{LA} but does not intersect the Δg_{mix} curve except at the tangency points. Note that $\Delta g_{mix1} > \Delta g_{mix2} > \Delta g_{mix3}$, and, therefore, Δg_{mix3} is a minimum. The single phase with the corresponding Δg_{mix1} has the highest Δg_{mix}. Outside the range $x_2^{L'A} \leq z_2 \leq x_2^V$, the tangent has one point of tangency and the system does not split into two phases. We will later show that for this range $(\partial \mu_2 / \partial x_2) < 0$, which ensures the stability of the single phase.

At $P = P_{3\phi}$, the plot of Δg_{mix} is shown in Fig. 5.2c. Any z_2 that is in the x_2^{LA} and x_2^V range $x_2^{LA} < z_2 < x_2^V$ may split into three phases: two liquid phases and one gas phase. The common tangent line implies that the chemical potentials are the same for all three phases for component 2, as well as for component 1. Outside the interval $x_2^{LA} \leq z_2 \leq x_2^V$, the tangent intersects the Δg_{mix} curve only at the point of tangency and therefore the single phase is stable. Note that at z_2, $\Delta g_{mix1} > \Delta g_{mix2}$ and, therefore, Δg_{mix2} corresponds to the more stable three phases.

Figure 5.2d shows the Δg_{mix} plot at P_4. For the overall composition z_2, there are several tangent lines. Three tangent lines are shown. Since Δg_{mix4} provides the lowest Δg_{mix} and does not intersect the Δg_{mix} curve, the corresponding tangent is the most stable solution giving liquid-liquid equilibria with composition $x_2^{L_A}$ for phase L_A and $x_2^{L_B}$ for phase L_B.

The above simple geometrical description for a binary system can be readily extended to a multicomponent mixture. For a multicomponent mixture ($n_c \geq 3$), the corresponding Δg_{mix} becomes a hypersurface instead of a curve, and the tangent line becomes a hyperplane. The criterion, however, remains the same. The tangent hyperplane corresponding to a stable equilibrium state cannot lie above the Gibbs free energy hypersurface for any composition. One approach to avoid false solutions is to search for all possible values of x_2 that provide the global minimum of Δg_{mix}. (In Fig. 5.2d, Δg_{mix4} is the global minimum for a fixed z_2 and all possible values of x_2.) This straightforward approach can be attacked mathematically using a minimization algorithm (as we will discuss later in this chapter), but it may be computationally expensive. A more practical approach is the method suggested by Michelsen (1982a) to determine certain minima of a distance function.

5.2 Tangent-Plane Distance (*TPD*) Analysis

Consider N_i mols of a homogeneous mixture at temperature T and pressure P. The overall mol fractions are $z_i = N_i / \sum_{i=1}^{n_c} N_i = N_i/N$. The Gibbs free energy of the homogeneous single-phase system (see Fig. 5.3a) is

$$G^{\mathrm{I}} = \sum_{i=1}^{n_c} N_i \mu_i(z, T, P) \tag{5.6}$$

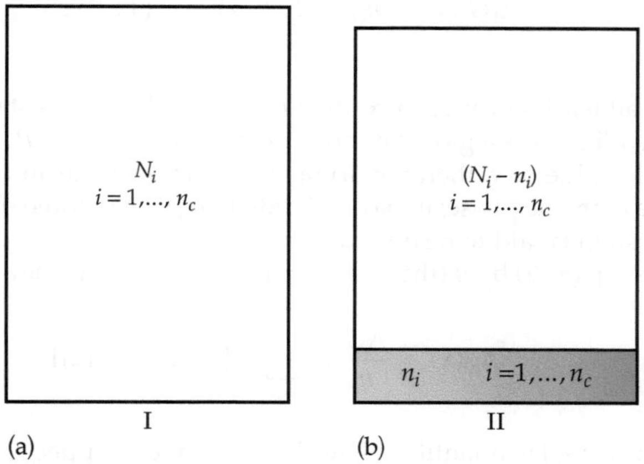

FIGURE 5.3 System with and without a second phase at temperature (T) and pressure (P).

where $z = (z_1, z_2, \ldots, z_{n_c-1})$. If a very small amount of a second phase (this may be referred to as the trial phase) is formed at the same pressure and temperature with mol numbers n_i, and $n_i << N_i$, then the Gibbs free energy of the new system II (see Fig. 5.3b) is

$$G^{\mathrm{II}} = G(\underline{n}, T, P) + G(\underline{N - n}, T, P) \tag{5.7}$$

where $G(\underline{n}, T, P)$ and $G(\underline{N - n}, T, P)$ are the Gibbs free energies of the two phases of system II.

$$G(\underline{n}, T, P) = \sum_{i=1}^{n_c} n_i \mu_i(x, T, P) \tag{5.8}$$

$$x_i = \frac{n_i}{\sum_{i=1}^{n_c} n_i} = \frac{n_i}{n} \tag{5.9}$$

and

$$G(\underline{N - n}, T, P) \approx G^{\mathrm{I}}(\underline{N}, T, P) - \sum_{i=1}^{n_c} \left(\frac{\partial G}{\partial N_i} \right) \Bigg|_{N_i} (n_i)$$

$$= G^{\mathrm{I}}(\underline{N}, T, P) - \sum_{i=1}^{n_c} n_i \mu_i(z, T, P) \tag{5.10}$$

Equation (5.10) is a Taylor's series expansion of $G(\underline{N - n}, T, P)$ around $G(\underline{N}, T, P)$. The change in Gibbs free energy from I to II is

$$\Delta G = G^{\mathrm{II}} - G^{\mathrm{I}} \tag{5.11}$$

$$\Delta G = \sum_{i=1}^{n_c} n_i [\mu_i(T, P, x) - \mu_i(T, P, z)] \tag{5.12}$$

If $\Delta G > 0$ for all feasible values of x, then the original state I is stable and G^{I} cannot be further reduced (i.e., G is a global minimum at constant T and P). However, if $\Delta G < 0$ for any feasible value of x, then the original state is unstable and the single phase will split into more than one phase; it may split into two phases, three phases, or more. When $\Delta G = 0$, the system is said to be neutral.

If we divide Eq. (5.12) by n (the total number of mols in the new phase),

$$\Delta g(x) = \frac{\Delta G}{n} = \sum_{i=1}^{n_c} x_i [\mu_i(x) - \mu_i(z)] \tag{5.13}$$

where $\Delta g(x)$ is a molar quantity. Note that we have dropped T and P dependencies since T and P are held constant. Similarly to Eq. (5.12), $\Delta g(x) > 0$ implies that system I is stable, and for $\Delta g(x) < 0$, system I is unstable. Equation (5.13) has a simple geometrical

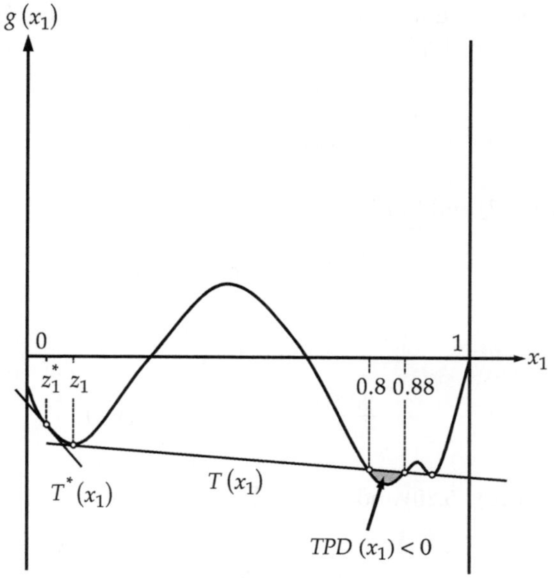

FIGURE 5.4 Plot of $g(x_1)$ vs. x_1.

interpretation. For simplicity, let us select a two-component system in which $x_2 = 1 - x_1$. Then Eq. (5.13) reduces to

$$\Delta g(x_1) = \sum_{i=1}^{2} x_i[\mu_i(x_1) - \mu_i(z_1)] \tag{5.14}$$

Figure 5.4 depicts the plot of $g(x_1)$,

$$g(x_1) = \sum_{i=1}^{2} x_i \mu_i(x_1) \tag{5.15}$$

vs. x_1; x_1 is the mol fraction of component 1, which can vary from 0 to 1. An EOS can be used to calculate $g(x_1)$. In Fig. 5.4, z_1 is the mol fraction of component 1, a fixed overall composition. Let us draw a tangent line at point z_1 to the curve $g(x_1)$. The equation for this tangent line is

$$T(x_1) = g(z_1) + \left.\frac{\partial g(x_1)}{\partial x_1}\right|_{z_1} (x_1 - z_1) \tag{5.16}$$

From Eq. (5.15) and the expression $x_2 = 1 - x_1$, we can evaluate the slope $\partial g(x_1)/\partial x_1$:

$$\frac{\partial g(x_1)}{\partial x_1} = \mu_1(x_1) - \mu_2(x_1) + \frac{x_1 \partial \mu_1(x_1)}{\partial x_1} + \frac{x_2 \partial \mu_2(x_1)}{\partial x_1} \tag{5.17}$$

From the Gibbs-Duhem Equation at constant T and P, [see Eq. (1.40)] $\sum_{i=1}^{2} x_i d\mu_i = 0$, which after dividing by ∂x_1 becomes

$$\sum_{i=1}^{2} x_i \left(\frac{\partial \mu_i}{\partial x_1}\right) = 0 \tag{5.18}$$

Combining Eqs. (5.17) and (5.18),

$$\frac{\partial g(x_1)}{\partial x_1} = \mu_1(x_1) - \mu_2(x_1) \tag{5.19}$$

at $x_1 = z_1$

$$\left.\frac{\partial g(x_1)}{\partial x_1}\right|_{z_1} = \mu_1(z_1) - \mu_2(z_1) \tag{5.20}$$

Combining Eqs. (5.16), (5.20) and $g(z_1) = \sum_{i=1}^{2} z_i \mu_i(z_1)$,

$$T(x_1) = \sum_{i=1}^{2} x_i \mu_i(z_1) \tag{5.21}$$

Now we define the tangent-plane distance (*TPD*) as the difference from $T(x_1)$ to $g(x_1)$ at any point x_1:

$$TPD(x_1) = g(x_1) - T(x_1) \tag{5.22}$$

From Eqs. (5.15), (5.20), and (5.21), one obtains

$$TPD(x_1) = \sum_{i=1}^{2} x_i[\mu_i(x_1) - \mu_i(z_1)] \tag{5.23}$$

Equation (5.23) is the same as Eq. (5.13) with $n_c = 2$. For a n_c-component system,

$$TPD(\boldsymbol{x}) = \sum_{i=1}^{n_c} x_i[\mu_i(\boldsymbol{x}) - \mu_i(\boldsymbol{z})] \tag{5.24}$$

which is the same as Eq. (5.13).

The criteria of the stability of system I in Fig. 5.3 now can be stated in terms of the tangent-plane distance. $TPD(\boldsymbol{x})$ should be positive over the whole range of $\boldsymbol{x}$. Note that in Fig. 5.4, the $TPD(\boldsymbol{x})$ from z_1 becomes negative for $0.80 < x_1 < 0.88$ and, therefore, the system of overall composition z_1 is not stable. On the other hand, the overall composition z_1^* has a tangent $T^*(x_1)$ that is below the curve $g(x_1)$ and does not have a $TPD(x_1)$ of less than zero; therefore, it is stable and the system cannot split into two phases.

Different methods have been proposed to search for the value(s) of $\boldsymbol{x}$ to test the stability of a system with a fixed composition $\boldsymbol{z}$. Michelsen (1982a) suggested to locate the stationary points (maxima, minima, or saddle points) of $TPD(\boldsymbol{x})$ rather than conducting an exhaustive search in $\boldsymbol{x}$-space for values of $\boldsymbol{x}$ where $TPD(\boldsymbol{x}) \geq 0$. Let us examine $g(x_1)$,

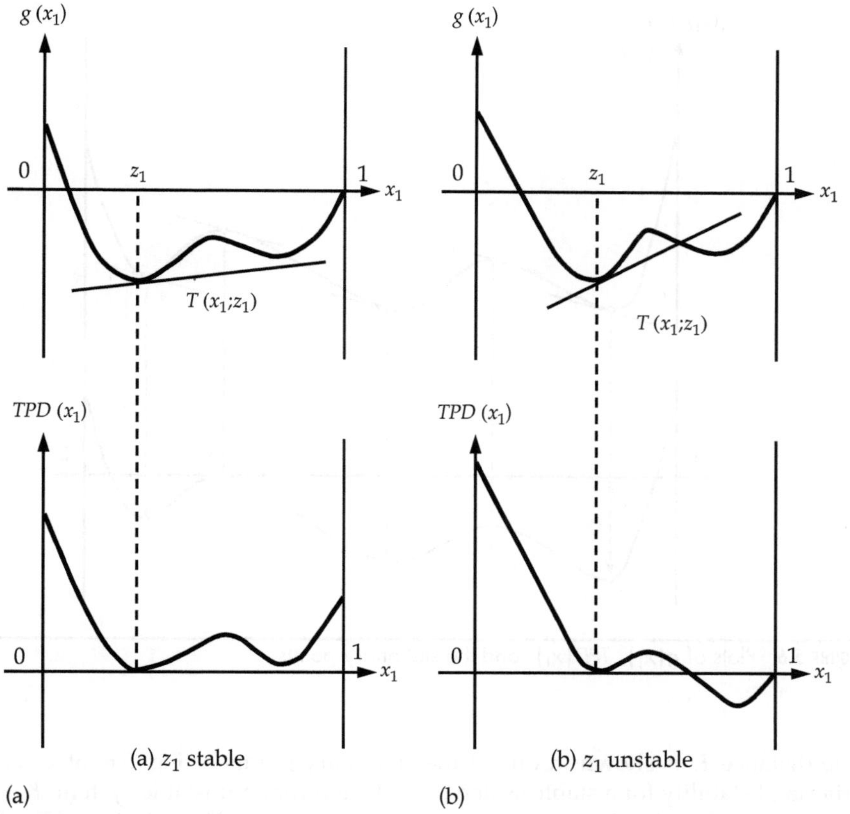

FIGURE 5.5 Plots of $g(x_1)$ and $TPD(x_1)$ vs. x_1.

the tangent $T(x_1)$ at point z_1, and the $TPD(x_1)$ shown in Fig. 5.5. The stationary points of $TPD(\boldsymbol{x})$ occur at points where

$$\frac{\partial}{\partial x_i}[TPD(\boldsymbol{x})] = 0 \quad i = 1, \ldots, n_c - 1 \tag{5.25}$$

In Eq. (5.25), the derivatives are with respect to x_1 to x_{n_c-1}. Substituting the expression for $TPD(\boldsymbol{x})$ from Eq. (5.24) and using the Gibbs-Duhem expression, $\sum_{i=1}^{n_c} x_i(\partial \mu_i/\partial x_1) = 0$, we obtain

$$\boxed{\mu_i(\boldsymbol{x}) - \mu_i(\boldsymbol{z}) = \mu_{n_c}(\boldsymbol{x}) - \mu_{n_c}(\boldsymbol{z}) = K \quad i = 1, \ldots, n_c - 1} \tag{5.26}$$

The $TPD(\boldsymbol{x})$ at the stationary points are obtained by substituting Eq. (5.26) into Eq. (5.24):

$$TPD(\boldsymbol{x}) = \sum_{i=1}^{n_c} x_i K = K. \tag{5.27}$$

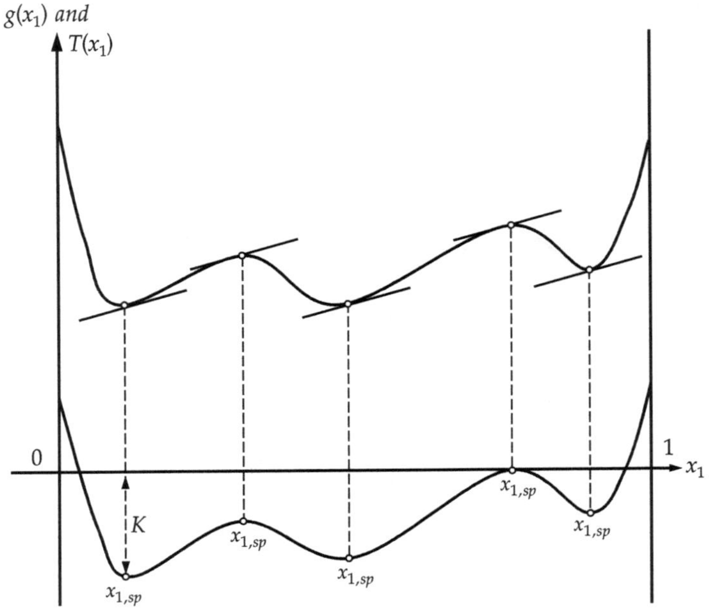

FIGURE 5.6 Plots of $g(x_1)$, $TPD(x_1)$, and the stationary points.

The distance K is shown in one of the stationary points of Fig. 5.6, at $x_{1,sp}$. From the criteria of stability for a stable system, $K \geq 0$, and for an unstable system, $K < 0$.

Equation (5.26) is a key equation for stability analysis. The solution of Eq. (5.26) provides the $\boldsymbol{x}$ of the stationary points of $TPD(\boldsymbol{x})$, which is not a trivial task to solve. The fugacity form of Eq. (5.26) can be obtained as follows.

From

$$\mu_i(\boldsymbol{x}, T, P) = \mu_i^0(T, P) + RT \ln f_i(\boldsymbol{x}, T, P)/f_i^0(T, P),$$
$$\mu_i(\boldsymbol{z}, T, P) = \mu_i^0(T, P) + RT \ln f_i(\boldsymbol{z}, T, P)/f_i^0(T, P)$$

Equation (5.26) and the definition of fugacity coefficients $\varphi_i(\boldsymbol{x}, T, P) = f_i(\boldsymbol{x}, T, P)/(x_i f_i^0(T, P))$ and $\varphi_i(\boldsymbol{z}, T, P) = f_i(\boldsymbol{z}, T, P)/(z_i f_i^0(T, P))$

$$\ln \varphi_i(\boldsymbol{x}) + \ln x_i - \ln z_i - \ln \varphi_i(\boldsymbol{z}) = k \quad i = 1, \ldots, n_c \tag{5.28}$$

where $k = K/RT$. Equation (5.28) provides the $\boldsymbol{x}$ at which $TPD(\boldsymbol{x})$ is at a stationary state. This equation is nonlinear and may be transformed to the following form by introducing a new variable X_i,

$$\ln X_i = \ln x_i - k \quad i = 1, \ldots, n_c \tag{5.29}$$

from which

$$x_i = X_i e^k \quad i = 1, \ldots, n_c \tag{5.30}$$

follows.

Since

$$\sum_{j=1}^{n_c} x_j = 1$$

then

$$e^k = \frac{1}{\sum_{j=1}^{n_c} X_j} \tag{5.31}$$

and, therefore,

$$x_i = \frac{X_i}{\sum_{j=1}^{n_c} X_j} \tag{5.32}$$

Equation (5.32) implies that the new variable X_i is similar to mol numbers for the new phase.

Combining Eqs. (5.28) and (5.29),

$$\boxed{\ln X_i + \ln \varphi_i(\underline{X}) - \ln z_i - \ln \varphi_i(z) = 0 \quad i = 1, \ldots, n_c} \tag{5.33}$$

Solution of the above nonlinear system of equations provides $\underline{X}$ that can be used to examine the stability analysis. From Eq. (5.31), $\sum_{j=1}^{n_c} X_j = 1/e^k = X_T$. Since for a stable system of fixed composition, z_i, when $k \geq 0$, the system is stable; therefore, in terms of $\underline{X}$, when $\sum_{j=1}^{n_c} X_j < 1$, the same system is stable. If $\sum_{j=1}^{n_c} X_j > 1$, the system is unstable.

The modified formulation of the *TPD* function in terms of *X* as suggested by Michelsen (1982b) takes the form

$$TPD^*(\underline{X}) = 1 + \sum_{i=1}^{n_c} X_i[\ln \varphi_i(\underline{X}) + \ln X_i - d_i(z) - 1] \tag{5.34}$$

where $d_i = \ln z_i + \ln \varphi_i(z)$. The TPD^* and *TPD* functions have the same stationary points. These two functions also have the same sign at the stationary points (see Example 5.3). Therefore, to determine the stability of the mixture, it is sufficient to check the positivity of the modified TPD^* function at all the stationary points. The modified formulation by Michelsen is in terms of unconstrained minimization when compared to $TPD(x) \geq 0$ and $\sum_{i=1}^{n_c} x_i = 1$. These treatments may bypass numerical difficulties of the constraint minimization (i.e., $\sum_{i=1}^{n_c} x_i = 1$).

Now we have the working equations to implement stability testing to be discussed next.

5.3 Stability Testing

The stability analysis of a given phase with a fixed overall composition z is a search for a trial phase of small amount that is taken out of the original single phase (i.e., the homogeneous phase). The trial phase when combined with the remainder of the original mixture gives a Gibbs free energy that should be lower than that of the original single-phase mixture for the single phase to be unstable.

Mathematically, the stability testing can be achieved by solving the nonlinear algebraic Eq. (5.33). There are many lines of attack. The combination of the successive substitution iterations (SSI) and the Newton method are often the method of choice. The quasi-Newton method with the Broyden–Fletcher–Goldfarb–Shanno (BFGS) update of the Hussian matrix (Dennis and Schnabel, 1996) is another choice. Both methods have been demonstrated to be effective in locating the local stationary points of the Gibbs free energy leading to the evaluation of the *TPD* (Hoteit and Firoozabadi, 2006; Li and Firoozabadi, 2012a). In the following, we will provide the algorithms based on the combined SSI and Newton methods.

5.3.1 SSI Method

In this method, Eq. (5.33) can be written as

$$X_i^{(k+1)} = \exp[d_i(z) - \ln \varphi_i(\underline{\mathbf{X}}^{(k)})] \quad i = 1, \ldots, n_c \tag{5.35}$$

where the index k denotes the iteration counter. The computational cost per iteration is very low. However, the number of iterations may be very high, in tens of thousands in the critical region, and close to the phase boundaries to a lesser degree.

5.3.2 Newton Method

In the Newton method, we write Eq. (5.33) as

$$F_i(\underline{\mathbf{X}}) = \ln X_i + \ln \varphi_i(\underline{\mathbf{X}}) - d_i(z) = 0 \quad i = 1, \ldots, n_c \tag{5.36}$$

The solution to Eq. (5.36) is given by

$$\boldsymbol{X}^{(k+1)} = \boldsymbol{X}^{(k)} - \boldsymbol{J}^{-1}\boldsymbol{F}^{(k)}(\underline{\mathbf{X}}) \tag{5.37}$$

where $\boldsymbol{X}$ is the column vector of X_i and $\boldsymbol{F}(\underline{\mathbf{X}})$ has a similar notation and $\boldsymbol{J}$ is the Jacobian matrix with elements

$$\frac{\partial F_i}{\partial X_j} = \frac{\delta_{ij}}{X_i} + \frac{\partial \ln \varphi_i(\underline{\mathbf{X}})}{\partial X_j}, \quad i = 1, \ldots, n_c, \; j = 1, \ldots, n_c \tag{5.38}$$

where

$$\frac{\partial \ln \varphi_i(\underline{X})}{\partial X_j} = \frac{1}{X_T}\sum_{k=1}^{n_c-1}\frac{\partial \ln \varphi_i(x)}{\partial x_k}(\delta_{kj} - x_k), \quad i = 1, \ldots, n_c, \ j = 1, \ldots, n_c \tag{5.39}$$

δ_{ij} and δ_{kj} are the Kronecker delta. Problem 5.1 provides the expressions for $\partial \ln \varphi_i/\partial x_k$.

The Newton methods may have quadratic convergence rate as shown in Example 5.4. Usually 2 to 4 iterations suffice convergence. However, there is a need for good enough initial guess, especially in the critical region or near the phase boundaries to a lesser extent. The combined SSI and Newton method is a powerful approach. The details of the algorithms for the combined approach and implementation of each method are outlined below.

1. We first need X_i. The initial K_i-values are required. For vapor-liquid equilibria, often the Wilson correlation (1965) is a good choice. The expression for the Wilson correlation is given by

$$K_i^{Wilson} = \frac{P_{c_i}}{P}\exp\left[5.37(1+\omega_i)\left(1 - \frac{T_{c_i}}{T}\right)\right] \tag{5.40}$$

where P_{ci}, T_{ci} and ω_i are the critical pressure, critical temperature, and acentric factor of component i. The Wilson correlation may not be adequate as the initial guess for liquid-liquid equilibria and even in the vapor-liquid equilibria in some cases. Li and Firoozabadi (2012a) present the following initial guesses for K_i to start the stability.

$$K_i^{stability} = \left\{K_i^{Wilson}, \frac{1}{K_i^{Wilson}}, \sqrt[3]{K_i^{Wilson}}, \frac{1}{\sqrt[3]{K_i^{Wilson}}}, K_i^{new}\right\} \tag{5.41}$$

The K_i^{new} is obtained from

$$K_i^{new} = \frac{0.9}{z_i} \tag{5.42a}$$

$$K_j^{new} = \frac{\left[\dfrac{0.1}{(n_c - 1)}\right]}{z_j} \qquad j \neq i \tag{5.42b}$$

Equation (5.42a) is motivated from the suggestion of Michelsen (1982a) that the trial phase composition can be initially assumed to be a pure species. Li and Firoozabadi (2012a) suggested that the composition of one of the components of the trial phase may be assigned a value of 90 mol % and the rest equally share the distribution as can be seen in Eq. (5.42b). Any species may assume the mol fraction of 0.90 in the trial phase.

In Eq. (5.41) there are $(n_c + 4)$ initial guesses. With these initial guesses, Li and Firoozabadi (2012a) report accurate results for an extensive set of mixtures in single-phase, in two-phase, and in three-phase systems.

2. Once K_i is available, X_i is calculated from $X_i = z_i K_i \quad i = 1, \ldots, n_c$.
3. The phase composition is then calculated from $x_i = X_i / \sum_{j=1}^{n_c} X_j$.
4. The corresponding compressibility factor Z and fugacity coefficients $\varphi_i(\mathbf{z})$ and $\varphi_i(\mathbf{x})$ are then computed.
5. Then $\mathbf{X}^{(k+1)}$ is computed from Eq. (5.35) and the convergence test is made; $\|\Delta\mathbf{X}\| < \varepsilon_{SSI}$ where $\Delta\mathbf{X} = \mathbf{X}^{(k+1)} - \mathbf{X}^{(k)}$. If there is no convergence, step 3 is returned to. When there is convergence, the next step is followed.
6. Newton method: $\mathbf{X}^{(k+1)}$ is computed from Eq. (5.37), and convergence is checked from $\|\mathbf{X}^{(k+1)} - \mathbf{X}^{(k)}\| < \varepsilon_N$.
7. TPD^* or TPD is evaluated ($TPD = -\ln X_T$).

Typical convergence criteria are: $\varepsilon_{SSI} = 10^{-2}$ and $\varepsilon_N = 10^{-9}$. If only SSI is used, then $\varepsilon_{SSI} = 10^{-9}$.

We have the following comments to make in relation to stability analysis. Based on our own experience, we have found a maximum of three nontrivial solutions to Eq. (5.33).

For one-phase stability testing (i.e., when examining the stability of a system in single-phase), the trivial solution is $\{K_i^{1p\ stability} = 1\}$. For two-phase stability testing (i.e., when examining the stability of one of the phases of a system in two-phase), the trivial solutions are $\{K_i^{2p\ stability} = 1\}$ and $\{K_i^{2p\ stability} = K_i^{2p\ split} = 1\}$ or $\{K_i^{2p\ stability} = 1/K_i^{2p\ split}\}$. We will define $K_i^{2p\ split}$ after a discussion of phase-split computations. We have also encountered trivial solutions of minima of TPD^* hypersurface. The true solution is stable if all nontrivial solutions have TPD^* higher than -10^{-10}. Figure 5.7 shows that one may encounter trivial solutions. Figure 5.8 shows the number of SSI and Newton iterations. Figure 5.9 shows the sign of TPD for gas-like and liquid-like states.

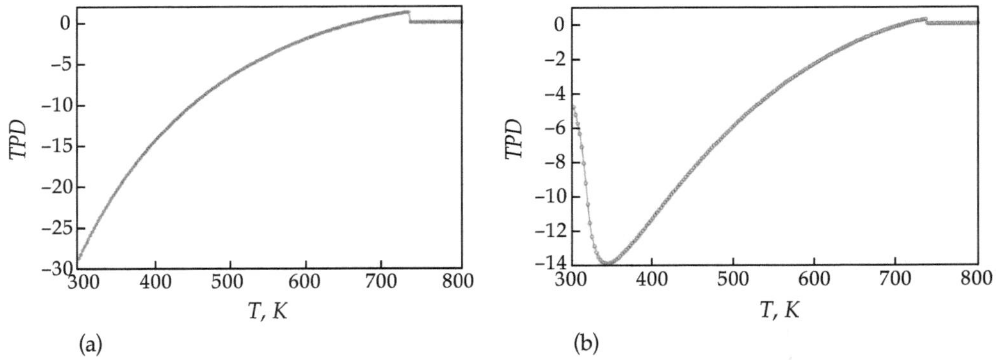

FIGURE 5.7 Minimum of the *TPD* vs. temperature: Mixture 4 of Hoteit and Firoozabadi (2006), (a) stability of gas-like phase at $P = 10$ bar, (b) stability of gas-like phase at $P = 50$ bar (adopted from Hoteit and Firoozabadi, 2006).

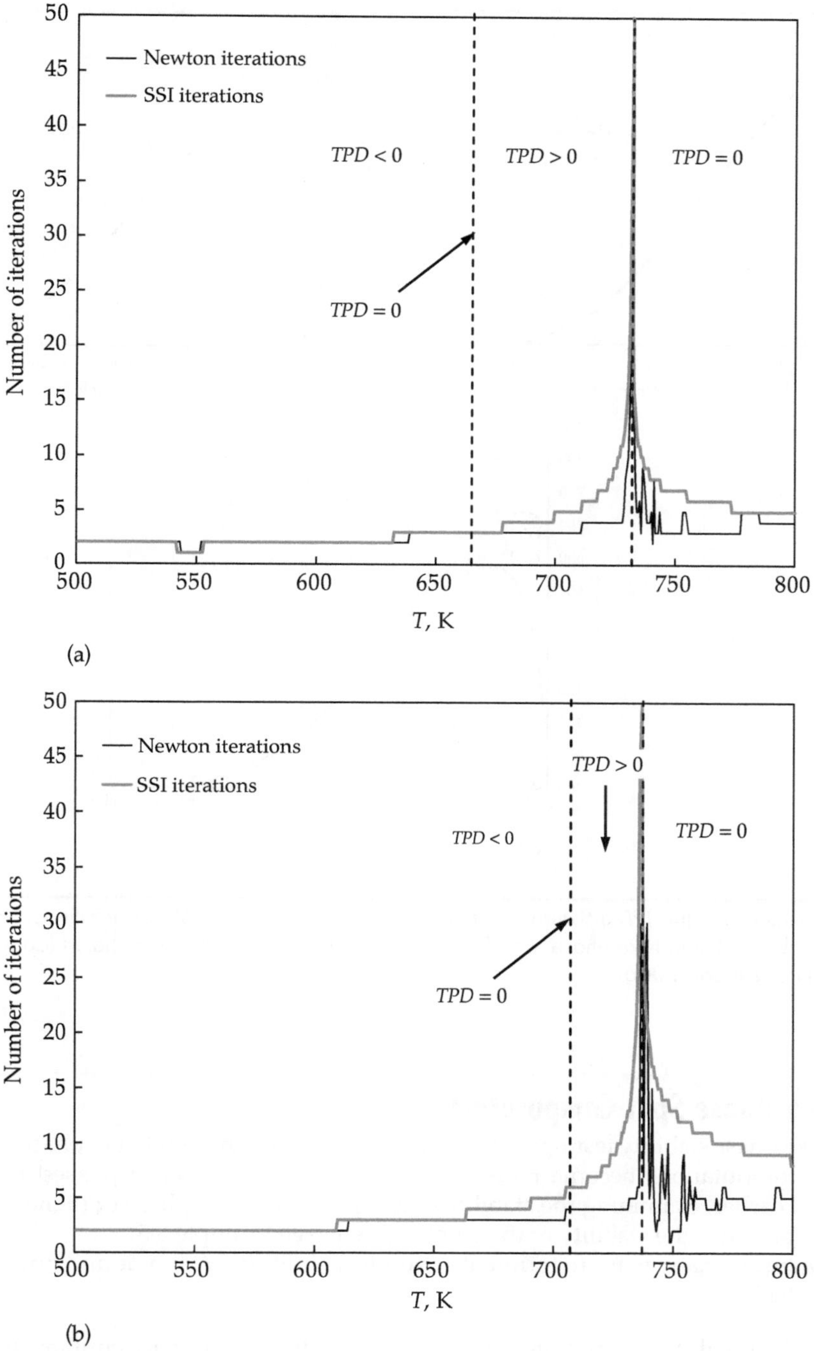

FIGURE 5.8 Number of SSI and Newton iterations in the SSI-Newton method versus temperature: Mixture 4 of Hoteit and Firoozabadi (2006), (a) stability of gas-like phase at $P = 10$ bar, (b) stability of gas-like phase at $P = 50$ bar (from Hoteit and Firoozabadi, 2006).

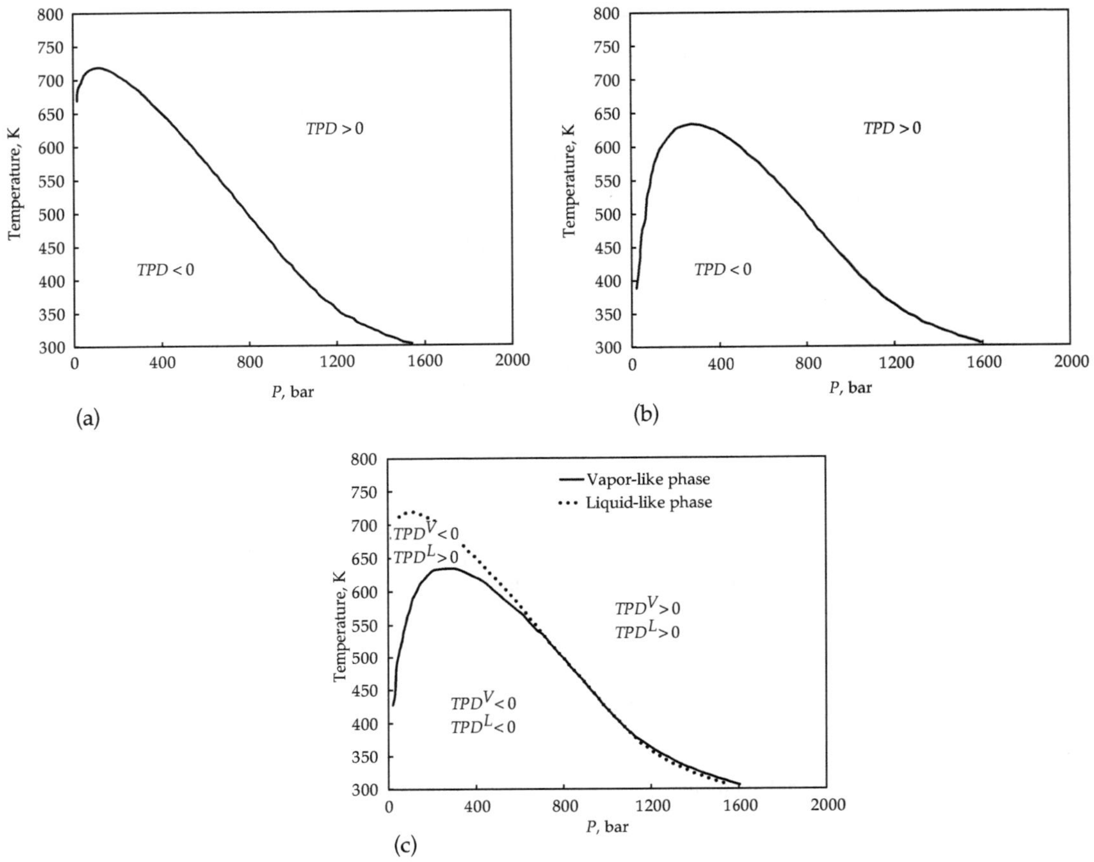

FIGURE 5.9 Minimum of the *TPD*: Mixture 4 of Hoteit and Firoozabadi (2006), (a) stability of gas-like phase, (b) stability of liquid-like phase, and (c) stability of gas-like and liquid-like phases (adapted from Hoteit and Firoozabadi, 2006).

5.4 Two-Phase Split Computations

Unlike phase stability testing which does not depend on the number of phases, the phase-split computations become more complicated as the number of phases increases. As stated earlier, examining the stability of any phase in a two-phase or higher-phase mixtures provides the stability of the whole system (see Example 5.5).

There are many ways to formulate two-phase splitting. A simple and direct approach includes:

1. Equality of fugacities to provide the conditions for phase equilibrium

$$f_i^L\left(T, P, x_1, \ldots, x_{n_c}\right) = f_i^V\left(T, P, y_1, \ldots, y_{n_c}\right) \tag{5.43}$$

2. Material balance of components to provide mass conservation

$$Fz_i = Lx_i + Vy_i \quad i = 1, \ldots, n_c \tag{5.44}$$

F, L, and V are the mols of feed, liquid, and vapor phases, respectively.

3. Constraint expressions

$$\sum_{i=1}^{n_c} x_i = 1 \tag{5.45}$$

$$\sum_{i=1}^{n_c} y_i = 1 \tag{5.46}$$

There are $2n_c$ unknowns in x_i and y_i and two unknowns in V and L. F and z_i are provided along with temperature and pressure. We have $2(n_c + 1)$ equations and $2(n_c + 1)$ unknowns. Various formulations of the phase-split calculations have been made over the years. It turns out that the formulation given above based on the equality of fugacities may not be numerically robust (see Example 5.6). There are various solution methods for the nonlinear algebraic equations. The Newton method (see Example 5.4) is the method of choice for rapid convergence. However, there is a need for good enough initial guesses. A clean and efficient formulation in terms of $\ln K_i$ (equilibrium ratios, the so-called K-values) is presented instead of Eq. (5.43). While K_i values may vary orders of magnitude, $\ln K_i$ variations are much less pronounced. From Eq. (1.114), $K_i = y_i/x_i = \varphi_i^L/\varphi_i^V$. We, therefore, write

$$F_i = \ln K_i - \ln \varphi_i^L + \ln \varphi_i^V = 0 \quad i = 1, \ldots, n_c \tag{5.47}$$

Next we write the mass balance starting with the relation between y_i and x_i.

$$y_i = K_i x_i \quad i = 1, \ldots, n_c \tag{5.48}$$

Combining Eqs. (5.44) and (5.48), we obtain

$$Fz_i = Lx_i + VK_i x_i \quad i = 1, \ldots, n_c \tag{5.49}$$

Taking the summation in Eq. (5.49),

$$L = F - V \tag{5.50}$$

Combining Eqs. (5.49) and (5.50), and defining $\beta = V/F$ (i.e., gas phase mol fraction)

$$\boxed{x_i = \frac{z_i}{1 + \beta(K_i - 1)}} \quad i = 1, \ldots n_c \tag{5.51}$$

Similarly, we obtain

$$\boxed{y_i = \frac{K_i z_i}{1 + \beta(K_i - 1)}} \qquad i = 1, \ldots n_c \tag{5.52}$$

From Eqs. (5.45), (5.46), (5.51), and (5.52),

$$\sum_{i=1}^{n_c} \frac{(K_i - 1)z_i}{1 + \beta(K_i - 1)} = 0 \tag{5.53}$$

We write Eq. (5.53), the Rachford-Rice (1952) expression, as

$$F_{n_c+1}(\beta) = RR = \sum_{i=1}^{n_c} \frac{(K_i - 1)z_i}{1 + \beta(K_i - 1)} = 0 \tag{5.54}$$

For fixed K_i and z_i, $F'_{n_c+1} = (dF_{n_c+1}/d\beta)$ is given by

$$F'_{n_c+1}(\beta) = -\sum_{i=1}^{n_c} \frac{(K_i - 1)^2 z_i}{[1 + \beta(K_i - 1)]^2} = 0 \tag{5.55}$$

Because F'_{n_c+1} is negative, $F'_{n_c+1}(\beta)$ is a monotonic decreasing function of β. This feature of the Rachford-Rice expression is important in solving Eq. (5.54).

Two-phase splitting calculations are more complex than stability analysis because of the need to include Eq. (5.54) in the computations. Efficient solution to the Rachford-Rice equation can become complex despite the fact that there is only one unknown.

In the formulation in terms of $\ln K_i$, the phase-split computations in two-phase contain $(n_c + 1)$ unknowns in $\ln K_i (i = 1, \ldots, n_c)$ and β. There are also $(n_c + 1)$ equations. The combination of two methods, the SSI and the Newton methods, has a definitive advantage over most other methods. We will first discuss the simple method of successive substitution iterations and then the Newton method.

5.4.1 SSI Method

In this iterative method only one unknown, the vapor mol fraction β, is searched. There are a few simple steps.

1. We start with the initial guess from the stability testing or from the correlations for K_i.
2. Then solve the Rachford-Rice equation for β. The Rachford-Rice equation is solved by the bisection method (see Example 5.7) or the Newton method, or the combination of the two. We may use $\beta = 0.5$ for the Newton method as the initial guess. Note that, based on stability analysis, we have a definitive knowledge of the two-phase state.

3. We then calculate x_i and y_i from Eqs. (5.51) and (5.52). The Newton method is used for the update of β values from [see Eq. (5.54)]

$$\beta^{(k+1)} = \beta^{(k)} - \frac{F_{n_c+1}(\beta^{(k)})}{F'_{n_c+1}(\beta^{(k)})} \tag{5.56}$$

Note that if $F'_{n_c+1}(\beta^{(k)})$ is close to zero, we need to use the bisection method. One may also need to use the bisection method for the initial guess of β (see Example 5.7).

5.4.2 Newton Method

For the solution of nonlinear algebraic equations, the general expression given by Eq. (5.37) is followed. The main step in the Newton method is the evaluation and inversion of the Jacobian matrix given by

$$J = \begin{pmatrix} \left[\dfrac{\partial F_i}{\partial \ln K_j}\right] & \left[\dfrac{\partial F_i}{\partial \beta}\right] \\ \left[\dfrac{\partial F_{n_c+1}}{\partial \ln K_j}\right] & \left[\dfrac{\partial F_{n_c+1}}{\partial \beta}\right] \end{pmatrix} \tag{5.57}$$

$[\partial F_i/\partial \ln K_j]$ represents $n_c \times n_c$ elements ($j = 1, \ldots, n_c$; $i = 1 \ldots, n_c$). $[\partial F_i/\partial \beta]$ represents n_c elements ($i = 1, \ldots, n_c$). $[\partial F_{n_{c+1}}/\partial \ln K_j]$ represents n_c elements ($j = 1, \ldots, n_c$) and $[\partial F_{n_{c+1}}/\partial \beta]$ represents a single element. Therefore, there are $(n_c + 1)(n_c + 1)$ elements in the matrix. In Example 5.8, we provide expressions for these elements. The solution procedure is outlined below.

1. Initial estimates of K_i is obtained from the SSI or from stability testing or from the previous time step in simulations.
2. The initial estimate of β is obtained from the solution of the Rachford-Rice equation.
3. The Newton method is then used. The method often converges in 2 to 3 iterations.

5.5 Three-Phase Split Calculations

In the three-phase split calculations, we designate phases by x, y, and z indices. The overall composition is represented by n_i. Similarly to two-phase, equilibrium is written in terms of fugacity coefficients (Haugen et al., 2011).

$$F_i^y = \ln K_i^y - \ln \varphi_i^x + \ln \varphi_i^y = 0 \quad i = 1, \ldots, n_c \tag{5.58a}$$

$$F_i^z = \ln K_i^z - \ln \varphi_i^x + \ln \varphi_i^z = 0 \quad i = 1, \ldots, n_c \tag{5.58b}$$

$K_i^y = y_i/x_i$ and $K_i^z = z_i/x_i$ are the equilibrium ratios of components in phases y and z with respect to reference phase x; x_i, y_i, z_i represent mol fractions in phases x, y, and z, respectively.

In a similar manner as in two-phase split, we write the material balance of species in three-phase and obtain the two Rachford-Rice equations:

$$F^y_{n_c+1} = RR^y = \sum_{i=1}^{n_c}(y_i - x_i) = \sum_{i=1}^{n_c} \frac{n_i\left(K_i^y - 1\right)}{1 + \beta^y\left(K_i^y - 1\right) + \beta^z\left(K_i^z - 1\right)} = 0 \qquad (5.59)$$

$$F^z_{n_c+1} = RR^z = \sum_{i=1}^{n_c}(z_i - x_i) = \sum_{i=1}^{n_c} \frac{n_i\left(K_i^z - 1\right)}{1 + \beta^y\left(K_i^y - 1\right) + \beta^z\left(K_i^z - 1\right)} = 0 \qquad (5.60)$$

β^y and β^z are the mol fractions of phases y and z, respectively. The functions $F^y_{n_c+1}$ and $F^z_{n_c+1}$ are monotonic with respect to β^y and β^z, respectively. For known K_i^y, K_i^z, β^y, and β^z one can calculate the mol fractions x_i, y_i, and z_i from the following expressions

$$x_i = \frac{n_i}{1 + \beta^y\left(K_i^y - 1\right) + \beta^z\left(K_i^z - 1\right)} \quad i = 1, \ldots, n_c \qquad (5.61)$$

$$y_i = K_i^y x_i \quad i = 1, \ldots, n_c \qquad (5.62)$$

$$z_i = K_i^z x_i \quad i = 1, \ldots, n_c \qquad (5.63)$$

In Eqs. (5.58a), (5.58b), (5.59), and (5.60), there are $2(n_c + 1)$ unknowns in $\ln K_i^y$, $\ln K_i^z$, β^y, and β^z. There are also $2(n_c + 1)$ equations. The combination of the SSI method and the Newton method is the natural approach to solve these equations. The SSI method is similar to two-phase calculations. In the Newton method for solving Eqs. (5.58a), (5.58b), (5.59), and (5.60), the unknowns are $\boldsymbol{x}$ defined by the column vector below,

$$\boldsymbol{x} = \begin{pmatrix} \left[\ln K_i^y\right] \\ \left[\ln K_i^z\right] \\ \left[\beta^y\right] \\ \left[\beta^z\right] \end{pmatrix} \qquad (5.64)$$

The elements of the column vector $\boldsymbol{F}$

$$\boldsymbol{F} = \begin{pmatrix} \left[F_i^y\right] \\ \left[F_i^z\right] \\ \left[F^y_{n_c+1}\right] \\ \left[F^z_{n_c+1}\right] \end{pmatrix} \qquad (5.65)$$

are used to calculate the elements of the Jacobian matrix.

$$J = \begin{pmatrix} \left[\dfrac{\partial F_i^y}{\partial \ln K_j^y}\right] & \left[\dfrac{\partial F_i^y}{\partial \ln K_j^z}\right] & \left[\dfrac{\partial F_i^y}{\partial \beta^y}\right] & \left[\dfrac{\partial F_i^y}{\partial \beta^z}\right] \\ \left[\dfrac{\partial F_i^z}{\partial \ln K_j^y}\right] & \left[\dfrac{\partial F_i^z}{\partial \ln K_j^z}\right] & \left[\dfrac{\partial F_i^z}{\partial \beta^y}\right] & \left[\dfrac{\partial F_i^z}{\partial \beta^z}\right] \\ \left[\dfrac{\partial F_{n_c+1}^y}{\partial \ln K_j^y}\right] & \left[\dfrac{\partial F_{n_c+1}^y}{\partial \ln K_j^z}\right] & \left[\dfrac{\partial F_{n_c+1}^y}{\partial \beta^y}\right] & \left[\dfrac{\partial F_i^y}{\partial \beta^z}\right] \\ \left[\dfrac{\partial F_{n_c+1}^z}{\partial \ln K_j^y}\right] & \left[\dfrac{\partial F_{n_c+1}^z}{\partial \ln K_j^z}\right] & \left[\dfrac{\partial F_{n_c+1}^z}{\partial \beta^y}\right] & \left[\dfrac{\partial F_{n_c+1}^z}{\partial \beta^z}\right] \end{pmatrix} \tag{5.66}$$

There are $2(n_c + 1) \times 2(n_c + 1)$ elements in the matrix in Eq. (5.66). In two-phase, there are $(n_c + 1) \times (n_c + 1)$ elements in the Jacobian matrix. We need initial guesses for $\ln K_j^y$, $\ln K_j^z$, β^y, and β^z. The SSI method is used to provide the initial guesses for the Newton method.

In the following, we will first discuss the SSI method and then the Newton method in the three-phase split computations.

5.5.1 SSI Method

The SSI method is based on availability of K_i^y, and K_i^z; n_i is known. There are only two unknowns, β^y and β^z in Eqs. (5.59) and (5.60). The K_i^y, and K_i^z are from the two-phase split computations and stability testing of one of the two phases of the two-phase split computations.

We solve the two *RR* equations, Eqs. (5.59) and (5.60), by the Newton method, which requires the initial guesses for β^y and β^z. Until recently, there has not been much progress despite the fact that various suggestions have been made. In recent years, there have been three suggestions (Okuno et al., 2010; Haugen et al., 2011; and Li and Firoozabadi, 2012b). Okuno et al. (2010) solved the two *RR* equations as a minimization of a convex function with gradients RR^y and RR^z, using the line-search Newton method. They suggest an approach to obtain the starting guesses of β^y and β^z. Haugen et al. (2011) used a geometric approach, employing the two-dimensional bisection method. This approach does not require starting guesses for β^y and β^z. The geometrical approach of Haugen et al. (2011) was later advanced further by Li and Firoozabadi (2012b). Li and Firoozabadi have suggested the use of the conventional Newton method to solve Eqs. (5.59) and (5.60) directly. The successful implementation critically depends on the starting guesses for β^y and β^z. The starting point is the two-phase split calculation. The phase with higher molecular weight is designated as x^{2p} and the phase with lower molecular weight is designated as y^{2p}.The equilibrium ratios, $K_i^{y\,2p}$, are with respect to phase x^{2p}. The phase molar fraction is, therefore, $\beta^{y\,2p}$. Then, the stability of phase x^{2p} is examined.

When x^{2p} is unstable, one should proceed to three-phase splitting. In the phase stability testing, the equilibrium ratios of the phase with the lower Gibbs *TPD* are selected and designated by $K_i^{z\,2p}$ (with respect to phase x^{2p}). The trial phase represents the third phase. With equilibrium ratios $K_i^{y\,2p}$ and $K_i^{z\,2p}$, and starting guesses β^y from two-phase stability and $\beta^z = 0$, the solution of Eqs. (5.59) and (5.60) by the Newton method proceeds nicely.

At each SSI step, K_i^y and K_i^z updated through

$$\ln K_i^y = \ln \varphi_i^x - \ln \varphi_i^y \tag{5.67}$$

$$\ln K_i^z = \ln \varphi_i^z - \ln \varphi_i^z \tag{5.68}$$

with x_i, y_i, and z_i are updated from Eqs. (5.61), (5.62), and (5.63), respectively. The update of the phase compositions provides the update of fugacity coefficients.

5.5.2 Newton Method

Li and Firoozabadi (2012a), based on the initial phase molar fractions from the SSI method and K_i^y and K_i^z described above, have implemented the Newton method for all the $2(n_c + 1)$ unknowns. There is generally convergence in two to three interations. To repeat, the initial K_i-values from the stability testing should correspond to the lowest *TPD* or *TPD**. The initial guess from stability provides initial estimates for the trial phase compositions that, when used with *G* minimization phase splitting algorithm, avoids trivial solutions (Michelsen and Mollerup, 2004). Other methods for phase-split computations, especially for two-phase, are discussed in detail by Michelsen and Mollerup (2004) and Whitson and Brulé (2000).

5.6 Direct Minimization of Gibbs Free Energy in Multiphase Split Calculation

The general problem of multicomponent, multiphase split calculation at constant temperature and pressure has been attacked by direct minimization of the Gibbs free energy (Pan and Firoozabadi, 1998). Using an algorithm that can provide the global minimum of Gibbs free energy, one can readily perform a multiphase-split to obtain the composition and the number of phases.

Consider the multiphase, multicomponent system at constant temperature and pressure sketched in Fig. 5.10. The stable equilibrium state consists of p phases; each phase j consists of $N_{j,1}, N_{j,2}, N_{j,3}, \ldots, N_{j,n_c}$ mols. The stable equilibrium state at which the Gibbs free energy of the system is a minimum is a necessary and sufficient condition. At constant temperature and pressure, the Gibbs free energy of the system in the right side of Fig. 5.10 can be written as

$$G = \sum_{j=1}^{p} G_j(N_{j,1}, N_{j,2}, \ldots, N_{j,n_c}) \tag{5.69}$$

when G_j is the Gibbs free energy of phase j and G is the total Gibbs free energy of the system. The stable equilibrium state corresponds to the global minimum of G with respect to $N_{j,i}$ mols ($i = 1, \ldots, n_c$; $j = 1, \ldots, p$), subject to the following constraints:

1. The material balance for component i

$$N_i = \sum_{j=1}^{p} N_{j,i} \tag{5.70}$$

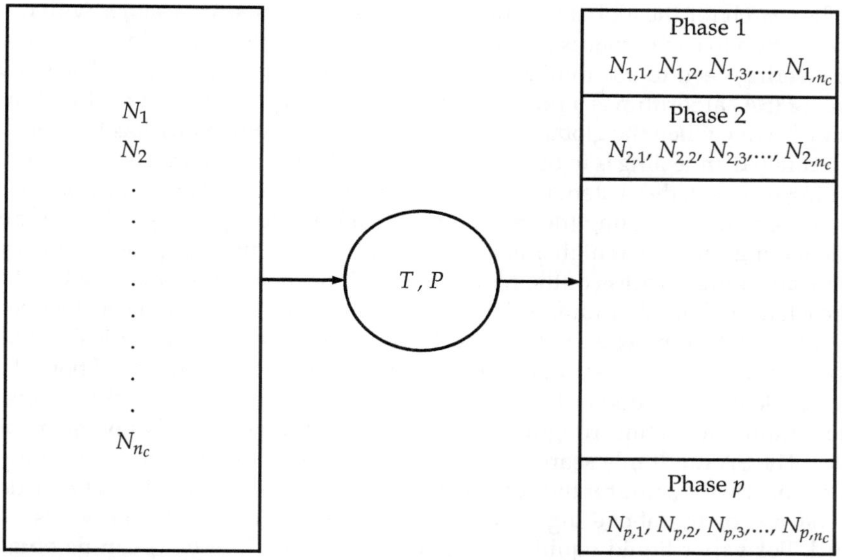

FIGURE 5.10 Schematic representation of multiphase, multicomponent flash at constant temperature and pressure (adapted from Pan and Firoozabadi, 1998).

2. Nonnegative mol numbers of component i in phase j

$$0 \leq N_{j,i} \leq N_i \quad i = 1, \ldots, n_c;\ j = 1, \ldots, p \tag{5.71}$$

The Gibbs free energy of phase j, G_j, in Eq. (5.69) is given by $G_j = \sum_{i=1}^{n_c} N_{j,i}\mu_{j,i}(\underline{\mathbf{N}}_j)$. Therefore, the Gibbs free energy of the system sketched in Fig. 5.10 can be written as

$$G = \sum_{j=1}^{p}\sum_{i=1}^{n_c} N_{j,i}\mu_{j,i}(\underline{\mathbf{N}}_j) \tag{5.72}$$

where $\mu_{j,i}(\underline{\mathbf{N}}_j) = \mu_i^o(T) + RT \ln f_{j,i}(\underline{\mathbf{N}}_j)/f_i^o(T)$. Therefore,

$$G = \left[\sum_{i=1}^{n_c} N_i\mu_i^o(T) - RT\sum_{i=1}^{n_c} N_i f_i^o(T)\right] + RT\sum_{j=1}^{p}\sum_{i=1}^{n_c} N_{j,i}\ln f_{j,i}(\underline{\mathbf{N}}_j) \tag{5.73}$$

In Eq. (5.73), the first term on the right side is constant, since T, P, and N_i are fixed. The second term on the right is a function of $N_{j,i}$. We can eliminate the mol numbers of one of the phases, say phase 1, from the material balance expression, Eq. (5.70):

$$N_{1,i} = N_i - \sum_{j=2}^{p} N_{j,i} \quad i = 1, \ldots, n_c \tag{5.74}$$

The above equations reduce the number of unknowns from $p(n_c)$ to $p(n_c - 1)$ in the minimization search.

The search for the global minimum of a function of many variables with minima that can be close to each other is a difficult problem. Disciplines ranging from economics to engineering need to use minimization (or optimization) algorithms. The simulated annealing (SA) algorithm is a powerful tool for this purpose. This algorithm is apparently very effective when the global extremum is hidden among many local extrema. The root of simulated annealing is in thermodynamics. When a molten metal is cooled slowly (annealing process), the system is able to reach a highly ordered crystalline state of the lowest energy. Rapid cooling (quenching process) leads to a polycrystalline or amorphous state having somewhat higher energy. The simulated annealing algorithm is analogous to the annealing process of the molten metal. It decreases the objective function slowly to reach its global minimum. A detailed description of the method can be found in Press et al. (1992), Corana et al. (1987), and Goffe et al. (1994). Mathematically, the SA algorithm globally optimizes an objective function with the constraints of bounds. The sole drawback of the method is the computational cost because of the "slow cooling process." The computational time roughly increases linearly with the number of independent variables. The SA randomly searches for the optimized point. Therefore, the computational time is nearly the same for the near-critical region and as far away from the critical region.

The use of the above algorithm for phase equilibrium calculations is straightforward. For vapor-liquid equilibria, initial estimates of $K_i = y_i/x_i$ can be obtained from the Wilson correlation. Close to the critical point, $K_i = 1$ is provided for light components and $K_i = 0.5$ for heavy components of a multicomponent mixture. Figure 5.11 depicts the

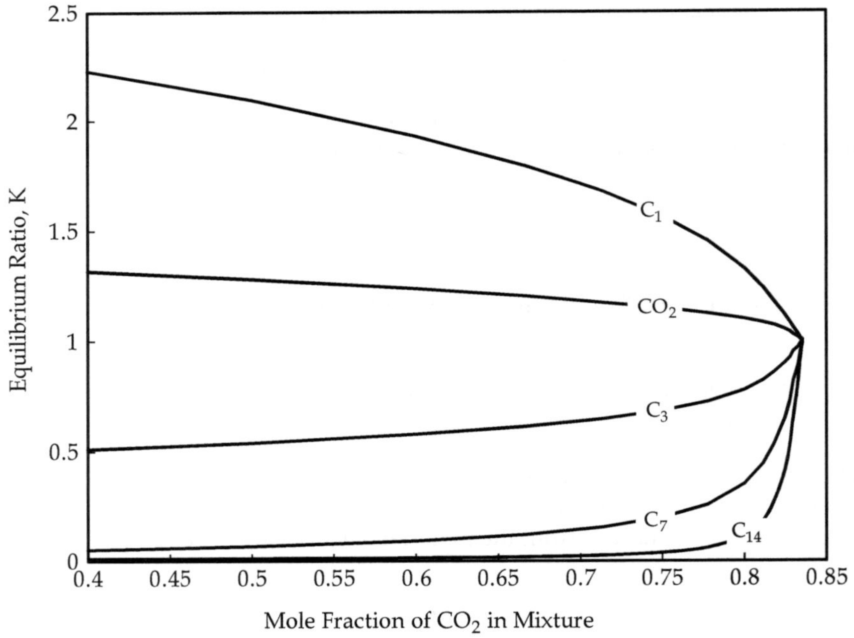

FIGURE 5.11 Equilibrium ratios vs. CO_2 concentration for the CO_2-synthetic oil mixture at 322.0 K and 105.35 bar (adapted from Pan and Firoozabadi, 1998).

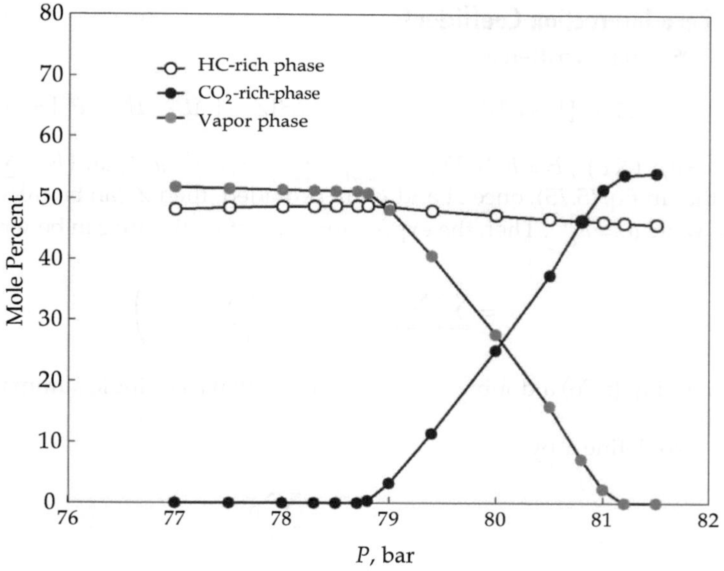

FIGURE 5.12 Mole percent of different phases versus pressure for the CO_2-reservoir oil system at 307.6 K (adapted from Pan and Firoozabadi, 1998).

example for the multicomponent mixture: $z_{C_1} = 0.35$, $z_{C_2} = 0.03$, $z_{C_3} = 0.04$, $z_{n\text{-}C_4} = 0.06$, $z_{n\text{-}C_5} = 0.04$, $z_{n\text{-}C_6} = 0.03$, $z_{n\text{-}C_7} = 0.05$, $z_{n\text{-}C_{10}} = 0.30$, $z_{n\text{-}C_{14}} = 0.05$, and varying amounts of CO_2. (The introduction of CO_2 in the mixture reduces the mol fraction of the other components proportionally.) Note that the calculation proceeds to the near-critical region where all K_i-values are close to one.

For V-L-L equilibrium calculations, one can select the light-liquid phase as the reference phase. Figure 5.12 plots the calculated results for CO_2-reservoir oil system at various pressures at fixed composition and 307.6 K. The compositions of oil B and CO_2 mixtures investigated by Shelton and Yarborough (1977) are used. Note that up to a pressure of 78.8 bar, two phases, a vapor phase and a hydrocarbon-rich liquid phase, coexist. From $P = 78.8$ to 81.2 bar, three phases co-exist and above $P = 81.2$ bar, the vapor phase disappears and two liquid phases are present. Details of various parameters can be found in Pan and Firoozabadi (1998). The direct minimization of Gibbs free energy can be used to perform wax and asphaltene precipitation calculations (Firoozabadi, 1999).

5.7 Phase-Split Calculations in Reduced Space

In the following, we will first present the expressions to perform phase-split calculations using the reduction method with zero interaction coefficients. The case with zero interaction coefficients results in 3 equations and 3 unknowns independent of the number of components (Michelsen, 1986). We will use the PR-EOS in our derivations.

5.7.1 Zero Interaction Coefficients

The PR-EOS can be written as:

$$Z^3 - [1 - \mathcal{B}]Z^2 + [\mathcal{A} - 3\mathcal{B}^2 - 2\mathcal{B}]Z - [\mathcal{A}\mathcal{B} - \mathcal{B}^2 - \mathcal{B}^3] = 0 \tag{5.75}$$

where $\mathcal{A} = ap/(RT)^2$, $\mathcal{B} = bP/RT$, $a = \sum_{i=1}^{n_c}\sum_{j=1}^{n_c} x_i x_j a_i^{1/2} a_j^{1/2}$, and $b = \sum_{j=1}^{n_c} x_j b_j$.

Note that in Eq. (5.75), once $\mathcal{A}$ and $\mathcal{B}$ are provided, then Z can be calculated.

Let us write $\alpha_i = a_i^{1/2}$. Then the expression for a of a mixture can be written as:

$$a = \sum_i \sum_j \alpha_i \alpha_j x_i x_j = \left(\sum_j \alpha_j x_j\right)^2 \tag{5.76}$$

Note that in Eq. (5.76) a double summation is written as a single summation because of symmetry.

Let us also define α by

$$\alpha = a^{1/2} = \sum_{j=1}^{n_c} \alpha_j x_j \tag{5.77}$$

Then

$$\sum_{j=1}^{n_c} a_i^{1/2} a_j^{1/2} x_j = \alpha_i \alpha \tag{5.78}$$

Combining Eqs. (5.76) and (5.78) with the definitions of a and b, and the expression for fugacity coefficient of component i from Eq. (3.32), one obtains

$$\ln \varphi_i = \frac{b_i}{b}[Z - 1] - \ln[Z - \mathcal{B}] - \frac{\alpha^2}{2\sqrt{2}RTb}\left[2\frac{\alpha_i}{\alpha} - \frac{b_i}{b}\right] \ln\left[\frac{Z + (1 + \sqrt{2})\mathcal{B}}{Z + (1 - \sqrt{2})\mathcal{B}}\right] \tag{5.79}$$

In terms of the variables b and α,

$$Z = Z(b, \alpha) \tag{5.80}$$

$$\varphi_i = \varphi_i(b, \alpha) \tag{5.81}$$

Note that P and T are assumed constant as in the case for the (isothermal) phase-split calculations.

In conventional two-phase split calculations, there are $(2n_c + 1)$ unknowns: V, y_i, and x_i (given T, P, and z_i). The equations are:

$$x_i \varphi_i^L = y_i \varphi_i^V \quad i = 1, \ldots, n_c \tag{5.82}$$

$$z_i = V y_i + (1 - V)x_i \quad i = 1, \ldots, n_c \tag{5.83}$$

$$\sum_{i=1}^{n_c} (y_i - x_i) = 0 \tag{5.84}$$

Note that in Eq. (5.83), V is the mol fraction of the gas phase. The symbol β is used in our earlier derivations.

In the reduction method, as we will see shortly, the number of equations is three. From Eq. (5.82),

$$K_i = \frac{\varphi_i^L(b^L, \alpha^L)}{\varphi_i^V(b^V, \alpha^V)} \tag{5.85}$$

Multiplying Eq. (5.83) by α_i and taking the summation, one obtains:

$$\sum_{i=1}^{n_c} \alpha_i z_i = V \sum_{i=1}^{n_c} \alpha_i y_i + (1 - V) \sum_{i=1}^{n_c} \alpha_i x_i \tag{5.86}$$

or

$$\boxed{\alpha^F = V\alpha^V + (1 - V)\alpha^L} \tag{5.87}$$

Similarly, multiplying Eq. (5.83) by b_i and taking the summation,

$$\boxed{b^F = Vb^V + (1 - V)b^L} \tag{5.88}$$

From Eqs. (5.87) and (5.88),

$$\alpha^L = \frac{\alpha^F - V\alpha^V}{(1 - V)} \tag{5.89a}$$

$$b^L = \frac{b^F - Vb^V}{(1 - V)} \tag{5.89b}$$

The subscripts, F, V, and L in the above equations represent feed, vapor, and liquid, respectively. If we now take the gas phase as the reference phase,

$$\boxed{K_i = K_i\left(\varphi_i^L, \varphi_i^V\right) = K_i(\alpha^L, b^L, \alpha^V, b^V) = K_i(\alpha^V, b^V, V)} \tag{5.90}$$

In other words, given α^V, b^V, and V, one can determine K_i.

Multiplying Eq. (5.52) by α_i and b_i, and taking the summation, we obtain:

$$\sum_{i=1}^{n_c} \frac{z_i K_i \alpha_i}{1 + V(K_i - 1)} - \alpha^V = 0 \tag{5.91}$$

$$\sum_{i=1}^{n_c} \frac{z_i K_i b_i}{1 + V(K_i - 1)} - b^V = 0 \tag{5.92}$$

The above two equations, and Eq. (5.53) repeated below,

$$\sum_{i=1}^{n_c} \frac{z_i(K_i - 1)}{1 + V(K_i - 1)} = 0 \tag{5.93}$$

give the three equations in the reduction method for the case of zero interaction coefficients. Note that in Eqs. (5.91) to (5.93), $K_i = K_i(\alpha^V, b^V, V)$, and therefore, the unknowns are α^V, b^V, and V. Once these three variables are known, we can calculate α^L

[from Eq. (5.89a)] and b^L [from Eq. (5.89b)]. The rest follows from knowledge of α^V, α^L, b^V, and b^L.

Next, we will outline the solution procedure using the SSI method.

Solution Procedure

Given T, P, and z_i $i = 1, \ldots, n_c$, we are interested in performing phase-split calculations for mixtures composed of n_c components. The following steps provide the solution procedure.

1. Calculate α^F and b^F; $\alpha^F = \sum_{i=1}^{n_c} \alpha_i z_i, b^F = \sum_{i=1}^{n_c} b_i z_i$.
2. Decide on a reference phase; here we select the gas phase.
3. Estimate α^V, b^V, and V.
4. Estimate α^L and b^L from Eqs. (5.89a) and (5.89b), respectively.
5. Calculate Z^L and Z^V.
6. Calculate $\varphi_i^V(\alpha^V, b^V)$ and $\varphi_i^L(\alpha^V, b^V, V)$, then $K_i = \varphi_i^L / \varphi_i^V$.
7. Calculate x_i and y_i; y_i from Eq. (5.52).
8. Check the convergence criteria for $\sum_{i=1}^{n_c}(y_i - x_i) = \varepsilon_1$, $\sum y_i \alpha_i - \alpha^V = \varepsilon_2$, and $\sum_{i=1}^{n_c} y_i b_i - b^V = \varepsilon_3$. In case the tolerances are not met, go to the next step.
9. Update V, α^V, and b^V and go back to step 4.

Initial estimates of $\alpha^V = \alpha_1$, $\alpha^V = \alpha_{n_c}$, $b^V = b_1$, $b^V = b_{n_c}$, and $V = 0.5$ can be assumed. A more efficient approach will be from stability analysis, which will be covered in the general case.

5.7.2 Nonzero Interaction Coefficients

The key element to extend the reduction method to the nonzero interaction coefficient case relates to the parameter "a" of the PR-EOS. The expression for "a" in a mixture is given by

$$a = \sum_{i=1}^{n_c} \sum_{j=1}^{n_c} a_i^{1/2} a_j^{1/2} (1 - \delta_{ij}) x_i x_j \tag{5.94}$$

When $\delta_{ij} = 0$, Eq. (5.94) can be written as a single summation [see Eq. (5.76)]. Let us define $\beta_{ij} = (1 - \delta_{ij})$ as the elements of the matrix $\boldsymbol{\beta}$. The matrix $\boldsymbol{\beta}$ is symmetric. We invoke the spectral theorem of linear algebra (see Strang, 1988) to diagonalize $\boldsymbol{\beta}$,

$$\boldsymbol{\beta} = \boldsymbol{SDS}^{-1} = \boldsymbol{SDS}^T \tag{5.95}$$

where $\boldsymbol{D}$ is the eigenvalue matrix defined by

$$\boldsymbol{D} = \begin{pmatrix} \lambda_1 & 0 & 0 & 0 & 0 & 0 \\ 0 & \lambda_2 & | & | & | & | \\ 0 & 0 & \lambda_3 & | & | & | \\ \vdots & \vdots & \vdots & \vdots & \vdots & \vdots \\ 0 & 0 & 0 & 0 & 0 & \lambda_{n_c} \end{pmatrix} \tag{5.96}$$

where $\lambda_1, \lambda_2, \ldots, \lambda_{n_c}$ are the eigenvalues of $\boldsymbol{\beta}$. The matrix S is given by

$$S = \left(\boldsymbol{q}'^{(1)}, \boldsymbol{q}'^{(2)}, \ldots, \boldsymbol{q}'^{(\alpha)}, \ldots, \boldsymbol{q}'^{(n_c)}\right) \tag{5.97}$$

In Eq. (5.97), $\boldsymbol{q}'^{(1)}, \boldsymbol{q}'^{(2)}, \ldots, \boldsymbol{q}'^{(n_c)}$ are the eigenvectors. Each eigenvector is given by

$$\boldsymbol{q}'^{(i)} = \left(q'_{i1}, q'_{i2}, \ldots, q'_{in_c}\right) \tag{5.98}$$

Note that in Eq. (5.97), $S^T = S^{-1}$ because S is in orthogonal matrix.

From expansion of Eq. (5.95),

$$\beta_{ij} = (1 - \delta_{ij}) = \sum_{k=1}^{n_c} \lambda_k q'_{ki} q'_{kj} \tag{5.99}$$

Combining Eqs. (5.94) and (5.99), one obtains

$$a = \sum_{k=1}^{n_c} \lambda_k \sum_{i=1}^{n_c} a_i^{1/2} x_i q'_{ki} \sum_{j=1}^{n_c} a_j^{1/2} x_j q'_{kj} \tag{5.100}$$

Let $q_{ki} = a_i^{1/2} q'_{k_i}$ and $q_{kj} = a_j^{1/2} q'_{kj}$, then Eq. (5.100) can be written as

$$\boxed{a = \sum_{k=1}^{n_c} \lambda_k Q_k^2} \tag{5.101}$$

where

$$\boxed{Q_k = \sum_{i=1}^{n_c} x_i q_{ki}} \tag{5.102}$$

Generally, $k > m$; $m < n_c$, then λ_k becomes negligible (Firoozabadi and Pan, 2002); therefore,

$$a \approx \sum_{k=1}^{m} \lambda_k Q_k^2 \tag{5.103}$$

Equation (5.103) implies that

$$a = a(Q_1, \ldots, Q_m) \tag{5.104}$$

and therefore we can write [from Eq. (5.75)],

$$Z = Z(Q_1, \ldots, Q_m, b) \tag{5.105}$$

The following term in the fugacity coefficient expression [see Eq. (3.32)] can be written in terms of new variables Q_k and q_{ki},

$$\sum_{j=1}^{n_c} a_i^{1/2} a_j^{1/2} (1 - \delta_{ij}) x_j = \sum_{k=1}^{m} \lambda_k q_{ki} Q_k \quad i = 1, \ldots, n_c \tag{5.106}$$

We now substitute Eq. (5.106) into Eq. (3.32) for the fugacity coefficient.

$$\ln \varphi_i = \frac{b_i}{b}(Z-1) - \ln(Z-\mathcal{B}) - \frac{\mathcal{A}}{2\sqrt{2}\mathcal{B}} \left[\frac{2\sum_{k=1}^{m} \lambda_k q_{ki} Q_k}{a} - \frac{b_i}{b} \right] \ln \frac{Z+2.414\mathcal{B}}{Z-0.414\mathcal{B}} \qquad (5.107)$$

In Eq. (5.107), $\varphi_i = \varphi_i(Q_1, \ldots, Q_m, b)$, because according to Eq. (5.103) $a = a(Q_1, \ldots, Q_m)$.

In the following, we present the equations for phase-split calculations.

Phase-Split Calculations

From Eq. (5.82),

$$\boxed{K_i = \frac{\varphi_i^L(Q_1^L, \ldots, Q_m^L, b^L)}{\varphi_i^V(Q_1^V, \ldots, Q_m^V, b^V)}} \qquad (5.108)$$

We establish the relation between Q_α^V and Q_α^L, and b^V and b^L. Toward that end, we multiply Eq. (5.83) by $q_{\alpha i}$ and by b_i, then take the summation. The results are

$$Q_\alpha^F = VQ_a^V + (1-V)Q_\alpha^L \quad \alpha = 1, \ldots, m \qquad (5.109)$$

$$b^F = Vb^V + (1-V)b^L \qquad (5.110)$$

From the above two equations, we obtain

$$Q_\alpha^L = \frac{Q_\alpha^F - VQ_\alpha^V}{(1-V)} \quad \alpha = 1, \ldots, m \qquad (5.111)$$

$$b^L = \frac{b^F - Vb^V}{(1-V)} \qquad (5.112)$$

Therefore, if the vapor phase is selected as the reference phase, from $K_i = \varphi_i^L / \varphi_L^V$

$$K_i = K_i\left(Q_1^V, \ldots, Q_m^V, b^V, V\right) \qquad (5.113)$$

We can now set up the equations for phase-split calculations:

$$\sum_{i=1}^{n_c} \frac{z_i K_i q_{\alpha i}}{1+V(K_i-1)} - Q_\alpha^V = 0 \quad \alpha = 1, \ldots, m \qquad (5.114)$$

$$\sum_{i=1}^{n_c} \frac{z_i K_i b_i}{1+V(K_i-1)} - b^V = 0 \qquad (5.115)$$

$$\sum_{i=1}^{n_c} \frac{z_i(K_i-1)}{1+V(K_i-1)} = 0 \qquad (5.116)$$

The above $(m+2)$ equations provide the $(m+2)$ unknowns V, b^V, and Q_α^V $(\alpha = 1, \ldots, m)$. In order to solve the above nonlinear equations, we need initial guesses for Q_α^V, b^V, and V. The estimate of Q_α^V and b^V should come from stability analysis.

Stability Analysis

The stability analysis in the reduction method is both simple and efficient. The main idea is to write the *TPD* function in terms of the independent variables [Q_α ($\alpha = 1, \ldots, m$)] and b instead of conventional variables X_i ($i = 1, \ldots, n_c$). The new variables will increase the computational speed from: (1) smoothness of the *TPD* function (Firoozabadi and Pan, 2002), and (2) reduction in the number of variables. The Newton method, even when the initial estimates of K_i-values are far from the solution, often converges in the reduction model (Firoozabadi and Pan, 2002). In some cases, we may need one or two steps of the SSI method. In the following, we will present a simple formulation and discuss the algorithm based on the work of Hoteit and Firoozabadi (2006).

The basic expression for the reduction method for stability testing starts from [see Eq. (5.102)]:

$$Q_\alpha = \sum_{i=1}^{n_c} x_i q_{\alpha i} \quad \alpha = 1, \ldots, m+1 \tag{5.117}$$

where $q_{m+1,i} = b_i$ ($i = 1, \ldots, n_c$), and $Q_{m+1} = b$.

We now substitute $x_i = X_i / \sum_{i=j}^{n_c} X_j$, and $X_j = \exp(d_j(z))\varphi_j/(\underline{Q})$ from Eq. (5.33) in Eq. (5.117) to obtain

$$h_\alpha(\underline{Q}) = \sum_{i=1}^{n_c} \frac{\exp[d_i(z)]}{\varphi_i(\underline{Q})}(Q_\alpha - q_{\alpha i}) = 0 \quad \alpha = 1, \ldots, m+1 \tag{5.118}$$

In the above equation,

$$\min q_{\alpha i} \le Q_{\alpha i} \le \max q_{\alpha i} \quad \alpha = 1, \ldots, m+1, \qquad i = 1, \ldots, n_c \tag{5.119}$$

Problem 5.7 provides a hint on the derivation of Eq. (5.119). Equation (5.118) is the basic expression of the reduction method for stability testing. The solution by the Newton method is:

$$Q_\alpha^{(k+1)} = Q_\alpha^{(k)} - \left(\frac{\partial h_\alpha^{(k)}(\underline{Q})}{\partial Q_\gamma}\right)^{-1} h_\alpha^{(k)}(\underline{Q}) \quad \alpha, \gamma = 1, \ldots, m+1 \tag{5.120}$$

where $(\partial h_\alpha / \partial Q_\gamma)$ are the elements of the Jacobian matrix. The calculation of these elements is given in Example 5.10. We are now set to present the algorithm.

Algorithm for the Reduction Method

There are five steps in the implementation of the reduction method.

1. Initialize the K_i-values from the Wilson correlation or from other procedures.
2. Initialize X_i from $X_i = z_i/K_i$ or $X_i = z_i K_i$. Note that there may be no need to use Eq. (5.41).
3. Repeat steps 3(a) and 3(b) (below) twice. This is the SSI step.
   a. $Q_\alpha = (1/X_T)\sum_{i=1}^{n_c} X_i q_{\alpha i} \quad \alpha = 1, \ldots, m+1$
   b. $X_i = \exp[d_i(z)]/\varphi_i(\underline{Q})$

4. Repeat
   a. Calculate the Jacobian matrix and obtain $Q_\alpha^{(k+1)}$ from Eq. (5.120).
   b. Check for feasibility.
      i. If $\min q_{\alpha i} \le Q_\alpha^{(k+1)} \le \max q_{\alpha i} \quad \alpha = 1, \ldots, m+1; i = 1, \ldots, n_c$, go to step 4(c).
      ii. Otherwise, update $Q_\alpha^{(k+1)}$ from $Q_\alpha = \sum_{i=1}^{n_c} X_i q_{\alpha i} / X_T \quad \alpha = 1, \ldots, m+1$
   c. Update X_i from $X_i = \exp[d_i(z)]/\varphi_i(\underline{Q}) \quad i = 1, \ldots, n_c$
   d. Check convergence: $\| Q^{k+1} - Q^k \| < \varepsilon$. If not satisfied, go to Step 4(c).
5. Check the sign of $TPD(\underline{X})$

Note that if the inequality expressed by Eq. (5.119) is not satisfied, the algorithm overpasses the Newton iteration [Step 4(a)] and instead, an update by the SSI is carried out (Step 3). Therefore, the algorithm couples the SSI method and the Newton method where in our experience only a single SSI step is performed in case of the failure of the Newton method.

The number of iterations in the reduction method is shown in Fig. 5.13. Close to the discontinuity, there is a substantial increase in the number of iterations, as expected. However, away from the discontinuity, the number of iterations is less than five. These results are remarkable in the sense that the combined effect of a small number of iterations in the reduction method and a small number of variables give rise to an efficient calculation.

Previously we showed the number of iterations for the SSI/Newton methods with conventional variables in Fig. 5.8. The CPU time, in the conventional-variable formulation, is an order of magnitude higher than in the reduction method, because of the large number of variables. In the reduction method, for the data presented in Fig. 5.13, the number of variables is 4. In the conventional method, the number of variables is the same as the number of components—it is 14. In the reduction method for stability

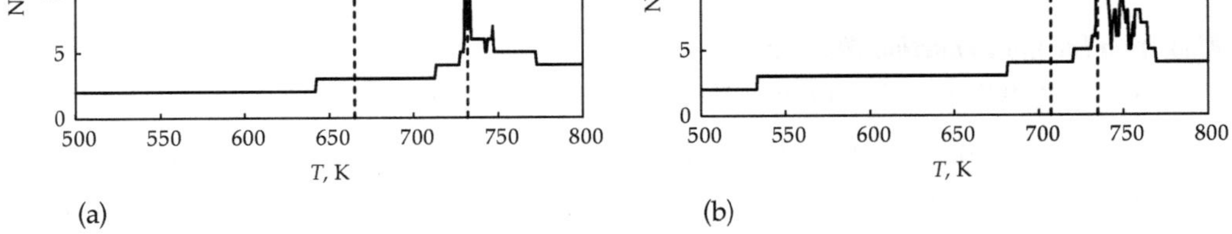

FIGURE 5.13 Number of iterations in the reduction method versus temperature: Mixture 4 of Hoteit and Firoozabadi (2006), (a) stability of gas-like phase at $P = 10$ bar, (b) stability of gas-like phase at $P = 50$ bar (from Hoteit and Firoozabadi, 2006).

testing, the results from Firoozabadi and Pan (2002) and Hoteit and Firoozabadi (2006) show clear advantage over the conventional variables. For phase-split calculations, there have been discussions in the literatures in efficiency of the reduction method over the conventional-variable formulations (Haugen and Becker, 2013; Gorucu and Johns, 2013; Michelsen, Yan and Stenby, 2013; Petitfrere and Nichita, 2015). There is agreement, however, as the number of species increases to over 20, the phase-split calculation by the reduction method becomes more efficient than the conventional-variable formulation.

5.8 Critical-Point Calculation

There are two main motivations for the calculation of the critical point in a multicomponent mixture. One motivation is to calculate the critical point to have an idea of the criticality and its features, such as vanishing of the interfacial tension. In compositional reservoir simulation, the main motivation for the estimation of the critical point is to use it for phase recognition, as we discussed in Chap. 3.

Hoteit et al. (2006) present a simple, robust, and efficient algorithm for critical-point calculation in multicomponent mixtures, which is as efficient as two-phase split calculations. We will closely follow their approach.

The two criteria for criticality based on the work of Heidemann and Khail (1980) were presented in Chap. 4 in the following forms.

$$\sum_{i=1}^{n_c}\sum_{j=1}^{n_c}(A_{ij})(\Delta N_i)(\Delta N_j) = 0 \tag{5.121}$$

$$\sum_{i=1}^{n_c}\sum_{j=1}^{n_c}\sum_{k=1}^{n_c}(A_{ijk})(\Delta N_i)(\Delta N_j)(\Delta N_k) = 0 \tag{5.122}$$

where $A_{ij} = (\partial^2 A/\partial N_i \partial N_j)$ and $A_{ijk} = (\partial^3 A/\partial N_i \partial N_j \partial N_k)$; A is the Helmholtz free energy. The expression for various derivatives of A, including A_i, A_{ij}, and A_{ijk} are given in Example 4.7.

Eq. (5.121) can be expressed by

$$\det(\boldsymbol{H}) = 0 \tag{5.123}$$

and

$$\boldsymbol{H}\Delta\boldsymbol{N} = 0 \tag{5.124}$$

where $\boldsymbol{H}$ is the $n_c \times n_c$ Hessian matrix with elements A_{ij} given above, and $\Delta\boldsymbol{N}$ is the column vector of nonzero perturbations in mol numbers $(\Delta N_1, \Delta N_2, \ldots \Delta N_{n_c})$. The derivatives in the above equations are at constant T, V, and mol numbers other than those used in the differentiation. There are $(n_c + 2)$ unknowns and $(n_c + 2)$ equations. The unknowns are T_c, V_c, and $\Delta N_i; i = 1, \ldots, n_c$. The $(n_c + 2)$ equations come from Eqs. (5.122) to (5.124). One of the equations is dependent since the system given by Eq. (5.124) is singular due to conditions given by Eq. (5.123). In order to evaluate $\Delta\boldsymbol{N}$, we need to impose an additional condition by setting the value of one of the elements in $\Delta\boldsymbol{N} = (\Delta N_1, \Delta N_2, \ldots, \Delta N_{n_c})$, say ΔN_1, or the magnitude of the vector $\Delta\boldsymbol{N}$.

Critical-point calculation is the solution to the nonlinear one-variable expressions given by Eqs. (5.122) and (5.123), and the system of linear equations expressed by

Eq. (5.124). Because for some mixtures there may not be a critical point, and for some mixtures there will be more than one critical point, we are interested in a method that is both robust and efficient.

A key equation to be solved is the volume objective function from Eq. (5.122), which can be simplified to

$$\sum_{i=1}^{n_c}\sum_{j=1}^{n_c}\sum_{k=1}^{n_c} A_{ijk}\Delta N_i \Delta N_j \Delta N_k = F_1\chi^3 + \left(F_2\sum_{i=1}^{n_c}\Delta N_i + F_3\sum_{i=1}^{n_c} A_i\Delta N_i\right)\chi^2$$
$$+F_4\left(2\sum_{i=1}^{n_c}\sum_{j=1}^{n_c} a_{ij}\Delta N_i\Delta N_j + \sum_{i=1}^{n_c} a_{ij}\Delta N_i^2\right)\chi \tag{5.125}$$
$$-RT\sum_{i=1}^{n_c}\frac{\Delta N_i^3}{N_i^2}$$

where $\chi = \sum_{i=1}^{n_c} b_i\Delta N_i$ and $A_i = (\partial A/\partial N_i)$; the coefficients F_1 to F_4 are independent of the indices i, j, and k. They are defined in the following:

$$F_1 = \frac{2RT}{(V-\mathcal{B})^3} + \frac{3\mathcal{A}F_6}{2\sqrt{2}\mathcal{B}^4} - \frac{2V\mathcal{A}(3V-2\mathcal{B})}{\mathcal{B}^2F_5^2} - \frac{8V\mathcal{A}(V-\mathcal{B})^2}{\mathcal{B}F_5^3} \tag{5.126a}$$

$$F_2 = \frac{3RT}{(V-\mathcal{B})^2} \tag{5.126b}$$

$$F_3 = \frac{3F_6}{2\sqrt{2}\mathcal{B}^3} - \frac{6V(\mathcal{B}-V)}{\mathcal{B}F_5^2} \tag{5.126c}$$

$$F_4 = \frac{3F_6}{2\sqrt{2}\mathcal{B}^2} \tag{5.126d}$$

F_5 and F_6 are given by:

$$F_5 = (V^2 + 2V\mathcal{B} - \mathcal{B}^2) \tag{5.126e}$$

$$F_6 = 2\ln\frac{V+(\sqrt{2}+1)\mathcal{B}}{V-(\sqrt{2}-1)\mathcal{B}} - \frac{4\sqrt{2}\mathcal{B}V}{F_5} \tag{5.126f}$$

The form of the triple-summation and its reduction given by Eq. (5.125), and the solution of this nonlinear equation by an appropriate technique, result in efficiency and robustness of the critical-point calculation. In the following, we will outline the numerical approach and present the algorithm for the critical-point calculation.

5.8.1 Numerical Solution

Equations (5.122) through (5.124) which are composed of two one-variable functions in nonlinear form, and a linear system of equations, are solved iteratively using two nested one-variable searches.

Suppose V_c is provided, then T_c is computed from Eq. (5.123). Next, we solve for ΔN from Eq. (5.124), which are the eigenvectors corresponding to the zero eigenvalues of $\boldsymbol{H}$.

The procedure for the solution based on given overall composition z_i is outlined in the following.

1. Provide an initial guess for V_c.
2. Provide an initial guess for T_c.
3. Calculate the elements of $\boldsymbol{H}$ based on V_c, T_c, and P_c from the EOS. If det($\boldsymbol{H}$), is greater than a given tolerance, assume a new value of T_c and repeat this step. Otherwise, continue to the next step.
4. Solve Eq. (5.124) for $\Delta \boldsymbol{N}$, calculate A_{ijk} $(i, j, k = 1, \ldots, n_c)$ and then evaluate Eq. (5.122). If it is greater than a given tolerance, assume a new V_c and return to Step 3. Otherwise, end.

In the following, we discuss the details of the solution to Eqs. (5.123) and (5.122) and updating of T_c and V_c.

5.8.2 One-D Search

Equation (5.123) provides T_c and Eq. (5.122) is the cubic function for V_c. Given V_c, Eq. (5.123) may be solved using the Newton method which may fail to converge when a derivative is very small or when the initial guess is not close enough to the solution. The Newton method also requires the evaluation of the derivatives. An alternative method for the solution of Eq. (5.123) is the use of the Brent method (Hoteit et al., 2006). The Brent approach combines three methods: (1) Bisection method, (2) Secant method, and (3) Inverse quadratic method (Press et al. 1992).

The bisection method is based on the idea that a root is bracketed in the interval (a, b) of $f(a)$ and $f(b)$ of function f have opposite signs. Once it is known that an interval contains a root, one can use the bisection method with confidence to find the root. The method consists of evaluating the function at the interval's midpoint and examining its sign. Use the midpoint to replace whichever limit has the same sign. After each iteration, the bounds containing the limit decreases by a factor of two. The method can be slow, but when combined with other methods, it can become efficient. The bisection method has a linear convergence rate. We have already commented on the use of the bisection method to solve the Rachford-Rice nonlinear equation in two-phase split calculations.

The secant method which is closely related to the Newton method replaces the function derivative by a finite difference approximation. It is faster than the bisection method due to a superlinear convergence rate of about 1.62 (Arfken, 1985).

The details of the procedure for the solution of Eq. (5.123) to obtain T_c are:

1. Obtain an initial bracketing interval $[T_{c,\text{inf}}, T_{c,\text{sup}}]$. This step will be discussed later.
2. Evaluate the objective function $f_{T,\text{inf}}$ and $f_{T,\text{sup}}$ corresponding to $T_{c,\text{inf}}$ and $T_{c,\text{sup}}$.
3. Set $T_c = T_{c,\text{inf}}$ and $f_T = f_{T,\text{inf}}$.
4. Obtain a new estimate of T_c:
   a. by using the inverse quadratic method. If T_c belongs to the bracketing interval, go to Step 5.
   b. by using the secant method. If T_c belongs to the bracketing interval, go to Step 5.
   c. by using the bisection method.

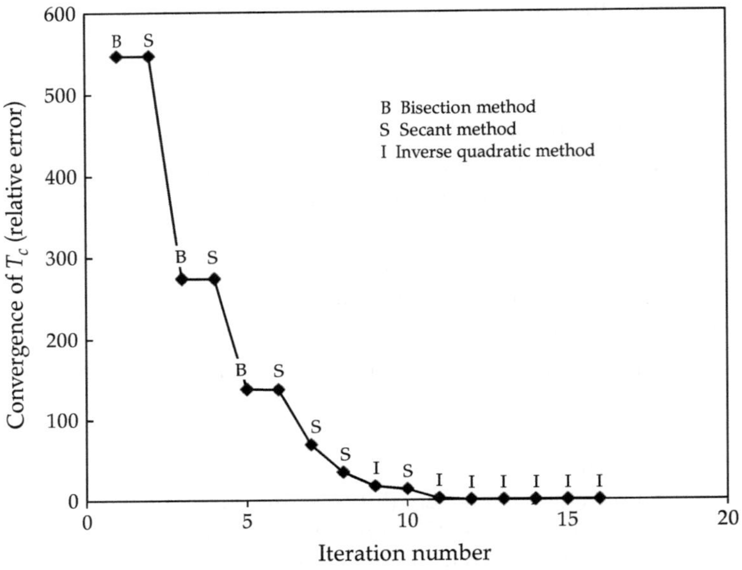

FIGURE 5.14 Switch between the three methods in each iteration to calculate critical temperature (from Hoteit et al., 2006).

5. Evaluate the objective function f_T corresponding to T_c.
6. Update the bracketing interval by checking the sign of f_T.
7. Update the relative error for $T_{c,\mathrm{inf}}$ and $T_{c,\mathrm{sup}}$. If it is greater than a predefined tolerance, go to Step 4. Otherwise, end. Relative tolerance for T_c can be $([T_c^{(k+1)} - T_c^{(k)}]/T_c^{(k)} = 10^{-12})$.

Figure 5.14 shows an example where the algorithm switches between the three methods to the critical temperature for a six-component mixture from Hoteit et al. (2006). The bisection method is used when the initial guess may be far from the solution. Then it may or may not switch to faster methods (secant and inverse gradient methods).

5.8.3 Initial Guess

The initial bracketing interval $[T_{c,\mathrm{inf}}, T_{c,\mathrm{sup}}]$ and $[V_{c,\mathrm{inf}}, V_{c,\mathrm{sup}}]$ for the critical temperature and critical volume, respectively, must be provided. The interval boundaries can be estimated from

$$V_{c,\mathrm{inf}} = m^b_{V,\mathrm{inf}}\mathcal{B} \quad \text{and,} \quad V_{c,\mathrm{sup}} = m^b_{V,\mathrm{sup}}\mathcal{B} \tag{5.127}$$

and

$$T_{c,\mathrm{inf}} = m_{T,\mathrm{inf}} \min(T_{ci}),\ T_{c,\mathrm{sup}} = m_{T,\mathrm{sup}} \max(T_{ci}) \tag{5.128}$$

The coefficients $m_{V,\inf}$, $m_{V,\sup}$, $m_{T,\inf}$, and $m_{T,\sup}$ are selected such that the intervals cover all possible physical solutions for the critical temperature and volume. Values of $m_{V,\inf} = 1.01$, $m_{V,\sup} = 4$, $m_{T,\inf} = 0.5$, and $m_{T,\sup} = 1.5$ have been suggested by Hoteit et al. (2006). The idea is to find a sign change bracketing interval to subdivide uniformly the interval into n_{sub} subintervals.

The following procedure is used to estimate the bracketing interval for T_c. The procedure to obtain an initial bracketing interval for V_c is similar to that of T_c with some differences which will be discussed shortly.

1. Subdivide interval $[T_{c,\inf}, T_{c,\sup}]$ into n_{sub} subintervals by defining a spacing step $\Delta T_c = (T_{c,\sup} - T_{c,\inf})/n_{sub}$.
2. Evaluate the objective function f_T corresponding to $T_{c,\inf}$.
3. Set $T_{c,\inf} = T_{c,\inf} + \Delta T$ and $f_{T,int} = f_T$.
4. Evaluate the objective function f_T corresponding to $T_{c,\inf}$.
5. If f_T and $f_{T,int}$ have the same sign, go to Step 3. Otherwise, end.

Solution of Eq. (5.122) using a similar procedure as the above should lead to all zeros of the volume objective function. There may be no critical point or there may be more than one critical point. One would specify all the bracketing intervals that have a sign change of the volume objective function. Relative tolerance for V_c can be $|V_c^{(k+1)} - V_c^{(k)}|/V_c^{(k)} = 10^{-10}$. Figure 5.15 shows a plot of the cubic objective function versus molar volume for Mixture 3 in Table 1 of Hoteit et al. (2006). Note that the function has three zeros indicating three critical points. Figure 5.16 depicts a plot of the cubic

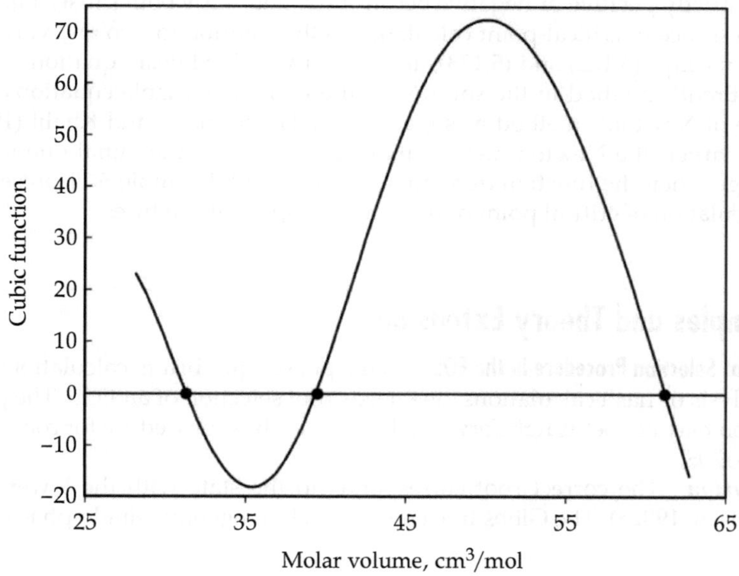

FIGURE 5.15 Behavior of the cubic function for Mixture 3 in Hoteit et al. (2006).

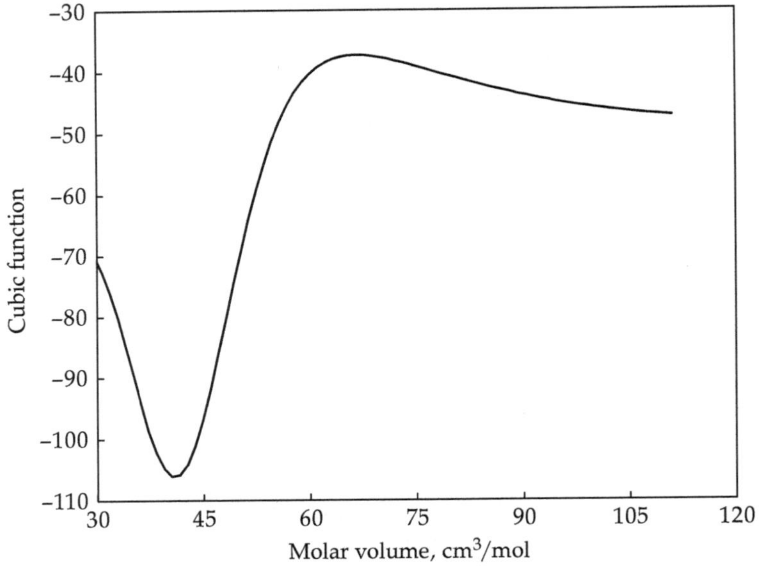

FIGURE 5.16 Behavior of the cubic function for Mixture 1 in critical volume calculation (Hoteit et al., 2006).

function for Mixture 1 in Table 1 of the same reference. Note that the function is strictly negative indicating no critical point. Depending on the choice of ΔN, the cubic function can be strictly positive or negative because ΔN and $-\Delta N$ both satisfy Eq. (5.121).

The essence of critical-point calculation is the solution to two one-variable nonlinear equations, Eqs. (5.122) and (5.123), and a multivariable linear equation, Eq. (5.124). The use of Brent's method in the solution of the two one-variable equations is superior to the use of Newton's method as suggested by Heidermand and Khalil (1980). As mentioned earlier, the Newton method may be sensitive to the initial guess, and may not converge when the function derivative is very small. Example 5.11 presents results for the calculation of critical point of two multicomponent mixtures.

5.9 Examples and Theory Extension

5.1: Root Selection Procedure in the EOS In the phase equilibrium calculations, either stability analysis or flash calculations, one needs root selection of an EOS. The process of root selection may not be straightforward. Present a robust procedure for root selection using the PR-EOS.

Solution The correct root corresponds to the state with the lower Gibbs energy (Michelsen, 1982a). The Gibbs free energy of a homogenous single-phase system is

$$g = \sum_{i=1}^{n_c} z_i \mu_i(z, T, P)$$

Let us assume that Z_1 and Z_2 are two different roots of EOS. We define

$$\Delta g = (g|_{Z_2} - g|_{Z_1})$$

From the above equations, Eq. (1.111), and Eq. (3.32) we obtain

$$\Delta g = \sum_{i=1}^{n_c} z_i[\mu_i(z, T, P)|_{Z=Z_2} - \mu_i(z, T, P)|_{Z=Z_1}]$$

$$= RT\sum_{i=1}^{n_c} z_i[\ln f_1(z, T, P)|_{Z=Z_2} - \ln f_1(z, T, P)|_{Z=Z_2}]$$

$$= RT\sum_{i=1}^{n_c} z_i \left[\frac{b_i}{b}(Z_2 - Z_1) - \ln\left(\frac{Z_2 - \mathcal{B}}{Z_1 - \mathcal{B}}\right) - \frac{\mathcal{A}}{2\sqrt{2}\mathcal{B}}\left[\frac{2\sum_{j=1}^{n_c} z_j a_{ij}}{a} - \frac{b_i}{b}\right] \times \ln\frac{(Z_2 + 2.414\mathcal{B})(Z_1 - 0.414\mathcal{B})}{(Z_2 - 0.414\mathcal{B})(Z_1 + 2.414\mathcal{B})}\right]$$

If $\Delta g > 0$, then Z_1 should be selected, otherwise, Z_2 is the true root.

5.2: Molar Gibbs Free Energy of Binary Mixtures over the Full Compositional Range Consider a binary mixture of C_1 and n-C_5 at $P = 100$ bar and $T = 350$ K. Plot Δg_{mix} vs. x_2 (mol fraction of n-C_5) over the whole range (i.e., $0 \leq x_2 \leq 1$), assuming that the C_1/n-C_5 mixture will stay in the hypothetical single-phase state. Then derive

$$\left(\frac{\partial^2 \Delta g_{mix}}{\partial x_2^2}\right)_{T,P} = -\frac{1}{x_2}\left(\frac{\partial \mu_1}{\partial x_2}\right)_{T,P}.$$

Based on stability criteria and the use of the above equation, show the unstable, stable, and metastable parts of the Δg_{mix} curve. Relate the curvatures to the stability.

Solution In order to plot Δg_{mix} [see Eq. (5.2)], the expression for $(\mu_i - \mu_i^0)$ is first evaluated [see Eq. (1.111)]:

$$\mu_i(T, P, x_2) - \mu_i^0(T, P) = RT\ln\frac{f_i(T, P, x_2)}{f_i^0(T, P)}$$

Using the above equation in Eq. (5.2),

$$\left(\frac{\Delta g_{mix}}{RT}\right) = \sum_{i=1}^{2} x_i \ln\frac{f_i}{f_i^0}$$

The PR-EOS can be used to evaluate f_i and f_i^0 for the C_1/n-C_5 mixture, and for C_1 and n-C_5 pure components at 100 bar and 350 K, respectively. The results are presented in Fig. 5.17.

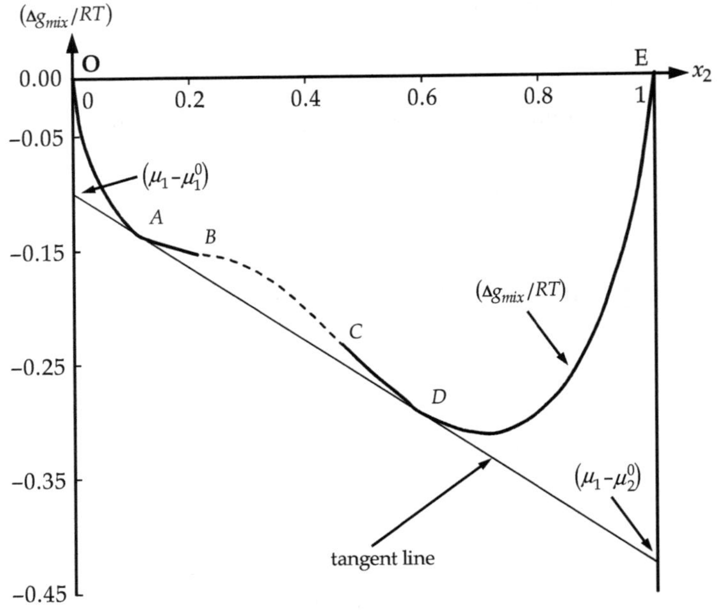

FIGURE 5.17 $(\Delta g_{mix}/RT)$ vs. x_2 for the C_1/n-C_5 mixture at 350 K and 100 bar.

Taking the derivative of Eq. (5.4), one readily obtains

$$\left(\frac{\partial^2 \Delta g_{mix}}{\partial x_2^2}\right)_{T,P} = -\frac{1}{x_2}\left(\frac{\partial \mu_1}{\partial x_2}\right)_{T,P}$$

For a two-component system to be stable, $(\partial \mu_1/\partial x_2)_{T,P}$ should be less than zero [see Eq. (4.61)]. Therefore for stability of a two-component system, $(\partial^2 \Delta g_{mix}/\partial x_2^2)_{T,P} > 0$. Points B and C in Fig. 5.17 represent $(\partial^2 \Delta g_{mix}/\partial x_2^2)_{T,P} = 0$. For points between B and C on the Δg_{mix} curve $(\partial^2 \Delta g_{mix}/\partial x_2^2)_{T,P} < 0$, and therefore are unstable; the curve is convex. In other words at points B and C the curvature of the Δg_{mix} changes. All the points on the curves OAB and CDE are stable; the curves are convex. The points between A and B and between C and D are of metastable nature.

5.3: Comparison of *TPD* and *TPD** Show that the stationary points of *TPD* and *TPD** have the same sign.

Solution The *TPD* at the stationary point is given by

$$TPD(\mathbf{x}^{sp}) = K = -\ln X_T \tag{5.129}$$

where $X_T = \sum_{i=1}^{n_c} X_i$. At the stationary points of TPD^* from Eq. (5.34)

$$\partial TPD^*(\underline{\mathbf{X}})/\partial X_i = 0 \quad i = 1, \ldots, n_c \tag{5.130}$$

$$\ln \varphi_i(\underline{\mathbf{X}}) + \ln X_i - d_i(z) = 0, \quad i = 1, \ldots, n_c \tag{5.131}$$

Comparing Eqs. (5.131) and (5.33) of the text shows the same stationary points for *TPD* and *TPD**.

Combining Eqs. (5.131) and (5.33) of the text, we obtain

$$TPD^*(\underline{\boldsymbol{X}}^{sp}) = 1 + \sum_{i=1}^{n_c} X_i[-1] = 1 - X_T = 1 - \exp[-TPD(\boldsymbol{x}^{sp})]$$

Therefore, *TPD* and *TPD** have the same sign at the stationary points.

5.4: Local Quadratic Convergence of Newton Method

a. Use the Newton method to show that $f(x) = x^2 - a$ has quadratic convergence (a is positive).

b. Find the square root of 3 by using the Newton method and present its rapid convergence.

Solution

a. The Newton method for a single-value function can be written as,

$$x^{(k+1)} = x^{(k)} - \frac{f(x^{(k)})}{f'(x^{(k)})} = x^{(k)} - \frac{x^{2(k)} - a}{2x^{(k)}} = \frac{x^{(k)}}{2} + \frac{a}{2x^{(k)}}$$

Subtracting $\sqrt{a}$ from both sides of the above equation (to write the expression for error),

$$x^{(k+1)} - \sqrt{a} = \frac{(x^{(k)} - \sqrt{a})^2}{2x^{(k)}}$$

In terms of relative error, the above equation becomes

$$\frac{x^{(k+1)} - \sqrt{a}}{\sqrt{a}} = \left(\frac{x^{(k)} - \sqrt{a}}{\sqrt{a}}\right)^2 \left(\frac{\sqrt{a}}{2x^{(k)}}\right)$$

In the above equation, as long as $|x^{(k)} - \sqrt{a}| < |2x^{(k)}|$, the new error $|x^{(k+1)} - \sqrt{a}|$ will be less than the old error $|x^{(k)} - \sqrt{a}|$. As a consequence, new errors will be much less than the previous errors. The error at each iteration will be approximately the square of the previous error.

b. $f(x) = x^2 - 3$

Let us use $x^{(0)} = 2$ as the initial guess, then from

$$x^{(k+1)} = x^{(k)} - \frac{f(x^{(k)})}{f'(x^{(k)})} = x^{(k)} - \frac{(x^{2(k)} - 3)}{2x^{(k)}}$$

$x^{(1)} = 1.75$, $x^{(2)} = 1.7321428$, $x^{(3)} = 1.7320508$, which has eight correct digits. Note that for a linear function, there is need for only one iteration.

5.5: Stability in Two-Phase and Higher Systems Show that in order to test the stability of the two-phase split, one can determine the stability of one of the phases only. The same applies to stability testing of multiphases where any one phase may be selected.

Solution Let us consider state I where phases x and y are at equilibrium. Let N_i^x and N_i^y be the mol numbers of component i in phases x and y, respectively. The corresponding mol fractions are z_i^x and z_i^y, and the chemical potentials are μ_i^x and μ_i^y ($\mu_i^x = \mu_i^y$ at two-phase equilibrium). The Gibbs free energy of state I is given by

$$G^I = G^x(\underline{N}^x, T, P) + G^y(\underline{N}^y, T, P) = \sum_{i=1}^{n_c} N_i^x \mu_i^x(\mathbf{z}^x, T, P) + \sum_{i=1}^{n_c} N_i^y \mu_i^y(\mathbf{z}^y, T, P)$$

If phases x and y are not stable, let us assume formation of infinitesimal amount of a new phase with mols n^x and n^y from phases x and y, respectively. The compositions of phases x and y are not affected. The Gibbs free energy of state II after formation of very small amount of a new trial phase is given by,

$$\begin{aligned} G^{II} &= G^x(\underline{N}^x - \underline{n}^x, T, P) + G^y(\underline{N}^y - \underline{n}^y, T, P) + G^{trial}(\underline{n}^x + \underline{n}^y, T, P) \\ &\approx G^x(\underline{N}^x, T, P) - \sum_{i=1}^{n_c} n_i^x \mu_i^x(\mathbf{z}^x, T, P) + G^y(\underline{N}^y, T, P) \\ &\quad - \sum_{i=1}^{n_c} n_i^y \mu_i^y(\mathbf{z}^y, T, P) + \sum_{i=1}^{n_c} (n_i^x + n_i^y) \mu_i^{trial}(\mathbf{z}^{trial}, T, P) \end{aligned}$$

where $n_i^x + n_i^y$ is the total number of mols of component i in the trial phase, and z_i^{trial} and μ_i^{trial} are the corresponding mol fractions and chemical potentials, respectively. The change in Gibbs free energy from states I to II, using $\mu_i^x = \mu_i^y$, is given by

$$\begin{aligned} \Delta G &= \sum_{i=1}^{n_c} (n_i^x + n_i^y) \left[\mu_i^{trial}(\mathbf{z}^{trial}, T, P) - \mu_i^x(\mathbf{z}^x, T, P)\right] \\ &= \sum_{i=1}^{n_c} (n_i^x + n_i^y) \left[\mu_i^{trial}(\mathbf{z}^{trial}, T, P) - \mu_i^y(\mathbf{z}^y, T, P)\right] \end{aligned}$$

Let us divide the last expression by $n^x + n^y$ to obtain

$$\begin{aligned} \Delta g &= \sum_{i=1}^{n_c} z_i^{trial} \left[\mu_i^{trial}(\mathbf{z}^{trial}, T, P) - \mu_i^x(\mathbf{z}^x, T, P)\right] \\ &= \sum_{i=1}^{n_c} z_i^{trial} \left[\mu_i^{trial}(\mathbf{z}^{trial}, T, P) - \mu_i^y(\mathbf{z}^y, T, P)\right] \end{aligned}$$

Therefore, the two-phase stability testing is equivalent to testing the stability of one of the two equilibrium phases from the two-phase split calculation.

5.6: Weakness in Phase-Split Formulation Based on Equality of Fugacities The equations based on equality of fugacities read

$$F_i = f_i^L(\underline{N}^L) - f_i^V(\underline{N}^V) = 0 \quad i = 1, \ldots, n_c \tag{5.132}$$

$$N_i = N_i^L + N_i^V \quad i = 1, \ldots, n_c \tag{5.133}$$

The above equations with $2n_c$ unknowns can be reduced readily to n_c unknown using Eq. (5.133). Here, we will use the mol numbers in the gas phase for demonstration. One can also use $N_i^L (i = 1, \ldots, n_c)$ as the independent variables. Newton's method for the solution of Eq. (5.132) in terms of N_i^V can be written as

$$N_i^{V(k+1)} = N_i^{V(k)} - (J_{ij})^{-1} F_i^{(k)} \quad i = 1, \ldots, n_c \tag{5.134}$$

where $N_i^{V(k+1)}$ is the value of N_i^V at $(k+1)^{\text{th}}$ iteration level. J_{ij} is the element of $n_c \times n_c$ Jacobian matrix, given by

$$\frac{\partial F_i}{\partial N_j^V} = \frac{1}{N^V} \sum_{k=1}^{n_c-1} \frac{\partial f_i^V}{\partial y_k} [\delta_{k,j} - y_k] + \frac{1}{N - N^V} \sum_{k=1}^{n_c-1} \frac{\partial f_i^L}{\partial y_k} [\delta_{k,j} - y_k] \quad i, j = 1, \ldots, n_c \tag{5.135}$$

where $N = \sum_{i=1}^{n_c} N_i$, $N^V = \sum_{i=1}^{n_c} N_i^V$. $N^L = \sum_{i=1}^{n_c} N_i^L$

Derive the above expression and discuss the weakness in formulation.

Solution Let us consider the expression of fugacity of vapor in terms of number of mols of vapor components, N_i^V as:

$$f_i^V = f_i^V (N_1^V, N_2^V, \ldots, N_{nc}^V) \quad i = 1, \ldots, n_c$$

Partial differentiation of the above equation with respect to N_j^V yields,

$$\frac{\partial f_i^V}{\partial N_j^V} = \sum_{k=1}^{n_c-1} \frac{\partial f_i^V}{\partial y_k} \frac{\partial y_k}{\partial N_j^V} \quad i, j = 1, \ldots, n_c$$

in which the term $\partial y_k / \partial N_j^V$ can be derived from:

$$\frac{\partial y_k}{\partial N_j^V} = \frac{1}{N^V} [\delta_{k,j} - y_k] \quad k = 1, \ldots, n_c - 1, \text{ and } j = 1, \ldots, n_c$$

where $\delta_{kj} = 1$ for $k = j$ and $\delta_{kj} = 0$ for $k \neq j$.

Combining the above two equations yields,

$$\frac{\partial f_i^V}{\partial N_j^V} = \frac{1}{N^V} \sum_{k=1}^{n_c-1} \frac{\partial f_i^V}{\partial y_k} [\delta_{k,j} - y_k] \quad i, j = 1, \ldots, n_c$$

Similarly, the partial differentiation of the fugacities in liquid can be written as

$$\frac{\partial f_i^L}{\partial N_j^L} = \frac{1}{N^L} \sum_{k=1}^{n_c-1} \frac{\partial f_i^L}{\partial x_k} [\delta_{k,j} - x_k] \quad i, j = 1, \ldots, n_c$$

in which $N^L = N - N^V$. Note that close to the gas phase boundary, these derivatives become very large. As a result, the method suffers from division by zero.

Partial differentiation of material balance equation [Eq. (5.133)] with respect to N_i^V yields:

$$\frac{\partial N_j^L}{\partial N_i^V} = -\delta_{ij} \quad i, j = 1, \ldots, n_c$$

Therefore,

$$\frac{\partial f_i^L}{\partial N_j^V} = \frac{\partial f_i^L}{\partial N_j^L}\frac{\partial N_j^L}{\partial N_j^V} = -\delta_{ij}\frac{\partial f_i^L}{\partial N_j^L} \quad i, j = 1, \ldots, n_c$$

Performing partial differentiation of Eq. (5.132) with respect to N_j^V with the help of above equation yields:

$$\frac{\partial F_i}{\partial N_j^V} = \frac{\partial f_i^V}{\partial N_j^V} - \frac{\partial f_i^L}{\partial N_j^V} = \frac{\partial f_i^V}{\partial N_j^V} + \delta_{ij}\frac{\partial f_i^L}{\partial N_j^L} \quad i, j = 1, \ldots, n_c$$

Plugging $\partial f_i^V/\partial N_j^V$ and $\partial f_i^L/\partial N_j^L$, minor manipulations, and substituting $N^L = N - N^V$ into the above equation provides the result. The weakness in the formulation in N^V or N^L is in the denominator. When either N^V or N^L is small, there is division by a small number.

5.7: Root Finding in the Rachford-Rice Equation The root finding in the Rachford-Rice equation can be complicated using only the Newton method in the near critical region. Use the bisection method (a very robust method) to solve the Rachford-Rice equation in phase-split calculations in the SSI method, and outline the procedure.

Solution

1. The initial value of β for the Rachford-Rice equation [see Eq. (5.54)] is based on the maximum and minimum value of K_i such that, $\beta_1 = 1/[1 - \max(K_i)]$ and $\beta_2 = 1/[1 - \min(K_i)]$.
2. The value of β is set as: $\beta = (\beta_1 + \beta_2)/2$.
3. The function $RR(\beta)$ is calculated from $RR(\beta) = \sum_{i=1}^{n_c} (K_i - 1)z_i/[1 + (K_i - 1)\beta]$.
4. β is set to β_1 when $RR(\beta) \geq 0$ and β is set to β_2 when $RR(\beta) < 0$.
5. Steps 2 to 5 are repeated until absolute value of $|\beta_1 - \beta_2| \leq 10^{-6}$ (i.e., convergence criterion).

5.8: Elements of the Matrix *J* in Eq. (5.57) Derive various derivatives of $F_i \quad i = 1, \ldots, n_c + 1$ in Eq. (5.57).

Solution Overall composition is given by $z_k = (1 - \beta)x_k + \beta y_k$, therefore,

$$\left.\frac{\partial z_k}{\partial \ln K_j}\right|_\beta = (1 - \beta)\left.\frac{\partial x_k}{\partial \ln K_j}\right|_\beta + \beta\left.\frac{\partial y_k}{\partial \ln K_j}\right|_\beta = 0 \quad k = 1, \ldots, n_{c-1}; \; j = 1, \ldots, n_c$$

We write

$$\left.\frac{\partial y_k}{\partial \ln K_j}\right|_\beta = -\frac{1-\beta}{\beta}\left.\frac{\partial x_k}{\partial \ln K_j}\right|_\beta \tag{5.136}$$

From

$$\left.\frac{\partial x_k}{\partial \ln K_j}\right|_\beta = K_j\left.\frac{\partial x_k}{\partial K_j}\right|_\beta$$

and from Eq. (5.51), we obtain

$$\left.\frac{\partial x_k}{\partial \ln K_j}\right|_\beta = K_j\left(-\frac{\delta_{k,j} z_k \beta}{[1+\beta(K_k-1)]^2}\right) \quad k=1,\ldots,n_{c-1};\ j=1,\ldots,n_c \tag{5.137}$$

Similarly,

$$\left.\frac{\partial y_k}{\partial \ln K_j}\right| = K_j\frac{(1-\beta)\delta_{k,j} z_k}{[1+\beta(K_k-1)]^2} \quad k=1,\ldots,n_{c-1};\ j=1,\ldots,n_c \tag{5.138}$$

From Eqs. (5.137) and (5.138), and $F_{n_c+1} = \sum_i^{n_c} y_i - \sum_{i=1}^{n_c} x_i$

$$\left.\frac{\partial F_{n_c+1}}{\partial \ln K_j}\right|_\beta = \sum_{k=1}^{n_c}\left(\left.\frac{\partial y_k}{\partial \ln K_j}\right|_\beta - \left.\frac{\partial x_k}{\partial \ln K_j}\right|_\beta\right) \quad j=1,\ldots,n_c \tag{5.139}$$

From Eq. (5.47),

$$\left.\frac{\partial F_i}{\partial \ln K_j}\right|_\beta = \left.\frac{\partial \ln K_i}{\partial \ln K_j}\right|_\beta - \left.\frac{\partial \ln \varphi_i^L}{\partial \ln K_j}\right|_\beta + \left.\frac{\partial \ln \varphi_i^V}{\partial \ln K_j}\right|_\beta \quad i,\ j=1,\ldots,n_c$$

where

$$\left.\frac{\partial \ln K_i}{\partial \ln K_j}\right|_\beta = \delta_{i,j} \quad i,\ j=1,\ldots,n_c \tag{5.140}$$

$$\left.\frac{\partial \ln \varphi_i^L}{\partial \ln K_j}\right|_\beta = \sum_{k=1}^{n_c-1}\left.\frac{\partial \ln \varphi_i^L}{\partial x_k}\right|_\beta \left.\frac{\partial x_k}{\partial \ln K_j}\right|_\beta \quad i,\ j=1,\ldots,n_c$$

From Eq. (5.137)

$$\left.\frac{\partial \ln \varphi_i^L}{\partial \ln K_j}\right|_\beta = \sum_{k=1}^{n_c-1}\left.\left(\frac{\partial \ln \varphi_i^L}{\partial x_k}\right)\right|_\beta \frac{-K_j\delta_{k,j} z_k \beta}{[1+\beta(K_k-1)]^2} \quad i,\ j=1,\ldots,n_c$$

Hence,

$$\left.\frac{\partial \ln \varphi_i^L}{\partial \ln K_j}\right|_\beta = \left.\left(\frac{\partial \ln \varphi_i^L}{\partial x_j}\right)\right|_\beta \frac{-K_j z_i \beta}{[1+\beta(K_j-1)]^2} \quad \text{if} \quad j \neq n_c \tag{5.141}$$

and

$$\left.\frac{\partial \ln \varphi_i^L}{\partial \ln K_j}\right|_\beta = 0 \ \text{ if } \ j = n_c \tag{5.142}$$

Similarly,

$$\left.\frac{\partial \ln \varphi_i^V}{\partial \ln K_j}\right|_\beta = \left.\left(\frac{\partial \ln \varphi_i^V}{\partial y_i}\right)\right|_\beta \frac{-K_j z_j(1-\beta)}{[1+\beta(K_j-1)]^2} \text{ if } j \neq n_c \tag{5.143}$$

and

$$\left.\frac{\partial \ln \varphi_i^V}{\partial \ln K_j}\right|_\beta = 0 \text{ if } j = n_c \tag{5.144}$$

From some of the above equations,

$$\left.\frac{\partial F_i}{\partial \ln K_j}\right|_\beta = \delta_{i,j} + \frac{z_j K_j}{[1+\beta(K_j-1)]^2}\left(\beta\left.\left(\frac{\partial \ln \varphi_i^L}{\partial x_j}\right)\right|_\beta + (1-\beta)\left.\left(\frac{\partial \ln \varphi_i^V}{\partial y_j}\right)\right|_\beta\right) \tag{5.145}$$

$i = 1, \ldots, n_c;\; j \neq n_c$

$$\left.\frac{\partial F_i}{\partial \ln K_j}\right|_\beta = \delta_{i,j} \quad i = 1, \ldots, n_c;\; j = n_c \tag{5.146}$$

Next, we derive the expression for $(\partial F_i/\partial \beta)$ $i = 1, \ldots, n_c$

From Eq. (5.47),

$$\left.\frac{\partial F_i}{\partial \beta}\right|_{\underline{\mathbf{K}}_j} = \left.\frac{\partial \ln K_i}{\partial \beta}\right|_{\underline{\mathbf{K}}_j} - \left.\frac{\partial \ln \varphi_i^L}{\partial \beta}\right|_{\underline{\mathbf{K}}_j} + \left.\frac{\partial \ln \varphi_i^V}{\partial \beta}\right|_{\underline{\mathbf{K}}_j} \quad i = 1, \ldots, n_c$$

Note that

$$\left.\frac{\partial \ln K_i}{\partial \beta}\right|_{\underline{\mathbf{K}}_j} = 0$$

As for the second term,

$$\left.\frac{\partial \ln \varphi_i^L}{\partial \beta}\right|_{\underline{\mathbf{K}}_j} = \sum_{k=1}^{n_c-1}\left.\left(\frac{\partial \ln \varphi_i^L}{\partial x_k}\right)\right|_{x_k}\left.\left(\frac{\partial x_k}{\partial \beta}\right)\right|_{\underline{\mathbf{K}}_j} \quad i = 1, \ldots, n_c.$$

Substituting for $\partial x_k/\partial \beta|_{\underline{\mathbf{K}}_j}$ using Eq. (5.51),

$$\left.\frac{\partial \ln \varphi_i^L}{\partial \beta}\right|_{\underline{\mathbf{K}}_j} = \sum_{k=1}^{n_c-1}\left.\left(\frac{\partial \ln \varphi_i^L}{\partial x_k}\right)\right|_{x_k}\left.\left(\frac{-z_k(K_k-1)}{[1+\beta(K_k-1)]^2}\right)\right|_{\underline{\mathbf{K}}_j}$$

Similarly,

$$\left.\frac{\partial \ln \varphi_i^V}{\partial \beta}\right|_{\underline{\mathbf{K}}_j} = \sum_{k=1}^{n_c-1}\left.\left(\frac{\partial \ln \varphi_i^V}{\partial y_k}\right)\right|_{y_k}\left(\frac{-z_k K_k(K_k-1)}{[1+\beta(K_k-1)]^2}\right)$$

Combining the above equations, we obtain

$$\left.\frac{\partial F_i}{\partial \beta}\right|_{\underline{\mathbf{K}}_j} = \sum_{k=1}^{n_c-1} \frac{z_k(K_k-1)}{[1+\beta(K_k-1)]^2}\left\{\left(\frac{\partial \ln \varphi_i^L}{\partial x_k}\right) - K_k\left(\frac{\partial \ln \varphi_i^V}{\partial y_k}\right)\right\}$$

$$i = 1, \ldots, n_c$$

The derivative $(\partial F_{nc+1}/\partial \beta)$ is easily computed.

5.9: Multiphase Split and Stability Analysis Discuss in detail the stability analysis of a three-component system of $C_1/CO_2/n\text{-}C_{16}$ at $T = 294.0$ K and $P = 67$ bar with $z_{C_1} = 0.05$, $z_{CO_2} = 0.90$, and $z_{n\text{-}C_{16}} = 0.05$.

Solution At fixed temperature and pressure, from the phase rule $F = n_c + 2 - p$, there can be a maximum of three phases when the interface between the phases is flat. The first question is what types of phases may exist: gas, liquid, or solid. Let us assume that a solid phase does not exist for the above system. Therefore one might expect: (1) a single gas phase or a single liquid phase, (2) gas and liquid phases, (3) liquid and liquid phases, or (4) gas-liquid-liquid phase separation. The difficulty in liquid-liquid (L-L) and vapor-liquid-liquid (V-L-L) and higher-phase equilibria (for more than three components) is how many phases should be considered for phase-split calculations. One approach is to determine whether one, two, or more phases are to be considered without prior knowledge of the true number of phases. One may, therefore, follow a sequential approach as outlined next for the $C_1/CO_2/n\text{-}C_{16}$ example.

1. Perform the stability analysis on the overall mixture to examine the stability of the single phase mixture. In case the original mixture is unstable, an estimate of equilibrium ratios (K_i-values) are then on hand for the two-phase split.
2. Perform the two-phase split using the K_i-values from the stability test as the initial guess.
3. Perform stability analysis on only one of the phases of the two-phase flash. If the selected phase is unstable, estimates of the equilibrium ratios for part of the three-phase flash are available.
4. Perform the three-phase flash using the estimates of equilibrium ratios from steps 2 and 3.
5. Perform the stability analysis on only one of the phases of the three-phase flash to check stability of the three phases. The three-phase split should be stable since single-phase and two-phases were unstable (this is true only for a three-component system).

In Step 1, we can use K_i-values from the Wilson correlation to proceed with the stability of the original phase and to calculate the composition of the trial second phase. The Wilson correlation may not be appropriate if the trial phase and the original phase are in liquid states. From the Wilson correlation, $K_{C_1} = 4.564$, $K_{CO_2} = 0.8709$, and $K_{n\text{-}C_{16}} = 0.000001$. We can use $x_i = K_i z_i$ and $x_i = z_i/K_i$ as estimates of the trial phase composition. If we minimize the Gibbs free energy from Eq. (5.13) with respect to x_i with the initial estimates above, $TPD = -0.4493$ and -0.026348. We use the K_i-values from the lowest TPD for two-phase split calculations. They are $K_{C_1} = 1.2264$, $K_{CO_2} = 1.8396$, and

$K_{n\text{-}C_{16}} = 0.05465$. The K_i-values corresponding to $TPD = -0.026438$ are $K_{C_1} = 2.2284$, $K_{CO_2} = 1.01650$, and $K_{n\text{-}C_{16}} = 0.0015$. Then, we perform flash computations. The K_i-values of two-phase split are: $K_{C_1} = 1.2845$, $K_{CO_2} = 1.7410$, and $K_{n\text{-}C_{16}} = 0.00469$. Note that the K_i-values of the two-phase split are close to the K_i-values of the lower TPD.

Next, we perform stability of both phases of the two-phase split. Only one is required, but here, for the purpose of demonstration, we examine the stability of both phases. We first examine the stability of the vapor phase with composition of $y_{C_1} = 0.0523$, $y_{CO_2} = 0.09382$, and $y_{n\text{-}C_{16}} = 0.0095$ (all mol fractions). The TPDs are -0.013075 and -2.84×10^{-7}. We use the K_i-values from the lowest TPD. A two-phase mixture is the outcome in two-phase computation. The liquid-phase of the two-phase is also unstable and its phase-split calculation results in two-phase.

In the final step, we perform three-phase split calculations based on K_i-values from the two-phase split calculation and stability analysis of one of the phases in two-phase split with lowest TPD results. The final results (composition in mol fractions) are

Feed	Vapor Phase	Liquid Phase 1	Liquid Phase 2
C_1, 0.05	0.072490	0.03444	0.030418
CO_2, 0.90	0.925058	0.96333	0.549997
n-C_{16}, 0.05	0.000060	0.00221	0.419584

In all the two-phase and three-phase split calculations, the number of SSI was equal or less than 7. The number of Newton iterations varied between 1 and 3. The number of iterations depends on the tolerances in our code.

5.10: $\partial h_\alpha / \partial Q_\gamma$ Derivatives in the Jacobian Matrix

Solution The calculations of the above derivatives are relted to $\partial \ln \varphi_i(\underline{Q})/\partial Q_\gamma$. Let us first calculate $\partial \ln \varphi_i(\underline{Q})/\partial \mathcal{B}$ ($Q_{m+1} = \mathcal{B}$). We have

$$\frac{\partial \ln \varphi_i(\underline{Q})}{\partial \mathcal{B}} = \frac{\partial \ln \varphi_i(\underline{Q})}{\partial Z}\frac{\partial Z}{\partial \mathcal{B}} + \left.\frac{\partial \ln \varphi_i(\underline{Q})}{\partial \mathcal{B}}\right|_Z$$

where we can readily find

$$\frac{\partial Z}{\partial \mathcal{B}} = \frac{-Z^2 + 2(3\mathcal{B}+1)Z + \mathcal{A} - 2\mathcal{B} - 3\mathcal{B}^2}{3Z^2 - 2(1-\mathcal{B})Z + (\mathcal{A} - 2\mathcal{B} - 3\mathcal{B}^2)}$$

$$\frac{\partial \ln \varphi_i(\underline{Q})}{\partial Z} = \frac{\mathcal{B}_i}{\mathcal{B}} - \frac{1}{Z-\mathcal{B}} - \frac{\mathcal{A}}{2\sqrt{2}\mathcal{B}}\left[\frac{2\sum_{k=1}^{m}\lambda_k Q_k q_{k,i}}{\mathcal{A}} - \frac{\mathcal{B}_i}{\mathcal{B}}\right]\left(\frac{2\sqrt{2}\mathcal{B}}{\mathcal{B}^2 - 2\mathcal{B}Z - Z^2}\right)$$

$$= \frac{\mathcal{B}_i}{\mathcal{B}} - \frac{1}{Z-\mathcal{B}} - \left(\frac{\mathcal{A}}{\mathcal{B}^2 - 2\mathcal{B}Z - Z^2}\right)\left[\frac{2\mathcal{A}_i}{\mathcal{A}} - \frac{\mathcal{B}_i}{\mathcal{B}}\right]$$

and

$$\left.\frac{\partial \ln \varphi_i(\underline{Q})}{\partial \mathcal{B}}\right|_Z = -\frac{\mathcal{B}_i}{\mathcal{B}^2}(Z-1) + \frac{1}{Z-\mathcal{B}} + \frac{(\mathcal{A}\mathcal{B}_i - 2\mathcal{B}\mathcal{A}_i)Z}{\mathcal{B}^2(-\mathcal{B}^2 + 2\mathcal{B}Z + Z^2)} + \frac{(\mathcal{B}\mathcal{A}_i - \mathcal{A}\mathcal{B}_i)}{\sqrt{2}\mathcal{B}^3} \ln \frac{Z + (1+\sqrt{2})\mathcal{B}}{Z + (1-\sqrt{2})\mathcal{B}} \tag{5.147}$$

with $\mathcal{A}_i = \sum_{k=1}^{m} \lambda_k Q_k q_{k,i}$.

Similarly, the derivatives $\partial \ln \varphi_i(\underline{Q})/\partial Q_\gamma;\ \gamma = 1, \ldots, m$ are

$$\frac{\partial \ln \varphi_i(\underline{Q})}{\partial Q_\gamma} = \frac{\partial \ln \varphi_i(\underline{Q})}{\partial Z}\frac{\partial Z}{\partial Q_\gamma} + \left.\frac{\partial \ln \varphi_i(\underline{Q})}{\partial Q_\gamma}\right|_Z \tag{5.148}$$

where,

$$\frac{\partial Z}{\partial Q_\gamma} = \frac{\partial Z}{\partial \mathcal{A}}\frac{\partial \mathcal{A}}{\partial Q_\gamma} = \frac{(\mathcal{B} - Z)}{3Z^2 - 2(1-\mathcal{B})Z + (\mathcal{A} - 2\mathcal{B} - 3\mathcal{B}^2)}(2\lambda_\gamma Q_\gamma)$$

The derivative $\partial \ln \varphi_i(\underline{Q})/\partial Z$ is given in Eq. (5.147). The remaining derivatives $\partial \ln \varphi_i(\underline{Q})/\partial Q_\gamma|_Z$ are given by:

$$\left.\frac{\partial \ln \varphi_i(\underline{Q})}{\partial Q_\gamma}\right|_Z = -\frac{1}{2\sqrt{2}\mathcal{B}}\left[2\lambda_\gamma q_{\gamma,i} - \frac{2\lambda_\gamma Q_\gamma \mathcal{B}_i}{\mathcal{B}}\right] \ln \frac{Z + (1+\sqrt{2})\mathcal{B}}{Z + (1-\sqrt{2})\mathcal{B}}$$

5.11: Critical-Point Calculation of Multicomponent Mixtures Consider Mixtures 23 and 25 of Peng and Robinson (1977) shown below. Calculate the critical pressure and critical temperature of these two mixtures using: (1) Baker and Luks (1980), and (2) Heidemann and Khalil (1980) formulations.

Mixture 23: $x_{\text{n-C}_5} = 0.2465$, $x_{\text{n-C}_6} = 0.2176$, $x_{\text{n-C}_7} = 0.1925$, $x_{\text{n-C}_8} = 0.1779$, $x_{\text{n-C}_9} = 0.1656$ (mol fraction)

Mixture 25: $x_{\text{C}_1} = 0.7057$, $x_{\text{C}_2} = 0.0669$, $x_{\text{C}_3} = 0.0413$, $x_{\text{n-C}_4} = 0.0508$, $x_{\text{n-C}_5} = 0.1353$ (mol fraction)

Solution The system of two equations [Eqs. (4.131) and (4.132)] and two unknowns (i.e., T_c and P_c) from the Baker and Luks formulation can be solved via the secant method. This is just the Newton method with numerical derivatives. Computation of the residual in Eq. (4.131) is fairly straightforward, because it only requires expressions for the second derivatives of the Helmhotz free energy A in terms of V and N_i. These derivatives are provided in Example 4.7. Equation (4.132) is somewhat more complicated; its determinant requires derivatives of Eq. (4.131) with respect to V and N_i. After both residuals are computed, the system of two equations and two unknowns are solved by the Newton method to convergence.

The Heidmann and Khalil formulation (1980) uses Eqs. (5.122), (5.123), and (5.124). The numerical algorithm for implementing the Heidemann and Khalil formulation is coded as follows. First, with an initial guess of critical volume, the critical temperature in Eq. (5.123) is determined via a 1-D Newton technique. Then, $\Delta N_i s$ in Eq. (5.124) are computed. In the next step, V_c is computed from Eq. (5.121) by the Newton method. The algorithm developed by Hoteit et al (2006), discussed in this chapter, is an improvement over Heidermann and Khalil (1980) in the solution of the same equations used by these authors.

The result from the two formulations are the same: Mixture 23, $T_c = 540K$, $P_c = 31$ bar, Mixture 25: $T_c = 309K$, $P_c = 152$ bar. The results for Mixture 25 are different from the work by Hoteit et al (2006) due to use of different interaction coefficients. The PR-EOS was used in the calculations. The Baker and Luks formulation was found to be more sensitive to the initial guess, requiring values to be within several degrees or bars of the final computed temperature and pressure. The Heidemann and Khalil formulation was found to be sufficiently insensitive to initial guess of $v_c = 4b$ and $T_c = 1.5\sum_{i=1}^{n_c} T_{ci}$. The run times were faster for the Heidemann and Khalil formulation by a factor of three for the 5-component systems. When very small amounts of C_1, C_2, and C_3 (less than 0.00001 mol fractions) were introduced to Mixture 23, the run time was 10 times faster for the Heidemann and Khalil formulation than Baker and Luks formulation.

5.12: Root Finding of the CPA-EOS Consider the mixture of CO_2 and H_2O at a given composition, temperature and pressure. Suppose the system is in single phase. Find the molar volume and the compressibility factor of the CPA-EOS from the following expression.

$$Z - \frac{Z}{Z-\mathcal{B}} + \frac{\mathcal{A}Z}{Z^2 + 2\mathcal{B}Z - \mathcal{B}^2} + \frac{4 + 4\eta - 2\eta^2}{2 - 3\eta + \eta^2}[x_1(1-\chi_1) + x_2(1-\chi_2)] = 0 \quad (5.149)$$

The symbols and the expression for the PR-EOS are from Chap. 3. The complexity is from noncubic form of Eq. (5.149) and the need for χ_1 (1 represents H_2O) and χ_2 (2 represents CO_2), which are defined below,

$$\chi_1 = \frac{Z}{Z + [4x_1\chi_1\Delta + 4x_2\chi_2\Delta](P/RT)} \quad (5.150)$$

$$\chi_2 = \frac{Z}{Z + (2x_1\chi_1\Delta)(P/RT)} \quad (5.151)$$

where the association strength is given by

$$\Delta = g\kappa b\left[\exp\left(\frac{\epsilon}{kT}\right) - 1\right] \quad (5.152)$$

There are three unknowns (Z, χ_1 , and χ_2) in the above 3 equations [that is, Eqs. (5.149), (5.150), and (5.151)]. The symbols of Eqs. (5.150) to (5.152) are defined in Chap. 3. Propose a method for the solution of the unknowns.

Solution When good initial guesses are not available, a fast solution technique, such as the Newton method, is not guaranteed to converge. The number of solutions is not known beforehand under general conditions.The bisection method is the most robust and will always converge, but at a relatively high computational root. For the aqueous phase, there is often one solution for the compressibility factor, slightly greater than the volume parameter of water, $\mathcal{B}$. Other solutions may exist with $Z < \mathcal{B}$, but these solutions have no physical meaning. The bisection method can be used to search for Z in the range $[\mathcal{B}, \mathcal{B} + \alpha]$ with α a small number (say $\alpha = 0.1$, but it can be increased when a solution is not found). χ_1 and χ_2 are found by the SSI method, followed by the Newton method (when tolerance is reached), using 0.5 as an initial guess for χ_1 and χ_2.

The bisection method for Z in Eq. (5.149) is considerably less efficient than the simultaneous solution for Z, χ_1, and χ_2. In reservoir simulation initial guesses from the previous time steps are available. As a result, the CPA-EOS can be used in reservoir simulations with comparable computational efficiency as the traditional PR-EOS. (See Moortgat et al., 2012.) The initial guess for Z can be also from the solution to the PR-EOS.

5.10 Problems

5.1 Derive the expression for $\partial \ln \varphi_i(x)/\partial x_k$ given below

$$\frac{\partial \ln \varphi_i(x)}{\partial x_k} = A_{ik} + B_{ik} + C_{ik} + D_{ik} \quad i = 1, \ldots, n_c, \quad k = 1, \ldots, n_{c-1}$$

where

$$A_{ik} = S_{bi}\left[\frac{\partial Z}{\partial x_k} - \left(S_{bk} - S_{n_c}\right)(Z-1)\right] - \frac{\left(\dfrac{\partial Z}{\partial x_k} - \dfrac{\partial \mathcal{B}}{\partial x_k}\right)}{Z - \mathcal{B}}$$

$$i = 1, \ldots, n_c, \quad k = 1, \ldots, n_{c-1}$$

$$B_{ik} = \frac{-1}{2\sqrt{2}\mathcal{B}}\left(\frac{\partial \mathcal{A}}{\partial x_k} - \frac{\mathcal{A}}{\mathcal{B}}\frac{\partial \mathcal{B}}{\partial x_k}\right)(2S_{ai} - S_{bi}) \ln\left[\frac{(Z + (\sqrt{2}+1)\mathcal{B})}{(Z - (\sqrt{2}-1)\mathcal{B})}\right]$$

$$i = 1, \ldots, n_c, \quad k = 1, \ldots, n_{c-1}$$

$$C_{ik} = \frac{-\mathcal{A}}{2\sqrt{2}\mathcal{B}}\left[\frac{2\left(a_{ik} - a_{n_c}\right)}{a} - 4S_{ai}\left(S_{ak} - S_{n_c}\right) + S_{bi}\left(S_{bk} - S_{n_c}\right)\right] \ln\left[\frac{(Z + (\sqrt{2}+1)\mathcal{B})}{(Z - (\sqrt{2}-1)\mathcal{B})}\right]$$

$$i = 1, \ldots, n_c, \quad k = 1, \ldots, n_{c-1}$$

$$D_{ik} = -\frac{\mathcal{A}}{\mathcal{B}}(2S_{ai} - S_{bi})\frac{\left(Z\dfrac{\partial \mathcal{B}}{\partial x_k} - \mathcal{B}\dfrac{\partial Z}{\partial x_k}\right)}{(Z^2 + 2Z\mathcal{B} - \mathcal{B}^2)} \quad i = 1, \ldots, n_c, \quad k = 1, \ldots, n_{c-1}$$

and

$$S_i = \sum_{j=1}^{n_c} x_j a_{ij} \quad i = 1, \ldots, n_c$$

$$S_{ai} = \frac{\sum_{j=1}^{n_c} x_j a_{ij}}{a} \quad i = 1, \ldots, n_c$$

$$S_{bi} = \frac{b_i}{b} \quad i = 1, \ldots, n_c$$

and

$$\frac{\partial Z}{\partial \mathcal{A}} = \frac{\mathcal{B} - Z}{3Z^2 - 2(1-\mathcal{B})Z + (\mathcal{A} - 2\mathcal{B} - 3\mathcal{B}^2)}$$

$$\frac{\partial Z}{\partial \mathcal{B}} = \frac{[-Z^2 + 2(3\mathcal{B}+1)] + (\mathcal{A} - 2\mathcal{B} - 3\mathcal{B}^2)}{3Z^2 - 2(1-\mathcal{B})Z + (\mathcal{A} - 2\mathcal{B} - 3\mathcal{B}^2)}$$

and

$$\frac{\partial Z}{\partial x_i} = \frac{\partial Z}{\partial \mathcal{A}}\frac{\partial \mathcal{A}}{\partial x_i} + \frac{\partial Z}{\partial \mathcal{B}}\frac{\partial \mathcal{B}}{\partial x_i} \qquad i = 1, \ldots, n_c - 1$$

$$\frac{\partial \mathcal{A}}{\partial x_i} = \frac{2P}{R^2T^2}\left(S_i - S_{n_c}\right) \quad i = 1, \ldots, n_c - 1$$

5.2 Consider the total Gibbs free energy of a multicomponent two-phase mixture given by $G = \sum_{i=1}^{n_c}\sum_{j=1}^{2} n_{ij}\mu_{ij}$, where n_{ij} is the number of mols of component i in phase j and μ_{ij} is the chemical potential of component i in phase j. Derive the basic equation of equilibrium $\mu_{i1} = \mu_{i2}$ $(i = 1, \ldots, n_c)$ by taking the derivative of G with respect to either n_{i1} or n_{i2} at constant T and P. Note that $n_{i1} + n_{i2} = n_i$ $(i = 1, \ldots, n_c)$ is a constant.
Hint: Use the Gibbs-Duhem Equation in your derivation.

5.3 The stability analysis can provide the initial guess for the K_i-values in phase-split calculations if the system is unstable. When the *TPD* is negative and the trial phase is gas-like, the K_i-values are updated from

$$K_i = \frac{\varphi_i(\underline{z})}{\varphi_i(\underline{X})} = \frac{X_i}{z_i}$$

When the *TPD* is negative and the trial phase is liquid-like, the K_i-values are updated from

$$K_i = \frac{\varphi_i(\underline{X})}{\varphi_i(\underline{z})} = \frac{z_i}{X_i}$$

When the *TPD* is negative for both liquid- and vapor-like trial phases, the K_i-values that correspond to the smaller *TPD* are selected. Note that the K_i-values should not be calculated from the normalized mol fractions y_i instead of the mol numbers in the above equations. As was demonstrated by Hoteit and Firoozabadi (2006), the use of x_i in the above equations results in numerical inefficiency. Offer your reasoning for the x_i.

5.4 Combine the Newton method and the bisection method to develop an efficient and robust algorithm for the solution of the Rachford-Rice equation.

5.5 Show that

$$\sum_{i=1}^{n_c}\sum_{j=1}^{n_c} a_{ij}\Delta N_i \Delta N_j = 2\sum_{i=1}^{n_c}\sum_{j=1}^{i-1} a_{ij}\Delta N_i \Delta N_j + \sum_{i=1}^{n_c} a_{ii}\Delta N_i^2$$

5.6 Show that

$$\sum_{j=1}^{n_c} a_i^{1/2} a_j^{1/2} (1-\delta_{ij}) x_j = \sum_{k=1}^{n_c} \lambda_k Q_k q_{ki}$$

5.7 The molar Gibbs free energy of mixing of a binary mixture, Δg_{mix}, is plotted vs. x_2 (mol fraction of component 2) in Fig. 5.17 at constant temperature and pressure. Show that the intercept of Δg_{mix} with $x_2 = 0$ and $x_2 = 1$ axes are $\mu_1 - \mu_1^0$ and $\mu_2 - \mu_2^0$, respectively; μ_1 and μ_2 are the chemical potentials at points A and D, respectively.

5.8 Based on Eq. (5.102) show that

$$\min q_{\alpha i} \leq Q_\alpha \leq \max q_{\alpha i} \quad \alpha = 1, \ldots, m+1, \quad i = 1, \ldots, n_c$$

5.9: Calculation of $(\partial c_{\alpha i}/\partial c_i)_{T,V,c_i}$ in Two-Phase The definition of terms are $c_{\alpha i} = n_{\alpha i}/V_\alpha$, $c_i = n_i/V$, where $n_{\alpha i}$ is the number of mols of component i in phase α; V_α is the volume of phase α; n_i is the number of mols of component i in the two-phase system; and, V is the volume of two-phase system. There are two phases, $\alpha = 1$ and $\alpha = 2$. The above derivative can appear in two-phase compositional modeling (Zidane and Firoozabadi, 2015) and can be calculated using the relations: $V = V_1 + V_2$, $n_i = n_{1i} + n_{2i}$, and $f_{1i} = f_{2i}$. Derive the working equation for the calculations.

5.10: Stability of a Phase with Curved Interface In Chaps. 2 and 3, we presented results of phase-split computations in two-phase gas-liquid systems with a curved interface. The working equations for phase-split calculations were also presented (See Examples 3.13 and 3.14). The expression for phase stability analysis for a system with a curved interface is given by,

$$TPD = \sum_{i=1}^{n_c} x_i \left[\ln x_i + \ln \varphi_i(x, P^L) + \ln P^L - d_i(z) \right]$$

where $d_i(z) = \ln z_i + \ln \varphi_i(z, P^V) + \ln P^V$. Note that (z) and P^V are known. Derive the above expression and take the derivatives with respect to x and P^L to obtain expression at the minimum. Note that in the expression above, the liquid phase is the trial phase.

5.11 Derive Eq. (5.125) of the text.

5.12 Give examples for which the Newton method does not have quadratic convergence.

5.11 References

Arfken, G., *Mathematical Methods for Physicists*. 1985, Academic Press.

Baker, L. E., A. C. Pierce, and K. D. Luks, Gibbs energy analysis of phase equilibria. *SPE Journal*, 1982. 22(5): pp. 731–742.

Baker, L. E. and K. D. Luks, Critical point and saturation pressure calculations for multipoint systems. *SPE Journal*, 1980. 20(1): pp. 15–24.

Corana, A., M. Marchesi, C. Martini, and S. Ridella, Minimizing multimodal functions of continuous variables with the 'Simulated Annealing' algorithm. *ACM Transactions on Mathematical Software*, 1987. 13: pp. 262–280.

Dennis, J. E. Jr., and R. B. Schnabel, Numerical methods for unconstrained optimization and nonlinear equations, *Society for Industrial and Applied Mathematics*, 1996.

Firoozabadi, A., *Thermodynamics of Hydrocarbon Reservoirs*. 1999, McGraw-Hill.

Firoozabadi, A., and H. Pan, Fast and robust algorithm for compositional modeling: Part I -- stability analysis testing. *SPE Journal*, 2002. 7(1), pp. 78–89.

Gibbs, J. W., Transaction of the Connecticut Academy (1876). As Reprinted in the Scientific Papers of J. Willard Gibbs.

Goffe, W. L., G. D. Ferrier, and J. Rogers, Global optimization of statistical functions with simulated annealing. *Journal of Econometrics*, 1994. 60(1–2): pp. 65–99.

Gorucu, S. E. and R. T. Johns, Comparison of reduced and conventional phase equilibrium calculations (Conference Paper). *Society of Petroleum Engineers - SPE Reservoir Simulation Symposium*, 18-20 February 2013. 1: The Woodlands, TX; United States.

Haugen, K. B., and B. L. Beckner, A critical comparison of reduced and conventional EOS algorithms. *SPE Journal*, 2013. 18(2): pp. 378–388.

Haugen, K. B., L. Sun, and A. Firoozabadi, Efficient and robust three-phase split computations. *AIChE Journal*, 2011. 57(9): pp. 2555–2565.

Heidemann, R. A., and A. M. Khalil, The calculation of critical points. *AIChE Journal*, 1980. 26(5): pp. 769–779.

Hoteit, H., and A. Firoozabadi, Simple phase stability-testing algorithm in the reduction method. *AIChE Journal*, 2006. 52(8): pp. 2909–2918.

Hoteit, H., E. Santiso, and A. Firoozabadi, An efficient and robust algorithm for the calculation of gas-liquid critical point of multicomponent petroleum fluids. *Fluid Phase Equilibria*, 2006. 241: pp. 186–195.

Li, Z., and A. Firoozabadi, General strategy for stability testing and phase-split calculation in two and three phases. *SPE Journal*, 2012a. 17(4): pp. 1096–1106.

Li, Z., and A. Firoozabadi, Initialization of phase fractions in Rachford-Rice equations for robust and efficient three-phase split calculation. *Fluid Phase Equilibria*, 2012b. 332(0): pp. 21–27.

Michelsen, M. L., Simplified flash calculations for cubic equations of state. *Industrial & Engineering Chemistry Process Design and Development*, 1986. 25(1): pp. 184–188.

Michelsen, M. L., The isothermal flash problem. Part I. Stability. *Fluid Phase Equilibria*, 1982a. 9(1): pp. 1–19.

Michelsen, M. L., The isothermal flash problem. Part II. Phase-split calculation. *Fluid Phase Equilibria*, 1982b. 9(1): pp. 21–40.

Michelsen, M. L. and J. M. Mollerup, *Thermodynamic Models: Fundamentals and Computational Aspects*. 2004, Tie-Line Publications.

Michelsen, M. L., W. Yan, and E. H. Stenby, A comparative study of reduced-variables-based flash and conventional flash. *SPE Journal*, 2013. 18(5): pp. 952–959

Mikyška, J., and A. Firoozabadi, New thermodynamic function for phase-splitting at constant temperature, moles, and volume. *AIChE Journal*, 2011. 57(7): pp. 1897–1904.

Moortgat, J., Z. Li, and A. Firoozabadi, Three-phase compositional modeling of CO_2 injection by higher-order finite element methods with CPA equation of state for the aqueous phase, *Water Resources Research*, 2012. 48:W12511.

Okuno, R., R. T. Johns, and K. Sepehrnoori, A new algorithm for Rachford-Rice for multiphase compositional simulation. *SPE Journal*, 2010. 15(2): pp. 313–325.

Pan, H., and A. Firoozabadi, Complex multiphase equilibrium calculations by direct minimization of Gibbs free energy using simulated annealing. *SPE Reservoir Evaluation & Engineering*, 1998. 1(1): pp. 36–42.

Peng, D. Y., and D. B. Robinson, A rigorous method for predicting the critical properties of multicomponent systems from an equation of state. *AIChE Journal*, 1977. 23(2): pp. 137–144.

Petitfrere, M., and D. V. Nichita., A comparison of conventional and reduction approaches for phase equilibrium calculations. *Fluid Phase Equilibria*, 2015. 386: pp. 30–46.

Press, W. H., S. A. Teukolsky, W. T. Vetterling, and B. P. Flannery, *Numerical Recipes: the Art of Scientific Computing*. 2nd ed. 1992, New York: Cambridge University Press.

Rachford, H. H. J., and J. D. Rice, Procedure for use of electronic digital computers in calculating flash vaporization hydrocarbon equilibrium. *Journal of Petroleum Technology*, 1952. 4(10): pp. 189–196.

Shelton, J. L., and L. Yarborough, Multiple phase behavior in porous media during CO_2 or rich-gas flooding. *Journal of Petroleum Technology*, 1977. 29(9): pp. 1171–1178.

Strang, G., *Linear Algebra and Its Applications*. 3rd ed. 1988, Harcourt Brace Jovanovich College Publishers.

Whitson, C. H. and M. R. Brulé, *Phase Behavior. Monograph Volume 20, Henry L. Doherty Series*. 2000, SPE, Richardson, TX.

Wilson, G. M., A Modified Redlich-Kwong Equation of State, Application to General Physical Data Calculations, in AIChE 65th National Meeting. 1965, Cleveland, OH.

Zidane, A., and A. Firoozabadi, An implicit numerical model for multicomponent compressible two-phase flow in porous media. *Paper in review*, 2015.

CHAPTER 6

Theory of Irreversibility and Diffusion in Multicomponent Mixtures

In all of our derivations for equilibrium in Chap. 1 to 5, we have made the fundamental assumption that the process is reversible. For an isolated system, the process is reversible when there is no entropy production in the system. A consequence of the reversibility assumption for all the equilibrium cases that we have studied (i.e., thermal, mechanical, and chemical) is that the temperature is the same throughout the composite system. In other words, the necessary condition to achieve equilibrium is the uniformity of temperature in the composite system. Under nonisothermal conditions, the process becomes irreversible and equilibrium cannot be achieved. In this chapter, we relax the assumption of reversibility and derive the expression for the rate of entropy production per unit volume, a key expression in irreversible thermodynamics. With a postulation that the fluxes are a linear function of driving forces, we derive the general expressions for various diffusion fluxes and the associated coefficients. The derivations from irreversible thermodynamics are superior to other approaches because we are able to evaluate various diffusion coefficients for nonideal mixtures. We use flux expressions to formulate a variety of problems, including, species distribution in hydrocarbon reservoirs, and interpretation of ice core data to relate them to abrupt past climate changes. There are 11 examples and 15 problems at the end of this chapter that cover various derivations for better understanding of concepts and numerical results to appreciate the significance of various effects.

6.1 Irreversibility in a Closed System

Let us examine the time variation of the entropy of a composite system (entropy change for the composite system from time t to $t + \mathrm{d}t$) $\mathrm{d}S$:

$$\mathrm{d}S = (\mathrm{d}S)_{ex} + (\mathrm{d}S)_i \tag{6.1}$$

where $(\mathrm{d}S)_{ex}$ is the change in entropy supplied to the composite system from the exterior and $(\mathrm{d}S)_i$ is the change in entropy produced inside the system (during time interval $\mathrm{d}t$).

An alternate statement of the second law (Postulates II, III, and IV of Chap. 1) is that $(\mathrm{d}S)_i$ must be zero for reversible processes and positive for irreversible processes:

$$(\mathrm{d}S)_i = 0 \quad \text{for reversible processes} \tag{6.2}$$

$$(\mathrm{d}S)_i > 0 \quad \text{for irreversible processes} \tag{6.3}$$

According to the second law, the entropy production within the system, $(\mathrm{d}S)_i$, cannot be negative. The entropy supplied to the system, $(\mathrm{d}S)_{ex}$, can be, however, positive, zero, or negative. For an isolated system, $(\mathrm{d}S)_{ex} = 0$, and therefore

$$\mathrm{d}S = (\mathrm{d}S)_i \tag{6.4}$$

Equation (6.1) implies that for a reversible process in an isolated system, $\mathrm{d}S = 0$, and for an irreversible process, $\mathrm{d}S > 0$. The mathematical expressions are

$$\mathrm{d}S = 0 \quad \text{isolated system, reversible process} \tag{6.5}$$

$$\mathrm{d}S > 0 \quad \text{isolated system, irreversible process} \tag{6.6}$$

In Chap. 1, the entropy supplied to the system was defined by the relationship $(\mathrm{d}S)_{ex} = đQ/T$ for a closed system, where $đQ$ is the heat supplied to the system at constant temperature T. Since the process was assumed reversible, then

$$\mathrm{d}S = \frac{đQ}{T}, \quad (\mathrm{d}S)_i = 0 \quad \text{closed system} \tag{6.7}$$

In summary, equilibrium thermodynamics addresses the reversible processes for which there is no entropy production within the system. In thermodynamics of irreversible processes, the entropy production, $(\mathrm{d}S)_i$, is formulated and then is related to the irreversible phenomena that may occur in the system.

Let us write the expression for the rate of entropy production in a system of volume V:

$$\left(\frac{\mathrm{d}S}{\mathrm{d}t}\right)_i = \int_V \sigma \mathrm{d}V \tag{6.8}$$

where σ is the entropy production per unit volume and time; it is also called the entropy production strength.

6.1.1 Entropy Production Strength

We will derive the expression for σ in a simple way.

Consider a small volume element $\Delta x \Delta y \Delta z$ within a given system. The entropy balance for this volume element can be stated as

Rate of Entropy Change = Net Entropy Influx + Rate of Entropy Production.

Let s denote the specific entropy (i.e., entropy per unit mass), and $\vec{J}^s$ the total entropy flux per unit time per unit surface area; then

$$\frac{\partial}{\partial t}[\rho s \Delta x \Delta y \Delta z] = J_x^s \Delta y \Delta z|_{x+\Delta x}^{x} + J_y^s \Delta x \Delta z|_{y+\Delta y}^{y} + J_z^s \Delta x \Delta y|_{z+\Delta z}^{z} + \sigma \Delta x \Delta y \Delta z \tag{6.9}$$

In the above equation J_x^s, J_y^s, and J_z^s are the total entropy fluxes in the x, y, and z directions, respectively, and ρ is the mass density. Dividing by $\Delta x \Delta y \Delta z$ and taking the limit as $\Delta x \to 0$, $\Delta y \to 0$, and $\Delta z \to 0$, and writing the result in vector notation,

$$\frac{\partial \rho s}{\partial t} = -\nabla \cdot (\vec{J}^s) + \sigma \tag{6.10}$$

Equation (6.10) is similar to any other balance expression, with the difference that there is a source term representing the production of entropy. A different form of the above equation can be presented in terms of total derivative of s:

$$\rho \frac{\mathrm{d}s}{\mathrm{d}t} = -\nabla \cdot (\vec{J}^s - \rho s \vec{v}) + \sigma \tag{6.11a}$$

In Eq. (6.11a), the total entropy flux $\vec{J}^s$ consists of three parts: (1) $\vec{q}^{cond}/T$, (2) $\sum_{i=1}^{n_c} \tilde{S}_i \vec{J}_i$, and (3) $\rho s \vec{v}$ which represent the contributions from conduction, diffusion, and convection, respectively ($\vec{q}^{cond}$ is the energy flux by conduction, $\vec{J}_i$ is the diffusion mass flux of component i to be described later in this chapter, and $\tilde{S}_i = \bar{S}_i / M_i$). Therefore, Eq. (6.11a) can also be written as

$$\rho \frac{\mathrm{d}s}{\mathrm{d}t} = -\nabla \cdot \left(\frac{\vec{q}^{cond}}{T} + \sum_{i=1}^{n_c} \tilde{S}_i \vec{J}_i \right) + \sigma \tag{6.11b}$$

The total derivative $(\mathrm{d}s/\mathrm{d}t)$ represents the variation of s with respect to time and position; $s = s(t, x, y, z)$. The partial derivative $(\partial s/\partial t)$ represents the variation of s with respect to time at a fixed position. For any scalar quantity such as s,

$$\rho \frac{\mathrm{d}s}{\mathrm{d}t} = \frac{\partial \rho s}{\partial t} + \nabla \cdot \rho s \vec{v} \tag{6.12}$$

The above equation can be obtained by combining the mass balance expression $(\partial \rho/\partial t) = -\nabla \cdot (\rho \vec{v})$ and the expression $(\mathrm{d}s/\mathrm{d}t) = (\partial s/\partial t) + \vec{v} \cdot \nabla s$. Combining Eqs. (6.10) and (6.12) results in the expression given by Eq.(6.11a). Now recall the expression for $\mathrm{d}S$ from Chap. 1,

$$T\mathrm{d}S = \mathrm{d}U + P\mathrm{d}V - \sum_{i=1}^{n_c} \left(\frac{\mu_i}{M_i} \right) \mathrm{d}m_i \tag{6.13}$$

where S, U, and V are the total entropy, internal energy, and volume, respectively, and m_i is the total mass of component i. Note that in the above equation, the chemical potential is defined as $\mu_i = (\partial U/\partial n_i)_{S,V,n_j}$; the molecular weight of component i, M_i, appears to make the chemical potential stay the same as defined in Chap. 1. The extensive quantities S, U, and V can be expressed in terms of specific quantities (quantity per unit mass): $S = ms$, $U = mu$, and $V = mv$, where $m = \sum_{i=1}^{n_c} m_i$. Substituting specific quantities in Eq. (6.13), assuming $m = 1$ and divide by $\mathrm{d}t$,

$$T\frac{\mathrm{d}s}{\mathrm{d}t} = \frac{\mathrm{d}u}{\mathrm{d}t} + P\frac{\mathrm{d}v}{\mathrm{d}t} - \sum_{i=1}^{n_c} \tilde{\mu}_i \frac{\mathrm{d}w_i}{\mathrm{d}t} \tag{6.14}$$

where $\tilde{\mu}_i = \mu_i / M_i$ and w_i is the mass fraction of component i. Eq. (6.14) is based on the assumption of equilibrium; it is assumed that at a local level, the process is still at equilibrium. When the volume element becomes large, the equilibrium assumption may not be valid. In irreversible thermodynamics, it is shown that for most transport processes not far from equilibrium, the equilibrium assumption is justified at the local level (de Groot and Mazur, 1984; Hasse, 1969).

The mass balance equation for species i and the energy balance are then used to provide the expression for σ. For the volume element, one can readily derive the mass-balance expression

$$\rho \frac{dw_i}{dt} = -\nabla \cdot \vec{J}_i \quad i = 1, \ldots, n_c \tag{6.15}$$

In the above equation, $\vec{J}_i$ is the diffusion mass flux of component i.

The expression for energy balance is taken from Bird et al. (see Table 18.3-1 of Bird et al., 1960). These authors provide the details of the derivation of the energy-balance expression

$$\rho \frac{du}{dt} = -\nabla \cdot \left(\vec{q}^{\,cond} + \sum_{i=1}^{n_c} \tilde{H}_i \vec{J}_i \right) - (\vec{\vec{\tau}} + P\vec{\vec{\delta}}) : \nabla \vec{v} - \sum_{i=1}^{n_c} \vec{J}_i \cdot \vec{g} \tag{6.16}$$

In the above equation, in addition to the energy flux by conduction, $\vec{q}^{\,cond}$, the energy flux by interdiffusion, $(\sum_{i=1}^{n_c} \tilde{H}_i \vec{J}_i)$, is also taken into account ($\tilde{H}_i = \bar{H}_i / M_i$).

Note that the sign of the last term is reversed from Bird et al. (1960); this is done to be consistent with our sign selection on the effect of gravity on equilibrium in Chap. 2, where we have assumed the upward direction to be positive. In Eq. (6.16), the symbol "$\vec{\vec{\ }}$" represents a tensor; both $\vec{\vec{\tau}}$ and $\vec{\vec{\delta}}$ are tensors. The stress tensor $\vec{\vec{\tau}}$ has nine components τ_{xx}, τ_{yy}, τ_{zz}, τ_{xy}, τ_{xz}, τ_{yx}, τ_{yz}, τ_{zx}, and τ_{zy}, whereas the velocity vector $\vec{v}$ has three components v_x, v_y, and v_z. The unit tensor $\vec{\vec{\delta}}$ has also nine components; it can be represented by a 3 × 3 matrix with diagonal elements unity and nondiagonal elements zero (see the Appendix for a brief description of vector and tensor operations). Note that $\nabla \vec{v}$ in Eq. (6.16) is the dyadic product of vector ∇ and vector $\vec{v}$; it is a tensor with nine components, $\partial v_x/\partial x$, $\partial v_x/\partial y$, $\partial v_x/\partial z$, $\partial v_y/\partial x$, $\partial v_y/\partial y$, $\partial v_y/\partial z$, $\partial v_z/\partial x$, $\partial v_z/\partial y$, and $\partial v_z/\partial z$. The double-dot operation ":" represents the scalar product of two tensors,

$$\vec{\vec{\tau}} : \nabla \vec{v} = \sum_{i=1}^{3} \sum_{j=1}^{3} \tau_{ij} \frac{\partial v_i}{\partial x_j} \tag{6.17}$$

where i and j take on the values of x, y, and z of the coordinates.

The term $P\vec{\vec{\delta}} : \nabla \vec{v}$ in Eq. (6.16) can be simplified considerably,

$$P\vec{\vec{\delta}} : \nabla \vec{v} = P(\nabla \cdot \vec{v}) \tag{6.18}$$

From the continuity equation, $(d\rho/dt) = -\rho(\nabla \cdot \vec{v})$. Therefore,

$$P\vec{\vec{\delta}} : \nabla \vec{v} = -\left(\frac{P}{\rho}\right) \frac{d\rho}{dt} \tag{6.19}$$

Since $\rho = 1/v$, where v is the specific volume,

$$\left(\frac{1}{\rho}\right)\left(\frac{d\rho}{dt}\right) = -\rho\left(\frac{dv}{dt}\right) \tag{6.20}$$

and Eq. (6.19) simplifies to

$$P\overrightarrow{\overrightarrow{\delta}} : \nabla\vec{v} = (P\rho)\frac{dv}{dt} \tag{6.21}$$

Combining Eqs. (6.16) and (6.21) gives

$$\rho\frac{du}{dt} = -\nabla\cdot\left(\vec{q}^{\,cond} + \sum_{i=1}^{n_c}\tilde{H}_i\vec{J}_i\right) - \overrightarrow{\overrightarrow{\tau}} : \nabla\vec{v} - P\rho\frac{dv}{dt} - \sum_{i=1}^{n_c}\vec{J}_i\cdot\vec{g} \tag{6.22}$$

The above form of the energy equation is combined with Eqs. (6.14) and (6.15) to provide the expression for ds/dt,

$$\rho T\frac{ds}{dt} = -\nabla\cdot\vec{q}^{\,cond} - \nabla\cdot\left(\sum_{i=1}^{n_c}\tilde{H}_i\vec{J}_i\right) - \overrightarrow{\overrightarrow{\tau}} : \nabla\vec{v} + \sum_{i=1}^{n_c}\tilde{\mu}_i(\nabla\cdot\vec{J}_i) - \sum_{i=1}^{n_c}\vec{J}_i\cdot\vec{g} \tag{6.23}$$

The first term on the right side of Eq. (6.11b) can be part of the following relationship,

$$\nabla\cdot\left(\frac{\vec{q}^{\,cond}}{T}\right) = \frac{1}{T}\nabla\cdot\vec{q}^{\,cond} - \left(\frac{1}{T^2}\right)(\vec{q}^{\,cond}\cdot\nabla T) \tag{6.24}$$

The second term on the right side of Eq. (6.11b) can be written as

$$-\nabla\cdot\left(\sum_{i=1}^{n_c}\tilde{H}_i\vec{J}_i - \sum_{i=1}^{n_c}\tilde{\mu}_i\vec{J}_i + \sum_{i=1}^{n_c}\tilde{\mu}_i\vec{J}_i\right) = -\nabla\cdot\left(T\sum_{i=1}^{n_c}\tilde{S}_i\vec{J}_i + \sum_{i=1}^{n_c}\tilde{\mu}_i\vec{J}_i\right) \tag{6.25}$$

Combining the above three expressions and using, $\nabla\cdot\sum_{i=1}^{n_c}(\tilde{\mu}_i\cdot\vec{J}_i) = \sum_{i=1}^{n_c}\vec{J}_i\cdot\nabla\tilde{\mu}_i + \sum_{i=1}^{n_c}\tilde{\mu}_i(\nabla\cdot\vec{J}_i)$ and a similar expression for $\nabla\cdot(T\sum_{i=1}^{n_c}\tilde{S}_i\vec{J}_i)$,

$$\begin{aligned}\rho\frac{ds}{dt} = &-\nabla\cdot\left[\left(\frac{\vec{q}^{\,cond}}{T}\right) + \sum_{i=1}^{n_c}\tilde{S}_i\vec{J}_i\right]\\ &-\frac{1}{T}\left[\left(\frac{\vec{q}^{\,cond}}{T}\right) + \sum_{i=1}^{n_c}\tilde{S}_i\vec{J}_i\right]\cdot\nabla T\\ &-\left(\frac{1}{T}\right)\overrightarrow{\overrightarrow{\tau}}:\nabla\vec{v} - \frac{1}{T}\sum_{i=1}^{n_c}\vec{J}_i(\nabla\tilde{\mu}_i + \vec{g})\end{aligned} \tag{6.26}$$

From the comparison of Eqs. (6.11b) and (6.26),

$$\boxed{\sigma = -\left(\frac{1}{T^2}\right)\left[\vec{q}^{cond} + T\sum_{i=1}^{n_c}\tilde{S}_i\vec{J}_i\right]\cdot\nabla T - \left(\frac{1}{T}\right)\vec{\vec{\tau}}:\nabla\vec{v} - \left(\frac{1}{T}\right)\sum_{i=1}^{n_c}\vec{J}_i\cdot(\nabla\tilde{\mu}_i + \vec{g})} \tag{6.27}$$

Note that the entropy production strength, according to the above equation consists of three terms. The first term arises from the temperature gradient, the second term is due to velocity gradients [see Eq. (6.30) below], and the third term is from the gradient of the chemical potentials. Equation (6.27) implies that at stationary states (in which the variables are independent of time), when temperature is time-independent but has a nonuniform value, σ is nonzero. We have shown that entropy production within the system is due to: (1) temperature gradient, (2) velocity gradients, and (3) chemical potential gradient (without gravity). One can also show that the terms on the right side of Eq. (6.27) are positive. Let us show that the second term is always ≥ 0.

In order to prove that $(-\vec{\vec{\tau}} : \nabla\vec{v})$ is positive, we need the expression for various stresses. For Newtonian fluids, the expression for stresses are (see Bird et al., 1960)

$$\tau_{ii} = -\frac{2}{3}\frac{\partial v_i}{\partial x_i} + \frac{2}{3}\mu(\nabla\cdot\vec{v}) \quad i = 1, 2, 3 \tag{6.28}$$

$$\tau_{ij} = \tau_{ji} = -\mu\left(\frac{\partial v_i}{\partial x_j} + \frac{\partial v_j}{\partial x_i}\right) \quad i = 1, 2, 3, \; j = 1, 2, 3, \; i \neq j \tag{6.29}$$

where i and j take on x, y, and z in the coordinates.

Combining Eqs. (6.17), (6.28), and (6.29) results in the following sum of the squares:

$$(-\vec{\vec{\tau}}:\nabla\vec{v}) = \frac{1}{2}\mu\sum_{i=1}^{3}\sum_{j=1}^{3}\left[\left(\frac{\partial v_i}{\partial x_j} + \frac{\partial v_j}{\partial x_i}\right) - \frac{2}{3}(\nabla\cdot\vec{v})\delta_{ij}\right]^2 \tag{6.30}$$

where $\delta_{ij} = 1$ for $i = j$, and $\delta_{ij} = 0$ for $i \neq j$. Therefore, it is established that the $(-\vec{\vec{\tau}} : \nabla\vec{v})$ term is always ≥ 0.

Let us also examine Eq. (6.27) for a convection-free system at isothermal conditions in the gravity field, where $\vec{v} = \vec{0}$, and $\nabla T = 0$. For the system to have no entropy production (i.e., to be at equilibrium state), $\nabla\tilde{\mu}_i = -\vec{g}$. Since in a gravity field $g_x = g_y = 0$ and $g_z = g$, then $d\tilde{\mu}_i/dz = -g$ or $(d\mu_i = -M_i g dz)_T$, which is the same as Eq. (2.13). This derivation of the sedimentation equation from irreversible thermodynamics is more rigorous than the derivation in Chap. 2.

6.2 Derivation of Diffusion Flux and Diffusion Coefficients in Multicomponent Mixtures

Let us rewrite Eq. (6.27) in the following form:

$$\sigma = +\left(\frac{1}{T^2}\right)\left[\vec{q}^{cond} + T\sum_{i=1}^{n_c}\tilde{S}_i\vec{J}_i\right]\cdot(-\nabla T) + \left(\frac{1}{T}\right)\vec{\vec{\tau}}:(-\nabla\vec{v}) + \left(\frac{1}{T}\right)\sum_{i=1}^{n_c}\vec{J}_i\cdot[-(\nabla\tilde{\mu}_i + \vec{g})] \tag{6.31}$$

In the above equation, $\nabla\tilde{\mu}_i$ is a function of ∇T as well as ∇P and ∇n_i. In order to eliminate the dependency of $\nabla\tilde{\mu}_i$ on ∇T, we write,

$$\nabla\tilde{\mu}_i = \nabla_T\tilde{\mu}_i + \left.\frac{\partial\tilde{\mu}_i}{\partial T}\right|_{P,\underline{n}}\nabla T \quad i = 1,\ldots,n_c \tag{6.32}$$

But from Eq. (1.125), $\partial\tilde{\mu}_i/\partial T|_{P,\underline{n}} = -\tilde{S}_i$. Therefore,

$$\nabla\tilde{\mu}_i = \nabla_T\tilde{\mu}_i - \tilde{S}_i\nabla T \tag{6.33}$$

Combining Eqs. (6.31) and (6.33),

$$\boxed{\sigma = +\left(\frac{1}{T^2}\right)\vec{q}^{cond}\cdot(-\nabla T) + \left(\frac{1}{T}\right)\vec{\vec{\tau}}:(-\nabla\vec{v}) + \left(\frac{1}{T}\right)\sum_{i=1}^{n_c}\vec{J}_i\cdot[-(\nabla_T\tilde{\mu}_i + \vec{g})]} \tag{6.34}$$

When there is no viscous dissipation, and with further simplification from $\sum_{i=1}^{n_c}\vec{J}_i = 0$ (see Prob. 6.1) and $\tilde{\mu}_i = \mu_i/M_i$, Eq. (6.34) simplifies to

$$\sigma = +\left(\frac{1}{T^2}\right)\vec{q}^{cond}\cdot(-\nabla T) + \left(\frac{1}{T}\right)\sum_{i=1}^{n_c-1}\vec{J}_i\cdot\left[-\nabla_T\left(\frac{\mu_i}{M_i} - \frac{\mu_{n_c}}{M_{n_c}}\right)\right] \tag{6.35a}$$

$$\sigma = +\,\vec{q}^{cond}\cdot\left(\nabla\left(\frac{1}{T}\right)\right) + \sum_{i=1}^{n_c-1}\vec{J}_i\cdot\left[-\nabla_T\left(\frac{\mu_i}{M_i} - \frac{\mu_{n_c}}{M_{n_c}}\right)\left(\frac{1}{T}\right)\right] \tag{6.35b}$$

The above equation can also be written as

$$\sigma = \sum_{i=1}^{n_c}\vec{\phi}_i\cdot\vec{X}_i \tag{6.36}$$

where $\vec{X}_i$ $(i = 1,\ldots,n_c)$ are the independent driving forces, $\nabla(1/T)$ for $i = n_c$, and $-\nabla_T(\mu_i/M_i - \mu_{n_c}/M_{n_c})(1/T)$ $(i = 1,\ldots,n_c - 1)$. The fluxes $\vec{\phi}_i$ are $\vec{q}^{cond}$ for $i = n_c$, and $\vec{J}_i$ $(i = 1,\ldots,n_c - 1)$.

6.2.1 Phenomenological Laws of Irreversible Thermodynamics

For a large class of problems, it is known empirically that the fluxes are linear functions of driving forces. This main postulate of irreversible thermodynamics assumes that for conditions not far from equilibrium, the various fluxes may be written as

$$\vec{\phi}_i = \sum_{k=1}^{n_c} \mathcal{L}_{ik}\vec{X}_k \quad i = 1, \ldots, n_c \tag{6.37}$$

where $\vec{\phi}_i$ is the independent flux and $\vec{X}_k$ is the driving force; $\mathcal{L}_{ik}$ are the phenomenological coefficients. When $k = i$, $\mathcal{L}_{ii}$ coefficients represent the direct effect to be described later. The coefficients $\mathcal{L}_{ik}$ are a function of temperature, pressure, and composition.

Combining Eqs. (6.36) and (6.37) provides

$$\sigma = \sum_{i=1}^{n_c}\sum_{k=1}^{n_c} \mathcal{L}_{ik}\vec{X}_k \cdot \vec{X}_i \tag{6.38}$$

Because $\sigma \geq 0$, in the above quadratic expression, all the diagonal elements, that is, phenomenological coefficients $\mathcal{L}_{ii}$ are positive (see Example 6.5).

Now let us write Eq. (6.37) in the following form

$$\vec{\phi}_i = \mathcal{L}_{i1}\vec{X}_1 + \sum_{k=2}^{n_c} \mathcal{L}_{ik}\vec{X}_k \quad i = 1, \ldots, n_c \tag{6.39}$$

Combining Eqs. (6.36) and (6.39) provides

$$\sigma = \left[\mathcal{L}_{1,1}\vec{X}_1 + \sum_{k=2}^{n_c} \mathcal{L}_{1k}\vec{X}_k\right] \cdot \vec{X}_1 + \sum_{i=2}^{n_c}\left[\mathcal{L}_{i1}\vec{X}_1 + \sum_{k=2}^{n_c} \mathcal{L}_{ik}\vec{X}_k\right] \cdot \vec{X}_i \tag{6.40}$$

Let us denote $\mathcal{L}_{1,1} = L_{qq}$, $\mathcal{L}_{1k} = L_{q,k-1}$ $(k = 2, \ldots, n_c)$, $\mathcal{L}_{i1} = L_{i-1,q}$ $(i = 2, \ldots, n_c)$ and $\mathcal{L}_{ik} = L_{i-1,k-1}$ $(i, k = 2, \ldots, n_c)$. Then Eq. (6.40) can be written as

$$\sigma = \left[L_{qq}\vec{X}_1 + \sum_{k=2}^{n_c} L_{q,k-1}\vec{X}_k\right] \cdot \vec{X}_1 + \sum_{i=2}^{n_c}\left[L_{i-1,q}\vec{X}_1 + \sum_{k=2}^{n_c} L_{i-1,k-1}\vec{X}_k\right] \cdot \vec{X}_i \tag{6.41a}$$

or

$$\sigma = \left[L_{qq}\vec{X}_1 + \sum_{k=1}^{n_c-1} L_{qk}\vec{X}_{k+1}\right] \cdot \vec{X}_1 + \sum_{i=1}^{n_c-1}\left[L_{iq}\vec{X}_1 + \sum_{k=1}^{n_c-1} L_{ik}\vec{X}_{k+1}\right] \cdot \vec{X}_{i+1} \tag{6.41b}$$

Comparison of Eqs. (6.35b) and (6.41a) provides

$$\vec{q}^{\,cond} = +\left(\frac{1}{T^2}\right) L_{qq}(-\nabla T) + \frac{1}{T}\sum_{k=1}^{n_c-1} L_{qk}\left[-\nabla_T\left(\frac{\mu_k}{M_k} - \frac{\mu_{n_c}}{M_{n_c}}\right)\right] \tag{6.42}$$

and

$$\vec{J}_i = +\left(\frac{1}{T^2}\right) L_{iq}(-\nabla T) + \frac{1}{T}\sum_{k=1}^{n_c-1} L_{ik}\left(-\nabla_T\left(\frac{\mu_k}{M_k} - \frac{\mu_{n_c}}{M_{n_c}}\right)\right) \quad i = 1, \ldots, n_c - 1 \quad (6.43)$$

Note that the Fourier's law of heat conduction flux (i.e., $-K\nabla T$) is a part of Eq. (6.42). From the Fourier law $K = L_{qq}/T^2$. The remaining terms on the right side of Eq. (6.42) are the additional contributions from diffusion which can occur in mixtures only. They may be small in most but not in all applications. The contribution of chemical potential gradients to conduction is called the Dufour-effect. For ideal liquid mixtures, the Dufour-effect is proportional to the concentration gradient. The Dufour-effect can be significant in some concentrations (Eckert and Drake, 1972).

In order to proceed further, we need to use the fundamental theorem of thermodynamics of irreversible processes due to Onsager (1931a, 1931b).

Onsager's Reciprocal Relations

In 1931, Onsager introduced the fundamental theorem of thermodynamics of irreversible processes in the form of reciprocal relations. According to this theorem, the phenomenological coefficients L_{ik} are symmetric,

$$L_{ik} = L_{ki} \quad i, k = 1, \ldots, n_c - 1 \quad (6.44)$$

These coefficients are selected such that at equilibrium all the fluxes are zero; they have a definite value.

In the following, we will use the reciprocal relations given by Eq. (6.44) to derive the expression for the multicomponent diffusion mass flux.

6.2.2 Diffusion Mass Flux

From $\mu_k = \mu_k(T, P, x_1, \ldots, x_{n_c-1})$ one can write at constant temperature,

$$\nabla_T \mu_k = \left.\frac{\partial \mu_k}{\partial P}\right|_{T,x} \nabla P + \sum_{j=1}^{n_c-1} \left.\frac{\partial \mu_k}{\partial x_j}\right|_{T,P,x_j} \nabla x_j, \quad k = 1, \ldots, n_c \quad (6.45)$$

From the Gibbs-Duhem equation [Eq. (1.40)] at constant temperature,

$$\nabla_T \mu_{n_c} = \frac{1}{x_{n_c}}\left(\frac{1}{c}\nabla P - \sum_{j=1}^{n_c-1} x_j \nabla_T \mu_j\right) \quad (6.46)$$

where c is the molar density.

From Eqs. (6.45) and (6.46),

$$\nabla_T\left(\frac{\mu_k}{M_k}-\frac{\mu_{n_c}}{M_{n_c}}\right)=\sum_{j=1}^{n_c-1}\left(\frac{x_j}{M_{n_c}x_{n_c}}+\frac{\delta_{jk}}{M_j}\right)\sum_{l=1}^{n_c-1}\left.\frac{\partial\mu_j}{\partial x_l}\right|_{T,P,x_l}\nabla x_l$$
$$+\left[\sum_{j=1}^{n_c-1}\frac{x_j}{M_{n_c}x_{n_c}}\left.\frac{\partial\mu_j}{\partial P}\right|_{T,x}+\frac{1}{M_k}\left.\frac{\partial\mu_k}{\partial P}\right|_{T,x}-\frac{1}{cM_{n_c}x_{n_c}}\right]\nabla P \tag{6.47}$$
$$k=1,\ldots,n_c-1$$

where $\delta_{jk}=0$ for $k\neq j$ and $\delta_{jk}=1$ for $k=j$.

Combining Eqs. (6.43) and (6.47), the expression for the diffusion mass flux is obtained,

$$\vec{J}_i=-L_{iq}\left(\frac{1}{T^2}\right)\nabla T-\frac{1}{T}\sum_{k=1}^{n_c-1}L_{ik}\sum_{j=1}^{n_c-1}\left(\frac{x_j}{M_{n_c}x_{n_c}}+\frac{\delta_{jk}}{M_j}\right)\sum_{l=1}^{n_c-1}\left.\frac{\partial\mu_j}{\partial x_l}\right|_{T,P,x_l}\nabla x_l$$
$$-\frac{1}{T}\sum_{k=1}^{n_c-1}L_{ik}\left[\frac{1}{M_k}\left.\frac{\partial\mu_k}{\partial P}\right|_{T,x}+\sum_{j=1}^{n_c-1}\frac{x_j}{M_{n_c}x_{n_c}}\left.\frac{\partial\mu_j}{\partial P}\right|_{T,x}-\frac{1}{cM_{n_c}x_{n_c}}\right]\nabla P \tag{6.48}$$
$$i=1,\ldots,n_c-1$$

From $\partial\mu_j/\partial x_l|_{T,P,x_l}=RT\partial\ln f_j/\partial x_l|_{T,P,x_l}$ ($j=1,\ldots,n_c;\, l=1,\ldots,n_c-1$) and $\partial\mu_k/\partial P|_{T,\underline{x}}=\bar{V}_k$ ($k=1,\ldots,n_c$), and Eq. (6.48), we obtain the general expression for the diffusion mass flux,

$$\vec{J}_i=-ca_{in_c}D_{in_c}\left\{Mk_{iT}\frac{\nabla T}{T^2}+\frac{M_ix_i}{L_{ii}}\sum_{j=1}^{n_c-1}\sum_{k=1}^{n_c-1}L_{ik}\sum_{l=1}^{n_c-1}\frac{M_jx_j+M_{n_c}x_{n_c}\delta_{jk}}{M_j}\right.$$
$$\left.\times\left.\frac{\partial\ln f_j}{\partial x_l}\right|_{T,P,x_l}\nabla x_l+\frac{M_ix_i}{RTL_{ii}}\sum_{k=1}^{n_c-1}L_{ik}\left[\frac{M_{n_c}x_{n_c}}{M_k}\bar{V}_k+\sum_{j=1}^{n_c-1}x_j\bar{V}_j-\frac{1}{c}\right]\nabla P\right\} \tag{6.49}$$
$$i=1,\ldots,n_c-1$$

where

$$a_{in_c}=\frac{M_iM_{n_c}}{M^2}\quad i=1,\ldots,n_c-1 \tag{6.50a}$$

$$D_{in_c}=\frac{M^2RL_{ii}}{cM_i^2M_{n_c}^2x_ix_{n_c}}\quad i=1,\ldots,n_c-1 \tag{6.50b}$$

$$k_{iT}=\frac{M_iM_{n_c}x_ix_{n_c}L_{iq}}{MRL_{ii}}=\alpha_{iT}x_ix_{n_c}\quad i=1,\ldots,n_c-1 \tag{6.50c}$$

In Eq. (6.50c), k_{iT} and α_{iT} denote the thermal diffusion ratio and the thermal diffusion factor of component i, respectively. Some authors have used an expression of the form $k_{iT}=\alpha_{iT}x_ix_j$ to relate thermal diffusion ratio and thermal diffusion factor in multicomponent systems. This form of thermal diffusion ratio expression is valid for two-component systems only. In three- and higher-component mixtures, Eq. (6.50c) should be used; x_{n_c} represents the reference component.

In order to facilitate the algebraic manipulations, we will write Eq. (6.49) in a vector form. Let us define the following column vectors and matrices.

$$\boldsymbol{J} \equiv [\vec{J}_i] = \begin{pmatrix} \vec{J}_1 \\ \vdots \\ \vdots \\ \vdots \\ \vec{J}_{n_c-1} \end{pmatrix} \tag{6.51a}$$

$$\boldsymbol{L} \equiv [L_{ij}] = \begin{pmatrix} L_{1,1} & \cdots\cdots & L_{1,n_c-1} \\ \vdots & \ddots & \vdots \\ \vdots & \ddots & \vdots \\ \vdots & \ddots & \vdots \\ L_{n_c-1,1} & \cdots\cdots & L_{n_c-1,n_c-1} \end{pmatrix} \tag{6.51b}$$

$$\boldsymbol{D} \equiv [D_{ij}] = \left[a_{in_c} D_{in_c} \delta_{ij}\right] \tag{6.51c}$$

$$\boldsymbol{M} \equiv [M_{ij}] = \left[\frac{M_{n_c} x_i}{L_{ii}} \delta_{ij}\right] \tag{6.51d}$$

$$\boldsymbol{W} \equiv [W_{ij}] = \left[\frac{M_i x_i + M_{n_c} x_{n_c} \delta_{ij}}{M_j}\right] \tag{6.51e}$$

$$\boldsymbol{F} \equiv [F_{ij}] = \left[\left.\frac{\partial \ln f_i}{\partial x_j}\right|_{T,P,x_j}\right] \tag{6.51f}$$

$$\nabla \boldsymbol{x} \equiv [\nabla x_i] \tag{6.51g}$$

$$\boldsymbol{K}_T \equiv [K_{iT}] = \left[\frac{Mk_{iT}}{T}\right] \tag{6.51h}$$

$$\boldsymbol{V} \equiv [V_i] = \left[\frac{1}{RT}\left(\sum_{j=1}^{n_c-1} x_j \bar{V}_j + \frac{M_{n_c} x_{n_c}}{M_i} \bar{V}_i - \frac{1}{c}\right)\right] \tag{6.51i}$$

In the above expressions Eqs. (6.51a) and (6.51b) are written in full column (a matrix in three directions) and matrix forms, respectively; also the rest of the expressions Eq. (6.51a). Note that in Eq. (6.51a) the flux of components also have three directions.

Equation (6.49) in the vector form becomes (we use dot between symbols for clarity)

$$\boldsymbol{J} = -c(\boldsymbol{D} \cdot \boldsymbol{M} \cdot \boldsymbol{L} \cdot \boldsymbol{W} \cdot \boldsymbol{F} \cdot \nabla \boldsymbol{x} + \boldsymbol{D} \cdot \boldsymbol{K}_T \nabla T + \boldsymbol{D} \cdot \boldsymbol{M} \cdot \boldsymbol{L} \cdot \boldsymbol{V} \nabla P) \tag{6.52}$$

Let

$$\boldsymbol{D}^M = \boldsymbol{D} \cdot \boldsymbol{M} \cdot \boldsymbol{L} \cdot \boldsymbol{W} \cdot \boldsymbol{F} \tag{6.53}$$

$$\boldsymbol{D}^T = \boldsymbol{D} \cdot \boldsymbol{K}_T \tag{6.54}$$

$$\boldsymbol{D}^P = \boldsymbol{D} \cdot \boldsymbol{M} \cdot \boldsymbol{L} \cdot \boldsymbol{V} \tag{6.55}$$

The form of various diffusion coefficients in Eqs. (6.53) to (6.55) can be used to study variations as we will discuss in the *Examples and Theory Extension* section. Then the diffusion mass flux expression can be written in a compact form,

$$\boxed{\boldsymbol{J} = -c(\boldsymbol{D}^M \cdot \nabla \boldsymbol{x} + \boldsymbol{D}^T \nabla T + \boldsymbol{D}^P \nabla P)} \tag{6.56}$$

In the above equation, the first, second, and third terms on the right side represent Fickian (molecular) diffusion, thermal diffusion, and pressure diffusion, respectively. The elements in the matrix $\boldsymbol{D}^M$ and the column vectors $\boldsymbol{D}^T$ and $\boldsymbol{D}^P$ are given by,

$$\boxed{D_{il}^M = a_{i,n_c} D_{i,n_c} \frac{M_i x_i}{L_{ii}} \sum_{k=1}^{n_c-1} L_{ik} \sum_{j=1}^{n_c-1} \frac{M_j x_j + M_{n_c} x_{n_c} \delta_{jk}}{M_j} \frac{\partial \ln f_j}{\partial x_l}\bigg|_{T,P,x_l}} \quad i, l = 1, \ldots, n_c - 1 \tag{6.57}$$

$$\boxed{D_i^T = a_{in_c} D_{in_c} M \frac{k_{iT}}{T}} \quad i = 1, \ldots, n_c - 1 \tag{6.58}$$

$$\boxed{D_i^P = a_{i,n_c} D_{i,n_c} \frac{M_{n_c} x_i}{RTL_{ii}} \sum_{k=1}^{n_c-1} L_{ik} \left[\sum_{j=1}^{n_c-1} x_j \bar{V}_j + \frac{M_{n_c} x_{n_c}}{M_j} - \frac{1}{c} \right]} \quad i = 1, \ldots, n_c - 1 \tag{6.59}$$

The expression for diffusion mass flux of component i from Eq. (6.56) can be written as,

$$\boxed{\vec{J}_i = -c \left[\sum_{j=1}^{n_c-1} D_{ij}^M \nabla x_j + D_i^T \nabla T + D_i^P \nabla P \right]} \quad i = 1, \ldots, n_c - 1. \tag{6.60}$$

The diffusion coefficients are given in Eqs. (6.57) to (6.59). Equation (6.57) reveals that the Fickian diffusion coefficient $D_{il}^M (i, l = 1, \ldots, n_c - 1)$ is a function of phenomenological coefficients $L_{ik}(i, k = 1, \ldots, n_c - 1)$ and of the quantities $(\partial \ln f_j / \partial x_l)_{x_l,T,P}$ as well as the mol fractions. The diagonal Fickian diffusion coefficient D_{ii}^M relates the concentration gradient ∇x_i of species i to the flux $\vec{J}_i$. The cross Fickian diffusion coefficient $D_{ij}(i \neq j)$ relates the concentration gradient $\nabla x_j (j \neq i)$ of species j to the flux $\vec{J}_i$. In a multicomponent mixture, the Fickian diffusion coefficient D_{ii}^M cannot be generally assumed to be the binary Fickian diffusion coefficient measured in binary mixtures of components i and n_c. Likewise, the thermal diffusion ratio k_{iT} for component i is not generally the same in binary and ternary mixtures, even at the same pressure, temperature, and mol fraction of component i.

The Fickian diffusion coefficients D_{il}^M given by Eq. (6.57) are not independent of each other. In the following, we will find the number of independent Fickian diffusion coefficients in a multicomponent mixture.

From Eq. (6.53),

$$\boldsymbol{L} = (\boldsymbol{D} \cdot \boldsymbol{M})^{-1} \cdot \boldsymbol{D}^M \cdot (\boldsymbol{W} \cdot \boldsymbol{F})^{-1} \tag{6.61}$$

Using the relation $[A \cdot B]^T = B^T \cdot A^T$ (A^T is the transpose of A) and the fact that L is symmetric and that $D \cdot M$ is a diagonal matrix, ($[D \cdot M]^T = D \cdot M$; D and M are both a diagonal matrix), one obtains the following relationship between D^M and $(D^M)^T$

$$(D^M)^T = (W \cdot F)^T \cdot D^M \cdot (W \cdot F)^{-1} \tag{6.62}$$

Equations (6.61) and (6.62) are valid provided $\det(W \cdot F) \neq 0$ which implies that $\det(W) \neq 0$ and $\det(F) \neq 0$. $[\det(W \cdot F) = \det(W) \cdot \det(F)]$. The former is nonzero but the latter may approach zero as the limit of stability, or criticality approaches as we demonstrated in Chap. 4.

From Eqs. (6.61) and (6.62),

$$L \cdot W \cdot F = D^{-1} \cdot M^{-1} \cdot D^M \tag{6.63}$$

and

$$(D^M)^T \cdot W \cdot F = (W \cdot F)^T \cdot D^M \tag{6.64}$$

respectively. From Eq. (6.63), one obtains the expression that relates the phenomenological coefficients to the Fickian diffusion coefficients,

$$\sum_{l=1}^{n_c-1} \sum_{k=1}^{n_c-1} \frac{M_k x_k + M_{n_c} x_{n_c} \delta_{lk}}{M_k} \left. \frac{\partial \ln f_j}{\partial x_l} \right|_{x_j, T, P} L_{li} = \frac{c M_{n_c} x_{n_c}}{R} D_{ij}^M \quad (i, j = 1, \ldots, n_c - 1) \tag{6.65}$$

Therefore, given D_{ij}^M, one can calculate the phenomenological coefficients L_{li} $(l, i = 1, \ldots, n_c - 1)$. From Eq. (6.64), one obtains,

$$\begin{aligned} &\sum_{l=1}^{n_c-1} \sum_{k=1}^{n_c-1} \frac{M_k x_k + M_{n_c} x_{n_c} \delta_{lk}}{M_k} \left. \frac{\partial \ln f_k}{\partial x_j} \right|_{x_j, T, P} D_{li}^M \\ &= \sum_{l=1}^{n_c-1} \sum_{k=1}^{n_c-1} \frac{M_k x_k + M_{n_c} x_{n_c} \delta_{lk}}{M_k} \left. \frac{\partial \ln f_k}{\partial x_j} \right|_{x_j, T, P} D_{lj}^M \quad (i, j = 1, \ldots, n_c - 1) \end{aligned} \tag{6.66}$$

The above equations provide the relationships among the Fickian diffusion coefficients. The $(n_c - 1)^2$ expressions in Eq. (6.66) are not independent. For $i = j$, the two sides of the equality are identical; which implies that the number of independent expressions reduces to $(n_c - 1)^2 - (n_c - 1)$. Additionally, by replacing i with j, in Eq. (6.66), one obtains the same expressions. Thus the number of independent expressions relating the Fickian diffusion coefficients reduces to $[(n_c - 1)^2 - (n_c - 1)]/2$, and the number of independent Fickian diffusion coefficients reduces to $(n_c - 1)^2 - [(n_c - 1)^2 - (n_c - 1)]/2 = n_c(n_c - 1)/2$. The following expression provides the number of independent Fickian diffusion coefficients in an n_c-component mixture,

$$\text{Number of independent Fickian diffusion coefficients} = \frac{n_c(n_c - 1)}{2} \tag{6.67}$$

We may comment that Kett and Anderson (1969) among others have verified the validity of the Onsager reciprocal relations in ternary systems of dodecane/hexadecane/hexane and toluene/chlorobenzene/bromobenzene. The measured Fickian diffusion coefficients for the dodecane/hexadecane/hexane at 25°C are $D_{1,1}^M = 0.968 \times 10^{-5}$ cm^2/s,

$D_{1,2}^M = 0.266 \times 10^{-5}$ cm^2/s, $D_{2,1}^M = 0.255 \times 10^{-5}$ cm^2/s , $D_{2,2}^M = 0.1031 \times 10^{-5}$ cm^2/s for $x_1 = 0.350$, and $x_2 = 0.317$ (the reference component is hexane).

Next, we will provide specific expressions for binary mixture diffusion flux.

Diffusion Flux in Binary Mixtures

The expression for mass diffusion flux in a binary system is readily obtained from Eq. (6.49) (after substitution for $a_{12} = M_1 M_2 / M^2$):

$$\vec{J}_1 = -\rho D_{1,2} \frac{M_1 M_2}{M^2} \left[\left. \frac{\partial \ln f_1}{\partial \ln x_1} \right|_{T,P} \nabla x_1 + \frac{x_1}{RT} \left(\bar{V}_1 - \frac{M_1}{\rho} \right) \nabla P + \frac{k_{1T}}{T} \nabla T \right] \tag{6.68}$$

Note that in a binary mixture, the phenomenological diffusion coefficient L_{11} does not appear in the expression for the diffusion flux. As a result, the expression for diffusion flux is very simple. In Eq. (6.68) the first, second, and third terms represent the Fickian, pressure, and thermal diffusion fluxes, respectively. In the *Examples and Theory Extension* section, we will present various forms of diffusion fluxes and diffusion coefficients.

Diffusion Flux and Diffusion Coefficients in the Critical Region

It is well known that some transport properties, as well as some thermodynamic properties, as discussed in Chap. 4, exhibit singular behavior near the vapor-liquid critical point (as well as liquid-liquid critical point) (Luettmer-Strathmann and Sengers, 1996). The variation of transport properties and thermodynamic properties may also be markedly different in single component and in multicomponent mixtures. Here we are mainly interested in the study of diffusion fluxes and diffusion coefficients at the critical point.

There are two methods in relation to the critical region. In one approach, the criteria of criticality are used to study various diffusion coefficients (i.e., Fickian, thermal, and pressure diffusion) in the critical region. In this approach the classical equations of state are used for fluid description both in the critical region and away from the critical region. An alternative approach relies on the description by the scaled equations with universal scaling functions and universal critical exponents. The second approach requires a crossover equation of state to link the scaled equations in the critical region, to the classical equations which are used for fluid description away from the critical region. The crossover equations cannot predict thermodynamic properties and phase behavior without a preliminary fit to the measured thermodynamic properties of mixtures under study.

We prefer the approach based on the criticality criterion (discussed in Chap. 4) due to its generality for nonideal fluids. This is the approach that is adopted in the following, where we first discuss the diffusion mass flux and diffusion coefficients in binary mixtures in the critical region because of simplicity and then move on to diffusion flux and diffusion coefficients in a multicomponent mixture.

Binary Mixtures

At isothermal and isobaric conditions, Eq. (6.68) simplifies to,

$$\vec{J}_1 = -\rho D_{1,2} \frac{M_1 M_2}{M^2} \left. \frac{\partial \ln f_1}{\partial \ln x_1} \right|_{T,P} \nabla x_1 = -\rho D_{1,1}^M \nabla x_1 \tag{6.69}$$

In Eq. (6.69), $D_{1,2} > 0$ because $L_{1,1} > 0$, see Eq. (6.50b). Therefore, the Fickian diffusion coefficient $D^M_{1,1}$ is positive except at the critical point where it is zero: at the critical point $\partial \ln f_1/\partial x_1|_{T,P} = 0$. For a stable fluid mixture $\partial \ln f_1/\partial x_1|_{T,P} > 0$. From this simple analysis, one demonstrates that the diffusion flux in a binary mixture is zero at the critical point provided ∇x_1 is finite.

Let us now consider a binary mixture at isothermal and nonisobaric conditions. At steady state with zero flux of species, Eq. (6.68) simplifies to

$$\left.\frac{\partial \ln f_1}{\partial \ln x_1}\right|_{T,P} \nabla x_1 + \frac{x_1}{RT}\left(\bar{V}_1 - \frac{M_1}{\rho}\right)\nabla P = 0 \tag{6.70}$$

At nonisobaric condition $\nabla P \neq 0$ (say, due to gravity, $dP/dz = -\rho g$; z is positive upward). In general, $(\bar{V}_1 - M_1/\rho)$ is nonzero, therefore, the second term has a finite nonzero value. As a result, the first term also has a nonzero finite value even at the critical point. At the critical point, $\partial \ln f_1/\partial \ln x_1|_{T,P} = 0$, which implies that ∇x_1 should be either $+\infty$ or $-\infty$. Figure 6.1 shows the variation of composition x_1 in a binary mixture in the gravity field at isothermal conditions. The inflection point corresponds to the critical point. From our simple analysis, the pressure diffusion coefficient which is given by

$$D^P_1 = D_{1,2}\frac{M_1 M_2}{M^2}\frac{x_1}{RT}\left(\bar{V}_1 - \frac{M_1}{\rho}\right) \tag{6.71}$$

is finite at the critical point.

A reasoning similar to the above can be made to analyze thermal diffusion coefficient and thermal diffusion flux at the critical point. At isobaric and nonisothermal conditions, Eq. (6.68) simplifies to

$$\left.\frac{\partial \ln f_1}{\partial \ln x_1}\right|_{T,P} \nabla x_1 + \frac{k_{1T}}{T}\nabla T = 0 \tag{6.72}$$

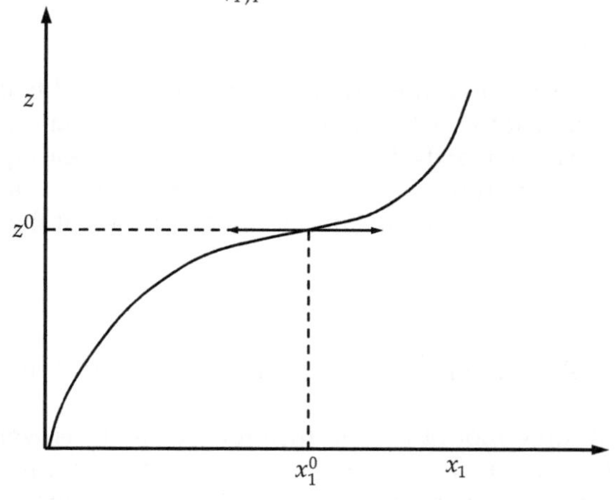

FIGURE 6.1 Composition variation in a vertical domain in the critical region (adapted from Ghorayeb and Firoozabadi, 2000a).

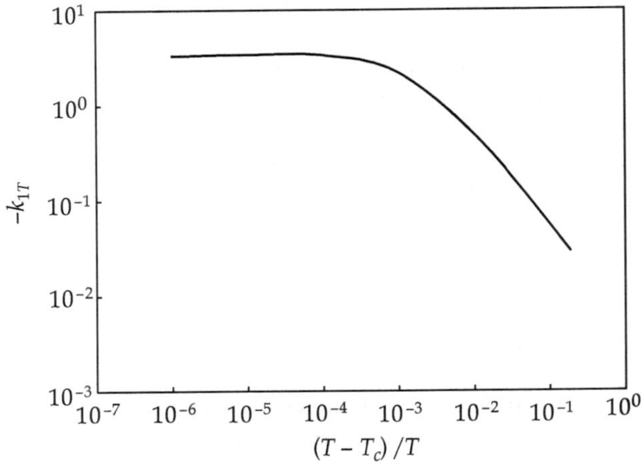

FIGURE 6.2 Thermal diffusion ratio vs. $(T - T_c)/T$ at the critical pressure for the binary mixture of CO_2/C_2, 25/75 mol% (adapted from Ghorayeb and Firoozabadi, 2009a).

at steady state with zero flux of species. As was demonstrated above, the term $\partial \ln f_1/\partial \ln x_1|_{T,P} \nabla x_1$ may have finite nonzero value at the critical point. For a finite ∇T, the thermal diffusion ratio k_{1T} is, therefore, finite at the critical point. In other words, thermal diffusion ratio does not exhibit a singular behavior at the critical point. Figure 6.2 shows the calculated thermal diffusion ratio vs. $(T - T_c)/T$ at the critical pressure for a binary mixture of CO_2/ethane (C_2) (25% CO_2, 75% C_2, mols). The negative sign of thermal diffusion ratio of CO_2 implies that CO_2 will segregate to the hot side (see Prob. 6.14).

Multicomponent Mixtures

For a binary mixture, we used the criteria of the limit of stability $\partial \ln f_1/\partial \ln x_1|_{T,P} = 0$ to analyze diffusion flux at the critical point. The other criterion for criticality has already been used in the derivation of the flux expression from the local equilibrium assumption. For multicomponent mixtures, we will primarily use the criterion $\det(\boldsymbol{F}) = 0$ from Eq. (4.131) which provides the following important expression at the critical point,

$$\det(\boldsymbol{D}^M) = 0 \tag{6.73}$$

Note that in general $\boldsymbol{D}^T$ and $\boldsymbol{D}^P$ have finite nonzero entries at the critical point [see Eqs. (6.54) and (6.55)].

Figure 6.3 shows the plot of four diffusion coefficients as well as the $\det(\boldsymbol{D})$ of the water-chloroform-acidic acid at temperature of 25°C and pressure of 1 atm. The five data points in Fig. 6.3 have a molar ratio of H_2O/chloroform of about 1.43 corresponding to the mol ratio of the critical point. The amount of acidic acid is varied. In the figure, as the critical point is approached, the $\det(\boldsymbol{D})$ decreases and approaches zero at the critical point. The reference component is acidic acid. Water is component 1, and chloroform is component 2.

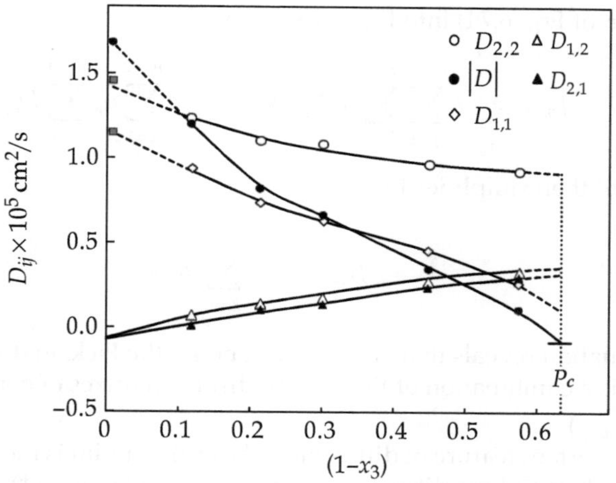

FIGURE 6.3 Fickian diffusion coefficients and det $|D|$ for the ternary mixture of $H_2O/CHCl_3/CH_3COOH$ at 25°C and 1 atm. P_c is the plait point. x_3 is mol fraction of acetic acid (adapted from Vitagliano et al., 1978).

According to data in Fig. 6.3, $D_{1,1}$ decreases more drastically than $D_{2,2}$; cross diffusion coefficients $D_{1,2}$ and $D_{2,1}$ become greater than $D_{1,1}$ as the critical point is approached.

Equation (6.73) implies that in a multicomponent mixture ($n_c \geq 3$), the determinant of the Fickian diffusion coefficients vanishes at the critical point. For a binary mixture there is only one Fickian diffusion coefficient which should vanish at the critical point.

From Eq. (6.73), one deduces that the rank of the $(n_c - 1)$ by $(n_c - 1)$ matrix of Fickian diffusion coefficients is $\leq (n_c - 2)$. The rank of a matrix counts the number of independent rows; for rank $= n_c - 2$, one of the rows of $\det(\mathbf{D}^M)$ is linearly dependent on the others. One way to obtain the rank of a matrix is to perform triangular factorization and then count the rows with zero entries (Strang, 1988). Every row with zero entries reduces the rank by one. Any row with zero entries also makes the determinant zero. Let us discuss the case of the rank of $\mathbf{D}^M = n_c - 2$ [a similar reasoning applies for the general case of rank $\mathbf{D}^M < (n_c - 2)$]. We designate the row that is not independent by $(n_c - 1)$. Then there exist constant coefficients $\alpha_1, \ldots, \alpha_{n_c-2}$ such that every entry of row $(n_c - 1)$ is given by,

$$D^M_{n_c-1,j} = \sum_{k=1}^{n_c-2} \alpha_k D^M_{kj} \quad j = 1, \ldots, n_c - 1 \tag{6.74}$$

At isothermal and isobaric conditions, Eq. (6.60) reduces to

$$\vec{J}_{n_c-1} = -c \sum_{j=1}^{n_c-1} D^M_{n_c-1,j} \nabla x_j \tag{6.75}$$

for the diffusion flux of component $(n_c - 1)$.

Substitution of Eq. (6.74) into Eq. (6.75) gives

$$\vec{J}_{n_c-1} = -c \sum_{j=1}^{n_c-1} \sum_{k=1}^{n_c-2} \alpha_k D^M_{k,j} \nabla x_j = -c \sum_{k=1}^{n_c-2} \alpha_k \sum_{j=1}^{n_c-1} D^M_{k,j} \nabla x_j \tag{6.76}$$

Equation (6.76) then simplifies to

$$\vec{J}_{n_c-1} = -c \sum_{k=1}^{n_c-2} \alpha_k \vec{J}_k \tag{6.77}$$

The above equation reveals that at the critical point, the Fickian diffusion flux of component $(n_c - 1)$ is a combination of the Fickian diffusion fluxes of components 1 to $(n_c - 2)$ (i.e., $\vec{J}_1, \ldots, \vec{J}_{n_c-2}$).

Another important feature of diffusion at the critical point is the reduction in the number of independent Fickian diffusion coefficients from $[n_c(n_c - 1)/2]$ to $[(n_c(n_c - 1)/2) - 1]$ as we will discuss next.

Let $\boldsymbol{D}_j = (D^M_{1,j}, D^M_{2,j}, \ldots, D^M_{n_c-2,j})$ represent the column vector and $\boldsymbol{A} = (\alpha_1\ \alpha_2 \ldots .\alpha_{n_c-2})$ represent the row vector. Then Eq. (6.74) can be written as,

$$D^M_{n_c-1,j} = \boldsymbol{A} \cdot \boldsymbol{D}_j \tag{6.78}$$

Let us define $\boldsymbol{D}_{n_c-1} = (D^M_{n_c-1,1}, \ldots, D^M_{n_c-1,n_c-1})$ and the matrix $\boldsymbol{D}^M_1 = [D^M_{ij}]$; $i, j = 1, \ldots, (n_c - 2)$. The column vector $\boldsymbol{D}_{n_c-1}$ is given by

$$\boldsymbol{D}_{n_c-1} = \boldsymbol{A} \cdot \boldsymbol{D}^M_1 \tag{6.79}$$

The rank of matrix $\boldsymbol{D}^M_1$ is $(n_c - 2)$, it is, therefore, invertible; that is, $\det(\boldsymbol{D}^M_1) \neq 0$ and one can write

$$\boldsymbol{A} = \left(\boldsymbol{D}^M_1\right)^{-1} \cdot \boldsymbol{D}_{n_c-1} \tag{6.80}$$

Now we write Eq. (6.78) for $j = n_c - 1$,

$$D^M_{n_c-1,n_c-1} = \boldsymbol{A} \cdot \boldsymbol{D}_{n_c-1} \tag{6.81}$$

Combining Eqs. (6.80) and (6.81),

$$D^M_{n_c-1,n_c-1} = \left[\left(\boldsymbol{D}^M_1\right)^{-1} \cdot \boldsymbol{D}_{n_c-1}\right] \cdot \boldsymbol{D}_{n_c-1} \tag{6.82}$$

From Eq. (6.82), the Fickian diffusion coefficient $D^M_{n_c-1,n_c-1}$ can be calculated from knowledge of other molecular diffusion coefficients. Therefore, the number of independent Fickian diffusion coefficients at the critical point is equal to $[(n_c(n_c - 1)/2)] - 1$ [see Eq. (6.67)].

Now we consider a multicomponent mixture ($n_c \geq 3$) at the critical point at isothermal and isobaric conditions. The diffusion flux of all the components is in general nonzero at unsteady state.

At isothermal and nonisobaric conditions, when $\vec{J} = 0$, one can write from Eq. (6.56),

$$\boldsymbol{D}^M . \nabla x + \boldsymbol{D}^P \nabla P = \boldsymbol{0} \tag{6.83}$$

The solution of Eq. (6.83) for ∇x_i is given by,

$$\nabla x_i = -\nabla P \frac{\det(\boldsymbol{B}_i)}{\det(\boldsymbol{D}^M)} \quad i = 1, \ldots, n_c - 1 \tag{6.84}$$

where $\boldsymbol{B}_i$ is the matrix $\boldsymbol{D}^M$ with i replaced by the entries in column vector $\boldsymbol{D}^P$. Both ∇P (say, from gravity) and $\det(\boldsymbol{B}_i)$ are in principle different from zero. Because $\det(\boldsymbol{D}^M)$ is zero at the critical point, therefore, ∇x_i ($i = 1, \ldots, n_c - 1$) is infinity. Figure 6.1 also applies to a multicomponent mixture ($n_c \geq 3$) at steady state, as well as to a binary mixture at steady state. One can apply a similar analysis at isobaric and nonisothermal conditions.

Our derivation of diffusion fluxes and diffusion coefficients is based on irreversible thermodynamics. An alternative approach is through the mechanism of molecular collision. Taylor and Krishna (1993) present excellent derivations through molecular collisions. This approach leads to the Maxwell-Stefan (MS) diffusion relations and the Maxwell-Stefan diffusion coefficients $\mathcal{D}_{ij}^{MS}$. One distinct feature of the matrix of Maxwell-Stefan diffusion coefficients is that they form a symmetric matrix. The Fickian diffusion coefficients form, in general, a nonsymmetrical matrix. Another important feature of the MS diffusion coefficients is that they do not vary as much as Fickian diffusion coefficients vary with composition, pressure, and temperature. As a result, composition dependency of the Maxwell-Stefan diffusion coefficients can be expressed by simple relations.

The diffusion flux in terms of MS diffusion coefficients is given (in vector notation) by

$$\vec{\boldsymbol{J}} = -c\boldsymbol{BF}\nabla \boldsymbol{x} \tag{6.85}$$

where the elements of matrix $\boldsymbol{B}$ are given by

$$B_{ii} = \frac{x_i}{\mathcal{D}_{in_c}} + \sum_{\substack{k=1 \\ i \neq k}}^{n_c} \frac{x_k}{\mathcal{D}_{ik}} \quad i = 1, \ldots, n_c - 1$$

$$B_{ij} = -x_i \left(\frac{1}{\mathcal{D}_{ij}} - \frac{1}{\mathcal{D}_{in_c}} \right) \quad i, j = 1, \ldots, n_c - 1, i \neq j \tag{6.86}$$

The $\mathcal{D}_{ij}$ are the MS diffusion coefficients. $\boldsymbol{F}$ is the matrix of nonideality with element F_{ij} given by

$$F_{ij} = \left(\frac{\partial \ln f_i}{\partial x_j} \right)_{T,P,x_j} \quad i, j = 1, \ldots, n_c - 1 \tag{6.87}$$

Comparison of Eqs. (6.85) and (6.56) without thermal and pressure diffusion provides

$$\boldsymbol{D}^M = [\boldsymbol{B}]^{-1} \boldsymbol{F} \tag{6.88}$$

Once $\mathcal{D}_{ij}$ are known, $\boldsymbol{D}^M$ can be computed using the above relationship. $\boldsymbol{F}$ can be evaluated from an EOS, discussed in Chap. 3. In order to complete the discussion on evaluation of Fickian diffusion coefficients, we briefly discuss how to obtain $\mathcal{D}_{ij}$ for hydrocarbon mixtures and petroleum fluids assuming that the species do not associate (nonpolar species).

Leahy-Dios and Firoozabadi (2007) provide a unified model to compute Fickian diffusion coefficients in gas and liquid mixtures. They relate Fickian diffusion coefficients to MS diffusion coefficients in terms of infinite dilution diffusion coefficients $\mathcal{D}_{ij}^{\infty}$. At infinite dilution, as the name implies, one of the species is infinitely diluted in the other species. Suppose we are interested in Fickian diffusion coefficients in a binary mixture. Then $\mathcal{D}_{1,2}$ can be related through the Vignes (1966) relation to the infinite dilution diffusion coefficients

$$\mathcal{D}_{1,2} = \left(\mathcal{D}_{1,2}^{\infty}\right)^{x_2} \left(\mathcal{D}_{2,1}^{\infty}\right)^{x_1} \tag{6.89}$$

$\mathcal{D}_{1,2}^{\infty}$ and $\mathcal{D}_{2,1}^{\infty}$ are the dilution diffusion coefficients of component 2 infinitely diluted in component 1, and component 1 infinitely diluted in component 2, respectively. x_1 and x_2 are mol fractions of components 1 and 2, respectively. Note that at infinite dilution $\mathcal{D}_{1,2}^{\infty} = D_{1,2}^{\infty}$, also $\mathcal{D}_{2,1}^{\infty} = D_{2,1}^{\infty}$. In mixtures of more than two components, $\mathcal{D}_{ij}$ are obtained from

$$\mathcal{D}_{ij} = \left(\mathcal{D}_{ij}^{\infty}\right)^{x_1} \left(\mathcal{D}_{ji}^{\infty}\right) \prod_{\substack{k=1 \\ k \neq i,\, j}}^{n_c} \left(\mathcal{D}_{ik}^{\infty}\mathcal{D}_{jk}^{\infty}\right)^{x_k/2} \quad i, j = 1, \ldots, n_c \quad i \neq j \tag{6.90}$$

Leahy-Dios and Firoozabadi (2007) provide a general expression for $\mathcal{D}_{ij}^{\infty}$ in terms of reduced pressure, reduced temperature, and the viscosity of the solvent. Note that the measurement of infinite dilution diffusion coefficients can be easily performed; however, at conditions far from high pressures and critical region. Clustering in the critical region and at high pressures may affect diffusion significantly and may require further research. Only recently, Mutoru et al. (2013) have studied the effect of clustering at higher pressures in self-diffusion coefficients of a number of normal alkanes.

The above derivations and discussions have set the stage for analysis and computations of species transport in which diffusion is important. In the following, we will first discuss in detail species distribution in hydrocarbon reservoirs and then move to abrupt past climate changes. In climate change, we require high precision in diffusion coefficients. However, the mixtures are ideal.

Let us start the discussion by writing the expression for the mass flux, $\vec{m}_i$:

$$\vec{m}_i = \rho w_i \vec{v} + \vec{J}_i \quad i = 1, \ldots, n_c \tag{6.91}$$

where ρ is the mass density, w_i is the mass fraction of component i, $\vec{v}$ is the mass-average velocity, and $\vec{J}_i$ is the diffusion mass flux (nonconvective) of component i. In the above equation, the first term on the right side is the convective mass flux, and the second term is the diffusive flux. We first start with the convection in Eq. (6.91).

6.2.3 Thermal Convection

One major goal of this section is to understand what drives flow in thermal convection at steady state; it may not be the buoyancy as one may believe. The benefit of buoyancy is that it will (if it is large enough) cause instabilities and these instabilities will cause a horizontal density gradient. Let us consider a single-component fluid in a two-dimensional cross-sectional (x, z) domain shown in Fig. 6.4. We assume that the process has been

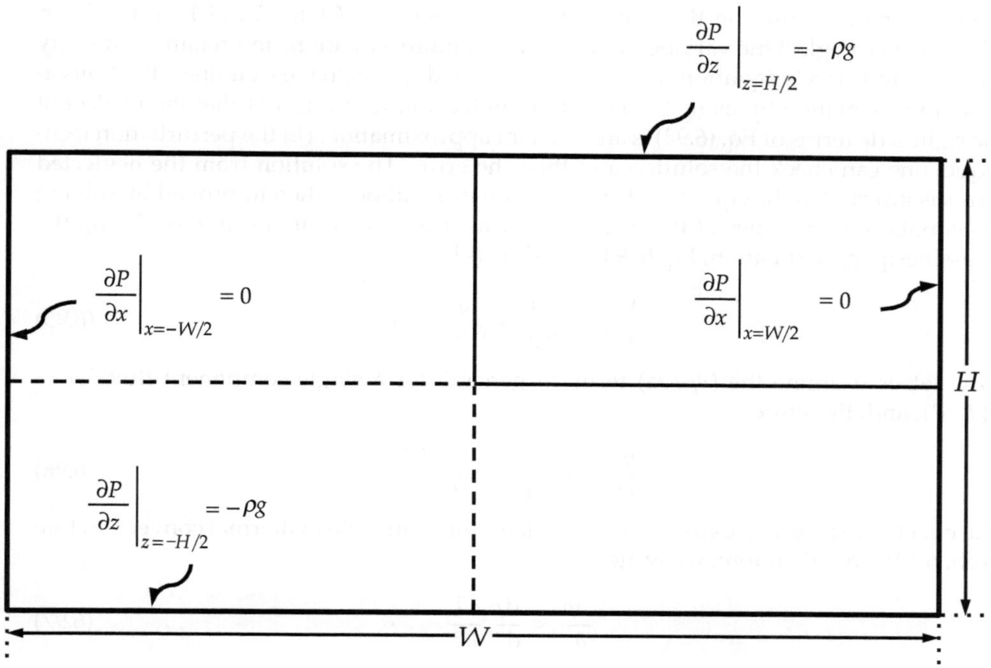

FIGURE 6.4 Geometry and boundary conditions for the 2D single-component fluid system used in the study of thermal convection.

going on long enough to have reached steady state. The domain boundaries are closed to flow and there is a steady-state temperature field imposed on the system.

The flow is governed by the continuity equation:

$$\nabla \cdot (\rho \vec{v}) = 0 \tag{6.92}$$

where the velocity is given by Darcy's law,

$$\vec{v} = -\left(\frac{k}{\mu}\right)(\nabla P + \rho \vec{g} \nabla z) \tag{6.93}$$

Note that in Eq. (6.93), $\vec{v}$ is not the true velocity; the true velocity is given by $\vec{v}/\phi$ where ϕ is the fractional porosity. Figure 6.4 provides the boundary conditions of the 2D geometry. The boundary conditions are derived from the assumption that fluid does not cross the boundaries; that is, the normal component of velocity is zero.

By combining the continuity equation, Eq. (6.92), and Darcy's equation, Eq. (6.93), one obtains,

$$\frac{\partial^2 P}{\partial x^2} + \frac{\partial^2 P}{\partial z^2} + g\frac{\partial \rho}{\partial z} = -\frac{1}{\rho}\left(\frac{\partial \rho}{\partial x}\right)\frac{\partial P}{\partial x} - \frac{1}{\rho}\left(\frac{\partial \rho}{\partial z}\right)\left(\frac{\partial P}{\partial z} + \rho g\right) \tag{6.94}$$

In the derivation of Eq. (6.94), it is assumed that the permeability is independent of x and z and μ is constant. In order to solve the problem of thermal convection in a simple manner, one may use the Boussinesq approximation (see Chap. 2 of Chandrasekhar, 1961), that is, neglect the variation of ρ in the continuity equation, and retain the density variation in Darcy's equation. Riley and Firoozabadi (1998) have examined the Boussinesq approximation by using the perturbation technique and found that the neglect of the right-side terms of Eq. (6.94) is an excellent approximation. (In the perturbation technique, one can check the solution and then the error. The solution from the neglected terms is inserted in the equation. The approximate solution is then improved by solving the equation with values of the neglected terms from the original solution). Using the Boussinesq approximation, Eq. (6.94) transforms into

$$\frac{\partial^2 P}{\partial x^2} + \frac{\partial^2 P}{\partial z^2} + g\frac{\partial \rho}{\partial z} = 0 \tag{6.95}$$

Now let us examine the $(\partial\rho/\partial z)$ term. In general, for a single-component fluid, $\rho = \rho(T, P)$, and, therefore,

$$\frac{\partial \rho}{\partial z} = \frac{\partial \rho}{\partial P}\frac{\partial P}{\partial z} + \frac{\partial \rho}{\partial T}\frac{\partial T}{\partial z} \tag{6.96}$$

The effect of pressure in density variation does not contribute to thermal convection (see Prob. 6.12), and, therefore, we write

$$\frac{\partial \rho}{\partial z} \approx \frac{d\rho}{dT}\frac{\partial T}{\partial z} \tag{6.97}$$

The temperature dependence of the density can be approximated using the Taylor series,

$$\rho(T) \approx \rho(T^0) + \left.\frac{d\rho}{dT}\right|_{T^0}(T - T^0) \tag{6.98}$$

where T^0 is the reference temperature. From the definition of thermal expansivity, $e = (1/V)(dV/dT)$ and $e = 1/(V/m)d(V/m)/dT = 1/(1/\rho)d(1/\rho)/dT = -(1/\rho)(d\rho/dT)$ (we have divided the numerator and denominator by the mass m of volume), then $d\rho/dT|_{T^0} = -\rho^0 e$, where ρ^0 is the density at temperature T^0. Therefore, Eq. (6.98) can be written as

$$\rho(T) \approx \rho^0[1 - e(T - T^0)] \tag{6.99}$$

Now let us choose a temperature variation in the domain according to

$$T = T^0 - BW\cos\frac{n\pi\left(x + \dfrac{W}{2}\right)}{W} + Cz \tag{6.100}$$

where parameter B influences the horizontal temperature variation and parameter C influences the vertical temperature variation. The constant n is an odd number; we will discuss its effect on the shape of the temperature variation in Example 6.4. Instead of Eq. (6.100), one may use the simpler expression,

$$T = T^0 + 2Bx + Cz \tag{6.101}$$

The factor "2" is included so that the two temperature profiles, from Eqs. (6.100) and (6.101), will have the same overall horizontal temperature change for the same value of B. Equation (6.101) describes very well the temperature variation in hydrocarbon reservoirs; it is also a solution to the energy equation where heat transfer is by pure conduction, that is, the solution to $\nabla^2 T = 0$ (see Example 6.3). The advantage of using the cosine temperature distribution is that $\cos[n\pi(x + W/2)/W]$ is orthogonal to the eigenfunctions of the problem, $\cos(m\pi(x + W/2)/W)$. This reduces the number of infinite sum cosines to two terms, $m = 0$ and $m = n$.

From Eqs. (6.97), (6.99), and (6.100), $\partial\rho/\partial z = -\rho^0 eC$, and then Eq. (6.95) transforms into

$$\frac{\partial^2 P}{\partial x^2} + \frac{\partial^2 P}{\partial z^2} - \rho^0 geC = 0 \tag{6.102}$$

and the boundary conditions are

$$\left.\frac{\partial P}{\partial x}\right|_{x=-\frac{W}{2}} = \left.\frac{\partial P}{\partial x}\right|_{x=\frac{W}{2}} = 0 \tag{6.103}$$

$$\left.\frac{\partial P}{\partial z}\right|_{z=-\frac{H}{2}} = -\left.\rho g\right|_{z=-\frac{H}{2}} = -\rho^0 g\left[1 + eBW\cos\frac{n\pi\left(x+\frac{W}{2}\right)}{W} + eC\frac{H}{2}\right] \tag{6.104}$$

$$\left.\frac{\partial P}{\partial z}\right|_{z=\frac{H}{2}} = -\left.\rho g\right|_{z=\frac{H}{2}} = -\rho^0 g\left[1 + eBW\cos\frac{n\pi\left(x+\frac{W}{2}\right)}{W} - eC\frac{H}{2}\right] \tag{6.105}$$

The solution of the above partial differential equation, Eq. (6.102), and the boundary conditions, Eqs. (6.103) to (6.105), are readily obtained:

$$P(x, z) = P^0 - \rho^0 gz + \frac{\rho^0 geCz^2}{2} - \frac{\rho^0 geW^2}{n\pi}B\frac{\sinh\left(\frac{n\pi z}{W}\right)}{\cosh\left(\frac{n\pi H}{2W}\right)}\cos\frac{n\pi\left(x+\frac{W}{2}\right)}{W} \tag{6.106}$$

In the above equation, P^0 is the pressure at the origin $x = 0$ and $z = 0$. T^0 and ρ^0 are also the temperature and density at the origin. The Darcy velocities in the x and z directions are obtained from $v_x = -(k/\mu)(\partial P/\partial x)$ and $v_z = -(k/\mu)(\partial P/\partial z + \rho g)$:

$$\boxed{v_x = -\frac{k\rho^0 geWB}{\mu}\frac{\sinh\left(\frac{n\pi z}{W}\right)}{\cosh\left(\frac{n\pi H}{2W}\right)}\sin\frac{n\pi\left(x+\frac{W}{2}\right)}{W}} \tag{6.107}$$

and

$$v_z = \frac{k\rho^0 geWB}{\mu}\left(\frac{\cosh\left(\frac{n\pi z}{W}\right)}{\cosh\left(\frac{n\pi H}{2W}\right)} - 1\right)\cos\frac{n\pi\left(x+\frac{W}{2}\right)}{W} \tag{6.108}$$

Note that in Eqs. (6.107) and (6.108), only the parameter of the horizontal temperature variation, B, appears. The vertical temperature variation parameter, C, is absent. In other words, the above two equations reveal that v_x and v_z are independent of the vertical temperature gradient and the vertical density difference; the thermal convection is proportional to the magnitude of the horizontal density gradient.

Now let us write the flow equation in terms of the stream function. In rotational flow in fluid mechanics, the stream function is often used instead of pressure. The stream function, ψ, is the complement of pressure, P. The value of the stream function is constant across a streamline; there is no flow across a streamline. The stream function is defined by (Lu, 1973):

$$v_x = -\frac{\partial\psi}{\partial z} \quad \text{and} \quad v_z = +\frac{\partial\psi}{\partial x} \tag{6.109}$$

Note that the stream function automatically satisfies the continuity equation, $\nabla\cdot\vec{v}=0$, for the Boussinesq case where ρ is assumed constant in the continuity equation. The equation that defines the stream function is from the curl of velocity (Lu, 1973); the curl of velocity is called the vorticity. The vorticity provides a measure of rotational flow. The curl of velocity, $\nabla\times\vec{v}$, is given by (see Borisenko and Tarapov, 1968)

$$\nabla\times\vec{v} = \frac{\partial}{\partial z}(v_x) - \frac{\partial}{\partial x}(v_z) \tag{6.110}$$

Substituting for $v_x = -(k/\mu)\partial P/\partial x$ and $v_z = -(k/\mu)(\partial P/\partial z + \rho g)$, assuming μ and k constant, and combining the results with Eq. (6.109),

$$\frac{\partial^2\psi}{\partial x^2} + \frac{\partial^2\psi}{\partial z^2} = -\frac{k}{\mu}g\frac{\partial\rho}{\partial x} \tag{6.111}$$

The expression for $\partial\rho/\partial x$ is $\partial\rho/\partial x \approx (\mathrm{d}\rho/\mathrm{d}T)(\partial T/\partial x)$, where the pressure dependence of ρ does not contribute to convection (see Prob. 6.12). Using Eqs. (6.99) and (6.100),

$$\left(\frac{\partial\rho}{\partial x}\right) = -(\rho^0 e)(n\pi B)\sin\frac{n\pi\left(x+\frac{W}{2}\right)}{W} \tag{6.112}$$

Combining Eqs. (6.111) and (6.112),

$$\frac{\partial^2\psi}{\partial x^2} + \frac{\partial^2\psi}{\partial z^2} = \left(\frac{k}{\mu}\right)(\rho^0 ge)(n\pi B)\sin\frac{n\pi\left(x+\frac{W}{2}\right)}{W} \tag{6.113}$$

The boundary conditions for Eq. (6.113) are $\partial\psi/\partial z = 0$ and $\partial\psi/\partial x = 0$. Therefore, $\psi(x=-W/2, z) = \psi(x=W/2, z) = \psi(x, z=-H/2)$, $\psi(x, z=H/2) = constant$. In other words, the boundary of the system is a streamline. We may assign $\psi = 0$, since one may choose any value for this streamline. Therefore, the boundary conditions of Eq. (6.113) are

$$\psi\left(x = -\frac{W}{2}, z\right) = \psi\left(x = \frac{W}{2}, z\right) = \psi\left(x, z = -\frac{H}{2}\right) = \psi\left(x, z = \frac{H}{2}\right) = 0 \quad (6.114)$$

The solution to Eqs. (6.113) and (6.114) is also in the form of Fourier series with only one term,

$$\psi(x, z) = +\frac{k\rho^0 g e W^2}{\mu \pi n} B \left(\frac{\cosh\left(\frac{n\pi z}{W}\right)}{\cosh\left(\frac{n\pi H}{2W}\right)} - 1 \right) \sin\left(\frac{n\pi\left(x + \frac{W}{2}\right)}{W} \right) \quad (6.115)$$

The velocities on the streamline can be readily calculated from Eq. (6.109); the results are the same as from Eqs. (6.107) and (6.108). The advantage of the stream function formulation is that streamlines are provided.

Example 6.4 at the end of this chapter provides numerical results for vertical and horizontal velocities in a 2D cross-sectional reservoir. Comments on the vertical and horizontal velocity profiles will be made in that example.

Next we add diffusion to the above problem with the objective of studying composition variation in hydrocarbon reservoirs where there always exists a thermal gradient.

6.2.4 Natural Convection and Diffusion in Porous Media

In the formulation of thermal convection in porous media at steady state, it was demonstrated that the horizontal gradient of temperature drives the thermal convection. In fact, the driving force for both thermal convection (i.e., the convection due to thermal gradient) and natural convection (i.e., the convection due to both thermal gradient and composition gradient) is governed by $(\partial\rho/\partial x)$ at steady state. The expression for $(\partial\rho/\partial x)$ is given by

$$\frac{\partial\rho}{\partial x} = \left(\frac{\partial\rho}{\partial T}\right)\left(\frac{\partial T}{\partial x}\right) + \sum_{i=1}^{n_c-1}\left(\frac{\partial\rho}{\partial x_i}\right)\left(\frac{\partial x_i}{\partial x}\right) \quad (6.116)$$

When there is only bulk flow and diffusive fluxes are zero, the temperature gradient $(\partial T/\partial x)$ is the sole contributor to density gradient $(\partial\rho/\partial x)$. With diffusion, the second term on the right side of Eq. (6.116) becomes effective. The two terms on the right side of Eq. (6.116) may have the same sign, or opposite signs and may have different magnitudes relative to each other. Therefore, convection may enhance composition variation due to the effect of the second term. Such a behavior is not in line with the common belief in the literature that convection always reduces composition variation in hydrocarbon reservoirs as we will demonstrate.

We now present the equations that describe the combined effect of convection and diffusion in porous media. Let us assume that there are two components in the mixture, and that there is a single phase, either gas or liquid; the geometry is a two-dimensional rectangle (see Fig. 6.4). It is also assumed that the temperature field is known. In hydrocarbon reservoirs, temperature data can be measured with modern tools. At steady state, one may write the continuity equation for components 1 and 2:

$$\nabla \cdot (\rho w_1 \vec{v} + \vec{J}_1) = 0 \tag{6.117}$$

$$\nabla \cdot (\rho w_2 \vec{v} + \vec{J}_2) = 0 \tag{6.118}$$

where $\vec{J}$ is the total diffusion mass flux. Adding Eqs. (6.117) and (6.118) results in the continuity equation for bulk flow given by Eq. (6.92), since $\vec{J}_1 + \vec{J}_2 = 0$. One of the three equations among Eqs. (6.92), (6.117), and (6.118) is a linear combination of the other two, we then use Eqs. (6.92) and (6.117). The velocity in porous media is given by Darcy's law [see Eq. (6.93)].

The diffusive mass flux of component 1 is given by the expression

$$\vec{J}_1 = -C^x(x_1, T, P)\nabla x_1 - C^P(x_1, T, P)\nabla P - C^T(x_1, T, P)\nabla T \tag{6.119}$$

which is an alternative form of Eq. (6.68). The coefficients of the above expression are

$$C^x = \rho \frac{dw_1}{dx_1} D_{1,2} \left(\frac{\partial \ln f_1}{\partial \ln x_1} \right)_{T,P} \tag{6.120}$$

$$C^P = \rho \frac{dw_1}{dx_1} D_{1,2} \frac{x_1}{RT} \left(\bar{V}_1 - \frac{M_1}{\rho} \right) \tag{6.121}$$

$$C^T = \rho \frac{dw_1}{dx_1} D_{1,2} \left(\frac{k_{1T}}{T} \right) \tag{6.122}$$

The mol fraction x_1 and the mass fraction w_1 are related to each other by

$$w_1 = \frac{x_1 M_1}{x_1 M_1 + (1 - x_1) M_2} \tag{6.123}$$

The boundary conditions for the 2D problem are

$$J_{1,x} = v_x = 0 \quad \text{at} \quad x = \pm \frac{W}{2} \tag{6.124}$$

$$J_{1,z} = v_z = 0 \quad \text{at} \quad z = \pm \frac{H}{2} \tag{6.125}$$

where H and W are the height and width of the domain (see Fig. 6.4 for geometry). We also need to specify pressure and composition in one point in the domain, say at the origin at, $x = 0$ and $z = 0$ because the boundary conditions are of Neumann type (i.e., derivatives are provided at the boundary). The problem formulation is then completed. The unknowns are pressure P and composition x_1, similarly to the 1D case, which does not allow for natural convection. The above system of equations in its general form has been solved by Riley and Firoozabadi (1998) using a method of successive approximation. In that work, the equations are first transformed into Poisson's equations. Riley and Firoozabadi studied the effect of convection on horizontal composition variation in a two-component mixture. Let us review some results from the work of these authors for the binary system C_1/n-C_4. The pressure, temperature, and composition at the origin ($x = 0$, $z = 0$) are fixed: $P^0 = 110$ atm, $T^0 = 339$ K, and $x_1^0 = 0.20$. The dimensions of the reservoir are $H = 150$ m and $W = 3000$ m. The horizontal and vertical temperature gradients are $\partial T/\partial x = 1$ K/300 m, and $\partial T/\partial z = -2$ K/30 m. Note that the geothermal temperature decreases toward the surface of the earth. The fluid viscosity is assumed constant in the domain, $\mu = 0.2$ cp—independent of pressure, temperature, and composition. Throughout the domain, the C_1/n-C_4 mixture remains a liquid for all cases studied by Riley and Firoozabadi. The fractional porosity of the domain $\phi = 0.20$, and the permeability is varied to examine its effect on composition variation. Constant Fickian diffusion coefficient $D_{1,1} = 1.02 \times 10^{-9} \text{m}^2/\text{s}$ was used in all the calculations. The thermal diffusion ratio $k_{1T} = 82{,}600[x_1(1 - x_1)/RT]$ was estimated from the work of Rutherford and Roof (1959), with R having units of atm·cm^3/mol·K, and T in K. No adjustments were made for the factor $(\partial \ln f_1/\partial \ln x_1)$ in k_{1T}. The Peng-Robinson equation of state (1976) was used for the estimation of volumetric and thermodynamic properties.

A contour plot of the methane mol fraction is shown for $k = 0$ md (i.e., no convection) in Fig. 6.5. The contour interval is 0.1 mol% in this and subsequent contour plots. The constant composition contours are essentially straight lines. The horizontal composition variation is 0.91 mol% at $z = 0$. Figure 6.6 shows the composition contours for $k = 0.2$ md. The surprise is that with a small permeability and the introduction of convection, the horizontal composition variation increases. The expectation is that convection would decrease the composition variation. Figure 6.7 shows that by increasing the permeability to 30 md, the trend of compositional gradient is reversed. There are other features in all the compositional contours: (1) they become more vertical; (2) except near the side boundaries, the curves seem to have the same shape and spacing, indicating $(\partial x_1/\partial x)$ being constant; (3) the curves develop a subtle "S" shape, which indicates that the vertical compositional gradient is not monotonic. Figure 6.8 provides the variation of horizontal methane composition gradient at the origin ($x = 0$, $z = 0$) versus permeability.

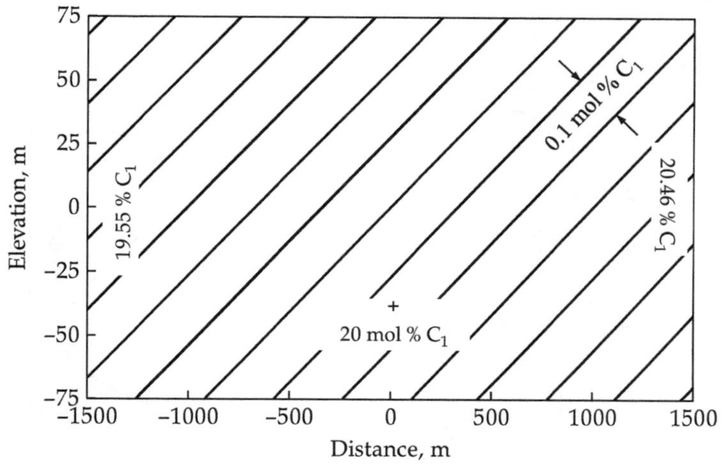

FIGURE 6.5 Methane mol fraction contours: $k = 0$ md (from Riley and Firoozabadi, 1998).

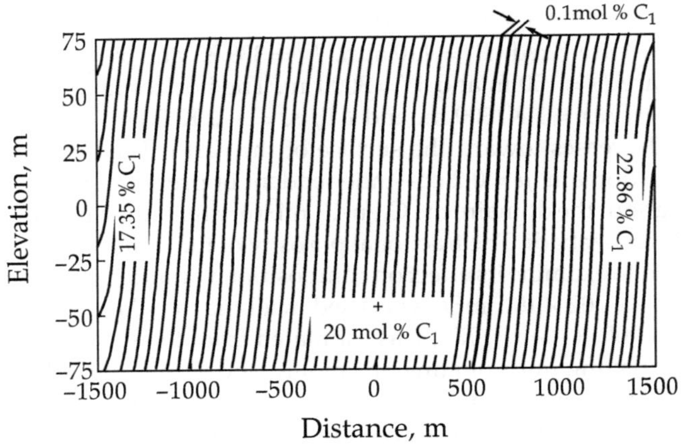

FIGURE 6.6 Methane mol fraction contours: $k = 0.2$ md (from Riley and Firoozabadi, 1998).

Figure 6.9 shows the horizontal velocities for $k = 10$ md. The vertical velocity is nearly zero except close to the vertical boundaries. This figure reveals that the horizontal velocity v_x varies linearly with z. (An explicit approximate expression for v_x will be derived shortly.)

Now that the features of the solution are available, we can make appropriate assumptions to derive an approximate analytical expression for v_x. Let us define a stream function that can accommodate compressible flow. The modified stream function can be defined in terms of mass flux:

$$\rho v_x = -\partial \psi_m / \partial z \quad \text{and} \quad \rho v_z = +\partial \psi_m / \partial x \tag{6.126}$$

Defined in this way, the stream function ψ_m satisfies the continuity equation, Eq. (6.92).

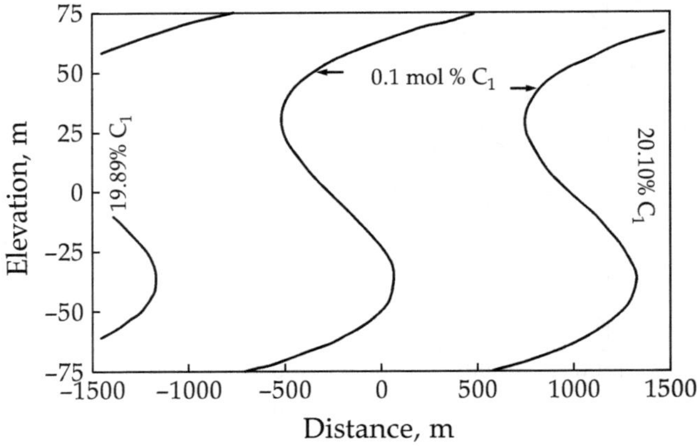

FIGURE 6.7 Methane mol fraction contours: $k = 30$ md (from Riley and Firoozabadi, 1998).

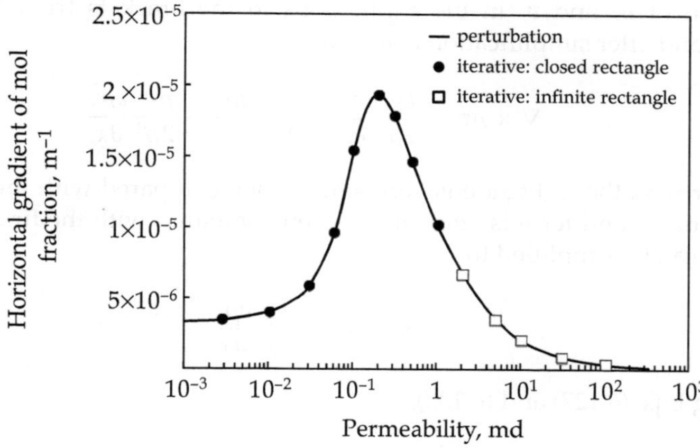

FIGURE 6.8 Horizontal composition gradient at $x = 0$, $z = 0$ vs. permeability (adapted from Riley and Firoozabadi, 1998).

Let us take the curl of $\rho\vec{v}$ and write it first in terms of the modified stream function and then in terms of $\vec{v}$:

$$\nabla \times \rho\vec{v} = \frac{\partial}{\partial z}(\rho v_x) - \frac{\partial}{\partial x}(\rho v_z) = -\frac{\partial^2 \psi_m}{\partial z^2} - \frac{\partial^2 \psi_m}{\partial x^2} \tag{6.127}$$

and

$$\nabla \times \rho\vec{v} = \rho\left[\frac{\partial v_x}{\partial z} - \frac{\partial v_z}{\partial x}\right] + v_x\frac{\partial \rho}{\partial z} - v_z\frac{\partial \rho}{\partial x} \tag{6.128}$$

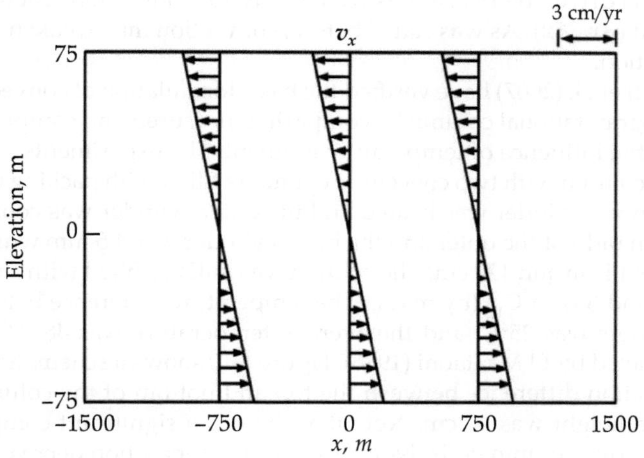

FIGURE 6.9 v_x vs. z at $x = -750$, 0, and 750 m: $k = 10$ md.

Substitution of v_x and v_z in the expressions in the brackets from Darcy's law [see Eq. (6.93)] and after simplification results in

$$\nabla \times \rho\vec{v} = \frac{kg}{2\mu}\frac{\partial \rho^2}{\partial x} + \frac{\rho v_x}{2\rho^2}\frac{\partial \rho^2}{\partial z} - \frac{\rho v_z}{2\rho^2}\frac{\partial \rho^2}{\partial x} \tag{6.129}$$

The last term on the right side is very small when compared with the second term on the right; the second term is very small when compared with the first term; therefore, Eq. (6.129) can be simplified to

$$\nabla \times \rho\vec{v} \approx \frac{kg}{2\mu}\frac{\partial \rho^2}{\partial x} \tag{6.130}$$

Combining Eqs. (6.127) and (6.130),

$$\frac{\partial^2 \psi_m}{\partial z^2} + \frac{\partial^2 \psi_m}{\partial x^2} \approx -\frac{kg}{2\mu}\frac{\partial \rho^2}{\partial x} \tag{6.131}$$

Since $\partial^2\psi_m/\partial x^2 = \partial(\rho v_z)/\partial x$ and since $\rho v_z \approx 0$ except around the side boundaries,

$$\frac{\partial^2 \psi_m}{\partial z^2} \approx -\frac{kg}{2\mu}\frac{\partial \rho^2}{\partial x} \tag{6.132}$$

Also note that $v_x = 0$ at $z = 0$ (see Fig. 6.9). Therefore, the integration of Eq. (6.132) provides the following simple expression for horizontal velocity:

$$v_x \approx \frac{kg}{\mu\phi}\frac{\partial \rho}{\partial x} z \tag{6.133}$$

Note that we have divided the Darcy velocity by fractional porosity in the last step to have true velocity. The $(\partial\rho/\partial x)$ term was previously given in terms of $(\partial T/\partial x)$ and $(\partial x_i/\partial x)$ [see Eq. (6.116)]. Equation (6.133) applies to both thermal convection, where the convection is driven by $(\partial T/\partial x)$ as well as natural convection where flow is driven by $(\partial T/\partial x)$ and $(\partial x_i/\partial x)$. As was stated before, convection may weaken or enhance composition variation.

Nasrabadi et al. (2007) have verified the basic formulation of convection and diffusion in a thermogravitational column by comparing measured and computed species separation under the influence of temperature gradient. The experiments were conducted in a cylindrical column with two concentric cylinders filled with packing of calibrated zircon balls. The inner cylinder was heated and the outer cylinder was cooled uniformly. The difference in radii of the outer and the inner cylinder was 4.5 mm while the height of the column was 14 cm and 120 cm. The column was initially filled with a uniform mixture of 15% n-C_{24} and 85% n-C_{12} (by mass). The temperature difference between the inner and outer cylinders was 25°C, and the average temperature was 48.5°C. The experiments were conducted by El Maataoni (1986). Figure 6.10 shows a schematic of the column and the composition difference between the top and bottom of the column at steady state. The column height was 40 cm. Note that there is a significant composition difference between the two column ends. Note also that the separation decreases with increase in permeability in the range of experiments. However, with decreasing permeability the trend reverses, similar to the results in Fig. 6.8.

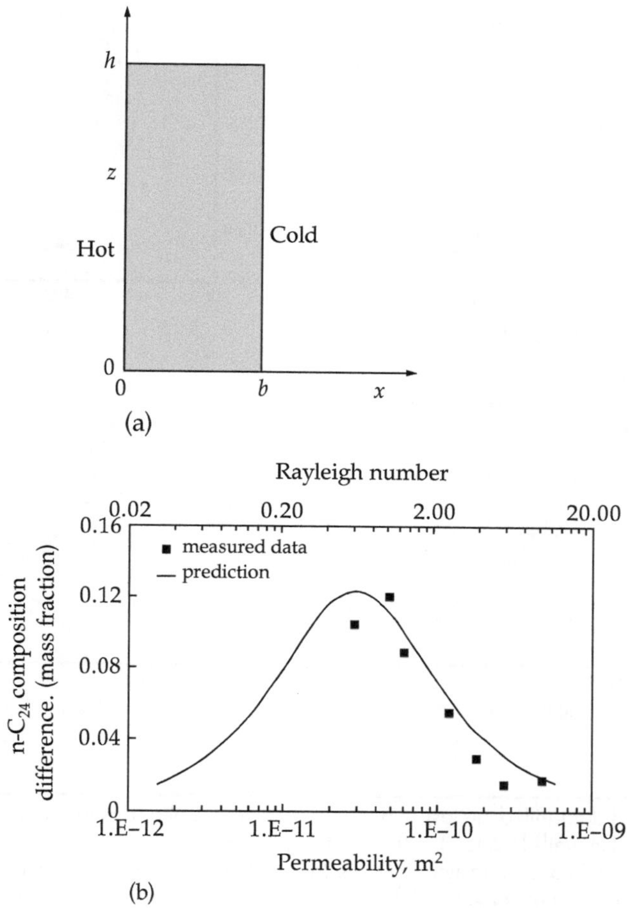

FIGURE 6.10 (a) Thermogravitational column and (b) measured and predicted separation in the column at steady state. Initial composition: 15% n-C_{24} and 85% n-C_{12} (mass) $\Delta T = 25°C$ (adapted from Nasrabadi et al., 2007).

In the following we will briefly discuss the variation of species in the subsurface in hydrocarbon reservoirs.

6.2.5 Compositional Variation in Hydrocarbon Reservoirs

The species distribution in hydrocarbon reservoirs may have a great impact on drilling of expensive wells in the offshore. It may also affect the production planning. Various authors have presented formulations of transport and diffusion in the subsurface (Ghorayeb and Firoozabadi, 2000a; Ghorayeb et al., 2003). In the following we present two unusual field examples. There are many indications that the species distribution in the Yufutsu gas condensate field in Japan is unusual due to the influence of thermal

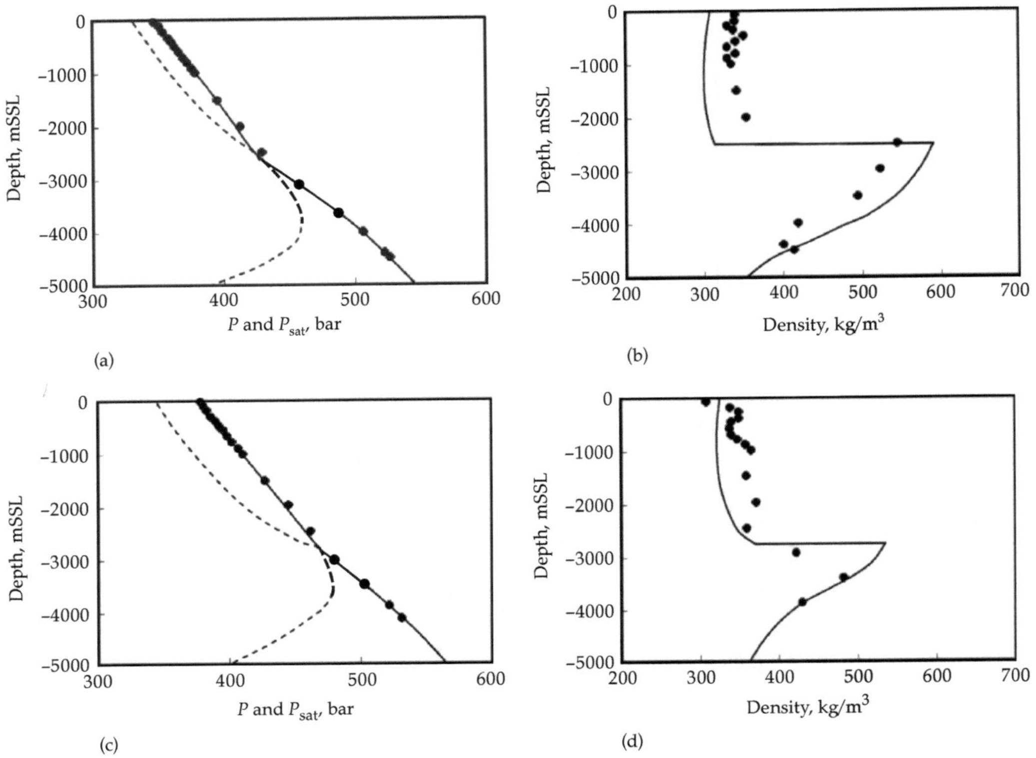

FIGURE 6.11 Measured data (filled circles) and model prediction (solid, dashed and long-dashed lines) from two different shut-in wells (a and c) in the Yufutsu gas condensate field in Japan. The dashed lines are the computed dewpoint. The long-dashed lines are the bubblepoint. Both dewpoint and bublepoint are denoted by saturation pressure, P_{sat}. In (a) and (c) pressure and saturation pressure are shown. In (b) and (d) density is shown (from Ghorayeb and Firoozabadi, 2003).

diffusion (Ghorayeb et al., 2003). Figure 6.11 shows the pressure data in two different wells, about 4 km apart. The wells were closed for a long time (some 2 years). Note that there is a sharp slope change in pressure data (Figs. 6.11a and 6.11c) around the depth of 2500 to 3000 m subsea level (SSL). The increase in slope is related to a significant increase in density. From the pressure data, one can evaluate the density, as depicted in Figs. 6.11b and 6.11d. The high density corresponds to liquid density (Ghorayeb et al., 2003). In these two wells, a liquid floats on top of a lighter fluid. The vertical temperature gradient in the subsurface is around 2 K/100 m. Without this temperature gradient, there will be segregation. Figure 6.12 shows the distribution of methane and heptane-plus as well as density and pressure with depth. Interestingly, heptane-plus which is the heaviest fraction is segregated at the top. The density of the fluid is also higher at the top than at the bottom. A dominant thermal diffusion has overcome gravity segregation from pressure diffusion. There are perhaps few other formations in the world where thermal diffusion has such a strong effect.

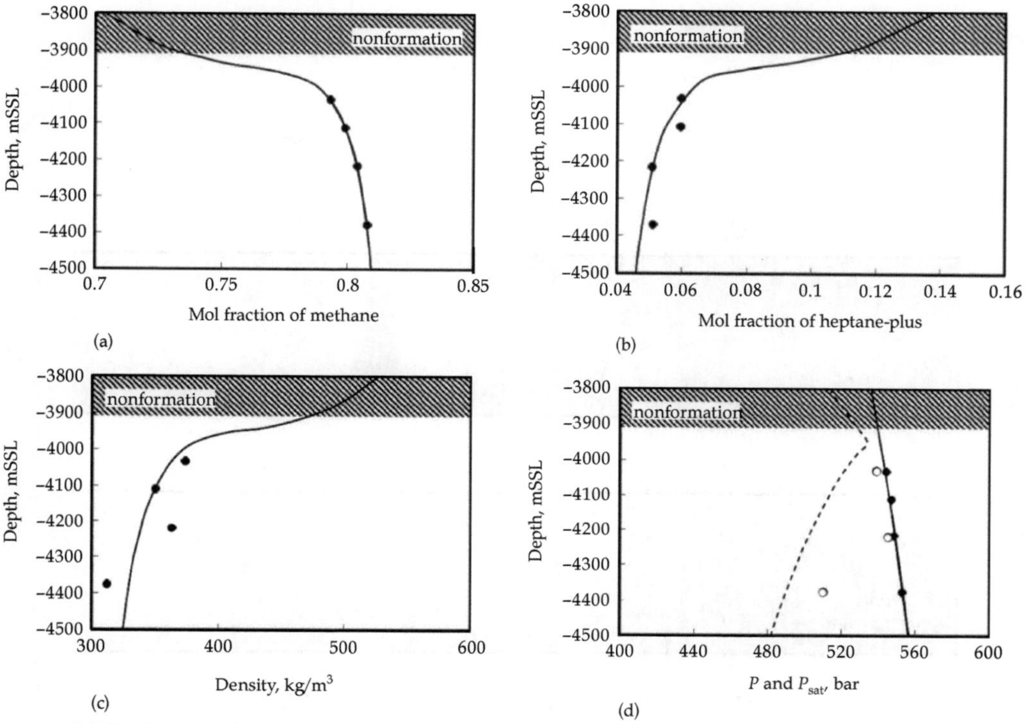

FIGURE 6.12 Measured data and model predictions for well AK1 in the Yufutsu gas condensate field in Japan; solid lines represent predicted results, and circles present measured data (adapted from Ghorayeb, Firoozabadi and Anraku, 2003).

The variation of composition in the two dimensional domain of width = 10 km and height = 1.5 km containing a near critical gas condensate fluid is presented next. The horizontal thermal gradient is 1.5 K/km. The vertical thermal gradient is 2.75 K/1000 m (Ghorayeb and Firoozabadi, 2000b). In this field which contains the near-critical fluid, there is very little variation of composition. Without thermal diffusion, the composition variation will be very significant. An oil column may be predicted by the Gibbs sedimentation expression. Figure 6.13 shows the composition and density contour plots when there is no convection (i.e., permeability is zero). When convection is allowed for permeability of 10 md, there is no significant change in composition and density profiles (see Fig. 6.14). However, when thermal diffusion is increased, there is a significant reduction in segregation of components in line with field data (see Fig. 6.15). The results presented in this section show that thermal diffusion can be a major influence on species distribution in hydrocarbon reservoirs.

Next we will discuss applications of thermal diffusion in another important problem in relation to past climate changes and the interpretation of ice core data.

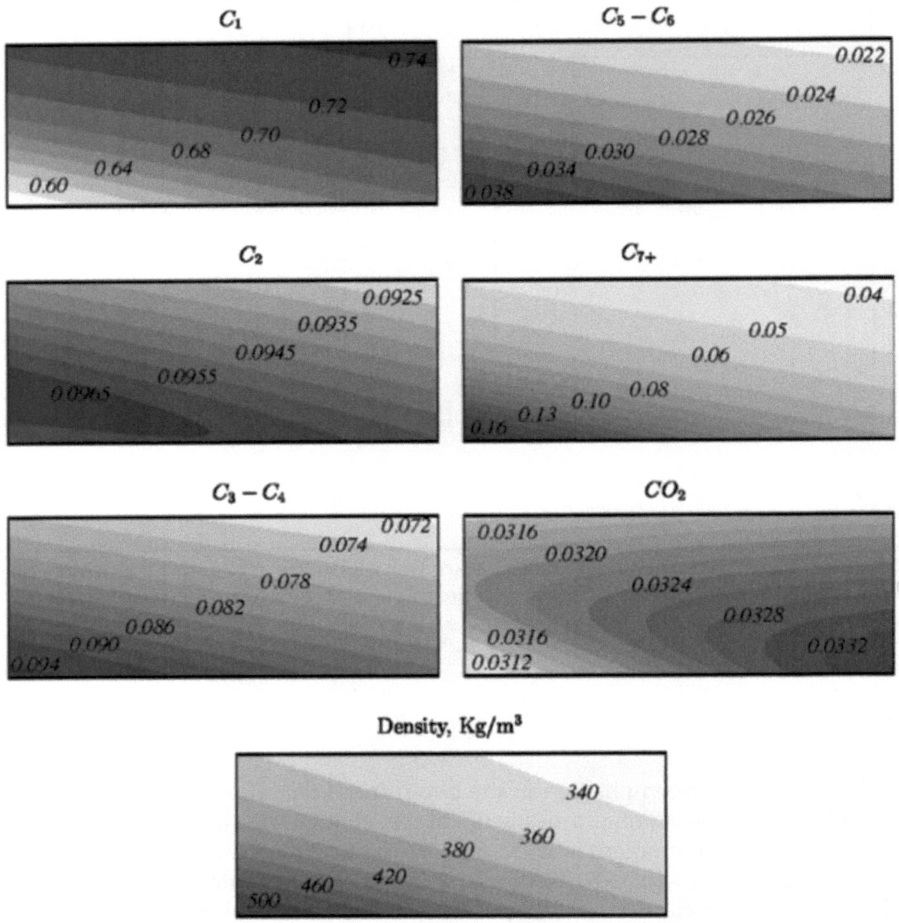

FIGURE 6.13 Composition and density contour plots (x, z) for the field example; $k = 0$ md (convection-free) (from Ghorayeb and Firoozabadi, 2000b).

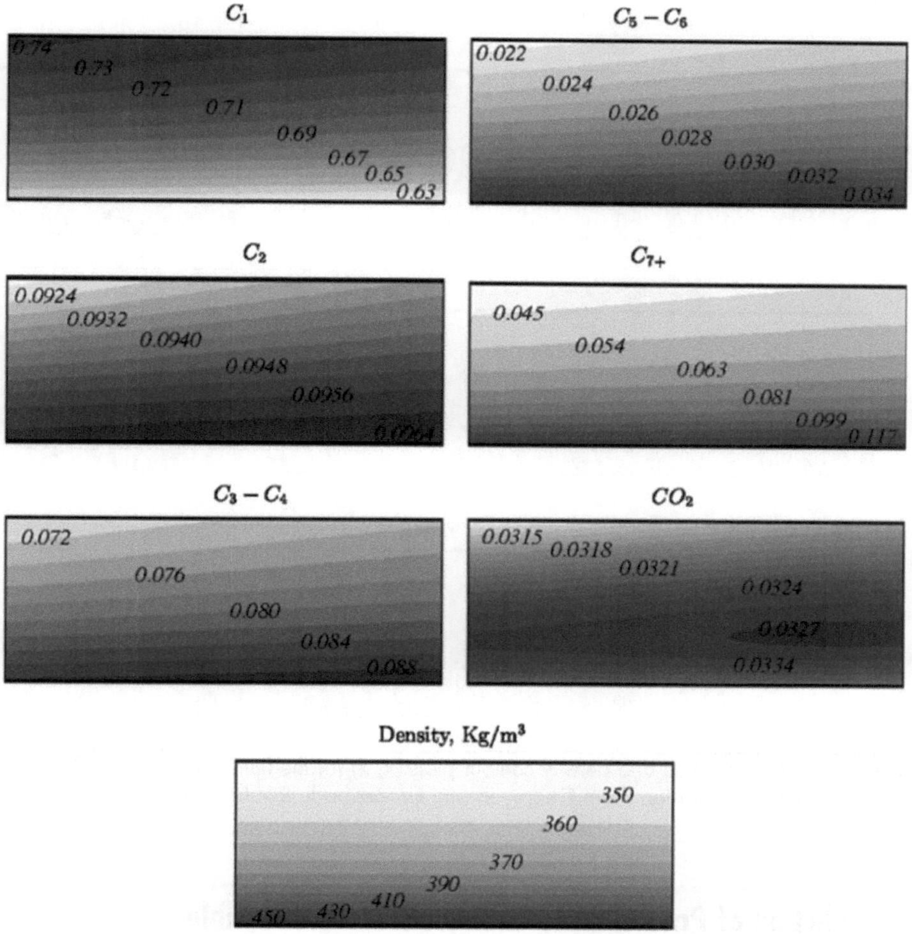

FIGURE 6.14 Composition and density contour plots (x, z) for the field example; $k = 10$ md (from Ghorayeb and Firoozabadi, 2000b).

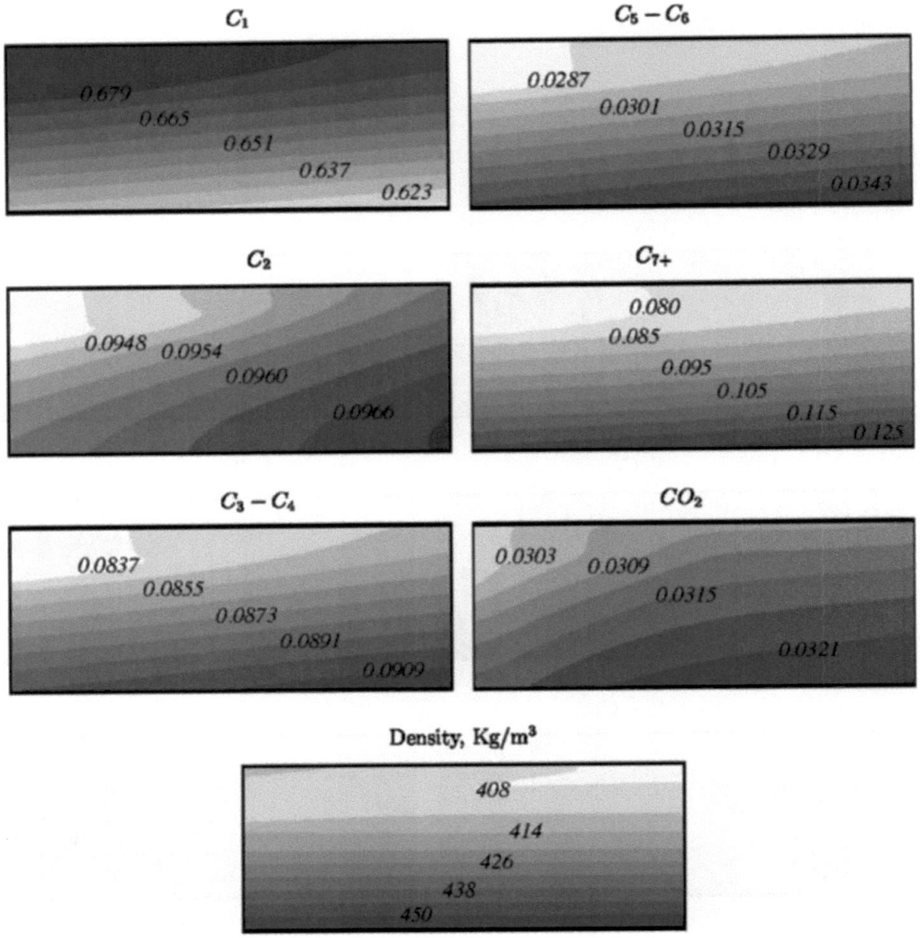

FIGURE 6.15 Composition and density contour plots (x, z) for the field example; $k = 0$ md with increased thermal diffusion (from Ghorayeb and Firoozabadi, 2000b).

6.3 Prediction of Past Climate Changes from Irreversible Thermodynamics

Figure 6.16 presents the composition of greenhouse gases CO_2 and methane (CH_4) in the atmosphere as well as the temperature of the past 440,000 years. The composition data are directly measured from ice cores in the glaciers. The upper horizontal axis shows the depth at which the samples are taken. The temperature is computed. There are two approaches to compute the past climate changes. While our focus is the approach based on irreversible thermodynamics for past climate change calculations introduced by Severinghaus et al. (1998), we will also briefly discuss the method used in the alternative approach.

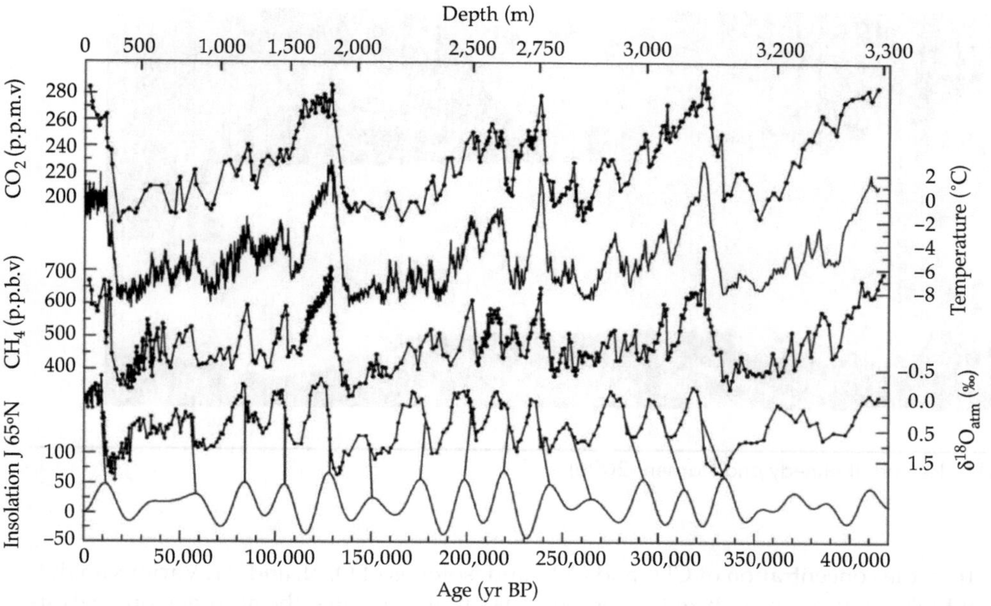

FIGURE 6.16 Greenhouse gases CO_2 and CH_4 and temperature of the past 440,000 years: CO_2; isotopic temperature of atmosphere; CH_4; $\delta^{18}O_{atm}$; and mid-June insolation at 65° N (in W/m^2). (Petit et al., 1999).

6.3.1 Ice Core Data

Ice core data have proven to be a powerful source for detailed information on the history of Earth's climate. As porous snow consolidates into ice, the air within it is trapped in bubbles at a close-off depth where porous snow transforms into glacial ice. The process preserves samples of the atmospheric air. CO_2, methane, and various isotopes can be analyzed in the gas bubbles. Because the isotopes of nitrogen and argon have had a constant composition in the atmosphere for more than 10^6 years, analysis of these isotopes is valuable in past climate change modeling. Therefore, in addition to knowledge of relative air-component concentration (to estimate the extent of greenhouse warming), the isotopic composition serves as traces for a different purpose. The isotopic composition can be used in modeling of past temperature changes. They can also help with other useful information. For example, CO_2 from fossil-fuel burning is relatively depleted in ^{13}C. Some aspects of ice core sampling will be discussed next.

6.3.2 Sampling

Deep drilling into hard ice using a hollow cylindrical drill is used to obtain ice core samples (Fig. 6.17). Collection of a large number of cores can be made from different depths below the snow layer. There is need for extreme care to avoid contamination of the core sample. The data gathered from the ice in the core include deuterium content of the ice D_{ice}, and dust content (aerosols in the ice). From the entrapped gas bubbles, one can

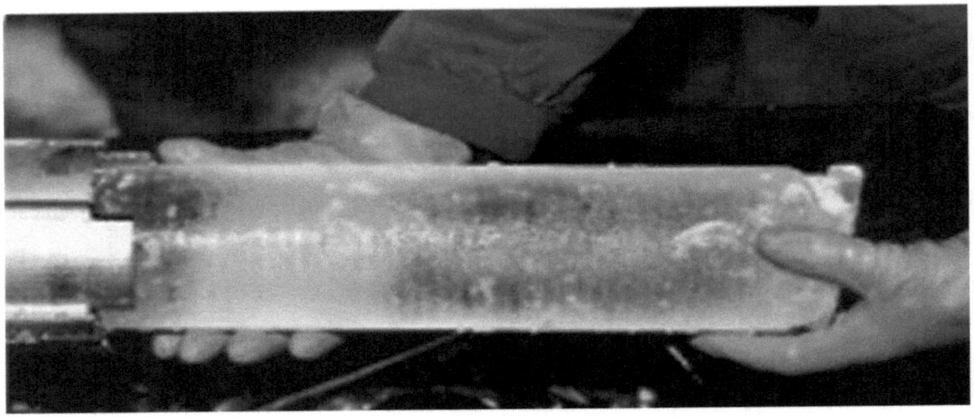

FIGURE 6.17 Ice core (Kennedy and Hansen, 2006).

extract the concentration of CH_4 and CO_2, and isotopes of O, N, and Ar. Various analytical tools are used to analyze the trapped gas. In recent years, the accuracy on analysis has increased in some cases by one order of magnitude. We will later comment in the *Examples and Theory Extension* section on the analysis accuracy of some isotopes. One needs to match the composition of the air in the bubbles and species of the ice in the ice core with dating of the air bubble and dating of the ice.

6.3.3 Ice Dating

As snow accumulates and is buried, it is compressed to form firn, a grainy material with a texture similar to granulated sugar. With increasing pressure at higher depths, the firn is compressed and transformed into glacial ice. The depth at which firn is transformed into ice may range between several tens of meter to 100 m in Antarctica. The transition from firn to ice is around firn density of 0.80 to 0.83 g/cm^3. To a certain depth, ice cores can be dated by counting individual layers, each layer representing the ice formation in one year as is shown in Fig. 6.18. This is similar to a growth ring in a tree. Layers can be visually distinguished in ice to significant depths. Deeper into the ice the layers become indistinguishable from increased pressure. In deeper locations, there is ice movement. Deep ice chronology is then constructed to model accumulations and ice flow (Lorius et al., 1985). Known volcanic eruptions are also identifiable in ice cores and are used as markers for calibration of past events.

6.3.4 Trapped Air Dating

As snow compresses into firn, the air in the firn column is in communication with the air above. With snow accumulation on top, the firn continues to densify with increasing depth. At firn density of about 0.80 to 0.83 g/cm^3, the pores close off and the air is trapped into glacial ice. As stated above, the firn-to-ice transition can be as deep as 90 to 100 m below the surface (at Vostok). Because of the diffusion processes (Fickian diffusion and pressure diffusion) across the snow column, the age of the ice and the age of the gas

FIGURE 6.18 19-cm long section of GISP 2 ice core from 1855 m showing annual layer structure illuminated from below by a fiber optic source. Section contains 11 annual layers with summer layers sandwiched between darker winter layer (Anthony Gow, United States Army Corps of Engineers, Cold Regions Research and Engineering Laboratory).

inside the ice core are not the same. They may be different by hundreds, and thousands of years. The difference in the age of the ice and that of the gas bubble inside the ice may be as high as 6000 years in glacial ice from Vostok (Barnola et al., 1991).

6.3.5 Temperature Modeling

There are two approaches in estimating past temperatures from isotopic composition data. In one approach the composition of the isotopes of ^{18}O and D in the ice are related to changes in the temperature where the precipitation has occurred. Through a process called **Rayleigh fractionation**, the temperature is computed (Dansgaard, 1964; Jouzel and Merlivat, 1984).

An alternative approach, which is much simpler, was introduced by Severinghaus et al. in 1998. In this approach, there is no need to calculate the difference between the age of the ice and the age of the gas bubble. Thermal diffusion is the main concept used in the alternative temperature calculation. In both approaches, gravity settling of the isotope is used to compute the thickness of the firn layer. In the following, we will present the theory upon which past climate changes are computed.

Gravitational Settling

The isotope sedimentation is often referred to as gravitational settling or gravitational fractionation in paleoenvironmental applications. Based on the Gibbs sedimentation expression derived in Eq. (2.15), we write

$$\frac{f_{iz}}{f_{i0}} = \exp\left[\frac{gzM_i}{RT}\right] \tag{6.134}$$

where the index 0 represents the surface of the firn. Note that in Eq. (6.134), z is positive downward to be compatible with the literature. For an ideal gas, $f_i = x_i P$ (from Chap. 1) and Eq. (6.134) can be written as

$$x_{iz} = x_{i0} \exp\left[\frac{gzM_i}{RT}\right] \tag{6.135}$$

Let us consider two isotopes comprising the gas mixture. Let $i=1$ stand for the heavy isotope and $i=2$ stand for the lighter isotope, then the ratio of the mol fraction of isotope 2 to isotope 1 at depth z to the top is given by

$$RI = \frac{x_{1z}}{x_{2z}} = \frac{x_{10}}{x_{20}} \exp\left[\frac{gz(M_1 - M_2)}{RT}\right] \tag{6.136}$$

We next define isotope deviation δ due to gravity settling by

$$\delta_G = \frac{RI - RI_0}{RI_0} = \frac{RI}{RI_0} - 1 \tag{6.137}$$

Let $\Delta M = M_1 - M_2$, then the isotope deviation is given by

$$\boxed{\delta_G = \exp\left[\frac{gz\Delta M}{RT}\right] - 1} \tag{6.138}$$

Given $\Delta M,\ T,$ and δ_G, one can calculate the thickness of the firn layer. The accuracy is of the order of 0.2 m, because of high accuracy of δ_G for some of the isotopes.

Thermal Diffusion Fractionation

In the sedimentation of isotopes in paleoenvironmental applications, the gravity fractionation is computed separately from thermal diffusion fractionation.

Let us rewrite the expression for the diffusion flux of species 1 from Fickian and thermal diffusion in a two-species system in the vertical direction [from Eq. (6.68)],

$$J_1 = -\rho D_{1,2} \frac{M_1 M_2}{M^2}\left[\frac{\partial \ln f_1}{\partial \ln x_1}\frac{\partial x_1}{\partial z} + \frac{k_{1T}}{T}\frac{\partial T}{\partial z}\right] \tag{6.139}$$

In Eq. (6.139), k_{1T} is the thermal diffusion ratio. It is related to thermal diffusion factor α_{1T} through $k_{1T} = \alpha_{1T} x_1 x_2$. At steady state, $J_1 = 0$. Assuming ideal gas behavior, $\partial \ln f_1 / \partial \ln x_1 = 1$, then we have

$$\frac{\partial x_1}{\partial z} = -\alpha_{1T} x_1 x_2 \frac{\partial \ln T}{\partial z} \tag{6.140}$$

Integrating the above equation between T_0 and T_z[$\int dx_1/(x_1 x_2) = \ln x_1/x_2 + constant$]

$$\ln \frac{x_{1z}/x_{2z}}{x_{10}/x_{20}} = \alpha_{1T}(T_{av}) \ln \frac{T_0}{T_z} \tag{6.141}$$

Equation (6.141) can also be written as

$$\left(\frac{x_{1z}}{x_{2z}}\right) \Big/ \left(\frac{x_{1,0}}{x_{2,0}}\right) = \left(\frac{T_0}{T_z}\right)^{\alpha_{1T}(T_{av})} \tag{6.142}$$

Let us define the separation factor from depth z to the reference point (0) by q,

$$q = \left(\frac{x_{1z}/x_{2z}}{x_{1,0}/x_{2,0}} \right) \tag{6.143}$$

The deviation factor for thermal diffusion can be then written as

$$\delta_T = (q-1) = \left(\frac{T_0}{T_z} \right)^{\alpha_{1T}(T_{av})} - 1 \tag{6.144}$$

Thermal diffusion factor α_{1T} which is a dimensionless parameter has been measured for various isotope pairs both in absence and presence of air. There is a small difference between the two measurements. One needs to use α_{1T} in air for high accuracy. Figure 6.19 shows the thermal diffusion factor of isotopes $^{15}N/^{14}N$ in air as a function of inverse of temperature. The data show linear dependency on $1/T$. A new measure of thermal diffusion isotope separation has been introduced in the interpretation of ice core data. The *thermal diffusion sensitivity parameter* Ω is defined as

$$\Omega(T_{av}) = \delta_T/(T_0 - T_z) = \left(\frac{\delta_T}{\Delta T} \right) \tag{6.145}$$

We write Eq. (6.144) in the following form using the Taylor series expansion

$$\delta_T \approx \alpha_{1T} \left(\frac{T_0 - T_z}{T_z} \right) = \alpha_{1T} \left(\frac{\Delta T}{T} \right) \tag{6.146}$$

provided $\Delta T/T \ll 1$, which is often the case.

Combining Eqs. (6.145) and (6.146), we obtain

$$\boxed{\Omega(T_{av}) = \alpha_{1T}(T_{av})/(T_{av})} \tag{6.147}$$

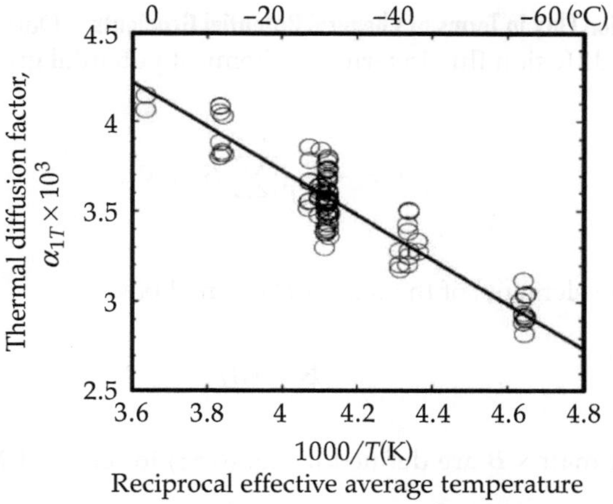

FIGURE 6.19 Thermal diffusion factor for $^{15}N/^{14}N$ in air (Grachev and Severinghaus, 2003).

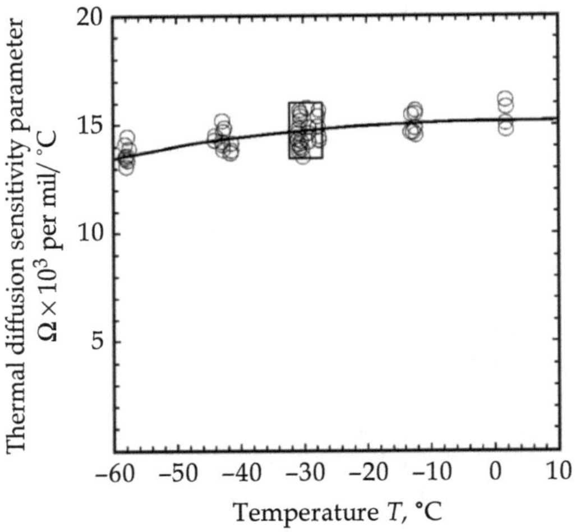

FIGURE 6.20 Variation of $\Omega(T)$ vs. temperature (Grachev and Severinghaus, 2003).

The expression for δ_T is then given by

$$\boxed{\delta_T = \Omega(T_{av})\Delta T} \tag{6.148}$$

Figure 6.20 shows that $\Omega(T)$ is nearly constant while $\alpha_{1T}(T)$ varies with temperature as shown in Fig. 6.19.

6.4 Examples and Theory Extension

6.1: Fickian Diffusion Flux in Terms of Chemical Potential Gradients Derive the following expression for Fickian diffusion flux in terms of chemical potential gradients

$$\vec{J}_i = -\frac{c}{RT}\sum_{k=1}^{n_c-1} \mathcal{B}_{ik} x_k \nabla \mu_k$$

where $\mathcal{B}_{ik}$ are the elements of the matrix $\boldsymbol{\mathcal{B}}$ defined below

$$\boldsymbol{\mathcal{B}} = (\boldsymbol{B})^{-1}$$

The elements of matrix $\boldsymbol{B}$ are defined in Eq. (6.86) in terms of MS diffusion coefficient $\mathcal{D}_{ij}$.

Solution The Fickian diffusion flux can be related to driving force at constant temperature and pressure based on $\vec{J} = \boldsymbol{L}\vec{Y}$ where $\vec{Y}$ is the driving force column vector.

($\underline{L}$ is the matrix of phenomenological coefficients.) The elements of the column vector $\vec{Y}$ are given by

$$Y_i = -\frac{1}{T}\sum_{k=1}^{n_c-1} A_{ik}\nabla\mu_k \tag{6.149}$$

A_{ik} are given by

$$A_{ik} = \left(\delta_{ik} + \frac{x_k}{x_{n_c}}\right) \tag{6.150}$$

We will derive the above expression in Example 6.2.

From $\mu_k = \mu_k(x_1, \ldots, x_{n_c-1})$ one readily writes

$$\nabla\mu_k = \sum_{l=1}^{n_c-1}\left(\frac{\partial\mu_k}{\partial x_l}\right)\nabla x_l \quad k = 1, \ldots, n_c \tag{6.151}$$

From $\mu_k = \mu_k^0(T, P) + RT\ln f_k$, we obtain

$$\frac{\partial\mu_k}{\partial x_l} = \frac{RT}{x_l}\Gamma_{kl} \quad k = 1, \ldots, n_c; l = 1, \ldots, n_c - 1 \tag{6.152}$$

where $\Gamma_{kl} = x_l\partial\ln f_k/\partial x_l$.

Combining Eqs. (6.151) and (6.152), the column vector of chemical potential gradients is written as

$$\nabla\boldsymbol{\mu} = RT\mathbf{X}^{-1}\boldsymbol{\Gamma}\nabla\boldsymbol{x} \tag{6.153}$$

where the element of diagonal matrix $\mathbf{X}$ are $X_{ii} = x_i$. Combining $\vec{J} = \boldsymbol{L}\vec{Y}$ and Eqs. (6.149) and (6.153), we obtain

$$\vec{J} = -R\boldsymbol{L}\boldsymbol{A}\boldsymbol{X}^{-1}\boldsymbol{\Gamma}\nabla\boldsymbol{X} \tag{6.154}$$

The Fickian diffusion flux (in vector notation) is given by

$$\vec{J} = -c\mathbf{D}\nabla\boldsymbol{x} \tag{6.155}$$

From $\mathbf{D} = \mathbf{B}\boldsymbol{\Gamma}$, and Eqs. (6.154) and (6.155), we obtain

$$\boxed{\mathbf{L} = \frac{c}{R}\mathcal{B}A^{-1}} \tag{6.156}$$

Combining $\vec{J} = \boldsymbol{L}\vec{Y}$ and Eq. (6.156), we obtain

$$\boxed{\vec{J} = -\frac{c}{RT}\mathcal{B}\nabla\boldsymbol{\mu}} \tag{6.157}$$

The above expression is the sought relationship. The usefulness of Eq. (6.157) is that it can be used to approximate Fickian diffusion flux between two different phases where the Fickian laws cannot be used (Hoteit, 2013; Moortgat and Firoozabadi, 2013).

6.2: Deriving Force in Terms of $\nabla\mu$ at Constant T and P Use the entropy production strength at constant temperature and pressure

$$\sigma = -\frac{1}{T}\sum_{i=1}^{n_c} \vec{J}_i \nabla\mu_i \geq 0$$

and $\sum_{i=1}^{n_c} \vec{J}_i = 0$ and the Gibbs-Duhem equation $\sum_{i=1}^{n_c} x_i \nabla\mu_i = 0$ to obtain

$$\sigma = -\frac{1}{T}\sum_{i=1}^{n_c-1} \vec{J}_i \sum_{k=1}^{n_c-1}\left(\delta_{ik} + \frac{x_k}{x_{n_c}}\right)\nabla\mu_k$$

Then use the following notation:

$$A_{ik} = \left(\delta_{ik} + \frac{x_k}{x_{n_c}}\right)$$

and show that the different driving forces are given by

$$Y_i = -\frac{1}{T}\sum_{k=1}^{n_c-1} A_{ik}\nabla\mu_k$$

Solution From $\sum_{i=1}^{n_c} \vec{J}_i = 0$, we obtain $\vec{J}_{n_c} = -\sum_{i=1}^{n_c-1} \vec{J}_i$.
From the Gibbs-Duhem expression $\sum_{i=1}^{n_c} x_i \nabla\mu_i = 0$, we obtain

$$\nabla\mu_{n_c} = -\sum_{i=1}^{n_c-1}(x_i/x_{n_c})\nabla\mu_i$$

Substituting the above equations in the expression for the entropy production strength, we obtain

$$\sigma = -\frac{1}{T}\sum_{i=1}^{n_c-1} \vec{J}_i \sum_{k=1}^{n_c-1}\left(\delta_{ik} + \frac{x_k}{x_{n_c}}\right) = \sum_{i=1}^{n_c-1} \vec{J}_i \vec{Y}_i$$

6.3: Basic Energy Expression in Permeable Media Derive the following equation, which describes temperature variation in porous media saturated with a single-phase fluid mixture at steady state. The equation is based on the negligible effect of inter-diffusion on heat transfer.

$$\nabla \cdot (K\nabla T) - \rho\vec{v}c_P\nabla T = 0$$

where K is the thermal conductivity of the porous medium, and ρ and c_P are the fluid mass density, and the heat capacity per unit mass of the fluid, respectively.

Solution We will derive the energy equation in a 2D $x-z$ system and then extend it to 3D.

Consider a volume element of dimensions Δz, 1, and Δx dimensions sketched in Fig. 6.21. At steady state, energy flux in at z − energy flux out at $(z + \Delta z)$ + energy flux in

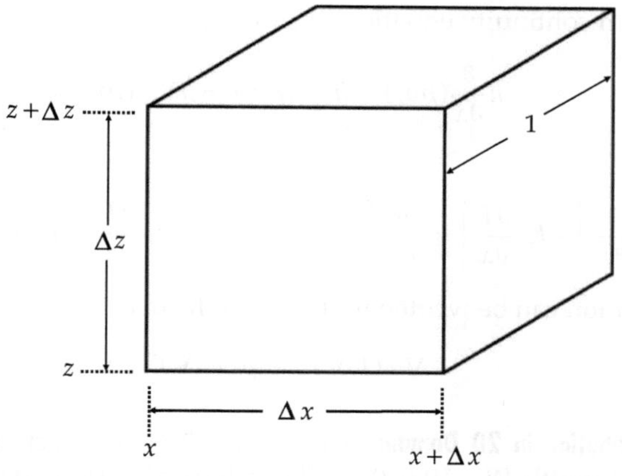

FIGURE 6.21 Volume element of Δx, Δz and 1.

at x energy − flux out at $(x + \Delta x) = 0$. The energy flux consists of two terms: conduction and convection. Note that we have neglected the energy flux due to diffusion.

Therefore

$$q_x^{cond}\Delta z \times 1\Big|_x - q_x^{cond}\Delta z \times 1\Big|_{x+\Delta x} + q_z^{cond}\Delta x \times 1\Big|_z - q_z^{cond}\Delta x \times 1\Big|_{z+\Delta z}$$
$$+q_x^{conv}\Delta z \times 1\Big|_x - q_x^{conv}\Delta z \times 1\Big|_{x+\Delta x} + q_z^{conv}\Delta x \times 1\Big|_z - q_z^{conv}\Delta x \times 1\Big|_{z+\Delta z} = 0$$

In the above equation, $\vec{q}^{cond}$ represents the heat flux by conduction through the porous medium, and $\vec{q}^{conv}$ represents the heat flux by convection of the fluid mixture. The y direction is assumed to be of unit length.

Dividing the above expression by $\Delta x \Delta z$ and taking the limit as $\Delta x \to 0$ and $\Delta z \to 0$, one obtains

$$\frac{\partial q_x^{cond}}{\partial z} + \frac{\partial q_z^{cond}}{\partial x} + \frac{\partial q_x^{conv}}{\partial z} + \frac{\partial q_z^{conv}}{\partial x} = 0$$

But $\vec{q}^{cond} = -K\nabla T$ and $\vec{q}^{conv} = \rho\vec{v}h$, where h is enthalpy per unit mass. The conduction and convection in the x-direction are $q_x^{cond} = -K_x\partial T/\partial x$ and $q_x^{conv} = \rho v_x h$, respectively. Therefore, the energy-flux expression in 2D is

$$\frac{\partial}{\partial x}\left[-K_x\frac{\partial T}{\partial x}\right] + \frac{\partial}{\partial z}\left[-K_z\frac{\partial T}{\partial z}\right] + \frac{\partial}{\partial x}[\rho v_x h] + \frac{\partial}{\partial z}[\rho v_z h] = 0$$

The last two terms can be expressed as

$$\frac{\partial}{\partial x}[\rho v_x h] + \frac{\partial}{\partial z}[\rho v_z h] = \rho v_x c_P\frac{\partial T}{\partial x} + \rho v_z c_P\frac{\partial T}{\partial z} + h\frac{\partial}{\partial x}(\rho v_x) + h\frac{\partial}{\partial z}(\rho v_z)$$

Note that from continuity equation, Eq. (6.92),

$$h\frac{\partial}{\partial x}(\rho v_x) + h\frac{\partial}{\partial z}(\rho v_z) = h\nabla \cdot (\rho \vec{v}) = 0$$

Therefore,

$$\frac{\partial}{\partial x}\left[-K_x\frac{\partial T}{\partial x}\right] + \frac{\partial}{\partial z}\left[-K_z\frac{\partial T}{\partial z}\right] + \rho v_x c_P\frac{\partial T}{\partial x} + \rho v_z c_P\frac{\partial T}{\partial z} = 0$$

The above equation can be written in the vector form as

$$\nabla \cdot (K\nabla T) - \rho \vec{v} c_P \nabla T = 0$$

6.4: Species Distribution in 2D Domain Consider a 2D cross-sectional reservoir of height $H = 1500$ ft and width $W = 10{,}000$ ft. The values of various parameters are 1) $C = -2°R/100$ ft, $B = 5°R/10{,}000$ ft [see Eqs. (6.100) and (6.101)], 2) $\mu = 0.2$ cp, $e = 1.67 \times 10^{-3}/°R$, $\rho^0 = 32.2\,\text{lb}_\text{m}/\text{ft}^3$, $T^0 = 150°F$, and 3) $k = 1000$ md, and $\phi = 0.25$.

First, assume $n = 1$ and $n = 3$ in Eq. (6.100) and compare the temperature profiles from Eqs. (6.100) and (6.101) by plotting T vs. x at $z = 0$. Then use Eqs. (6.107) and (6.108) to calculate v_x and v_z and plot the velocity contours for $n = 1$ from Eq. (6.100) and also plot v_x vs. z at $x = -5000$ ft, 0 ft, and +5000 ft, and v_z vs. x at $z = -125$ ft, 0 ft, and 125 ft.

In the next step, assume $n = 3$ in Eq. (6.100), and then calculate v_x and v_z and plot the velocity contours.

Solution The solution for velocity requires the evaluation of $(k/\mu)(\rho^0 geWB)$ in Eqs. (6.107) and (6.108). The rest is straightforward.

The multiplier $(k/\phi\mu)\,(\rho^0 geWB) = 90$ ft/year. Note that with this multiplier one obtains the true velocity, not the Darcy velocity.

Figure 6.22 depicts the temperature plots, indicating that Eq.(6.101), and Eq. (6.100) with $n = 1$ give comparable results. Figure 6.23 depicts the horizontal velocity vs. z

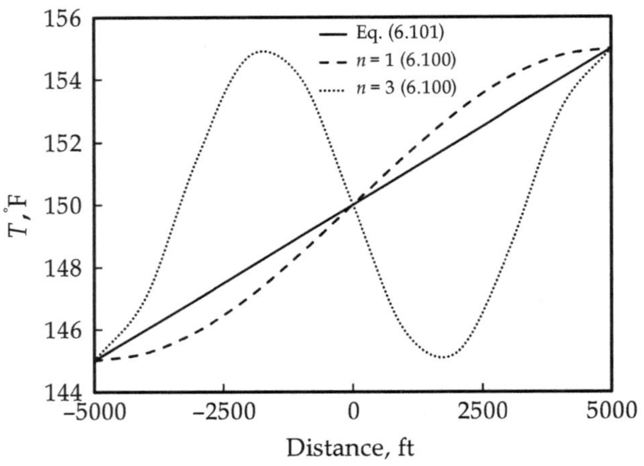

FIGURE 6.22 Temperature (at $z = 0$) versus distance using the linear and cosine functions.

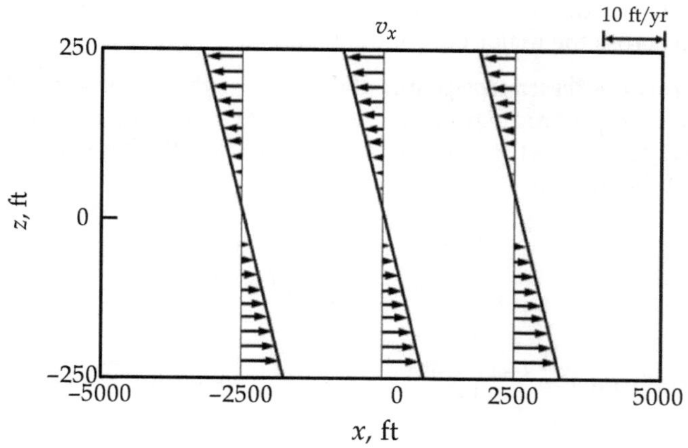

FIGURE 6.23 v_x vs. z at $x = -2500, 0$, and 2500 ft: [$n = 1$ in Eq. (6.100)].

at $x = -2500$ ft, 0 ft, and +2500 ft. An interesting feature of the plot is that v_x varies linearly with z, which was not transparent from Eq. (6.107).

Figure 6.22 also shows the temperature profiles for $n=3$. The factor n in the cosine term dictates the number of cells in the x-direction. This is because the flowlines will turn whenever $\partial\rho/\partial x$ is zero. Figure 6.24 provides the velocity contours for $n=3$, showing that there are indeed three cells. Note the alternation in the velocity directions of the neighboring cells.

The results from this thermal-convection example and the results from the natural convection-diffusion problem presented in the text have a basic difference: the vertical velocity not shown. For the former, the vertical velocity in the whole distance across the

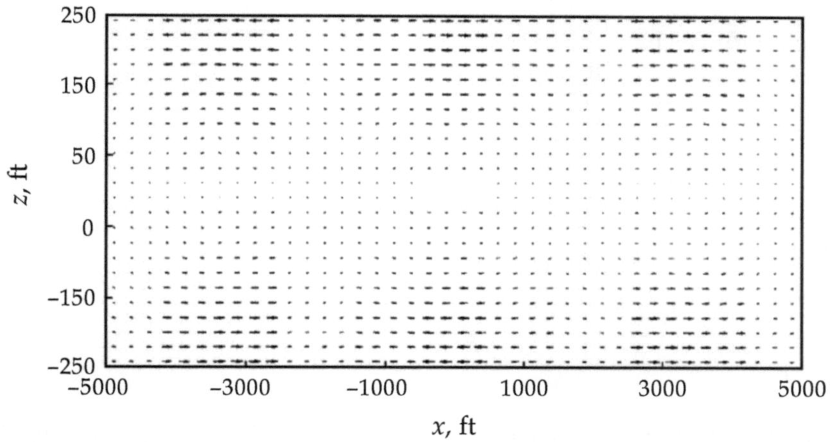

FIGURE 6.24 Velocity contours due to thermal convection: [$n = 3$, Eq. (6.100)].

reservoir is observed. For the latter, it is significantly close to the side boundaries and is close to zero across the major part of the reservoir.

6.5: Relation between Phenomenological Coefficients Use the reciprocity relation and the positivity of the entropy production strength to show that diagonal elements of the matrix of phenomenological coefficients are positive and the off-diagonal elements must satisfy conditions of the form

$$(L_{ii}L_{kk}) \geq \left(\frac{1}{4}\right)(L_{ik} + L_{ki})^2$$

Solution From $\sigma = \sum_{i=1}^{n_c}\sum_{k=1}^{n_c} L_{ik}X_kX_i \geq 0$, we write

$$\sigma = \sum_{i=1}^{n_c}\left(L_{i1}X_1X_i + L_{i2}X_2X_i + \cdots L_{i,n_c}X_{n_c}X_i\right) = \left(X_1 \; X_2 \ldots X_{n_c}\right)$$

$$= \begin{pmatrix} L_{1,1} & L_{1,2} & \ldots & L_{1,n_c} \\ L_{2,1} & L_{2,2} & \ldots & L_{2,n_c} \\ \vdots & & & \\ L_{n_c,1} & L_{n_c,2} & \ldots & L_{n_c,n_c} \end{pmatrix} \begin{pmatrix} X_1 \\ X_2 \\ \vdots \\ X_{n_c} \end{pmatrix} = \boldsymbol{X}^T\boldsymbol{L}\boldsymbol{X}$$

where $\boldsymbol{X}^T$ is the transpose of the column vector $\boldsymbol{X}$. The expression for σ can be written as,

$$\sigma = \sum_{i=1}^{n_c} R_iX_i^2 + \sum_{i=1}^{n_c}\sum_{j>i} \| L_{ij} \| (X_i + S_{ij}X_j)^2$$

where R_i is given by

$$R_i = L_{ii} - \sum_{j\neq i}^{n_c} \| L_{ij} \|$$

and

$$S_{ij} = \frac{L_{ij}}{\| L_{ij} \|}$$

In the above expression, $\| L_{ij} \|$ is the absolute value of L_{ij}, and $\sum_{i\neq j}^{n_c} \| L_{ij} \|$ is the sum of the absolute values of the off-diagnonal elements of row i of the matrix of phenomenological coefficients. Because $\sigma \geq 0$, then $R_i \geq 0$, therefore, $L_{ii} > 0$.

Now, let $L_{1,1}^{(0)} = L_{1,1}$

$$L_{2,2}^{(1)} = \frac{\begin{vmatrix} L_{1,1} & L_{1,2} \\ L_{2,1} & L_{2,2} \end{vmatrix}}{L_{1,1}}$$

$$\vdots$$

$$L_{(n_c,n_c)}^{(n_c-1)} = \frac{\begin{vmatrix} L_{1,1} & L_{1,2} & \dots & L_{1,n_c} \\ \vdots & & & \\ L_{n_c,1} & L_{n_c,2} & \dots & L_{n_c,n_c} \end{vmatrix}}{\begin{vmatrix} L_{1,1} & \dots & L_{1,n_c-1} \\ \vdots & & \\ L_{n_c-1,1} & \dots & L_{n_c-1,n_c-1} \end{vmatrix}}$$

and

$$L_{ik}^{(l)} = L_{ik}^{(l-1)} - 1/L_{ll}^{(l-1)} \left[L_{li}^{(l-1)} L_{lk}^{(l-1)} \right], \, L_{ik}^{(0)} = L_{ik}$$

$$Z_i = X_i + \frac{1}{L_{ii}^{(i-1)}} \sum_{k=i+1}^{n_c} L_{ki}^{(i-1)} X_k \quad i = 1, \dots, n_c$$

Therefore, σ can be expressed as

$$\sigma = L_{1,1} Z_1^2 + \frac{\begin{vmatrix} L_{1,1} & L_{1,2} \\ L_{2,1} & L_{2,2} \end{vmatrix}}{L_{1,1}} Z_2^2 + \frac{\begin{vmatrix} L_{1,1} & L_{1,2} & L_{1,3} \\ L_{2,1} & L_{2,2} & L_{2,3} \\ L_{3,1} & L_{3,2} & L_{3,3} \end{vmatrix}}{\begin{vmatrix} L_{1,1} & L_{1,2} \\ L_{2,1} & L_{2,2} \end{vmatrix}} Z_3^2 + \dots$$

where $Z_1, Z_2, Z_3, \dots$ are all independent. The above expression based on $\sigma \geq 0$ provides,

$$\boldsymbol{L} = \begin{vmatrix} L_{1,1} & L_{1,2} & \dots & L_{1,n_c} \\ \vdots & & & \\ L_{n_c,1} & L_{n_c,2} & \dots & L_{n_c,n_c} \end{vmatrix} > 0$$

If we exchange two columns or two rows at the same time, the sign of the determinant of $\boldsymbol{L}$ will not change.

$$\ell' = \begin{vmatrix} L_{ii} & L_{ik} & \dots & L_{i,n_c} \\ L_{ki} & L_{kk} & \dots & L_{k,n_c} \\ \vdots & & & \\ L_{n_c,i} & L_{n_c,k} & \dots & L_{n_c,n_c} \end{vmatrix} > 0$$

Then we can write

$$\begin{vmatrix} L_{ii} & L_{ik} \\ L_{ki} & L_{kk} \end{vmatrix} > 0$$

from which $L_{ii}L_{kk} - L_{ik}L_{ki} > 0$ or $L_{ii}L_{kk} > L_{ik}^2 = (1/4)[L_{ik} + L_{ki}]^2$. Note that, $L_{ik} = L_{ki}$.

6.6: Interface Composition in Three-Component Mixtures Consider a flat interface between gas and liquid phases at given pressure P and given temperature T. Write the equations that can be used to find the composition of the gas phase and the liquid phase at the interface for the following two cases.

(a) Two-component mixtures

(b) Three-component mixtures

If criteria of local equilibrium do not provide adequate equations (for b), make suggestions for the information that is needed to solve the problem.

Solution

(a) For a two-component mixture, the problem is defined by the following equations:

$$\mu_1^L(T, P, x_1^L) = \mu_1^G(T, P, y_1^G)$$
$$\mu_2^L(T, P, x_1^L) = \mu_2^G(T, P, y_1^G)$$

We have two equations and two unknowns, x_1^L and y_1^G, so we have sufficient number of equations.

(b) For three-component mixtures, the local equilibrium is defined by the following equations:

$$\mu_1^L(T, P, x_1^L, x_2^L) = \mu_1^G(T, P, y_1^G, y_2^G)$$
$$\mu_2^L(T, P, x_1^L, x_2^L) = \mu_2^G(T, P, y_1^G, y_2^G)$$
$$\mu_3^L(T, P, x_1^L, x_2^L) = \mu_3^G(T, P, y_1^G, y_2^G)$$

We have three equations and four unknowns, x_1^L, x_2^L, y_1^G, and y_2^G. Additional equations can be obtained from mass balance across the interface:

$$c^L x_1^L v_z^L + J_{1z}^L = c^G y_1^G v_z^G + J_{1z}^G$$
$$c^L x_2^L v_z^L + J_{2z}^L = c^G y_2^G v_z^G + J_{2z}^G$$
$$c^L x_3^L v_z^L + J_{3z}^L = c^G y_3^G v_z^G + J_{3z}^G$$

where c is the molar density, v_z is mol-averaged vertical velocity, and J_{iz} is the vertical diffusive molar flux of component i. The above equations introduce two additional unknowns, v_z^L and v_z^G (the diffusive fluxes are not unknowns since they relate to composition through Fick's laws).

Therefore, we have 6 equations and 6 unknowns, and the solution is fully specified (Haugen and Firoozabadi, 2009).

6.7: Net Heat of Transport of Component *i* In thermodynamics of irreversible processes, the net heat of transport of component *i* is defined by,

$$Q_i^* = Q_i - \bar{H}_i$$

where Q_i is the energy transported across a given reference plane per mol of diffusing component *i*, in an isothermal process. Show that Q_i^* corresponds to the absorption of heat.

Solution We will follow Denbigh (1951), and consider a certain bounded region at constant temperature and pressure. Suppose dn_i moles of component *i* diffuse across the boundary. The energy flux across the boundary will be $Q_i dn_i$. In order to keep the pressure and temperature of region constant, $đQ$ heat may be absorbed by the system from the surrounding and $đW$ work may be done by the region. The change in internal energy is

$$dU = -Q_i dn_i + đQ - đW$$

The change in internal energy is $-\bar{U}_i dn_i$, and the change in volume is $-\bar{V}_i dn_i$, where $\bar{U}_i$ and $\bar{V}_i$ are the partial molar internal energy, and partial molar volume, respectively, and $đW = -P\bar{V}_i dn_i$. Therefore, $đQ = (Q_i - \bar{U}_i - P\bar{V}_i)dn_i$. From $\bar{H}_i = \bar{U}_i + P\bar{V}_i$,

$$đQ = (Q_i - \bar{H}_i)dn_i = Q_i^* dn_i$$

or

$$Q_i^* = đQ/dn_i$$

which is the net heat that must be absorbed per mol of diffusing component *i* to keep the pressure and temperature of the region constant.

6.8: Expression of Gravitational Settling At thermal equilibrium, the difference in isotope ratios of the snow (RI) at z and the atmosphere (RI_0) at $z = 0$ is caused exclusively by gravitational fractionation,

$$\delta_G = \frac{RI}{RI_0} - 1 = \exp\left[\frac{(M_1 - M_2)gz}{RT}\right] - 1$$

where M_1 and M_2 are the molar mass of the heavy and light isotopes, respectively, g is gravity acceleration, and z is depth, R is the universal gas constant, and T is temperature. Show that this expression reduces to

$$\delta_G \cong \left[\frac{(M_1 - M_2)gz}{RT}\right]$$

when $(M_1 - M_2)gz/RT \ll 1$

Solution We can expand the exponential function using $e^x = 1 + x + x^2/2! + x^3/3! + \ldots$ (Maclaurin series).

When $x \ll 1$, we write

$$\exp\left[\frac{(M_1 - M_2)gz}{RT}\right] \cong 1 + \frac{(M_1 - M_2)gz}{RT}$$

Therefore,

$$\delta_G = \exp\left[\frac{(M_1 - M_2)gz}{RT}\right] - 1 \cong \frac{(M_1 - M_2)gz}{RT}$$

6.9: Expression for Thermal Fractionation, Thickness of Firn Layer, and Estimation of Atmospheric Temperature When there is a temperature difference between the snow (T at z) and the atmosphere (T_0 at $z=0$), thermal diffusion leads to a difference in isotope ratios RI and RI_0,

$$\delta_T = \frac{RI}{RI_0} - 1 = \left[\frac{T_0}{T}\right]^{\alpha_{1T}} - 1$$

where α_{1T} is the thermal diffusion ratio of the heavy isotope.

(a) Show that this expression reduces to

$$\delta_T \cong \alpha_{1T}\left(\frac{T_0 - T}{T}\right)$$

when $T_0 - T \ll T$

(b) A scientist makes measurements of the isotope ratios of $^{15}N/^{14}N$ and $^{40}Ar/^{36}Ar$ in air bubbles trapped in an ice core from Greenland. He finds that at a given depth the deviation from atmospheric isotope ratios are given by

Nitrogen: $\delta = 4.1 \times 10^{-4}$

Argon: $\delta = 1.6 \times 10^{-3}$

Assuming thermal equilibrium at $T = -40°C$, estimate the thickness of the snow layer at the corresponding time.

(c) The scientist finds that the above deviations remain constant for a large section of the ice-core, supporting the assumption of thermal equilibrium. However, in a small interval he finds increased deviations

Nitrogen: $\delta = 5.7 \times 10^{-4}$

Argon: $\delta = 2.1 \times 10^{-3}$

He concludes that the change must be due to a sudden change in atmospheric temperature, a change in the thickness of the snow layer, or both. Assuming the temperature of the snow remains constant, estimate the atmospheric temperature and the thickness of the snow layer at the corresponding time. The thermal diffusion ratio of ^{15}N is $\alpha_{1T} = 5.0 \times 10^{-3}$ and the thermal diffusion ratio of ^{40}Ar is $\alpha_{1T} = 1.0 \times 10^{-2}$.

Solution

(a) Writing $\delta_T = RI/RI_0 - 1 = [T_0/T]^{\alpha_{1T}} - 1$ and adding and subtracting 1 in the bracket, we get

$$\delta_T = \left[1 + \frac{T_0 - T}{T}\right]^{\alpha_{1T}} - 1$$

Note that $(T_0 - T)/T \ll 1$. Applying the binomial theorem $(1 + x)^n \approx 1 + nx$ where $x \ll 1$, we get:

$$\delta_T \cong 1 + \alpha_{1T}\left(\frac{T_0 - T}{T}\right) - 1 = \alpha_{1T}\left(\frac{T_0 - T}{T}\right)$$

(b) Assuming thermal equilibrium,

$$\delta_G \cong \left[\frac{(M_1 - M_2)gz}{RT}\right]$$

or

$$z = \frac{RT\delta_G}{(M_1 - M_2)g}$$

From data of $^{15}N/^{14}N$,

$$z = \frac{RT\delta_G}{(M_1 - M_2)g} = \frac{(8.3145\ \text{J/mol/K})(233.15\ \text{K})(4.1 \times 10^{-4})}{(1 \times 10^{-3}\ \text{kg/mol})(9.8\ \text{m/s}^2)} = 81.1\ \text{m}$$

From data of $^{40}Ar/^{36}Ar$,

$$z = \frac{RT\delta_G}{(M_1 - M_2)g} = \frac{(8.3145\ \text{J/mol/K})(233.15\ \text{K})(1.6 \times 10^{-3})}{(4 \times 10^{-3}\ \text{kg/mol})(9.8\ \text{m/s}^2)} = 79.1\ \text{m}$$

From the results above, it is found that the calculated firn thicknesses are close to each other.

(c) There are two contributions to δ: gravitational and thermal. To solve this problem, we must solve a system of two equations simultaneously. Namely, we need to solve

$$\delta = \alpha_{1T}\left(\frac{T_0 - T}{T}\right) + \frac{(M_1 - M_2)gz}{RT}$$

for both $^{15}N/^{14}N$ and $^{40}Ar/^{36}Ar$, that is,

$$\delta_N = \alpha_{1T}\left(\frac{T_0 - T}{T}\right) + \frac{(M_1 - M_2)gz}{RT}$$

$$\delta_{Ar} = \alpha_{1T}\left(\frac{T_0 - T}{T}\right) + \frac{(M_1 - M_2)gz}{RT}$$

Combining the two equations, and solving using the given data, we obtain $T_0 = 237.7$ K and $z = 94.9$ m.

6.10: Estimation of Abrupt Global Warming: 15,000 Years Ago Use the argon and nitrogen isotope composition measurements from the gas bubbles in the ice cores from Greenland (GISPs) to calculate both the firn thickness as well as the abrupt warming of about 15,000 years ago. The measured isotope data versus the deduced ice age are presented in Fig. 6.25. The gas age is estimated from an independent approach. You may first solve parts (a), (b), and (c) and move on to parts (d) and (e).

(a) What is the gravity settling for $^{40}Ar/^{36}Ar$? Note that the settling can be compared with ^{36}Ar. Specifically, express your answer in ‰ (per mil) in a firn of thickness 92-m deep for a stagnant air temperature of 226 K.

(b) What is the gravity settling in the $^{15}N/^{14}N$ system for the same conditions as above?

(c) What is the maximum added settling from thermal diffusion if there is a sharp temperature increase of 10 K? What would the maximum thermal diffusion settling be for a sharp temperature decrease of 10 K?

(d) What was the thickness of the firn layer around 14.6 kyr BP (before present)? Use the data in Fig. 6.25.

(e) What was the global warming in degrees K around 14.6 kyr BP?

Note: In order to solve some parts of the above problem, you need data on thermal diffusion ratio of $^{15}N/^{14}N$ and $^{40}Ar/^{36}Ar$. For $^{15}N/^{14}N$ you may use Fig. 6.20. For $^{40}Ar/^{36}Ar$, use $\Omega = 0.0388/\text{K}$ ‰.

Solution

(a) We use the expression

$$\delta_G = \exp\left(\frac{\Delta Mgz}{RT}\right) - 1 \cong \frac{\Delta Mgz}{RT}$$

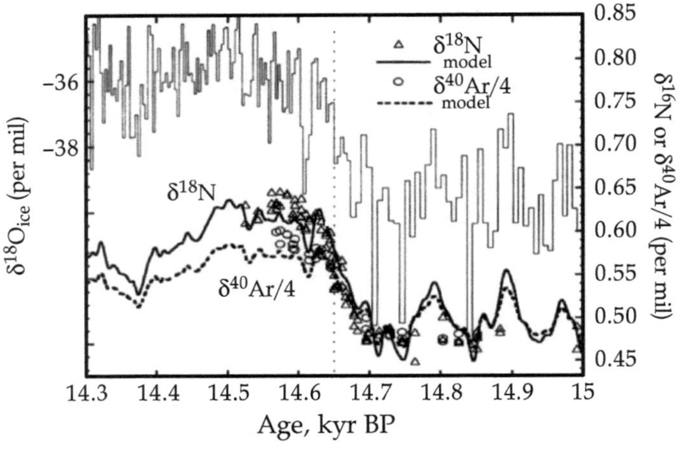

FIGURE 6.25 A profile of $\delta^{15}N$ and $\delta\,^{40}Ar$ in fossil air extracted from a ~15,000-year-old section of GISP2 ice core (Grachev and Severinghaus, 2003).

to compute the gravitational settling in the $^{40}Ar/^{36}Ar$ system. The parameters are $\Delta M = 4\times10^{-3}$ kg/mol, $z = 92$ m, $g = 9.8$ m/s^2, $T = 226$ K, and $R = 8.314$ J/mol·K. The computed gravitational settling is

$$\delta_G = 1.92 \text{ ‰}$$

(b) Using the same expression and conditions as above, except $\Delta M = 1 \times 10^{-3}$ kg/mol, the computed gravitational settling in the $^{15}N/^{14}N$ system is

$$\delta_G = 0.48 \text{ ‰}$$

(c) The settling due to thermal diffusion is given by $\delta_T = \Omega(T_{ave})\Delta T$. For the $^{40}Ar/^{36}Ar$ system, $\Omega(T_{ave}) = 0.0388$ ‰/K.
Thus, for a temperature increase of $\Delta T = 10$ K,

$$\delta_T = 0.388 \text{ ‰}$$

For $^{15}N/^{14}N$, as shown in Fig. 6.20, $\Omega = 8.656/T - 1232/T^2 = 0.0142$ ‰ K^{-1} at $T_{ave} = 226$ K. Thus, for a temperature increase of $\Delta T = 10$ K,

$$\delta_T = 0.142 \text{ ‰}$$

For a temperature decrease of $\Delta T = -10$ K, $\delta_T = -0.388$ ‰ and $\delta_T = -0.142$ ‰ for the $^{40}Ar/^{36}Ar$ and $^{15}N/^{14}N$ systems, respectively. The maximum settling is $\delta = \delta_G + \delta_T$

$$\delta = 1.92 - 0.388 = 1.532 \text{ ‰ for } ^{40}Ar/^{36}Ar$$

$$\delta = 0.48 - 0.142 = 0.338 \text{ ‰ for } ^{15}N/^{14}N$$

(d), (e) From Fig. 6.25, we obtain $\delta^{15}N = 0.62$ ‰ and $\delta^{40}Ar = 0.58 \times 4 = 2.32$ ‰. The total δ for both gravitational and thermal diffusion is given by

$$\delta = \delta_G + \delta_T = \frac{\Delta M g z}{RT} + \Omega \Delta T$$

Thus,

$$\delta^{15}N = \frac{\Delta M_N g z}{RT} + \Omega_N (T_0 - T)$$

and

$$\delta^{40}Ar = \frac{\Delta M_{Ar} g z}{RT} + \Omega_{Ar} (T_0 - T)$$

where $\delta^{15}N = 0.62$ ‰, $\delta^{40}Ar = 2.32$ ‰, $\Delta M_N = 1 \times 10^{-3}$ kg/mol, $\Delta M_{Ar} = 4 \times 10^{-3}$ kg/mol, $\Omega_N = 0.0142$ ‰ K^{-1}, $\Omega_{Ar} = 0.0388$ ‰ K^{-1}, $g = 9.8$ m/s^2, $T = 226$ K, $R = 8.314$ J/mol·K

We have two equations and two unknowns. Solving the system for z and T_0, we obtain

$$z = 94.7 \text{ m}$$

$$T_0 = 234.9 \text{ K}$$

$$\Delta T = T_0 - T = 234.9 - 226 = 8.9 \text{ K}$$

Therefore, the global warming around 14.6 kyrBP was around 9°C.

6.11: Invariance of Eigenvalue of Fickian Diffusion Coefficient Matrix in Multicomponent Mixtures
Show that the eigenvalues of the Fickian diffusion coefficient matrix in multicomponent fluid mixture should remain independent of the reference component. As one notes, the values of the diffusion coefficients in three- and higher-component mixtures will not be generally the same when the reference component changes. (See data in Prob. 6.15.)

Solution We first illustrate the derivations for the ternary and quaternary mixtures, for which the diffusion coefficient matrix is represented by 2 × 2 and 3 × 3 elements. We then illustrate the transformation for the general n_c-component case and demonstrate that the eigenvalues of the matrix of diffusion coefficient do not change with order of the component.

The diffusion flux of components 1 and 2 in a ternary mixture can be expressed by,

$$\begin{pmatrix} \vec{J}_1 \\ \vec{J}_2 \end{pmatrix} = \begin{pmatrix} D_{1,1} & D_{1,2} \\ D_{2,1} & D_{2,2} \end{pmatrix} \begin{pmatrix} \nabla x_1 \\ \nabla x_2 \end{pmatrix} \tag{6.158}$$

In the above expression, the reference component is species 3. We now change the reference component to 2. Since $\sum_{i=1}^{n_c} \vec{J}_i = 0$, then $\vec{J}_3 = -\vec{J}_1 - \vec{J}_2$ and $x_3 = 1 - x_1 - x_2$, we write

$$\begin{pmatrix} \vec{J}_1 \\ -\vec{J}_1 - \vec{J}_2 \end{pmatrix} = \begin{pmatrix} a & b \\ c & d \end{pmatrix} \begin{pmatrix} \nabla x_1 \\ -\nabla x_1 - \nabla x_2 \end{pmatrix} \tag{6.159}$$

in which $a,\ b,\ c$, and d can be determined by simple algebra, leading to

$$\begin{pmatrix} \vec{J}_1 \\ -\vec{J}_1 - \vec{J}_2 \end{pmatrix} = \begin{pmatrix} D_{1,1} - D_{1,2} & -D_{1,2} \\ -(D_{1,1} + D_{2,1}) + (D_{1,2} + D_{2,2}) & D_{1,2} + D_{2,2} \end{pmatrix} \begin{pmatrix} \nabla x_1 \\ -\nabla x_1 - \nabla x_2 \end{pmatrix} \tag{6.160}$$

Let us denote the diffusion coefficient matrices in Eqs. (6.158) and (6.160) by $\boldsymbol{D}_3$ and $\boldsymbol{D}_2$, respectively. The index i is chosen to designate the reference component. These two matrices are related by

$$\boldsymbol{D}_2 = \boldsymbol{A}_2 \boldsymbol{D}_3 \boldsymbol{A}_2 \tag{6.161}$$

where

$$\boldsymbol{A}_2 = \begin{pmatrix} 1 & 0 \\ -1 & -1 \end{pmatrix} \tag{6.162}$$

We proceed in the same way for the quaternary mixture. Let component 4 be selected as the reference component. The fluxes are then given by,

$$\begin{pmatrix} \vec{J}_1 \\ \vec{J}_2 \\ \vec{J}_3 \end{pmatrix} = \begin{pmatrix} D_{1,1} & D_{1,2} & D_{1,3} \\ D_{2,1} & D_{2,2} & D_{2,3} \\ D_{3,1} & D_{3,2} & D_{3,3} \end{pmatrix} \begin{pmatrix} \nabla x_1 \\ \nabla x_2 \\ \nabla x_3 \end{pmatrix} \tag{6.163}$$

We now select component 3 as the reference. The diffusion fluxes are

$$\begin{pmatrix} \vec{J}_1 \\ \vec{J}_2 \\ -\vec{J}_1 - \vec{J}_2 - \vec{J}_3 \end{pmatrix} = \begin{pmatrix} a & b & c \\ d & e & f \\ g & h & i \end{pmatrix} \begin{pmatrix} \nabla x_1 \\ \nabla x_2 \\ -\nabla x_1 - \nabla x_2 - \nabla x_3 \end{pmatrix} \tag{6.164}$$

where

$$\begin{pmatrix} a & b & c \\ d & e & f \\ g & h & i \end{pmatrix} = \begin{pmatrix} D_{1,1} - D_{1,3} & D_{1,2} - D_{1,3} & -D_{1,3} \\ D_{2,1} - D_{2,3} & D_{2,2} - D_{2,3} & -D_{2,3} \\ -(D_{1,1} + D_{2,1} + D_{3,1}) + E & -(D_{1,2} + D_{2,2} + D_{3,2}) & E \end{pmatrix} \tag{6.165}$$

where $E = D_{1,3} + D_{2,3} + D_{3,3}$. Let $\boldsymbol{D}_4$ and $\boldsymbol{D}_3$ be the matrices in Eqs. (6.163) and (6.165), respectively. The two matrices are related by,

$$\boldsymbol{D}_3 = \boldsymbol{A}_3 \boldsymbol{D}_4 \boldsymbol{A}_3 \tag{6.166}$$

where

$$\boldsymbol{A}_3 = \begin{pmatrix} 1 & 0 & 0 \\ 0 & 1 & 0 \\ -1 & -1 & -1 \end{pmatrix} \tag{6.167}$$

For the n_c-component system, the diffusion fluxes are $\vec{J}_i = \sum_{j=1}^{n_c-1} D_{ij} \nabla x_j$ $(i = 1, \ldots, n_c - 1)$ when the reference component is n_c. When the reference component is $n_c - 1$, the fluxes and diffusion coefficients are given below,

$$\begin{pmatrix} \vec{J}_1 \\ \vec{J}_2 \\ \vdots \\ \vec{J}_{n_c-2} \\ -\sum_{i=1}^{n_c-1} \vec{J}_i \end{pmatrix} = \boldsymbol{D}_{n_c-1} \begin{pmatrix} \nabla x_1 \\ \nabla x_2 \\ \vdots \\ \nabla x_{n_c-2} \\ -\sum_{i=1}^{n_c-1} \nabla x_i \end{pmatrix} \tag{6.168}$$

where

$$
D_{n_c-1} = \begin{pmatrix} D_{1,1} - D_{1,n_c-1} & D_{2,1} - D_{1,n_c-1} & \cdots & -D_{1,n_c-1} \\ D_{2,1} - D_{2,n_c-1} & D_{2,2} - D_{2,n_c-1} & \cdots & D_{2,n_c-1} \\ \vdots & \vdots & \ddots & \\ D_{n_c-2,1} - D_{n_c-2,n_c-1} & D_{n_c-2,2} - D_{n_c-2,n_c-1} & \cdots & -D_{n_c-2,n_c-1} \\ -\sum_{i=1}^{n_c-1} D_{i,1} + \sum_{i=1}^{n_c-1} D_{i,n_c-1} & \sum_{i=1}^{n_c-1} D_{i,2} + \sum_{i=1}^{n_c-1} D_{i,n_c-1} & \cdots & \sum_{i=1}^{n_c-1} D_{i,n_c-1} \end{pmatrix} \tag{6.169}
$$

The matrices $\boldsymbol{D}_{n_c-1}$ and $\boldsymbol{D}_{n_c}$ are related by,

$$
\boldsymbol{D}_{n_c} = \boldsymbol{A}_{n_c-1}\boldsymbol{D}_{n_c-1}\boldsymbol{A}_{n_c-1} \tag{6.170}
$$

where $\boldsymbol{A}_{n_c-1}[(n_{c-1}) \times (n_c - 1)]$ is given by

$$
\boldsymbol{A}_{n_c-1} = \begin{pmatrix} 1 & 0 & 0 & 0 & \cdots & 0 \\ 0 & 1 & 0 & 0 & \cdots & 0 \\ 0 & 0 & 1 & 0 & \cdots & 0 \\ \vdots & \vdots & \vdots & \vdots & \vdots & \vdots \\ -1 & -1 & -1 & -1 & \cdots & -1 \end{pmatrix} \tag{6.171}
$$

For the above matrix, $\boldsymbol{A}_{n_c-1}\boldsymbol{A}_{n_c-1}^{-1} = \boldsymbol{I}_{n_c-1}$, where $\boldsymbol{A}_{n_c-1}^{-1}$ is the inverse of $\boldsymbol{A}_{n_c-1}$ and $\boldsymbol{I}_{n_c-1}$ is the identity matrix of rank $n_c - 1$.

The determinant of $\boldsymbol{A}_{n_c-1}$ is

$$
\det(\boldsymbol{A}_{n_c-1}) = -1 \tag{6.172}
$$

The eigenvalues λ of $\boldsymbol{D}_{n_c}$ satisfy

$$
\det(\boldsymbol{D}_{n_c} - \lambda \boldsymbol{I}_{n_c-1}) = 0 \tag{6.173}
$$

Using the expressions from Eq. (6.170) and below, and the associated property of matrix manipulation,

$$
\begin{aligned} \det(\boldsymbol{D}_{n_c} - \lambda \boldsymbol{I}_{n_c-1}) &= \det(\boldsymbol{A}_{n-1}\boldsymbol{D}_{n_c-1}\boldsymbol{A}_{n_c-1} - \lambda \boldsymbol{I}_{n_c-1}) \\ &= \det(\boldsymbol{A}_{n_c-1}(\boldsymbol{D}_{n_c-1} - \lambda \boldsymbol{I}_{n_c-1})\boldsymbol{A}_{n_c-1}) \\ &= [\det(\boldsymbol{A}_{n_c-1})]^2 \det(\boldsymbol{D}_{n_c-1} - \lambda \boldsymbol{I}_{n_c-1}) \\ &= \det(\boldsymbol{D}_{n_c-1} - \lambda \boldsymbol{I}_{n_c-1}) \end{aligned} \tag{6.174}
$$

Equation (6.174) shows that the diffusion coefficient matrices $\boldsymbol{D}_{n_c-1}$ and $\boldsymbol{D}_{n_c}$ share the same eigenvalues. In other words, the eigenvalues of the Fickian diffusion coefficient matrix in multicomponent mixtures must be invariant with respect to the choice of the reference component. Problem 6.15 provides data in a ternary mixture with different reference components. The eigenvalues are about the same.

6.5 Problems

6.1 Show that in a multicomponent system the summation of the diffusive mass flux of all the components is zero:

$$\sum_{i=1}^{n_c} \vec{J_i} = 0$$

Hint: Use the diffusive mass-flux expression $\vec{J_i} = \rho_i(\vec{v_i} - \vec{v})$, where ρ_i is the mass density of component i (i.e., mass of component i per unit volume) and $\vec{v}$ is the velocity of component i, and the definition of mass average velocity $\vec{v} = \sum_{i=1}^{n_c} \rho_i \vec{v_i} / \rho$.

6.2 Derive the following expression in an isothermal gravity field.

$$\left.\frac{\partial \ln f_1}{\partial \ln x_1}\right|_{T,P} \nabla x_1 + \frac{x_1}{RT}\left(\bar{V}_1 - \frac{M_1}{\rho}\right) \nabla P = 0$$

Then demonstrate that x_1 increases with depth when $\bar{V}_1$ is negative. Show that the concentration gradient increases when a fluid approaches the critical point.

6.3 Consider a binary mixture at constant T and P. Is the diffusion flux always from a higher to lower concentration? Is the diffusion coefficient always positive? Use thermodynamic stability analysis to provide the answer for positivity of the Fickian diffusion coefficient.

6.4 For multiphase flow in permeable media, the expression for Fickian diffusion flux is often written as

$$\vec{J}_{\alpha,i} = -\phi S_\alpha c_\alpha \sum_{i=1}^{n_c-1} D_{\alpha,i,k}^{Fick} \nabla x_{\alpha,k}$$

where ϕ is the porosity, S_α is the saturation of phase α, c_α is the molar density of phase α, $D_{\alpha,i,k}^{Fick}$ is the Fickian diffusion coefficient of component i in phase α with component k, and $x_{\alpha,k}$ is the mol fraction of component k in phase α. What is the justification for the above expression?

6.5 Consider a ternary mixture at constant T and P. Is the diffusion flux always from the higher to the lower concentration? Is the determinant of diffusion coefficients always positive? Is there any sign restriction on any of elements of the diffusion coefficient matrix in multicomponent mixtures? How about sign restriction on phenomenological coefficients (see Mutoru and Firoozabadi, 2011)?

6.6 Derive Eqs. (6.42) and (6.43).

6.7 Derive Eqs. (6.61), (6.62), and (6.63).

6.8 What is the basic mathematical difference between pressure diffusion flux and the other two diffusion fluxes (Fickian and thermal)?

6.9 The mass diffusion flux in the mass-averaged frame ($\vec{J}_i$) and the molar diffusion flux in the mass averaged frame ($\vec{j}_i$) are related by $\vec{J}_i = \vec{j}_i / M_i$. In this case, show that $\sum_{i=1}^{n_c} \vec{J}_i = 0$ but $\sum_{i=1}^{n_c} \vec{j}_i \neq 0$.

6.10 The mass diffusion flux of a component i relative to the mass-average velocity is given by

$$\vec{j}_i = \rho_i(\vec{v}_i - \vec{v})$$

where $\vec{v}$ is the barycentric velocity; $\vec{v} = \sum_{i=1}^{n_c} \rho_i \vec{v}_i$.

The molar diffusion flux relative to the molar average velocity $\vec{v}^*$ is given by

$$\vec{J}_i^* = c_i(\vec{v}_i - \vec{v}^*)$$

where $\vec{v}^*$ is the molar averaged velocity; $\vec{v}^* = \sum_{i=1}^{n_c} c_i \vec{v}_i$.

Show that

$$\vec{j}_i = \sum_{k=1}^{n_c-1} \left\{ \delta_{ik} + \left[w_i \left(\frac{x_k}{x_i} \right) - w_k \right] \frac{w_i}{w_k} \right\} M_k \vec{J}_k^*$$

6.11 The mass and molar diffusion fluxes are given by

$$\vec{j} = -\rho \boldsymbol{D}^m \cdot \nabla \omega$$

$$\vec{J}^* = -c\boldsymbol{D}^M \cdot \nabla \boldsymbol{x}$$

where $\boldsymbol{D}^m$ and $\boldsymbol{D}^M$ are $(n_c - 1)$ square matrices. Show that $\boldsymbol{D}^m$ and $\boldsymbol{D}^M$ are related by

$$\boldsymbol{D}^m = \boldsymbol{K}^M \boldsymbol{w} \boldsymbol{x}^{-1} \boldsymbol{D}^M \boldsymbol{x} \boldsymbol{w}^{-1}$$

where $\boldsymbol{w}$ and $\boldsymbol{x}$ are $(n_c - 1)$ column vectors of mass and mol fractions, respectively. The elements of matrix K^M are given by

$$K_{ik}^M = \delta_{ik} - \omega_i \left(1 - \frac{w_{n_c} x_k}{w_k x_{n_c}} \right), \quad i, k = 1, \ldots, n_c - 1$$

6.12 King Hubert (1956) has shown that so long as the fluid density is constant or is a function of the pressure only, there will be no convective circulation in porous media. However, when ρ is a function of temperature, the fluid may have a convective circulation, which is referred to as thermal convection. In other words, the so-called free convection in porous media cannot occur because of pressure variation of density. For multicomponent systems, it is only the temperature and/or composition variations of density that result in natural convection.

Use the rotational and irrotational criteria for a vector field to prove the above.

Hint: Write the Darcy velocity of porous media in the following form: $\vec{v} = -(k/\mu)\rho[(\nabla P/\rho) + \vec{g}]$. Then take the curl of $[(\nabla P/\rho) + \vec{g}]$:

$$\nabla \times [(\nabla P/\rho) + \vec{g}] = \nabla \times (\nabla P/\rho) = \nabla P \times \nabla(1/\rho)$$

since $\nabla \times g = 0$. Show that $\nabla P \times \nabla(1/\rho) = 0$ if $\rho = \rho(P)$ and $\nabla P \times \nabla(1/\rho) \neq 0$ if $\rho = \rho(T)$.

6.13 Duncan and Toor (1962) have performed diffusion measurements in ternary ideal gas mixtures composed of hydrogen, nitrogen, and carbon dioxide at 35.2°C and atmospheric pressure. A two-bulb diffusion cell was used in the measurements. In one experiment (mixture 1), the concentration of nitrogen was 0.50 (mol fraction) in the two bulbs. In bulb 1, $x_{H_2} = 0.0$, $x_{N_2} = 0.50$, and $x_{CO_2} = 0.50$. In bulb 2, $x_{H_2} = 0.50$, $x_{N_2} = 0.50$, and $x_{CO_2} = 0.0$. In another experiment (mixture 2), the concentration of CO_2 was 0.50 (mol fraction) in the two bulbs. In bulb 1, $x_{H_2} = 0.50$, $x_{N_2} = 0.0$, and $x_{CO_2} = 0.50$; in bulb 2, $x_{H_2} = 0.0$, $x_{N_2} = 0.50$, and $x_{CO_2} = 0.50$. The time variation of composition in bulbs 1 and 2 are plotted in Fig. 6.26. Note that despite the fact that the initial composition gradient for N_2 is zero for mixture 1, and it is zero for CO_2 in mixture 2, there is composition variation for N_2 in mixture 1, and for CO_2 in mixture 2. Note also that for both mixtures, there is an initial increase in N_2 and CO_2 concentration in bulb 1, followed by a decrease. Explain the reason for N_2 concentration variation in mixture 1, and CO_2 variation in mixture 2.

Hint: The explanation can be made in terms of cross diffusion coefficients (see Hoteit, 2013).

6.14 Use thermodynamic stability analysis to show that when the thermal diffusion coefficient is positive, the component should go to the cold region in the binary mixture. There is no need to adopt a sign convention for thermal diffusion coefficients as often recommended in the literature (see Firoozabadi et al., 2000).

6.15 In a comprehensive work, Larrañaga et al. (2014) report Fickian diffusion coefficients in a ternary mixture of tetralin (THN), isobutyl benzene (IBB), and dodecane (n-C_{12}) at 1 atm pressure and 25°C. They report diagonal and off-diagonal diffusion coefficients for the three possible orders of components. They present results with different ordering of components as shown below (unit is m^2/s).

Order	$D_{1,1} \times 10^{-10}$	$D_{1,2} \times 10^{-10}$	$D_{2,1} \times 10^{-10}$	$D_{2,2} \times 10^{-10}$
THN/IBB/n-C_{12}	6.16	−4.61	0.37	11.40
n-C_{12}/IBB/THN	6.54	−0.51	−0.37	11.60
THN/n-C_{12}/IBB	11.30	5.08	0.14	6.65

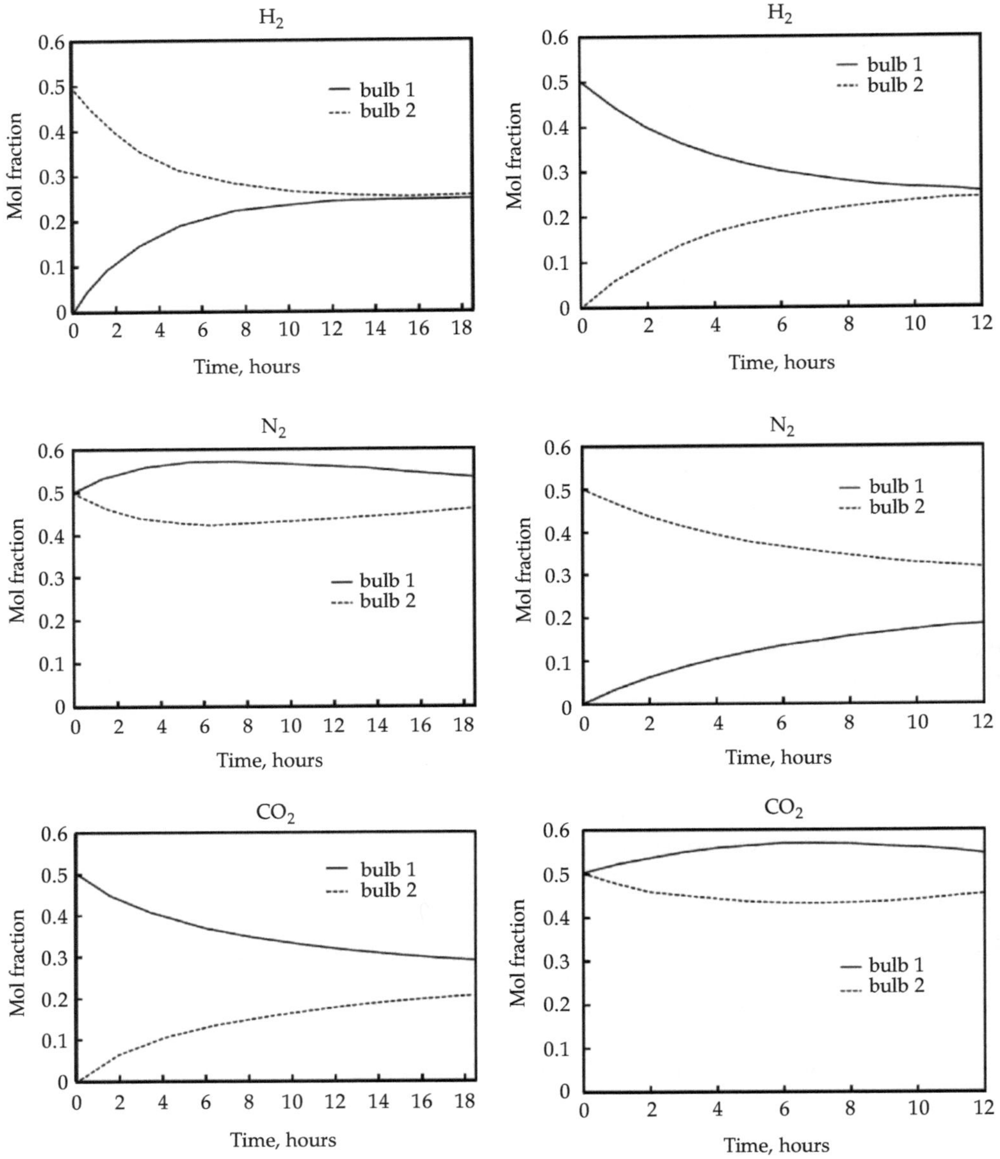

FIGURE 6.26 (Left) Variation of composition with time for ideal gas ternary mixture 1 (initial $x_{N_2} = 0.50$), and (right) variation of composition with time for ideal gas ternary mixture 2 (initial $x_{CO_2} = 0.50$).

Compute the eigenvalues (unit m^2/s) of the diffusion coefficient matrix which are given below.

Order	$\lambda_1 \times 10^{-10}$	$\lambda_2 \times 10^{-10}$
THN/IBB/n-C_{12}	6.51	11.10
THN/n-C_{12}/IBB	6.49	11.50
n-C_{12}/IBB/THN	6.50	11.10

Note that the third component is the reference.

6.6 Appendix

6.6.1 Scalars, Vectors, Tensors, and their Products

A scalar is defined as a quantity which has only magnitude, such as T, V, t, and P. A vector is defined to have a given magnitude and direction, such as velocity, momentum, and force.

Let us examine velocity $\vec{v}$ which can be written as:

$$\vec{v} = \sum_i \vec{\delta}_i v_i = \begin{pmatrix} v_1 \\ v_2 \\ v_3 \end{pmatrix}$$

where $\vec{\delta}_i$ is the unit normal vector in the direction i. Tensors are somewhat more complicated to define. Let us define a tensor such as stress tensor $\vec{\vec{\tau}}$

$$\vec{\vec{\tau}} = \sum_i \sum_j \vec{\delta}_i \vec{\delta}_j \tau_{ij} = \begin{pmatrix} \tau_{11} & \tau_{12} & \tau_{13} \\ \tau_{21} & \tau_{22} & \tau_{23} \\ \tau_{31} & \tau_{32} & \tau_{33} \end{pmatrix}$$

Note that the product $\vec{\delta}_i$ and $\vec{\delta}_j$ in the above expression is the dyadic product of two vectors as we will define shortly. Scalars τ_{ij} are referred to as the components of the tensor $\vec{\vec{\tau}}$.

6.6.2 Multiplication of a Vector by a Scalar

When a vector is multiplied by a scalar s, the magnitude is altered but the direction stays the same.

$$s\vec{v} = \begin{pmatrix} sv_1 \\ sv_2 \\ sv_3 \end{pmatrix} = \vec{v}s$$

6.6.3 Scalar Product or Dot Product (·) of Two Vectors

Vectors $\vec{v}$ and $\vec{w}$ can be written as

$$\vec{v} = \vec{\delta}_1 v_1 + \vec{\delta}_2 v_2 + \vec{\delta}_3 v_3$$
$$\vec{w} = \vec{\delta}_1 w_1 + \vec{\delta}_2 w_2 + \vec{\delta}_3 w_3$$

The dot product of vectors $\vec{v}$ and $\vec{w}$ are written as

$$\vec{v} \cdot \vec{w} = \left(\sum_i \vec{\delta}_i v_i\right) \cdot \left(\sum_j \vec{\delta}_j v_j\right) = v_1 w_1 + v_2 w_2 + v_3 w_3 = \sum_i v_i w_i$$

where we have used

$$\vec{v} \cdot \vec{w} = vw \cos\phi_{\vec{v}\vec{w}} \qquad \vec{\delta}_i \cdot \vec{\delta}_j = 1 \times 1 \cos\phi_{\vec{\delta}_i \vec{\delta}_j}$$
$$\vec{\delta}_i \cdot \vec{\delta}_j = 0 \quad i \neq j \qquad \vec{\delta}_i \cdot \vec{\delta}_j = 1 \quad i = j$$

$\phi_{\vec{v}\vec{w}}$ and $\phi_{\vec{\delta}_i \vec{\delta}_j}$ are the angles between vectors $\vec{v}$ and $\vec{w}$, and $\vec{\delta}_i$ and $\vec{\delta}_j$, respectively.

Scalar Product or Double Dot Product (:) of Two Tensors

One can readily show that

$$\vec{\vec{\sigma}} : \vec{\vec{\tau}} = \left(\sum_i \sum_j \vec{\delta}_i \vec{\delta}_j \sigma_{ij}\right) : \left(\sum_l \sum_k \vec{\delta}_l \vec{\delta}_k \tau_{lk}\right) = \sum_i \sum_j \sigma_{ij} \tau_{ji}$$

Differential Operator

The vector differential operator ∇, known as "nabla" or "del" is defined in rectangular coordinates as

$$\nabla = \vec{\delta}_1 \frac{\partial}{\partial x_1} + \vec{\delta}_2 \frac{\partial}{\partial x_2} + \vec{\delta}_3 \frac{\partial}{\partial x_3} = \sum_i \vec{\delta}_i \frac{\partial}{\partial x_i}$$

It may operate on a scalar, vector, or tensor.

Gradient of a Scalar

The gradient of a scalar is a vector. Let s be a scalar, then

$$\nabla s = \sum_i \vec{\delta}_i \frac{\partial s}{\partial x_i}$$

Divergence of a Vector

The divergence of a vector $\vec{v}$ is given as

$$\nabla \cdot \vec{v} = \sum_i \frac{\partial v_i}{\partial x_i}$$

The divergence of a vector is a scalar.

Dyadic Product of Two Vectors

The dyadic product of two vectors is a tensor. Therefore, for a velocity vector $\vec{v}$,

$$\nabla\vec{v} = \sum_i \sum_j \vec{\delta}_i \vec{\delta}_j \frac{\partial}{\partial x_i} v_j = \begin{pmatrix} \frac{\partial v_1}{\partial x_1} & \frac{\partial v_2}{\partial x_1} & \frac{\partial v_3}{\partial x_1} \\ \frac{\partial v_1}{\partial x_2} & \frac{\partial v_2}{\partial x_2} & \frac{\partial v_3}{\partial x_2} \\ \frac{\partial v_1}{\partial x_3} & \frac{\partial v_2}{\partial x_3} & \frac{\partial v_3}{\partial x_3} \end{pmatrix}$$

is a tensor. Most tensors are symmetric. The product $\vec{\delta}_i \vec{\delta}_j$ above is also a matrix with elements all zero except one element with unity. It is a unit dyad.

Let us now write the dyadic product of two vectors $\vec{v}$ and $\vec{w}$.

$$\vec{v}\vec{w} = \begin{pmatrix} v_1 w_1 & v_1 w_2 & v_1 w_3 \\ v_2 w_1 & v_2 w_2 & v_2 w_3 \\ v_3 w_1 & v_3 w_2 & v_3 w_3 \end{pmatrix}$$

and, therefore, for $\vec{\vec{\tau}} : \nabla\vec{v}$, one would obtain

$$\vec{\vec{\tau}} : \nabla\vec{v} = \sum_i \sum_j \tau_{ij} \frac{\partial v_i}{\partial x_j}$$

6.7 References

Aavatsmark, I., *Mathematiche Einfuhrug in die Thermodynamik der Gemische*. 1995, Berlin: Akademie Verlag.

Barnola, J. M., D. Raynaud, Y. S. Korotkevich, and C. Lorius, Vostok ice core provides 160,000-year record of atmospheric CO_2. *Nature*, 1987. 329(6138): pp. 408–414.

Barnola, J., P. Pimienta, D. Raynaud, and Y. Korotkevich, CO_2-climate relationship as deduced from the vostok ice core: A re-examination based on new measurements and on a re-evaluation of the air dating. *Tellus B*, 1991. 43(2): pp. 83–90.

Bird, R. B., W. E. Stewart, and E. N. Lightfoot, *Transport Phenomena*. 1960, New York: John Wiley & Sons.

Borisenko, A. I., and I. E. Tarapov, *Vector and Tensor Analysis with Applications*. 1968, New York: Dover Publications Inc.

Chandrasekhar, S., *Hydrodynamic and Hydrodynamic Stability*. 1961, New York: Dover Publications Inc.

Dansgaard, W., Stable isotopes in precipitation. *Tellus*, 1964. 16(4): pp. 436–468.

de Groot, S. R., and P. Mazur, *Non-Equilibrium Thermodynamics*. 1984, New York: Dover Publications Inc.

Denbigh, K., *The Thermodynamics of Steady State*. 1951, New York: John Wiley & Sons.

Duncan, J. B., and H. L. Toor, An experimental study of three component gas diffusion. *AIChE Journal*, 1962. 8(1): pp. 38–41.

Eckert, E. R., and R. M. Drake. *Analysis of Heat and Mass Transfer*. 1972, New York: McGraw-Hill.

El Maataoui, M., *Consequences de la Thermodiffusion en Milieu Poreux sur l'Hydrolyse des Solutions de Chlorures Ferriques et sur les Migrations d'Hydrocarbures dans les Melanges de n-Alcanes et dans Un Petrole Brut: Implications Geochimiques*. 1986, University of Paul Sabatier: Toulouse, France.

Firoozabadi, A., K. Ghorayeb, and K. Shukla, Theoretical model of thermal diffusion factors in multicomponent mixtures. *AIChE Journal*, 2000. 46(5): pp. 892–900.

Ghorayeb, K., and A. Firoozabadi, Molecular, pressure, and thermal diffusion in nonideal multicomponent mixtures. *AIChE Journal*, 2000. 46(5): pp. 883–891.

Ghorayeb, K., and A. Firoozabadi, Modeling multicomponent diffusions and convection in porous media. *SPE Journal*, 2000. 5(2): pp. 158–171.

Ghorayeb, K., A. Firoozabadi, and T. Anraku, Interpretation of the unusual fluid distribution in the yufutsu gas-condensate field. *SPE Journal*, 2003. 8(2): pp. 114–123.

Grachev, A. M., and J. P. Severinghaus, Laboratory determination of thermal diffusion constants for $^{29}N_2/^{28}N_2$ in air at temperatures from −60 to 0°C for reconstruction of magnitudes of abrupt climate changes using the ice core fossil–air paleothermometer. *Geochimica et Cosmochimica Acta*, 2003. 67(3): pp. 345–360.

Grachev, A. M., and J. P. Severinghaus, Determining the thermal diffusion factor for $^{40}Ar/^{36}Ar$ in air to aid paleoreconstruction of abrupt climate change. *Journal of Physical Chemistry A*, 2003. 107(23): pp. 4636–4642.

Haase, R., *Thermodynamics of Irreversible Processes*. 1969, New York: Dover Publications Inc.

Haase, R., H.-W. Borgmann, K. H. Ducker, and W.P. Lee, Thermodiffusion im kritischen verdampfungsgebiet binarer systeme, *Naturforsch*, 1971. 26a: pp. 1224–1227.

Haugen, K. B., and A. Firoozabadi, Composition at the interface between multicomponent nonequilibrium fluid phases. *Journal of Chemical Physics*, 2009. 130(6): p. 064707.

Haugen, K. B., and A. Firoozabadi, Mixing of two binary nonequilibrium phases in one dimension. *AIChE Journal*, 2009. 55(8): pp. 1930–1936.

Hoteit, H., and A. Firoozabadi, Numerical modeling of diffusion in fractured media for gas-injection and -recycling schemes. *SPE Journal*, 2009. 14(2): pp. 323–337.

Hoteit, H., Modeling diffusion and gas-oil mass transfer in fractured reservoirs. *Journal of Petroleum Science and Engineering*, 2013. 105(0): pp. 1–17.

Hubbert, M. K., Darcy's law and the field equations of the flow of underground fluids. *International Association of Scientific Hydrology*. Bulletin, 1957. 2(1): pp. 23–59.

Jouzel, J., and L. Merlivat, Deuterium and oxygen 18 in precipitation: modeling of the isotopic effects during snow formation. *Journal of Geophysical Research: Atmospheres*, 1984. 89(D7): pp. 11749–11757.

Jouzel, J., N. I. Barkov, J. M. Barnola, et al., Extending the vostok ice-core record of palaeoclimate to the penultimate glacial period. *Nature*, 1993. 364(6436): pp. 407–412.

Kempers, L. J. T. M., A Thermodynamic theory of the soret effect in a multicomponent liquid. *Journal of Chemical Physics*, 1989. 90(11): pp. 6541–6548.

Kennedy, D., and B. Hanson, Ice and History. *Science*, 2006. 311(5768): pp. 1673.

Kett, T. K., and D. K. Anderson, Ternary isothermal diffusion and the validity of the onsager reciprocal relations in nonassociating systems. *Journal of Physical Chemistry*, 1969. 73(5): pp. 1268–1274.

Kingston, P. E., and H. Niko, Development planning of the brent field. *Journal of Petroleum Technology*, 1975: pp. 1190–1198.

Larrañaga, M., D. A. S., Rees, and M. M. Bou-Ali, Determination of the molecular diffusion coefficients in ternary mixtures by the sliding symmetric tubes technique. *Journal of Chemical Physics*, 2014. 140(054201): pp. 1–9.

Leahy-Dios, A., and A. Firoozabadi, Unified model for nonideal multicomponent molecular diffusion coefficients. *AIChE Journal*, 2007. 53(11): pp. 2932–2939.

Lorius, C., Jouzel, J., Raynaud, D., Hansen, J., and H. Le Treut, The ice-core record: climate sensitivity and future greenhouse warming. *Nature*, 1990. 347(6289): pp. 139–145.

Lorius, C., J., Jouzel, C. Ritz, L. Merlivat, N.I. Barkov, Y.S. Korotkevich, and V. M. Kotlyakov, A 150,000-year climatic record from antarctic ice. *Nature*, 1985. 316(6029): pp. 591–596.

Lu, P.-C., *Introduction to the Mechanics of Viscous Fluids*. 1973, New York: Holt, Pinehart, and Winston.

Luettmer-Strathmann, J., and J. V. Sengers, The transport properties of fluid mixtures near the vapor-liquid critical line. *Journal of Chemical Physics*, 1996. 104(8): pp. 3026–3047.

Moortgat, J., and A. Firoozabadi, Fickian diffusion in discrete-fractured media from chemical potential gradients and comparison to experiment. *Energy & Fuels*, 2013. 27(10): pp. 5793–5805.

Mutoru, J., and A. Firoozabadi, Form of multicomponent fickian diffusion coefficients matrix. *Journal of Chemical Thermodynamics*, 2011. 43(8): pp. 1192–1203.

Mutoru, J. W., W. Smith, C.S. O'Hern, and A. Firoozabadi, Molecular dynamics simulations of diffusion and clustering along critical isotherms of medium-chain n-alkanes. *Journal of Chemical Physics*, 2013. 138(2): pp. 024317.

Nasrabadi, H., H. Hoteit, and A. Firoozabadi, An analysis of species separation in a thermo-gravitational column filled with a porous medium. *Transport in Porous Media*, 2007. 67(3): pp. 473–486.

Onsager, L., Reciprocal relations in irreversible processes. I. *Physical Review*, 1931. 37(4): pp. 405–426.

Onsager, L., Reciprocal relations in irreversible processes. II. *Physical Review*, 1931. 38(12): pp. 2265–2279.

Peng, D.-Y., and D. B. Robinson, A new two-constant equation of state. *Industrial & Engineering Chemistry Fundamentals*, 1976. 15(1): pp. 59–64.

Petit, J. R., Jouzel, J., Raynaud, D., et al., Climate and atmospheric history of the past 420,000 years from the vostok ice core, Antarctica. *Nature*, 1999. 399(6735): pp. 429–436.

Riley, M. F., and A. Firoozabadi, Natural Convection in Hydrocarbon Reservoirs. 1996, Quarterly Report of Reservoir Engineering Research Institute.

Riley, M. F., and A. Firoozabadi, Compositional variation in hydrocarbon reservoirs with natural convection and diffusion. *AIChE Journal*, 1998. 44(2): pp. 452–464.

Rutherford, W. M., and J. G. Roof, Thermal diffusion in methane-n-butane mixtures in the critical region. *Journal of Physical Chemistry*, 1959. 63(9): pp. 1506–1511.

Rutherford, W. M., Calculation of thermal diffusion factors for the methane-n-butane system in the critical and liquid regions. *AIChE Journal*, 1963. 9(6): pp. 841–843.

Salamatin, A. N., V. Y. Lipenkov, N. I. Barkov, J. Jouzel, J. R. Petit, and D. Raynaud, Ice core age dating and paleothermometer calibration based on isotope and temperature profiles from deep boreholes at vostok station (East Antarctica). *Journal of Geophysical Research: Atmospheres*, 1998. 103(D8): pp. 8963–8977.

Severinghaus, J. P., T. Sowers, E. J. Brook, R. B. Alley, and M. L., Bender, Timing of abrupt climate change at the end of the younger dryas interval from thermally fractionated gases in polar ice. *Nature*, 1998. 391(6663): pp. 141–146.

Severinghaus, J. P., A. Grachev, B. Luz, and N. Caillon, A method for precise measurement of argon 40/36 and krypton/argon ratios in trapped air in polar ice with applications to past firn thickness and abrupt climate change in Greenland and at Siple Dome, Antarctica. *Cosmochim. Acta*, 2003. 67: pp. 325–343.

Shukla, K., and A. Firoozabadi, A new model of thermal diffusion coefficients in binary hydrocarbon mixtures. *Industrial & Engineering Chemistry Research*, 1998. 37(8): pp. 3331–3342.
Taylor, R., and R. Krishna, *Multicomponent Mass Transfer*. 1993, New York: John Wiley & Sons.
Vignes, A., Diffusion in binary solutions. variation of diffusion coefficient with composition. *Industrial & Engineering Chemistry Fundamentals*, 1966. 5(2): pp. 189–199.
Vitagliano, V., R. Sartorio, S. Scala, and D. Spaduzzi, Diffusion in a ternary system and the critical mixing point. *Journal of Solution Chemistry*, 1978. 7(8): pp. 605–622.

CHAPTER 7

Interfacial Thermodynamics

Properties and structures of matter can be altered by changes in the bulk or by changes on the surface. The alteration in the bulk requires large amounts of substance. Changes in the surface may require much less. Knowledge and description of surfaces (the interface between gas and liquid) and interfaces (the interface between liquid and solid, and liquid and liquid) relate to surface or interfacial thermodynamics. In this chapter, we use surfaces and interfaces often interchangeably.

Nanoparticles, capillary phenomena, adsorption onto surfaces, nucleation (formation of a new phase), colloidal systems, wettability, micelles, and microemulsions are among the problems which have a foundation in interfacial thermodynamics. In this chapter, unlike in Chap. 2, we assign properties to the interface. As a result, thermodynamic functions will have surface properties, as added variables.

This chapter starts with the definition of surface energy and surface energy density of a flat surface, and the Gibbs adsorption expression for a flat surface. We then derive the expression of work for a system with a curved interface, where the variation of interfacial tension across the interface is accounted for. Thermodynamic functions in a composite system containing an interface are then derived. We will show that in a composite system consisting of bulk phase prime and double prime, and the interface (say a system with a gas bubble of pressure P' in a liquid of pressure P''), $G = A + P''V$, where G, A, and V are the Gibbs free energy, Helmholtz free energy, and volume of the total system, respectively, and P'' is the pressure of the continuous bulk liquid phase. We will derive the expression which shows the effect of curvature on the interfacial tension. Next, we present expressions for the work of cluster formation in single and multicomponent mixtures. The derivation of the work of cluster formation in a two or multicomponent mixture is much more complicated than that of a single-component system. An interesting demonstration in this chapter is the change of Gibbs free energy of a composite system when a cluster is formed within it. The Gibbs free energy change in the total system has a maximum when plotted versus the size of a newborn cluster of the new phase. We learn for the first time, a physical meaning for the maximum of Gibbs free energy.

This chapter is supplemented with a set of examples. One fascinating example discusses the effect of surface charge on vapor pressure. Interestingly, this problem was fully covered in the 1888 book by Lord Kelvin (Thomson, 1888). A plot of ΔG versus the size of a droplet shows first a minimum and then a maximum as the size of the droplet increases for a fixed driving force. Another fascinating example covers the shape of the interface in weightless state. In this example, the interface energy fully defines the system. The bulk phase does not enter into the formulation. In addition to ten examples, about 24 problems are presented at the end of this chapter.

There are two approaches for studying the interface between two distinguishable bulk phases. In one approach, the applicability of the laws of macroscopic hydrostatics to the nonhomogeneous interface zone is assumed. In another approach, Gibbs proposed to study the properties of the interface by using a mathematical surface of separation which he referred to as the dividing surface. In the Gibbsian approach, no thickness is assigned to the interface. We will use the Gibbsian approach in our analysis of the interface and follow the work of Ono and Kondo (1960) assuming equilibrium state. Our derivations include an analytical expression for the interfacial tension as a function of the curvature. Toward the end of this chapter we will relax the chemical equilibrium assumption and follow Defay and Prigogine (1966) with some generalization.

Let us consider the two bulk phases and the interface sketched in Fig. 7.1. We divide the volume V into the volumes V' and V''. The dividing surface shown by the dashed line in Fig. 7.1c is selected such that it is located in the interface zone (in some cases the dividing surface may not be in the interface zone). Volume V' contains the bulk phase prime and a small volume in the interface region. Volume V'' includes the bulk phase double-prime with the rest of the interface region. The system consists of two hypothetical phases "′" and "″" which are homogenous up to the dividing surface. Let N_i' and N_i'' represent the number of mols (or molecules) of species i in the hypothetical bulk phases, "′" and "″", respectively. Let N_i represent the total number of mols (or molecules) of species i in the total system. Then,

$$N_i = N_i' + N_i'' + N_i^s \qquad i = 1, \ldots n_c \tag{7.1}$$

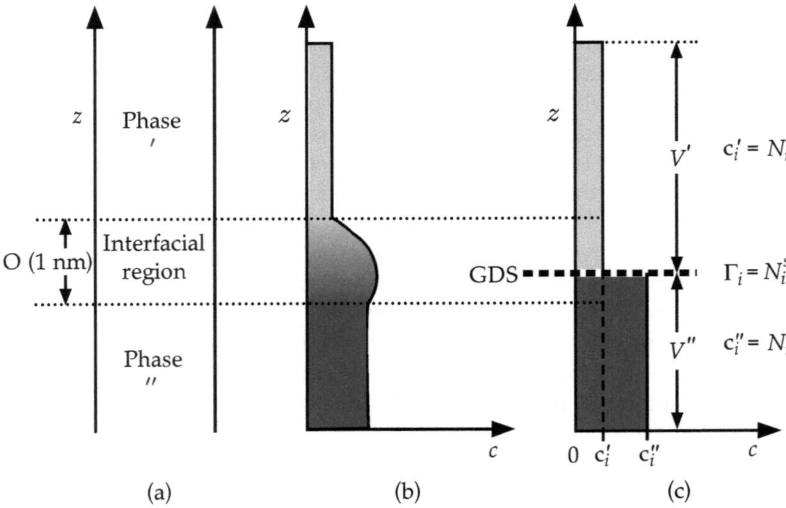

FIGURE 7.1 Concentration variation of component i and the Gibbs dividing surface: (GDS) (a) interfacial region, (b) concentration variation, and (c) idealized concentration variation based on GDS.

where N_i^s is the number of mols (or molecules) of species i in the interface. Note that N_i^s is not independent of the location of the dividing surface. The quantities describing the interface are, therefore, superficial values. In a system with a single species, the location of the dividing surface can be readily chosen such that $N^s = 0$. This surface is referred to as the equimolar (or equimolecular) dividing surface (EDS). In thermodynamics of interfaces, thermodynamic functions such as internal energy, entropy, and Helmholtz free energy are defined for the interface. Before we proceed further, some of these functions will be defined for the interface.

The Helmholtz free energy of the system in Fig. 7.1c, A, may be defined by

$$A = A' + A'' + A^s \tag{7.2}$$

where A' and A'' are the Helmholtz free energies of the hypothetical primed and double-primed phases, and A^s is the superficial (also referred to as excess) free energy of the interface. Similarly, the superficial internal energy U^s and the superficial entropy S^s of the interface are defined from

$$U = U' + U'' + U^s \tag{7.3}$$

$$S = S' + S'' + S^s \tag{7.4}$$

respectively. Using the relation $A = U - TS$ (assuming that T is constant throughout the system; i.e., the system is in thermal equilibrium), one readily derives from $A = U - TS$ and Eqs. (7.2) to (7.4),

$$A^s = U^s - TS^s \tag{7.5}$$

which has the same form as that of a bulk phase. Note that we may not write $G = A + PV$ for the whole system. In all the derivations in this chapter, the assumption of thermal equilibrium is an integral part of interfacial thermodynamics. Mechanical equilibrium will also be assumed in the sense that there is neither pressure driving force for bulk flow in a given phase nor pressure discontinuity other than that given by the Young-Laplace equation.

The chemical equilibrium assumption will also be made in most derivations (the chemical potentials of species i are equal in the bulk phases and in the interface). Later we will relax the chemical equilibrium assumption to derive the expression for chemical potentials in a nonequilibrium process for a heterogeneous system.

In the following, the interfacial thermodynamics for a system with a flat interface will be presented first. A flat interface simplifies the derivations considerably, mainly because the interfacial tension is independent of the location of the dividing surface. For a curved surface, the interfacial tension varies as the location of the dividing surface changes. The derivations in the next two sections are similar to the work of Ono and Kondo (1960), with some subtle differences in the interfacial thermodynamics of curved interfaces.

7.1 Plane Interface

We derived the expression for the incremental work (see Fig. 7.2) when the volume and area are increased in Chap. 2,

$$đW = -PdV + \sigma d\mathcal{A} \tag{7.6}$$

In the above equation, the interfacial tension σ can be interpreted as the work done on the system to increase the unit area of the interface [Eq. (7.6) assumes mechanical equilibrium]. The interfacial tension σ can also be interpreted as the force across the unit length at the interface. Note that in Eq. (7.6), the interfacial tension is independent of the location of the dividing surface.

Now, let us write the expression for the change of the internal energy of the total system (closed) in Fig. 7.2,

$$dU = TdS - PdV + \sigma d\mathcal{A} \tag{7.7}$$

In Eq. (7.7) we have assumed reversibility by writing $đQ = TdS$. The expression for dA is given by

$$dA = -SdT - PdV + \sigma d\mathcal{A} \tag{7.8}$$

For an open system with the assumption of chemical equilibrium,

$$dA = -SdT - PdV + \sigma d\mathcal{A} + \sum_{i=1}^{n_c} \mu_i dN_i \tag{7.9}$$

where N_i is given by Eq. (7.1). Because A is a homogeneous function of degree one, we can readily derive from Eq. (7.9) the following relationship,

$$A = -PV + \sigma\mathcal{A} + \sum_{i=1}^{n_c} \mu_i N_i \tag{7.10}$$

Note that Eq. (7.10) is based on the assumption of thermal, mechanical, and chemical equilibrium.

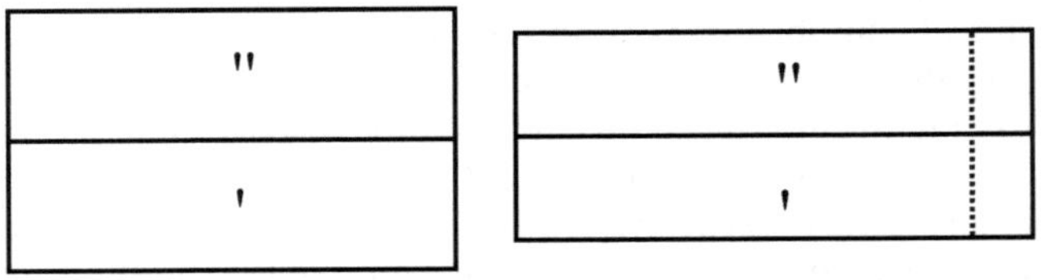

FIGURE 7.2 Change in interface area and volume: flat interface.

For the bulk phases prime and double-prime, one can write,

$$A' = -PV' + \sum_{i=1}^{n_c} N_i' \mu_i \tag{7.11}$$

$$A'' = -PV'' + \sum_{i=1}^{n_c} N_i'' \mu_i \tag{7.12}$$

From Eqs. (7.2), (7.5), and (7.10) to (7.12), one readily derives,

$$A^s = \sigma \mathcal{A} + \sum_{i=1}^{n_c} N_i^s \mu_i \tag{7.13}$$

According to the above equation, A^s is a function of the location of the dividing surface through N_i^s. If we select the location of the dividing surface such that

$$\sum_{i=1}^{n_c} N_i^s \mu_i = 0 \tag{7.14}$$

then from Eq. (7.13),

$$\boxed{\sigma = \frac{A^s}{\mathcal{A}}} \tag{7.15}$$

Eq. (7.13) shows that σ is equal to the superficial density of the Helmholtz free energy when the Gibbs dividing surface is determined by $\sum_{i=1}^{n_c} N_i^s \mu_i = 0$. For this choice of the dividing surface, σ is often referred to as the surface energy density in the nucleation literature. For a single component system, with $N^s = 0$, Eq. (7.14) defines the EDS.

Next we derive the Gibbs adsorption equation for a flat interface with the thermal, mechanical, and chemical equilibrium assumptions. The differential of the Helmholtz free energy for the bulk phases prime and double-prime are given by,

$$dA' = -S'dT - PdV' + \sum_{i=1}^{n_c} \mu_i dN_i' \tag{7.16}$$

$$dA'' = -S''dT - PdV'' + \sum_{i=1}^{n_c} \mu_i dN_i'' \tag{7.17}$$

Combining Eqs. (7.1), (7.2), (7.9), (7.16), and (7.17), one obtains,

$$dA^s = -S^s dT + \sigma d\mathcal{A} + \sum_{i=1}^{n_c} \mu_i dN_i^s \tag{7.18}$$

A comparison of the differentials of Eq. (7.13) and Eq. (7.18) leads to

$$d\sigma = -s^s dT - \sum_{i=1}^{n_c} \Gamma_i d\mu_i \tag{7.19}$$

where $s^s = S^s/\mathcal{A}$ and $\Gamma_i = N_i^s/\mathcal{A}$; Γ_i is the adsorption "inventory" of component i at the interface. In a constant temperature process,

$$d\sigma = -\sum_{i=1}^{n_c} \Gamma_i d\mu_i \tag{7.20}$$

Equation 7.20 is the Gibbs-adsorption isotherm. Among other applications, it can be used to study the effect of pressure on adsorption (see Probs. 7.3 and 7.4 at the end of this chapter).

7.2 Curved Interface

We will consider a curved interface with a spherical shape due to the widespread application in nucleation and many other phase-transition processes. In the derivations, a classical thermodynamic approach will be used. The majority of authors use hydrostatics in their derivations.

Let us consider the sketch portrayed in Fig. 7.3. The volume of the interior phase "′", and the exterior phase "″" are given by,

$$V' = \frac{\omega}{3}[r^3 - r'^3] \tag{7.21}$$

$$V'' = \frac{\omega}{3}[r''^3 - r^3] \tag{7.22}$$

respectively; r' and r'' are the radii of the inner and the outer phases, r is the radius of the interface and ω is the solid angle. The solid angle of a cone with apex angle 2θ is the area of a spherical cap with a unit-radius sphere with $\omega = 2\pi(1 - \cos\theta)$. The interface area is given by

$$\mathcal{A} = \omega r^2 \tag{7.23}$$

Note that r is a function of the dividing surface position.

Let us first derive the work term for constant r' and r'' due to change in ω by $d\omega$, which is given by

$$đW(\omega) = -\eta d\omega \tag{7.24}$$

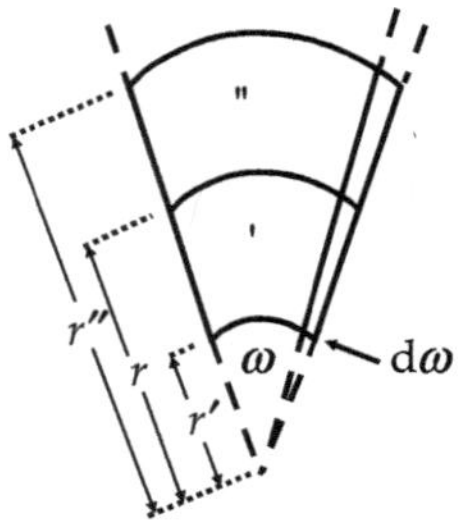

FIGURE 7.3 Change in interface area and volume of a cone-curved surface (r' and r'' constant).

The constant of proportionality η will be determined shortly. The expression for $đW(\omega)$ is also given by,

$$đW(\omega) = -P'\mathrm{d}V' - P''\mathrm{d}V'' + \gamma d\mathcal{A} \tag{7.25}$$

In Eq. (7.25), the first two terms represent the effect of volume expansion (compression) and the last term is work done on the system due to increase in the interface area. Note that instead of the symbol σ we have used γ, which is a function of the location of the dividing surface (to be shown soon). The minimum value of $\gamma = \sigma$, which is to be shown later. The differentials $\mathrm{d}V'$, $\mathrm{d}V''$, and $\mathrm{d}\mathcal{A}$ in Eq. (7.25) with r' and r'' fixed are given by

$$\mathrm{d}V' = \left(\frac{\partial V'}{\partial \omega}\right)_{r'} \mathrm{d}\omega \tag{7.26a}$$

$$\mathrm{d}V'' = \left(\frac{\partial V''}{\partial \omega}\right)_{r''} \mathrm{d}\omega \tag{7.26b}$$

$$\mathrm{d}\mathcal{A} = \left(\frac{\partial \mathcal{A}}{\partial \omega}\right)_{r',r''} \mathrm{d}\omega \tag{7.26c}$$

Combining Eqs. (7.21) to (7.26) and realizing that $(\partial\omega/\partial r) = 0$, because ω is independent of the location of the dividing surface

$$đW(\omega) = -\eta \mathrm{d}\omega = -\frac{1}{3}P'[r^3 - r'^3]\mathrm{d}\omega + \frac{1}{3}P''[r^3 - r''^3]\mathrm{d}\omega + \gamma r^2 \mathrm{d}\omega \tag{7.27}$$

Equation 7.27 readily provides

$$\eta = \frac{1}{3}P'[r^3 - r'^3] - \frac{1}{3}P''[r^3 - r''^3] - \gamma r^2 \tag{7.28}$$

Note that η is independent of r despite the apparent dependency from the above equation. The work $đW(\omega)$ does not depend on where we assign the location of the dividing surface. From Eq. (7.28), one can write,

$$\gamma = \frac{1}{3}r(P' - P'') + \left(\frac{1}{r^2}\right)K \tag{7.29}$$

where,

$$K = -\eta + \frac{1}{3}(P''r''^3 - P'r'^3) \tag{7.30}$$

According to Eq. (7.30), the parameter K is independent of the location of the dividing surface. Therefore, from Eq. (7.29), the interfacial tension γ depends on the location of the dividing surface. Let us multiply Eq. (7.28) by ω, and combine the result with Eqs. (7.21) to (7.23) to obtain,

$$-\eta\omega = -P'V' - P''V'' + \gamma\mathcal{A}. \tag{7.31}$$

Now let us multiply Eq. (7.29) by r^2 and differentiate with respect to r to obtain,

$$\gamma + \frac{1}{2} r \left[\frac{\partial \gamma}{\partial r}\right] = \frac{1}{2} r (P' - P'') \tag{7.32a}$$

An alternative form of the above question is

$$\boxed{(P' - P'') = \frac{2\gamma}{r} + \left[\frac{\partial \gamma}{\partial r}\right]} \tag{7.32b}$$

where $[\partial\gamma/\partial \mathrm{r}]$ represents the change in the interfacial tension with a change in the location of the dividing surface. In other words, $[\partial\gamma/\partial r]\mathrm{d}r$ is the change in the interfacial tension by the change in the location of the dividing surface. Eq. (7.32b) is one of the key expressions in interfacial thermodynamics of a curved interface.

Now we are ready to derive the general expression for $đW$ from changes in r', r'', as well as the solid angle ω, which is given by

$$đW = -\eta \mathrm{d}\omega - P' \left(\frac{\partial V'}{\partial r'}\right)_{\omega, r''} \mathrm{d}r' - P'' \left(\frac{\partial V''}{\partial r''}\right)_{\omega, r'} \mathrm{d}r'' \tag{7.33}$$

Substitution of $(\partial V'/\partial r')_{\omega,r''}$ and $(\partial V''/\partial r'')_{\omega,r'}$ from Eqs. (7.21) and (7.22) into Eq. (7.33) gives,

$$đW = -\eta \mathrm{d}\omega + P' \omega r'^2 \mathrm{d}r' - P'' \omega r''^2 \mathrm{d}r'' \tag{7.34}$$

By combining Eqs. (7.28), (7.34), and the differentials of Eqs. (7.21) to (7.23), one obtains,

$$đW = -P'\mathrm{d}V' - P''\mathrm{d}V'' + \gamma \mathrm{d}\mathcal{A} - \left[(P'' - P') + \frac{2\gamma}{r}\right] \mathcal{A}\mathrm{d}r \tag{7.35}$$

The above equation when combined with Eq. (7.32b) provides the final expression for $đW$,

$$\boxed{đW = -P'\mathrm{d}V' - P''\mathrm{d}V'' + \gamma \mathrm{d}\mathcal{A} + \left[\frac{\partial \gamma}{\partial r}\right] \mathcal{A}\mathrm{d}r} \tag{7.36}$$

Let us now derive another important relationship in interfacial thermodynamics of a curved interface. The difference in bulk-phase pressures, that is $(P' - P'')$, should be independent of the location of the dividing surface. Therefore,

$$\frac{\partial}{\partial r}(P' - P'') = 0 \tag{7.37}$$

Application of the above differentiation in Eq. (7.32b) provides,

$$\left[\frac{\partial^2 \gamma}{\partial r^2}\right] + \frac{2}{r}\left[\frac{\partial \gamma}{\partial r}\right] - \frac{2\gamma}{r^2} = 0 \tag{7.38}$$

Note that in Eq. (7.38), the differentiation is carried out with respect to r when the physical interface is fixed. The general solution to the differential equation given by the above equation is of the form

$$\gamma = k_1 r + \frac{k_2}{r^2} \tag{7.39}$$

We now define the interface of tension (i.e., the surface of tension) as the surface for which

$$\left[\frac{\partial \gamma}{\partial r}\right]_{r=r_S} = 0 \tag{7.40}$$

and denote γ by σ for this particular dividing surface. For this particular choice of the dividing surface, Eq. (7.32b) transforms to the familiar Young-Laplace equation,

$$P' - P'' = \frac{2\sigma}{r_S} \tag{7.41}$$

In Chap. 2, when the interior phase "prime" was assumed to be part of the bulk phase, we simply used r for the spherical radius for the interface.

The constants k_1 and k_2 with our choice of the dividing surface are given by,

$$k_1 = \frac{2\sigma}{3r_S} \tag{7.42}$$

$$k_2 = \frac{\sigma r_S^2}{3} \tag{7.43}$$

Substitution of the above two constants into Eq. (7.39) provides another important expression,

$$\boxed{\gamma = \sigma \left[\frac{2r}{3r_S} + \frac{r_S^2}{3r^2}\right]} \tag{7.44}$$

In the above expression, the interfacial tension is related to the surface tension. According to Eq. (7.38), the second derivative of γ with respect to r is positive at $r = r_S$, $(\partial^2\gamma/\partial r^2)\,|_{r=r_S} > 0$, which implies that the surface tension σ corresponds to the minimum of γ. We will use Eq. (7.44) to later obtain the variation of σ with r_S as the radius of curvature of the interface changes. Our derivation of Eq. (7.44) is simpler and more general than the derivation presented by Ono and Kondo (1960).

In the following, the thermodynamic functions for the systems with a curved interface will be derived. For certain thermodynamic functions such as the Gibbs free energy, there are some important differences between the derivation of G in a bulk-phase and in a heterogeneous system with a curved interface. This difference will be seen clearly in Chap. 8.

7.3 Thermodynamic Functions

The basic expression for the differential of the total internal energy in a heterogeneous system is given by

$$dU = TdS + đW + \sum_{i=1}^{n_c} \mu_i dN_i \tag{7.45}$$

where it is assumed that the process is reversible, and chemical equilibrium is assumed between the primed, double-primed phases, and the interface. One can combine Eqs. (7.36) and (7.45), to provide the expression for dU,

$$dU = TdS - P'dV' - P''dV'' + \gamma d\mathcal{A} + \left[\frac{\partial \gamma}{\partial r}\right] \mathcal{A}dr + \sum_{i=1}^{n_c} \mu_i dN_i \tag{7.46}$$

where thermal and mechanical equilibrium are assumed in addition to chemical equilibrium. From the definition $A \equiv U - TS$ one obtains,

$$dA = -SdT - P'dV' - P''dV'' + \gamma d\mathcal{A} + \left[\frac{\partial \gamma}{\partial r}\right] \mathcal{A}dr + \sum_{i=1}^{n_c} \mu_i dN_i \tag{7.47}$$

Combining Eqs. (7.32b) and (7.46), one obtains,

$$dU = TdS - P''dV - \left(\frac{2\gamma}{r} + \left[\frac{\partial \gamma}{\partial r}\right]\right) dV' + \gamma d\mathcal{A} + \left[\frac{\partial \gamma}{\partial r}\right] \mathcal{A}dr + \sum_{i=1}^{n_c} \mu_i dN_i \tag{7.48}$$

where we have substituted for P' from Eq. (7.32b) and $V = V' + V''$. Eq. (7.48) for a sphere of radius r simplifies to

$$\boxed{dU = TdS - P''dV + \sum_{i=1}^{n_c} \mu_i dN_i} \tag{7.49}$$

which is obtained using $\mathcal{A} = 4\pi r^2$ and $V' = (4\pi/3)r^3$. The above expression is useful in nucleation and in many other systems with a curved interface. Interestingly, Eq. (7.49) has the same form as that of a bulk homogenous phase when the pressure P is replaced by P''. Note that dU represents the internal energy change of the total system composed of the primed and double-primed phases as well as the interface. The expression for dA can be readily derived from $A = U - TS$ and Eq. (7.49),

$$dA = -SdT - P''dV + \sum_{i=1}^{n_c} \mu_i dN_i \tag{7.50}$$

Both Eqs. (7.49) and (7.50) provide a guideline for the derivation of the Gibbs free energy in a composite system composed of primed and double-primed phases and the interface,

$$G = U - TS + P''V = A + P''V \tag{7.51}$$

Equation (7.51) will be used in the derivation of work of cluster formation later. The differential of the Gibbs free energy of the composite system is given by

$$dG = -SdT + VdP'' + \sum_{i=1}^{n_c} \mu_i dN_i \tag{7.52}$$

Note that $G = G(T, P'', N_1, \ldots, N_{n_c})$. The Legendre transformation can be used to derive Eq. (7.52) from Eq. (7.49). From Eq. (7.52), one can readily derive (using the property of homogeneity of degree one)

$$\boxed{G = \sum_{i=1}^{n_c} \mu_i N_i} \tag{7.53}$$

where N_i is given by Eq. (7.1). Note that Eq. (7.53) is different from the expression derived in Example 2.7. Some authors include the term $\sigma\mathcal{A}$ in the additional expression for G. Alberty (1994) and Everett (1972) provide detailed discussions.

From Eqs. (7.46) and (7.47), one can obtain the following two expressions,

$$U = TS - P'V' - P''V'' + \gamma\mathcal{A} + \sum_{i=1}^{n_c} \mu_i N_i \tag{7.54}$$

$$A = -P'V' - P''V'' + \gamma\mathcal{A} + \sum_{i=1}^{n_c} \mu_i N_i \tag{7.55}$$

Using Eqs. (7.1), (7.2), the expressions $A' = -P'V' + \sum_{i=1}^{n_c} \mu_i N_i'$ and $A'' = -P''V'' + \sum_{i=1}^{n_c} \mu_i N_i''$, one obtains

$$\boxed{A^s = \gamma\mathcal{A} + \sum_{i=1}^{n_c} N_i^s \mu_i} \tag{7.56}$$

which is similar to Eq. (7.13) of the planar interface with the difference that γ is a function of the location of the dividing surface. When we assign $\sum_{i=1}^{n_c} N_i^s \mu_i = 0$ to the surface of separation

$$\gamma = \frac{A^s}{\mathcal{A}} \tag{7.57}$$

which implies that the interfacial tension becomes equal to the superficial density of the Helmholtz free energy, often referred to as the surface energy density.

Next, we derive the expression for the Gibbs adsorption isotherm. Subtracting Eqs. (7.16) and (7.17) from Eq. (7.47), and using Eqs. (7.1) and (7.2) provides,

$$dA^s = -S^s dT + \gamma d\mathcal{A} + \sum_{i=1}^{n_c} \mu_i dN_i^s + \left[\frac{\partial \gamma}{\partial r}\right] \mathcal{A} dr \tag{7.58}$$

A comparison of the above equation with the differential of Eq. (7.56) provides

$$\mathcal{A}d\gamma + S^s dT + \sum_{i=1}^{n_c} N_i^s d\mu_i = \left[\frac{\partial \gamma}{\partial r}\right] \mathcal{A}dr \tag{7.59}$$

Equation 7.59 can be viewed as the Gibbs-Duhem relation for a spherical interface. In terms of superficial densities, one rewrites Eq. (7.59) as

$$d\gamma = -s^s dT - \sum_{i=1}^{n_c} \Gamma_i d\mu_i + \left[\frac{\partial \gamma}{\partial r}\right] dr \tag{7.60}$$

The equation above differs from Eq. (7.19) only in the additional term from the location effect of the dividing surface. Under isothermal conditions, the first term on the right of Eq. (7.60) drops and we have the Gibbs adsorption equation,

$$\boxed{d\gamma = -\sum_{i=1}^{n_c} \Gamma_i d\mu_i + \left[\frac{\partial \gamma}{\partial r}\right] dr} \tag{7.61}$$

7.4 Effect of Curvature on Interfacial Tension

When the size of a bubble or a droplet decreases to 100 nm or less, the interfacial tension becomes a function of size, or curvature (Santiso and Firoozabadi, 2006). In this section, we will derive the expression that can be used to account for the effect of curvature on the interfacial tension at equilibrium conditions (i.e., thermal, mechanical, and chemical equilibrium). The curvature effect is important in the segregation of small clusters in a bulk liquid (Larson and Garside, 1986a and b) and in micellar formation, as well as in many other interfacial processes.

At chemical equilibrium, the chemical potential of species i is the same in the primed and double-primed phases, and in the interface. Let us write the differential of the chemical potentials in terms of the variables of the double-primed phase,

$$d\mu_i = \left(\frac{\partial \mu_i}{\partial T}\right)_{P'',\underline{x}''} dT + \left(\frac{\partial \mu_i}{\partial P''}\right)_{T,\underline{x}''} dP'' + \sum_{j=1}^{n-1} \left(\frac{\partial \mu_i}{\partial x_j''}\right)_{T,P'',x_j''} dx_j'' \quad i = 1, \ldots, n_c \tag{7.62}$$

Note that we have used the mol fractions rather than the mol numbers as independent variables. After substitution for $(\partial\mu_i/\partial T)_{P'',\underline{x}''} = -\bar{S}_i''$ and $(\partial\mu_i/\partial P'')_{T,\underline{x}''} = \bar{V}_i''$ in Eq. (7.62) and combining the result with Eq. (7.60) one obtains

$$d\gamma = \left(-s^s + \sum_{i=1}^{n_c} \Gamma_i \bar{S}_i''\right) dT - \left(\sum_{i=1}^{n_c} \Gamma_i \bar{V}_i''\right) dP''$$
$$- \sum_{j=1}^{n_c-1} \left[\sum_{i=1}^{n_c} \Gamma_i \left(\frac{\partial \mu_i}{\partial x_j''}\right)_{T,P'',x_j''}\right] dx_j'' + \left[\frac{\partial \gamma}{\partial r}\right] dr \tag{7.63}$$

Let us consider the dividing surface defined by

$$\sum_{i=1}^{n_c} \Gamma_i \bar{V}_i'' = 0 \tag{7.64}$$

For this dividing surface, the interfacial tension and the radius are denoted by γ_V and r_V, respectively, and Eq. (7.63) takes the form

$$d\gamma_V = \left(-s^s + \sum_{i=1}^{n_c} \Gamma_i \bar{S}_i'' \right) dT - \sum_{j=1}^{n_c-1} \left[\sum_{i=1}^{n_c} \Gamma_i \left(\frac{\partial \mu_i}{\partial x_j''} \right)_{T,P'',x_j''} \right] dx_j'' + \left[\frac{\partial \gamma}{\partial r} \right]_{r=r_V} dr_V \tag{7.65}$$

From Eq. (7.65), one obtains

$$\left[\frac{\partial \gamma_V}{\partial r_V} \right]_{T,\underline{x}''} = \left[\frac{\partial \gamma}{\partial r} \right]_{r=r_V} \tag{7.66}$$

The right side can be evaluated from Eq. (7.44):

$$\left[\frac{\partial \gamma}{\partial r} \right]_{r=r_V} = \sigma \left[\frac{2}{3r_S} - \frac{2r_S^2}{3r_V^3} \right] \tag{7.67}$$

One can also write from Eq. (7.44) (at $r = r_V$)

$$\gamma_V = \sigma \left[\frac{2r_V}{3r_S} + \frac{r_S^2}{3r_V^2} \right] \tag{7.68}$$

The derivative of γ_V with respect to r_V (at constant $T, \underline{x}''$) using the above relation is given by

$$\left[\frac{\partial \gamma_V}{\partial r_V} \right]_{T,\underline{x}''} = \sigma \left[\frac{2}{3r_S} - \frac{2r_S^2}{3r_V^3} \right] + \left(\frac{2r_V}{3r_S} + \frac{r_S^2}{3r_V^2} \right) \left(\frac{\partial \sigma}{\partial r_V} \right)_{T,\underline{x}''} + \left[\frac{-2r_V}{3r_S^2} + \frac{2r_S}{3r_V^2} \right] \sigma \left(\frac{\partial r_S}{\partial r_V} \right) \tag{7.69}$$

Combining Eqs. (7.66), (7.67), and (7.69) provides

$$\left(\frac{\partial \ln \sigma}{\partial \ln r_S} \right)_{T,\underline{x}''} = \frac{2(r_V^3 - r_S^3)}{2r_V^3 + r_S^3} \tag{7.70}$$

Let us define

$$\delta \equiv r_V - r_S \tag{7.71}$$

Then Eq. (7.70) can be written as,

$$\left(\frac{\partial \ln \sigma}{\partial \ln r_S} \right)_{T,\underline{x}''} = 2 \left(\frac{\delta}{r_S} \right) \frac{1 + \left(\frac{\delta}{r_S} \right) + \left(\frac{1}{3} \right) \left(\frac{\delta}{r_S} \right)^2}{1 + \left(\frac{2\delta}{r_S} \right) \left[1 + \left(\frac{\delta}{r_S} \right) + \left(\frac{1}{3} \right) \left(\frac{\delta}{r_S} \right)^2 \right]} \tag{7.72}$$

Integration of Eq. (7.72) from $r_S=\infty$ (flat interface) with $\sigma=\sigma_\infty$ to any r_S provides,

$$\ln\left(\frac{\sigma}{\sigma_\infty}\right)_{T,\underline{x}''}=2\int_\infty^{r_S}\left(\frac{\delta}{r_S}\right)\frac{1+\left(\frac{\delta}{r_S}\right)+\left(\frac{1}{3}\right)\left(\frac{\delta}{r_S}\right)^2}{1+\left(\frac{2\delta}{r_S}\right)\left[1+\left(\frac{\delta}{r_S}\right)+\left(\frac{1}{3}\right)\left(\frac{\delta}{r_S}\right)^2\right]}d\ln r_S \tag{7.73}$$

For the special case that $\delta \ll r_S$ and, $\delta\equiv\delta_\infty=$ constant for the value of a plane interface, Eq. (7.73) provides an analytical solution

$$\boxed{\sigma=\sigma_\infty\exp\left[-\frac{2\delta}{r_S}\right]} \tag{7.74}$$

Various authors, including Tolman (1949), Kirkwood and Buff (1949), Ono and Kondo (1960), and Laaksonen and McGraw (1996), and McGraw and Laaksonen (1997) have studied the effect of interface curvature on the interfacial tension. The derivation presented above is, perhaps, simpler than some of the derivations in the literature.

7.5 Work of Cluster Formation: Single-Component Systems

In this section, we will derive the expression for the work of formation of spherical and cap-shaped clusters in a single component system which is considerably simpler than the work of cluster formation in multicomponent systems. In a multicomponent mixture, cross chemical potentials complicate the study of a heterogeneous system with an interface. In our derivations we will follow the approach based on the surface energy concept derived earlier. An alternative method is the use of the concept of availability as will be discussed in Chap. 8. Some of our derivations are similar to those of Kashchiev (2000).

The work of cluster formation, W, is a major parameter in the nucleation rate expression $J_s=a\exp[-W^*/kT]$ where W^* is the work of the cluster formation at the critical size (the size at which the change in Gibbs free energy is a maximum), J_s is the stationary rate of nucleation, a is the kinetic factor, T is the absolute temperature, and k is the Boltzmann constant [Kashchiev (2000)]. The work of cluster formation is also directly useful in understanding of a large number of phenomena including the delay period, the memory effects (lack of reproducibility), and the most probable location for the formation of a new phase. Work of cluster formation is also a key parameter in the understanding of wettability and its alteration.

In the following, we will first derive the work of cluster formation for the spherical shape cluster, and then proceed to the cap-shaped clusters. The work of cluster formation in lens-shaped clusters is discussed in the Problems section (see Prob. 7.10).

7.5.1 Spherical Clusters

Consider Fig. 7.4 which depicts the state of N molecules in the initial state (state I) of uniform density at temperature T and pressure P (Fig. 7.4a). The final state (state II) corresponds to the homogeneous formation of a spherical cluster of n molecules at

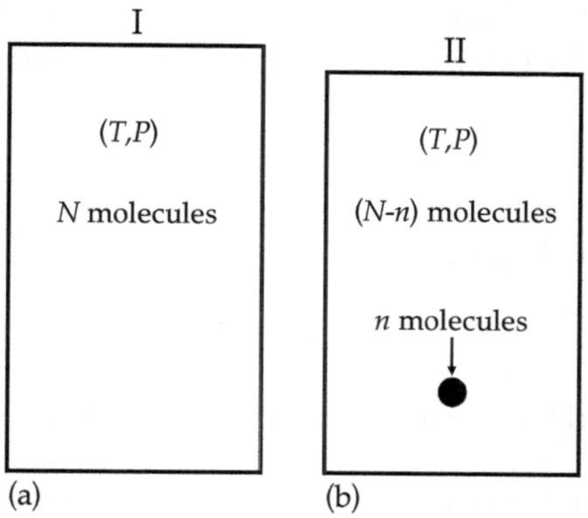

FIGURE 7.4 The states before and after formation of one spherical cluster at constant T and P. Note that the pressure in the cluster (P_n) is generally different from P.

temperature T and pressure P in the bulk continuous phase. Therefore, from the beginning, we are assuming thermal equilibrium.

Let us write the expression for the Gibbs free energy of the total system at state I,

$$G^{\mathrm{I}} = N\mu^{\mathrm{I}} \tag{7.75}$$

In the above equation, μ is the chemical potential per molecule. The Gibbs free energy of state II is given by

$$G^{\mathrm{II}} = PV + A^{\mathrm{II}} \tag{7.76}$$

Note that in Eq. (7.76), $V = V_{N-n} + V_n$; [V_{N-n} and V_n are defined based on the equimolecular dividing surface (EDS)] and A^{II} is the total Helmholtz free energy at state II. We have shown earlier [see Eq. (7.51)] that in Eq. (7.76), P is the pressure of the continuous bulk phase (note change of notation). The Helmholtz free energy at state II is given by

$$A^{\mathrm{II}} = \left[(N-n)\mu^{\mathrm{I}} - PV_{N-n}\right] + \left[n\mu^{\mathrm{II}}(P_n) - P_nV_n\right] + \phi(V_n) \tag{7.77}$$

In Eq. (7.77), $\phi(V_n)$ is the surface energy of the spherical cluster and $\mu^{\mathrm{II}}(P_n)$ is the chemical potential of the cluster per molecule. Equations (7.76) and (7.77) are combined to provide,

$$G^{\mathrm{II}} = -V_n(P_n - P) + n\left[\mu^{\mathrm{II}}(P_n) - \mu^{\mathrm{I}}\right] + N\mu^{\mathrm{I}} + \phi(V_n) \tag{7.78}$$

The difference between G^{II} and G^{I} can be obtained from Eqs. (7.75) and (7.78),

$$\Delta G = G^{\mathrm{II}} - G^{\mathrm{I}} = -(P_n - P)V_n + n\left[\mu^{\mathrm{II}}(P_n) - \mu^{\mathrm{I}}\right] + \phi(V_n) \tag{7.79}$$

The work of cluster formation $W = \Delta G$ is, therefore,

$$W = -(P_n - P)V_n + \left[\mu^{\mathrm{II}}(P_n) - \mu^{\mathrm{I}}\right]n + \phi(V_n) \tag{7.80}$$

Note that three terms contribute to the work of cluster formation. The first term is from expansion, the second term is from the nonequilibrium effect, and the third term is from the interfacial effect.

Next, we will assume that there is mechanical equilibrium. The condition for mechanical equilibrium is given by $(\partial G^{\mathrm{II}}/\partial V_n)_{n,P_n}=0$. Therefore, from Eq. (7.78),

$$P_n = P + \frac{\mathrm{d}\phi}{\mathrm{d}V_n} \tag{7.81}$$

For an equimolecular dividing surface (EDS),

$$\phi(V_n) = \sigma_n \mathcal{A}_n \tag{7.82}$$

where σ_n is the interfacial tension and $\mathcal{A}_n$ is the surface area of the cluster. For spherical clusters, $\mathcal{A}_n=(36\pi)^{1/3}V_n^{2/3}$.

Let us assume that the interfacial tension is independent of the size of the clusters, then $\sigma_n=\sigma=constant$, and

$$\phi(V_n) = (36\pi)^{\frac{1}{3}}\sigma V_n^{\frac{2}{3}} \tag{7.83}$$

From Eqs. (7.81) and (7.83),

$$P_n = P + \left[\frac{2}{3}(36\pi)^{\frac{1}{3}}\right]\frac{\sigma}{V_n^{\frac{1}{3}}} \tag{7.84}$$

For a sphere, $V_n=(4/3)\pi r^3$ (r is the cluster radius); an alternative form of Eq. (7.84) is

$$P_n = P + \frac{2\sigma}{r} \tag{7.85}$$

Equation (7.85) is the Young-Laplace equation.

In nucleation theory, instead of the true driving force $[\mu^{\mathrm{I}}(P)-\mu^{\mathrm{II}}(P_n)]$, the driving force in the form of $[\mu^{\mathrm{I}}(P)-\mu^{\mathrm{II}}(P)]$ is used. This choice does not affect W. We can write Eq. (7.80) in the following form,

$$W = -(P_n - P)V_n + [\mu^{\mathrm{II}}(P_n) - \mu^{\mathrm{II}}(P)]n + [\mu^{\mathrm{II}}(P) - \mu^{\mathrm{I}}(P)]n + \phi(V_n) \tag{7.86}$$

Note that in the third term on the right of Eq. (7.86) although the pressure and temperature of states I and II are the same, the chemical potentials are generally different because the fluid (and solid) states are not the same. The difference between $\mu^{\mathrm{II}}(P)$ and $\mu^{\mathrm{I}}(P)$ in Eq. (7.86) is shown by $\Delta\mu=+\,[\mu^{\mathrm{I}}(P)-\mu^{\mathrm{II}}(P)]$ and is defined in the nucleation literature as the driving force. With this definition, Eq. (7.86) can be rewritten as

$$\boxed{W = -(P_n - P)V_n + \left[\mu^{\mathrm{II}}(P_n) - \mu^{\mathrm{II}}(P)\right]n - n\Delta\mu + \phi(V_n)} \tag{7.87}$$

The second term on the right side of Eq. (7.87) can be evaluated using $[d\mu=(V_n/n)\mathrm{d}P]_T$,

$$\mu^{\mathrm{II}}(P_n) = \mu^{\mathrm{II}}(P) + \left(\frac{1}{n}\right)\int_P^{P_n} V_n(P)\mathrm{d}P \tag{7.88}$$

For an incompressible liquid or solid cluster, Eq. (7.88) becomes

$$\mu^{\text{II}}(P_n) = \mu^{\text{II}}(P) + v_o(P_n - P) \tag{7.89}$$

where v_o is the molecular volume.

For an ideal gas, $PV_n = nkT$. Substitution of $V_n = nkT/P$ in Eq. (7.88) provides

$$\mu^{\text{II}}(P_n) = \mu^{\text{II}}(P) + kT\ln\frac{P_n}{P} \tag{7.90}$$

We can now write the expression for the work of cluster formation.

7.5.2 Incompressible Clusters

For an incompressible cluster, the first and second terms on the right side of Eq. (7.87) cancel out [using Eq. (7.89)] and, therefore,

$$W = -n\Delta\mu + \phi(V_n) \tag{7.91}$$

Equation (7.91) can be expressed in various forms in terms of the cluster volume, V_n, number of molecules in the cluster, n, and the radius of the cluster, r. In terms of cluster volume, the work of cluster formation is given by

$$W(V_n) = -\left(\frac{V_n}{v_o}\right)\Delta\mu + (36\pi)^{\frac{1}{3}}\sigma V_n^{\frac{2}{3}} \tag{7.92}$$

The work of cluster formation in terms of n and r are

$$W(n) = -n\Delta\mu + \left(36\pi v_o^2\right)^{\frac{1}{3}}\sigma n^{\frac{2}{3}} \tag{7.93}$$

and

$$W(r) = -(4\pi/3v_o)r^3\Delta\mu + 4\pi r^2\sigma \tag{7.94}$$

respectively.

7.5.3 Gas Clusters

When a gas cluster follows an ideal gas behavior, the second expression in the right side of Eq. (7.87) can be simplified using Eq. (7.90). However, before proceeding to this step, let us simplify Eq. (7.87) by substituting Eqs. (7.84) and (7.83) for its first and fourth terms,

$$W = \left[\mu^{\text{II}}(P_n) - \mu^{\text{II}}(P) - \Delta\mu\right]n + \frac{1}{3}(36\pi)^{\frac{1}{3}}\sigma V_n^{\frac{2}{3}} \tag{7.95}$$

The second term on the right side of Eq. (7.95) is equal to $(1/3)\sigma\mathcal{A}$ where $\mathcal{A}$ is the surface area. The terms in the bracket are simply $[\mu^{\text{II}}(P_n) - \mu^{\text{I}}(P)]$, which becomes zero when $\mu^{\text{II}}(P_n) = \mu^{\text{I}}(P)$; this is the condition of equilibrium, but it is an unstable equilibrium due to the fact that it corresponds to a maximum in ΔG. In Example 7.4, there will be further discussion of ΔG and its maximum.

With the substitution of Eq. (7.90) for $\mu^{\mathrm{II}}(P_n) - \mu^{\mathrm{II}}(P)$ and $n = P_n V_n / kT$ and P_n from Eq. (7.84), Eq. (7.95) transforms into

$$W = -\left\{\left(\frac{\Delta\mu}{kT}\right) - \ln\left[1 + \frac{2}{3}(36\pi)^{\frac{1}{3}}\frac{\sigma}{\left(PV_n^{\frac{1}{3}}\right)}\right]\right\}$$

$$\times\left[P + \frac{2}{3}(36\pi)^{\frac{1}{3}}\frac{\sigma}{V_n^{\frac{1}{3}}}\right]V_n + \frac{1}{3}(36\pi)^{\frac{1}{3}}\sigma V_n^{\frac{2}{3}} \tag{7.96}$$

Equation (7.96) can also be expressed as a function of n and r,

$$W(n) = -n\left[\Delta\mu - kT\ln\left(\frac{P_n}{P}\right)\right] + \frac{1}{3}(36\pi)^{\frac{1}{3}}\sigma\left(\frac{kT}{P_n}\right)^{\frac{2}{3}} n^{\frac{2}{3}} \tag{7.97}$$

and

$$W(r) = -\left(\frac{4\pi}{3}\right)\left(\frac{\Delta\mu}{kT}\right) - \ln\left[1 + \frac{2\sigma}{Pr}\right]\left(P + \frac{2\sigma}{r}\right)r^3 + \frac{4}{3}\pi r^2\sigma \tag{7.98}$$

respectively.

7.6 Cap-Shaped Clusters

Consider Fig. 7.5; in state I, there is no cluster formation, in state II, a cap-shaped cluster is formed on the substrate. We are interested in deriving the expression for the work of cluster formation.

The Gibbs free energy of the system at state I is given by

$$G^{\mathrm{I}} = N\mu^{\mathrm{I}} + \phi_{so} + G_s \tag{7.99}$$

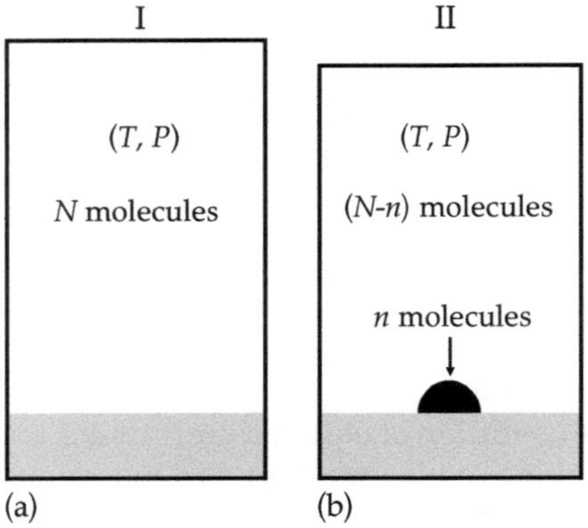

FIGURE 7.5 The states before and after formation of one cap-shaped cluster at constant T and P.

where ϕ_{so} is the surface energy of the substrate-continuous phase interface and G_s is the Gibbs free energy of the bulk-phase substrate. $\phi_{so} = \sigma_s \mathcal{A}_s$ where σ_s is the specific surface energy of the continuous substrate interface, and $\mathcal{A}_s$ is the surface area of the substrate.

The Gibbs free energy of the system at state II is given by

$$G^{\text{II}} = A^{\text{II}} + PV \tag{7.100}$$

where the Helmholtz free energy is obtained from,

$$A^{\text{II}} = \left[(N-n)\mu^{\text{I}} - PV_{N-n}\right] + \left[n\mu(P_n) - P_n V_n\right] + \phi + \phi_s + G_s \tag{7.101}$$

In Eq. (7.101), ϕ is the surface energy of the interface between: (1) the continuous fluid phase and the cluster, and (2) the cluster and the substrate. In the same equation, ϕ_s is the surface free energy of the interface between the continuous fluid phase and the substrate minus the portion of the base of the cluster. Substituting Eq. (7.101) into Eq. (7.100) and using $PV = P(V_{N-n} + V_n)$,

$$G^{\text{II}} = \left[\mu^{\text{II}}(P_n) - \mu^{\text{I}}(P)\right] n - \left[P_n - P\right] V_n + N\mu^{\text{I}} + \phi + \phi_s + G_s \tag{7.102}$$

The work of cluster formation is obtained by subtracting Eq. (7.102) from Eq. (7.99),

$$W = G^{\text{II}} - G^{\text{I}} = \Delta G = \left[\mu^{\text{II}}(P_n) - \mu^{\text{I}}(P)\right] n - \left[P_n - P\right] V_n + (\phi + \phi_s - \phi_{so}) \tag{7.103}$$

An alternative form of Eq. (1.103) is obtained by defining $\Delta\mu = [\mu^{\text{I}}(P) - \mu^{\text{II}}(P)]$

$$W = \left[\mu^{\text{II}}(P_n) - \mu^{\text{II}}(P)\right] n - n\Delta\mu - \left[P_n - P\right] V_n + (\phi + \phi_s - \phi_{so}) \tag{7.104}$$

The condition for mechanical equilibrium of the system II is given by $\partial G^{\text{II}}/\partial V_n = 0$ where the number of molecules in the cluster is held constant. As a result, from Eq. (7.102) one obtains,

$$P_n = P + \frac{\mathrm{d}(\phi + \phi_s)}{\mathrm{d}V_n} \tag{7.105}$$

In order to proceed further, we will obtain explicit expressions for surface free energies ϕ, ϕ_s, and the Young equation.

7.7 Derivation of the Young Equation

Young's equation is the condition for equilibrium shape of a liquid drop on a substrate. Such a shape is often called a cap. Let us consider a cap-shaped cluster sketched in Fig. 7.6. The specific surface energies are also shown in the same figure. The surface area of the cap dome is given by $2\pi r^2(1 - \cos\theta)$. The surface area of the cluster base is given by $\pi r^2 \sin^2\theta$.

The surface energy of the cluster (base and dome) is given by

$$\phi(r) = 2\pi r^2(1 - \cos\theta)\sigma + (\pi r^2 \sin^2\theta)\sigma_i \tag{7.106}$$

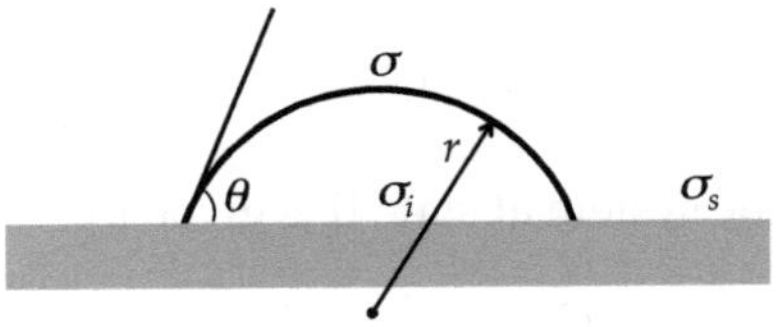

FIGURE 7.6 Cap-shaped liquid cluster on a flat substrate.

The surface energy of the substrate (minus the portion from the base of the cluster) is given by

$$\phi_s(\theta) = [\mathcal{A}_s - \pi r^2 \sin^2\theta]\sigma_s \tag{7.107}$$

where $\mathcal{A}_s$ is the total surface area of the substrate. The total surface free energy of the system in Fig. 7.6 is, therefore, given by

$$\phi_t = \phi + \phi_s = 2\pi r^2 \left[(1-\cos\theta)\sigma + \frac{1}{2}(1-\cos^2\theta)(\sigma_i - \sigma_s)\right] + \sigma_s \mathcal{A}_s \tag{7.108}$$

The volume of the cluster is given by

$$V_n = [\psi(\theta)]\frac{4}{3}\pi r^3 \tag{7.109}$$

where $\psi(\theta)$ is the shape factor,

$$\psi(\theta) = \left(\frac{1}{4}\right)(2+\cos\theta)(1-\cos\theta)^2 = \left(\frac{1}{4}\right)(2 - 3\cos\theta + \cos^3\theta) \tag{7.110}$$

In a cluster of n molecules, for a given volume of V_n, the shape of the cluster is determined by the minimum of ϕ_t with respect to variations of r and $x = \cos\theta$, then

$$\phi_t = 2\pi r^2 \left[(1-x)\sigma + \frac{1}{2}(1-x^2)(\sigma_i - \sigma_s)\right] + \sigma_s \mathcal{A}_s \tag{7.111}$$

From $d\phi_t = 0$

$$2\left[(1-x)\sigma + \frac{1}{2}(1-x^2)(\sigma_i - \sigma_s)\right]\frac{dr}{r} - [\sigma + x(\sigma_i - \sigma_s)]dx = 0 \tag{7.112}$$

The volume of the cluster is given by

$$V_n = \frac{1}{3}\pi r^3(2 - 3x + x^3) \tag{7.113}$$

From $dV_n = 0$ (i.e., constant volume)

$$(2 - 3x + x^3)\frac{dr}{r} - (1-x^2)dx = 0 \tag{7.114}$$

Combining Eqs. (7.112) and (7.114), one obtains the Young equation,

$$\boxed{\sigma\cos\theta = \sigma_s - \sigma_i} \tag{7.115}$$

The above equation is also valid for a crystal cap. A similar approach can be used to derive Young's equation for a cluster on a conical substrate.

Combining Eqs. (7.104), (7.108), (7.110), and (7.115) one obtains,

$$\boxed{W = \left[\mu^{\mathrm{II}}(P_n) - \mu^{\mathrm{II}}(P)\right]n - n\Delta\mu - \left[P_n - P\right]V_n + 4\psi(\theta)\pi r^2\sigma} \tag{7.116}$$

Note that Eq. (7.116) is similar to Eq. (7.87).

7.8 Driving Force ($\Delta\mu$)

We have defined the thermodynamic driving force for the first-order phase transition as,

$$\Delta\mu(T, P) = \mu^{\mathrm{I}}(T, P) - \mu^{\mathrm{II}}(T, P) \tag{7.117}$$

In the following, basic expressions for $\Delta\mu$ are derived.

7.8.1 Condensation from Vapor at Constant *T*

Condensation from vapor at temperature T occurs when $P > P_e$ where P_e is the equilibrium pressure at T. From $\mathrm{d}\mu = v\mathrm{d}P$ (state I is vapor)

$$\mu^{\mathrm{I}}(T, P) - \mu^{\mathrm{I}}(T, P_e) = -\int_P^{Pe} v\mathrm{d}p \tag{7.118}$$

For an ideal gas $v = kT/P$,

$$\mu^{\mathrm{I}}(T, P) - \mu^{\mathrm{I}}(T, P_e) = -\int_P^{Pe} \frac{kT}{P}\mathrm{d}P = kT\ln\frac{P}{P_e} \tag{7.119}$$

Therefore

$$\mu^{\mathrm{I}}(T, P) = \mu^{\mathrm{I}}(T, P_e) + kT\ln\frac{P}{P_e} \tag{7.120}$$

Next, the expression for $\mu^{\mathrm{II}}(T, P)-\mu^{\mathrm{II}}(T, P_e)$, from Eq. (7.89) is used with the assumption of liquid incompressibility. It is rewritten in the following form

$$\mu^{\mathrm{II}}(T, P) = \mu^{II}(T, P_e) - v_o(P_e - P) \tag{7.121}$$

At equilibrium (T and P_e),

$$\mu^{\mathrm{I}}(T, P_e) = \mu^{\mathrm{II}}(T, P_e) \tag{7.122}$$

Therefore, combining Eqs. (7.117) and (7.120) to (7.122)

$$\Delta\mu(T, P) = kT\ln\frac{P}{P_e} + v_o(P_e - P) \tag{7.123}$$

7.8.2 Evaporation from Liquid at Constant T

Evaporation from a liquid can occur at constant T when $P < P_e$. The expression for chemical potential of state I (i.e., liquid state) is

$$\mu^{\mathrm{I}}(T, P) = \mu^{\mathrm{I}}(T, P_e) + v_o(P - P_e) \tag{7.124}$$

where it is assumed that the liquid is incompressible. The chemical potential in the vapor (state II) is given by

$$\mu^{\mathrm{II}}(T, P) = \mu^{\mathrm{II}}(T, P_e) - kT \ln \frac{P_e}{P} \tag{7.125}$$

Combining Eqs. (7.117), (7.124), and (7.125)

$$\Delta\mu = kT \ln \frac{P_e}{P} + v_o(P - P_e) \tag{7.126}$$

Next, we will derive the expressions for thermodynamic functions when there is no chemical equilibrium between the bulk phases and the interface. Then the results are used to derive the expression for the work of cluster formation in multicomponent systems. As will be pointed out briefly in the *Examples and Theory Extension* section, there have been discussions in the literature in relation to the expression for the work of cluster formation in multicomponent mixtures.

7.9 Thermodynamic Functions and Gibbs Adsorption Equation without Chemical Equilibrium

Our derivations are based on the material from Chapters II to V of Defay and Prigogine (1966) and our generalization that the dividing surface position should be an independent variable of superficial thermodynamic functions and interfacial tension. Let us consider a multicomponent mixture with two bulk phases, prime and double prime, and the interface. We assume thermal and mechanical equilibrium in the system. We define

$$c_i' = \frac{N_i'}{V'} \tag{7.127}$$

$$c_i'' = \frac{N_i''}{V''} \tag{7.128}$$

where c_i' and c_i'' are the molar concentrations of component i in phases prime and double prime. Other symbols have the usual definition.

We have already defined the adsorption of component "i" in the interface by,

$$\Gamma_i = \frac{N_i^s}{\mathcal{A}} \tag{7.129}$$

Let us also define specific functions for the two bulk phases

$$u' = \frac{U'}{V'} \tag{7.130}$$

$$u'' = \frac{U''}{V''} \tag{7.131}$$

$$s' = \frac{S'}{V'} \tag{7.132}$$

$$s'' = \frac{S''}{V''}, \tag{7.133}$$

The superficial internal energy, entropy, and Helmholtz free energy of the interface per unit area are defined by,

$$u^s = \frac{U^s}{\mathcal{A}} \tag{7.134}$$

$$s^s = \frac{S^s}{\mathcal{A}} \tag{7.135}$$

$$a^s = \frac{A^s}{\mathcal{A}} \tag{7.136}$$

The Helmhotz free energy of the system composed of two bulk phases and a curved interface is defined by,

$$A = A(T, V', V'', \mathcal{A}; N_1', \ldots, N_{nc}'; N_1'', \ldots N_{nc}''; N_1^s, \ldots, N_{nc}^s; r) \tag{7.137}$$

Note that in the above expression we have assumed thermal and mechanical equilibrium in the system. Now we are ready to define the fundamental equation for A in an open system.

7.10 Helmholtz Free Energy of an Open System

The differential of A from Eq. (7.137) can be written as,

$$\begin{aligned} dA = {} & \left(\frac{\partial A}{\partial T}\right) dT + \left(\frac{\partial A}{\partial V'}\right) dV' + \left(\frac{\partial A}{\partial V''}\right) dV'' + \left(\frac{\partial A}{\partial \mathcal{A}}\right) d\mathcal{A} \\ & + \sum_{i=1}^{n_c} \left(\frac{\partial A}{\partial N_i'}\right) dN_i' + \sum_{i=1}^{n_c} \left(\frac{\partial A}{\partial N_i''}\right) dN_i'' + \sum_{i=1}^{n_c} \left(\frac{\partial A}{\partial N_i^s}\right) dN_i^s + \left(\frac{\partial A}{\partial r}\right) dr \end{aligned} \tag{7.138}$$

In Eq. (7.138), for the sake of brevity, the quantities that are held constant are not shown.

In a closed system when the mass of all the components remains unchanged from a transformation in T, V', V'', $\mathcal{A}$, or r one can write,

$$\frac{\partial A}{\partial T} = -S \tag{7.139}$$

$$\frac{\partial A}{\partial V'} = -P' \tag{7.140}$$

$$\frac{\partial A}{\partial V''} = -P'' \tag{7.141}$$

$$\frac{\partial A}{\partial \mathcal{A}} = \gamma \tag{7.142}$$

$$\frac{\partial A}{\partial r} = \frac{\partial A}{\partial \gamma}\left[\frac{\partial \gamma}{\partial r}\right] = \mathcal{A}\left[\frac{\partial \gamma}{\partial r}\right] \tag{7.143}$$

Note that the above are the first five derivatives in Eq. (7.47). Let us now define chemical potentials based on the total Helmholtz free energy A,

$$\tilde{\mu}_i' = \left(\frac{\partial A}{\partial N_i'}\right) \tag{7.144}$$

$$\tilde{\mu}_i'' = \left(\frac{\partial A}{\partial N_i''}\right) \tag{7.145}$$

$$\tilde{\mu}_i^s = \left(\frac{\partial A}{\partial N_i^s}\right) \tag{7.146}$$

The first two chemical potentials above are, in general, not equal to the bulk phase chemical potentials of the form $\mu_i' = (\partial A'/\partial N_i')$ and $\mu_i'' = (\partial A''/\partial N_i'')$. We will later establish the relation between $\tilde{\mu}_i'$, μ_i', $\tilde{\mu}_i''$ and μ_i''.

Combining Eqs. (7.138) to (7.146), we obtain the differential of Helmholtz free energy in an open system,

$$dA = -SdT - P'dV' - P''dV'' + \gamma d\mathcal{A} + \sum_{i=1}^{n_c} \tilde{\mu}_i' dN_i' + \sum_{i=1}^{n_c} \tilde{\mu}_i'' dN_i''$$
$$+ \sum_{i=1}^{n_c} \tilde{\mu}_i^s dN_i^s + \left[\frac{\partial \gamma}{\partial r}\right]\mathcal{A}dr \tag{7.147}$$

7.11 Relation between $\tilde{\mu}_i'$, $\tilde{\mu}_i''$, $\tilde{\mu}_i^s$ and μ_i', μ_i'', μ_i^s

First, we write the dependency of bulk phase Helmholtz free energies A', A'', and the surface free energy A^s on various independent variables.

$$A' = A'(T, V', N_1', \ldots, N_{nc}') \tag{7.148}$$

$$A'' = A''(T, V'', N_1'', \ldots, N_{nc}'') \tag{7.149}$$

$$A^s = A^s(T, V', V'', \mathcal{A}; N_1', \ldots, N_{nc}'; N_1'', \ldots, N_{nc}''; N_1^s, \ldots, N_{nc}^s; r) \tag{7.150}$$

Differentiating $A = A' + A'' + A^s$ with respect to N_i', N_i'', and N_i^s one obtains,

$$\tilde{\mu}_i' = \mu_i' + \frac{\partial A^s}{\partial N_i'} \tag{7.151}$$

$$\tilde{\mu}_i'' = \mu_i'' + \frac{\partial A^s}{\partial N_i''} \tag{7.152}$$

$$\tilde{\mu}_i^s = \mu_i^s = \frac{\partial A^s}{\partial N_i^s} \tag{7.153}$$

μ_i^s is analogous to μ_i' and μ_i''; it is the chemical potential of component i adsorbed at the interface.

To proceed further, let us write the expression for $a^s = A^s/\mathcal{A}$ in terms of different independent variables,

$$\tilde{a}^s = \tilde{a}^s(T, V', V'', \mathcal{A}; N_1', \ldots, N_{n_c}'; N_1'', \ldots, N_{n_c}''; N_1^s, \ldots, N_{n_c}^s; r) \tag{7.154}$$

$$a^s = a^s(T; c_1', \ldots, c_{n_c}'; c_1'', \ldots, c_{n_c}''; \Gamma_1, \ldots, \Gamma_{n_c}; r) \tag{7.155}$$

Note that $\tilde{a}^s$ and a^s are the same, the symbol "~" is used to denote the functional dependency. The derivatives of $\tilde{a}^s$ with respect to T, V', V'', $\mathcal{A}$; N_i', N_i'', N_i^s, and r are related to the derivatives of a^s with respect to T, c_i', c_i'', Γ_i, and r.

$$\frac{\partial \tilde{a}^s}{\partial T} = \frac{\partial a^s}{\partial T} \tag{7.156}$$

$$\frac{\partial \tilde{a}^s}{\partial V'} = \sum_{i=1}^{n_c} \frac{\partial a^s}{\partial c_i'} \frac{\partial c_i'}{\partial V'} = -\frac{1}{V'} \sum_{i=1}^{n_c} c_i' \frac{\partial a^s}{\partial c_i'} \tag{7.157}$$

$$\frac{\partial \tilde{a}^s}{\partial V''} = \sum_{i=1}^{n_c} \frac{\partial a^s}{\partial c_i''} \frac{\partial c_i''}{\partial V''} = -\frac{1}{V''} \sum_{i=1}^{n_c} c_i'' \frac{\partial a^s}{\partial c_i''} \tag{7.158}$$

$$\frac{\partial \tilde{a}^s}{\partial \mathcal{A}} = \sum_{i=1}^{n_c} \frac{\partial a^s}{\partial \Gamma_i} \frac{\partial \Gamma_i}{\partial \mathcal{A}} = -\frac{1}{\mathcal{A}} \sum_{i=1}^{n_c} \Gamma_i \frac{\partial a^s}{\partial \Gamma_i} \tag{7.159}$$

$$\frac{\partial \tilde{a}^s}{\partial N_i'} = \frac{1}{V'} \frac{\partial a^s}{\partial c_i'} \tag{7.160}$$

$$\frac{\partial \tilde{a}^s}{\partial N_i''} = \frac{1}{V''} \frac{\partial a^s}{\partial c_i''} \tag{7.161}$$

$$\frac{\partial \tilde{a}^s}{\partial N_i^s} = \frac{1}{\mathcal{A}} \frac{\partial a^s}{\partial \Gamma_i} \tag{7.162}$$

$$\frac{\partial \tilde{a}^s}{\partial r} = \frac{\partial a^s}{\partial r} \tag{7.163}$$

Let us now define cross chemical potentials ε_i' and ε_i'' by

$$\varepsilon_i' = \frac{\partial a^s}{\partial c_i'} \tag{7.164}$$

$$\varepsilon_i'' = \frac{\partial a^s}{\partial c_i''} \tag{7.165}$$

The cross chemical potentials of component i represent the effect of the bulk phase concentration on the surface free energy density on either side of the interface.

The interfacial chemical potential μ_i^s can be expressed by [from Eq. (7.153)],

$$\mu_i^s = \frac{\partial a^s}{\partial \Gamma_i} \tag{7.166}$$

The three chemical potentials μ_i^s, ε_i', and ε_i'' are intensive quantities.

Equations (7.160) to (7.161), (7.164) to (7.165), and $A^s = \mathcal{A}\tilde{a}^s$ lead to,

$$\frac{\partial A^s}{\partial N_i'} = \frac{\mathcal{A}}{V'}\frac{\partial a^s}{\partial c_i'} = \frac{\mathcal{A}}{V'}\varepsilon_i' \tag{7.167}$$

$$\frac{\partial A^s}{\partial N_i''} = \frac{\mathcal{A}}{V''}\frac{\partial a^s}{\partial c_i''} = \frac{\mathcal{A}}{V''}\varepsilon_i'' \tag{7.168}$$

By combining Eqs. (7.151), (7.152) and (7.167) to (7.168) one obtains

$$\boxed{\tilde{\mu}_i' = \mu_i' + \frac{\mathcal{A}}{V'}\varepsilon_i'} \tag{7.169}$$

$$\boxed{\tilde{\mu}_i'' = \mu_i'' + \frac{\mathcal{A}}{V''}\varepsilon_i''} \tag{7.170}$$

7.12 Fundamental Equation of the Interface

For each of the bulk phases, one can write

$$\frac{\partial A'}{\partial T} = -S'; \frac{\partial A'}{\partial V'} = -P'; \frac{\partial A'}{\partial N_i'} = \mu_i' \tag{7.171}$$

$$\frac{\partial A''}{\partial T} = -S''; \frac{\partial A''}{\partial V''} = -P''; \frac{\partial A''}{\partial N_i''} = \mu_i'' \tag{7.172}$$

Combining Eqs. (7.139) to (7.143) and (7.171) to (7.172),

$$\frac{\partial A^s}{\partial T} = -S^s; \frac{\partial A^s}{\partial V'} = 0; \frac{\partial A^s}{\partial V''} = 0; \frac{\partial A^s}{\partial \mathcal{A}} = \gamma; \frac{\partial A}{\partial r} = \mathcal{A}\left[\frac{\partial \gamma}{\partial r}\right] \tag{7.173}$$

The other derivatives of A^s can be obtained from rearranging Eqs. (7.151) to (7.153),

$$\frac{\partial A^s}{\partial N_i'} = \tilde{\mu}_i' - \mu_i'; \frac{\partial A^s}{\partial N_i''} = \tilde{\mu}_i'' - \mu_i''; \frac{\partial A^s}{\partial N_i^s} = \mu_i^s \tag{7.174}$$

The derivatives of a^s with respect to c_i', and c_i'', and Γ_i are given by Eqs. (7.164) to (7.166), respectively. The derivative of a^s with respect to T is [from the first expression in Eq. (7.173)]

$$\frac{\partial a^s}{\partial T} = -s^s \tag{7.175}$$

Now we combine $A^s = \mathcal{A}\tilde{a}^s$, the fourth expression in Eq. (7.173), and Eqs. (7.159) and (7.166) to obtain

$$\gamma = a^s - \sum_{i=1}^{n_c} \Gamma_i \mu_i^s \tag{7.176}$$

Equation (7.176) is the general expression for interfacial tension γ. If we select the dividing surface such that $\sum_{i=1}^{n_c} \Gamma_i \mu_i^s = 0$, then

$$\gamma = a^s \tag{7.177}$$

Note that both Eqs. (7.176) and (7.177) apply to chemical equilibrium as well as non-chemical equilibrium conditions.

Next we replace A^s with $\mathcal{A}\tilde{a}^s$ in the second and third expressions in Eq. (7.173) and use Eqs. (7.156), (7.157), (7.164), and (7.165) to obtain,

$$\sum_{i=1}^{n_c} \varepsilon_i' c_i' = 0 \tag{7.178}$$

$$\sum_{i=1}^{n_c} \varepsilon_i'' c_i'' = 0 \tag{7.179}$$

The above two equations show that in a one component system $\varepsilon' = 0$ and $\varepsilon'' = 0$; the cross chemical potentials in a one component system composed of bulk-prime and double prime phases and an interface are zero.

As a final expression in this section, from Eq. (7.176) one obtains

$$A^s = \gamma \mathcal{A} + \sum_{i=1}^{n_c} N_i^s \mu_i^s \tag{7.180}$$

Therefore, from $A^s = U^s - TS^s$

$$U^s = TS^s + \gamma \mathcal{A} + \sum_{i=1}^{n_c} N_i^s \mu_i^s \tag{7.181}$$

For the bulk phases,

$$U' = TS' - P'V' + \sum_{i=1}^{n_c} N_i' \mu_i' \tag{7.182}$$

$$U'' = TS'' - P''V'' + \sum_{i=1}^{n_c} N_i'' \mu_i'' \tag{7.183}$$

Therefore

$$U = TS - P'V' - P''V'' + \gamma A + \sum_{i=1}^{n_c} N_i^s \mu_i^s + \sum_{i=1}^{n_c} N_i' \mu_i' + \sum_{i=1}^{n_c} N_i'' \mu_i'' \tag{7.184}$$

$$A = -P'V' - P''V'' + \gamma A + \sum_{i=1}^{n_c} N_i^s \mu_i^s + \sum_{i=1}^{n_c} N_i' \mu_i' + \sum_{i=1}^{n_c} N_i'' \mu_i'' \tag{7.185}$$

At equilibrium Eq. (7.184) reduces to $(\mu_i = \mu_i' = \mu_i'' = \mu_i^s)$

$$U = TS - P'V' - P''V'' + \gamma A + \sum_{i=1}^{n_c} N_i \mu_i \tag{7.186}$$

7.12.1 Adsorption at the Interface

Let us write the differential of the interfacial tension from Eq. (7.176)

$$d\gamma = da^s - \sum_{i=1}^{n_c} \Gamma_i d\mu_i^s - \sum_{i=1}^{n_c} \mu_i^s d\Gamma_i \tag{7.187}$$

The total differential of a^s from Eq. (7.155) is,

$$da^s = \frac{\partial a^s}{\partial T} dT + \sum_{i=1}^{n_c} \frac{\partial a^s}{\partial c_i'} dc_i' + \sum_{i=1}^{n_c} \frac{\partial a^s}{\partial c_i''} dc_i'' + \sum_{i=1}^{n_c} \frac{\partial a^s}{\partial \Gamma_i} d\Gamma_i + \frac{\partial a^s}{\partial r} dr \tag{7.188}$$

Using Eqs. (7.175), (7.164), (7.165), (7.166), and the last expression in Eq. (7.173) the above expression becomes

$$da^s = -s^s dT + \sum_{i=1}^{n_c} \varepsilon_i' dc_i' + \sum_{i=1}^{n_c} \varepsilon_i'' dc_i'' + \sum_{i=1}^{n_c} \mu_i^s d\Gamma_i + \left[\frac{\partial \gamma}{\partial r}\right] dr \tag{7.189}$$

Substituting Eq. (7.189) into Eq. (7.187) we obtain,

$$d\gamma = -s^s dT + \sum_{i=1}^{n_c} \varepsilon_i' dc_i' + \sum_{i=1}^{n_c} \varepsilon_i'' dc_i'' - \sum_{i=1}^{n_c} \Gamma_i d\mu_i^s + \left[\frac{\partial \gamma}{\partial r}\right] dr \tag{7.190}$$

The above equation can provide the interfacial tension at nonequilibrium conditions. At equilibrium $\varepsilon_i' = \varepsilon_i'' = 0$, $(i = 1, \ldots, n_c)$, therefore,

$$\mu_i' = \mu_i^s = \mu_i'' = \mu_i \quad i = 1, \ldots, n_c \tag{7.191}$$

Equation (7.190) reduces to

$$d\gamma = -s^s dT - \sum_{i=1}^{n_c} \Gamma_i d\mu_i + \left[\frac{\partial \gamma}{\partial r}\right] dr. \tag{7.192}$$

Equation (7.192) is the Gibbs adsorption equation which applies at chemical equilibrium. Gibbs did not include the nonchemical equilibrium effects in his derivation.

7.12.2 Work of Cluster Formation in a Multicomponent Mixture

The expression for the work of cluster formation in a multicomponent mixture is more complicated than the expression for a single component system. There are a limited number of papers on the derivation and there have been discussions on the proper expression for the work of cluster formation and inconsistency in some derivations even when the final results are correct (Wilmeski, 1984, 1987; Nishioka and Kusaka, 1992; Debenedetti and Reiss, 1998). We will use Eq. (7.185) for the Helmholtz free energy of a heterogeneous system with two bulk phases and a curved interface to derive the expression for the work of cluster formation in a multicomponent mixture. The approach is simple.

Let us consider the formation of a cluster of a new phase within a continuous bulk phase at temperature T and pressure P. In the initial state (i.e., state I), the closed system contains n_c species with N_i molecules for each species. In state II, the system contains a spherical cluster with n_i molecules of each species i. The Gibbs free energy of the system at state I is given by

$$G^{\mathrm{I}} = \sum_{i=1}^{n_c} N_i \mu_i^{\mathrm{I}}(P) \tag{7.193}$$

The Gibbs free energy of state II is given by

$$G^{\mathrm{II}} = PV^{\mathrm{II}} + A^{\mathrm{II}} \tag{7.194}$$

The Helmoholtz free energy A^{II} is given [from Eq. (7.185)] by,

$$A^{\mathrm{II}} = -PV_{N-N^s-n} - P_n V_n + \gamma \mathcal{A} + \sum_{i=1}^{n_c} (N_i - N_i^s - n_i)\mu_i^{\mathrm{I}}(P, composition)$$
$$+ \sum_{i=1}^{n_c} N_i^s \mu_i^s + \sum_{i=1}^{n_c} n_i \mu_i^{\mathrm{II}}(P_n, composition) \tag{7.195}$$

In the above equation, the chemical potentials are functions of composition. The work of cluster formation is given by $\Delta G = G^{\mathrm{II}} - G^{\mathrm{I}}$. Combining Eqs. (7.193) to (7.195) and $V^{\mathrm{II}} = V_{N-N^s-n} + V_n$ where V_{N-N^s-n} is the volume of the bulk continuous phase and V_n is the volume of the cluster, we obtain

$$W = -(P_n - P)V_n + \gamma \mathcal{A} + \sum_{i=1}^{n_c} N_i^s \left[\mu_i^s - \mu_i^{\mathrm{I}}(P, compostion)\right]$$
$$+ \sum_{i=1}^{n_c} n_i \left[\mu_i^{\mathrm{II}}(P_n, composition) - \mu_i^{\mathrm{I}}(P, composition)\right] \tag{7.196}$$

Equation (7.196) can be regarded as the general expression for the work of cluster formation in a multicomponet mixture.

Under the condition that the interface chemical potential is approximately the same as the continuous bulk phase,

$$W = -(P_n - P)V_n + \gamma A + \sum_{i=1}^{n_c} n_i \left[\mu_i^{\text{II}}(P_n, \textit{composition}) - \mu_i^{\text{I}}(P, \textit{composition})\right] \qquad (7.197)$$

Equation (7.197) is similar to the expression derived earlier for the work of cluster formation in a single component system. We will use Eq. (7.196) to compute the work of cluster formation in multicomponent mixtures in terms of the thermodynamic driving forces in a multicomponent mixture in Example 7.2.

7.13 Examples and Theory Extension

7.1: Relation between Chemical Potential Driving Forces in Binary Mixtures Show that for droplet formation in a vapor, one can write

$$\frac{\Delta\mu_1}{\bar{V}_1^L} = \frac{\Delta\mu_2}{\bar{V}_2^L}$$

in a binary mixture for the critical size. What are the assumptions in the derivation of the above equation? The driving force is defined by

$$\Delta\mu_i = \mu_i^G(P^G) - \mu_i^L(P^G) \quad i = 1, 2$$

Solution In a given phase, one can write

$$d\mu_i = (\bar{V}_i dP)_{T,n}$$

$$\mu_i^L(P^L) = \mu_i^L(P^G) + \int_{P^G}^{P^L} \bar{V}_i^L dP = \mu_i^L(P^G) + \bar{V}_i^L(P^L - P^G)$$

The above equation is based on the assumption that $\bar{V}_i^L$ is independent of pressure in the range P^L to P^G or $\mu_i^L(P^L) - \mu_i^L(P^G) = \bar{V}_i^L(P^L - P^G)$.

At the critical size (unstable equilibrium)

$$\mu_i^L(P^L) = \mu_i^G(P^G)$$

Therefore

$$\mu_i^G(P^G) - \mu_i^L(P^G) = \bar{V}_i^L(P^L - P^G),$$

or

$$\Delta\mu_i = \bar{V}_i^L(P^L - P^G).$$

Writing the above equation for components 1 and 2, one obtains

$$\frac{\Delta\mu_1}{\Delta\mu_2} = \frac{\bar{V}_1^L}{\bar{V}_2^L}$$

7.2: Relation between Driving Force and Critical Radius In the expression for the calculation of the driving force for drop nucleation in multicomponent mixtures from vapor there has been confusion regarding the correctness of the following equation (Doyle,1961; Wilemski,1984; Nishioka and Kusaka,1992; Oxtoby and Kashchiev,1994; Debenedetti and Reiss,1998; Laaksonen et al.,1999),

$$-\Delta\mu_i + \frac{2\gamma\bar{V}_i^L}{r^*} = 0 \tag{7.198}$$

where r^* is the radius of the critical nucleus and $\bar{V}_i^L$ is the partial molar volume of component i in the droplet. $\Delta\mu_i$ is the driving force for the new phase formation. Use the work of cluster formation, Eq. (7.196), to obtain the above equation.

Solution We first add and subtract $\mu_i^L(P^G)$ in the brackets in the last term of Eq. (7.196) in order to have $\Delta\mu_i = \mu_i^G(P^G) - \mu_i^L(P^G)$ in the same expression. The result after rearrangement of terms is

$$W = (P^G - P^L)V^L + \gamma\mathcal{A} - \sum_{i=1}^{n_c} n_i^L\Delta\mu_i + \sum_{i=1}^{n_c} n_i^L\left[\mu_i^L(P^L) - \mu_i^L(P^G)\right] + \sum_{i=1}^{n_c} n_i^s\left[\mu_i^s - \mu_i^G(P^G)\right] \tag{7.199}$$

Note that in Eq. (7.199), the temperature and compositional dependencies of the chemical potentials are not shown for the sake of brevity. The pressure in the gas phase is designated by P^G and n_i^L is the number of molecules of component i in the liquid drop. We show the volume of the liquid drop by V^L.

Next, we take the partial derivatives of W with respect to n_j^L (at constant T, P, $\boldsymbol{n}_j$) and set it to zero (see Anklam and Firoozabadi, 2004).

$$\begin{aligned}
\left(\frac{\partial W}{\partial n_j^L}\right)_{T,P^G,\boldsymbol{n}_j} &= -V^L\left(\frac{\partial P^L}{\partial n_j^L}\right) + (P^G - P^L)\left(\frac{\partial V^L}{\partial n_j^L}\right) + \gamma\left(\frac{\partial \mathcal{A}}{\partial n_j^L}\right) \\
&+ \mathcal{A}\left(\frac{\partial\gamma}{\partial n_j^L}\right) - \Delta\mu_j - \sum_{i=1}^{n_c} n_i^L\frac{\partial\mu_i^G(P^G)}{\partial n_j^L} + \sum_{i=1}^{n_c} n_i^L\frac{\partial\mu_i^L(P^G)}{\partial n_j^L} \\
&+ \left[\mu_j^L(P^L) - \mu_j^L(P^G)\right] \\
&+ \sum_{i=1}^{n_c} n_i^L\frac{\partial\mu_i^L(P^L)}{\partial n_j^L} - \sum_{i=1}^{n_c} n_i^L\frac{\partial\mu_i^L(P^G)}{\partial n_j^L} + \sum_{i=1}^{n_c} n_i^s\frac{\partial\mu_i^s}{\partial n_j^L} - \sum_{i=1}^{n_c} n_i^s\frac{\partial\mu_i^G(P^G)}{\partial n_j^L} \\
&+ \sum_{i=1}^{n_c}\left(\frac{\partial n_i^s}{\partial n_j^L}\right)\left[\mu_i^s - \mu_i^s(P^G)\right] = 0 \quad j = 1, \ldots, n_c
\end{aligned} \tag{7.200}$$

The first and ninth terms on the right side cancel out due to the Gibbs-Duhem equation. Likewise, the fourth and eleventh terms cancel out based on the Gibbs absorption isotherm expression. Because there is no change in the composition of the gas phase (it is very large), the sixth and twelfth terms are zero. Further, at the critical nucleus size $\mu_i^G(P^G) = \mu_i^s$, (unstable equilibrium). Therefore, Eq. (7.200) reduces to

$$(P^G - P^L)\left(\frac{\partial V^L}{\partial n_j^L}\right) + \gamma\left(\frac{\partial \mathcal{A}}{\partial n_j^L}\right) + \left[\mu_j^L(P^L) - \mu_j(P^G)\right] - \Delta\mu_j = 0 \quad j = 1, \ldots, n_c \tag{7.201}$$

From

$$\left(d\mu_j^L = \bar{V}_j^L dP^L\right)_{T,\underline{n}^L} \tag{7.202}$$

and

$$\bar{V}_j^L = \left(\frac{\partial V^L}{\partial n_j^L}\right)_{T,P^L,n^L} \tag{7.203}$$

and assuming that $\bar{V}_j^L$ to be independent of pressure, one obtains

$$\mu_j^L(P^L) - \mu_j^L(P^G) = -\left(\frac{\partial V^L}{\partial n_j^L}\right)_{T,P^L,n_j^L}(P^G - P^L) \tag{7.204}$$

Combining Eqs. (7.201) and (7.204)

$$\gamma\left(\frac{\partial \mathcal{A}}{\partial n_j^L}\right) - \Delta\mu_j = 0$$

For a spherical nucleus $(\partial \mathcal{A}/\partial n_j^L) = (2/r)\bar{V}_j^L$. Therefore,

$$\frac{2\gamma \bar{V}_j^L}{r^*} - \Delta\mu_j = 0 \tag{7.205}$$

The main assumption made in the derivation of Eq. (7.205) is that $\bar{V}_j^L$ is independent of pressure.

7.3: Critical Nucleus Work in Multicomponent Mixtures Use the following expression (derived in Example 7.2),

$$\Delta\mu_i - \frac{2\gamma \bar{V}_i^L}{r^*} = 0 \tag{7.206}$$

to obtain,

$$r^* = +\frac{2\gamma v^L}{\sum_{i=1}^{n_c} x_i \Delta\mu_i} \tag{7.207}$$

and the expression for the Gibbs free energy of the formation of a critical-size liquid cluster (nucleus)

$$W^* = \Delta G^* = \frac{4}{3}(\pi r^{*2})\gamma = \frac{1}{3}\gamma\mathcal{A} \tag{7.208}$$

Note that Eq. (7.208) is valid for both a single and a multicomponent mixture.

Solution Multiply Eq. (7.206) by x_i (composition of bulk liquid cluster) and then take the sum over all the components, and use the expression $v^L = \sum_{i=1}^{n_c} x_i \bar{V}_i^L$. The result is Eq. (7.207). The expression for the Gibbs free energy of the formation of a critical-size liquid cluster in gas is given by [from Eq. (7.199) when $\mu_i^s = \mu_i^L(P^L) = \mu_i^G(P^G)$]

$$\Delta G^* = (P^G - P^L)V^L + \gamma\mathcal{A} \tag{7.209}$$

From $-(P^G - P^L) = 2\gamma/r^*$ and Eq. (7.209)

$$\Delta G^* = -\frac{2\gamma V^L}{r^*} + \gamma\mathcal{A} = \frac{1}{3}\gamma\mathcal{A}$$

7.4: Work of Critical-Size Cluster Formation and Critical Cluster Size Consider formation of liquid clusters in a gas phase. Plot the work of the cluster formation as a function of the number of molecules in the liquid cluster for differing driving forces of $\Delta\mu/kT = 1$, $\Delta\mu/kT = 2$, and $\Delta\mu/kT = 3$. Basic data are $v_0 = 5 \times 10^{-23}$ cm^3, $\sigma = 100$ erg/cm^2, $T = 293$ K, where v_0 is the molecular volume. Note that both the size of the critical nucleus and the height of the energy barrier decreases with increasing supersaturation (driving force).

Solution The work for the formation of an incompressible spherical liquid cluster is given by Eq. (7.93).

$$\frac{W(n)}{kT} = -\frac{n\Delta\mu}{kT} + \frac{\left(36\pi v_0^2\right)^{1/3}\sigma n^{2/3}}{kT} \tag{7.210}$$

For $v_0 = 5 \times 10^{-23}$ cm^3, $\sigma = 100$ erg/cm^2, $T = 293$ K, and $k = 1.38 \times 10^{-16}$ erg/K

$$\frac{W(n)}{kT} = -\frac{n\Delta\mu}{kT} + 16.23n^{\frac{2}{3}}$$

Figure 7.7 shows a plot of $W(n)$ as a function of n for $\Delta\mu/kT = 1, 2, 3$: At the critical nucleus, $\partial W(n)/\partial n = 0$. Therefore, based on differentiation of Eq. (7.210)

$$\boxed{n^* = \left(\frac{2}{3}\right)^3 \frac{36\pi v_0^2\sigma^3}{(\Delta\mu)^3}},$$

and

$$W^* = \boxed{\frac{16\pi}{3}\frac{v_0^2\sigma^3}{(\Delta\mu)^2}}.$$

The above two simple equations reveal that as the interfacial tension increases, the barrier to nucleation increases; that is, it will be more challenging to nucleate a new phase. The phase change is carried out with ease as we approach the limit of stability and the critical point, because of the decrease in interfacial tension. Both expressions show the strong effect of driving force on size of crucial nuclei and on work of forming crucial nuclei. The figure shows that as $\Delta\mu$ increases n^* decreases. The size of the critical nucleus are $n^* = 1270$, 158, and 47 molecules at $\Delta\mu/kT = 1$, 2, and 3, respectively.

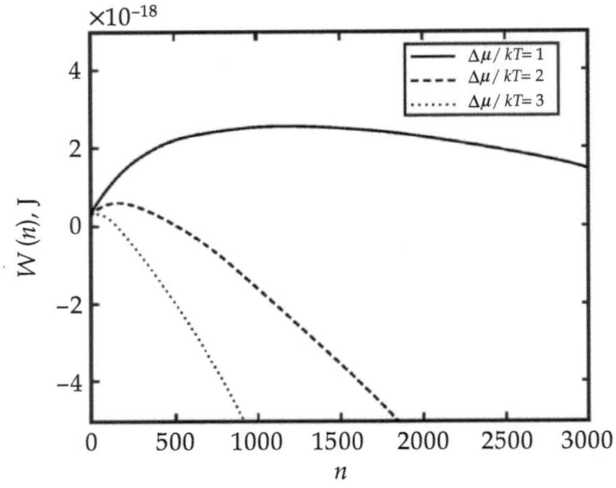

FIGURE 7.7 $W(n)$ as a function of n for $\Delta\mu/kT = 1, 2, 3$.

7.5: Driving Force of Freezing in a Subcooled Liquid Derive the following expression for the driving force of freezing in a subcooled liquid,

$$\Delta\mu = \frac{\Delta h_f \Delta T}{T_e} \tag{7.211}$$

where $\Delta T = (T_e - T)$ is the subcooling, Δh_f is the enthalpy of fusion, and T_e is the equilibrium freezing point temperature.

Solution We consider a liquid cooled to a temperature T, below the melting temperature T_e at constant external pressure P. The driving force of freezing at pressure P is defined by

$$\Delta\mu = \mu^L(T) - \mu^S(T) \tag{7.212}$$

where the superscript stands for the phase state. The pressure of both solid (S) and liquid (L) is assumed to be the same, P.

Adding and subtracting $\mu^L(T_e)$ on the right side of Eq. (7.212),

$$\Delta\mu = \mu^L(T) - \mu^L(T_e) + \mu^L(T_e) - \mu^S(T) \tag{7.213}$$

For a pure substance at constant pressure, $d\mu = -sdT$, therefore

$$\mu^L(T) - \mu^L(T_e) = -\int_{T_e}^{T} s^L(T)dT \tag{7.214}$$

$$\mu^S(T) - \mu^S(T_e) = -\int_{T_e}^{T} s^S(T)dT \tag{7.215}$$

However, at equilibrium, $\mu^L(T_e) = \mu^S(T_e)$. Subtracting Eq. (7.215) from Eq. (7.214),

$$\Delta\mu = -\int_{T_e}^{T} s^L(T)dT + \int_{T_e}^{T} s^S(T)dT \tag{7.216}$$

At constant pressure, $dh = T ds$, and at temperature T, $h^L - h^S = T(s^L - s^S)$. Let $\Delta h_f =$ enthalpy of fusion $= h^L - h^S$. Then $s^L = s^S + \Delta h_f / T$. Eq. (7.216) can now be written as

$$\Delta\mu = -\int_{T_e}^{T} \left(\frac{\Delta h_f}{T}\right) dT \tag{7.217}$$

Assume $\Delta h_f(T)$ to be independent of temperature (when T_e is close to T). One derives

$$\Delta\mu = \Delta h_f \ln \frac{T_e}{T} \tag{7.218}$$

or

$$\Delta\mu = \Delta h_f \ln \frac{T_e - T + T}{T} = \Delta h_f \ln\left(\frac{\Delta T}{T} + 1\right) \tag{7.219}$$

Equation (7.219) can be approximated by $[\ln(\Delta T/T + 1) \approx \Delta T/T \approx \Delta T/T_e]$. Therefore, the sought expression is obtained.

7.6: Effect of Surfactants on Interfacial Tension Surfactants when added to a solution move to the interface. Show that the adsorption of surfactants at the interface results in the reduction of interfacial tension.

Solution Consider a mixture of water and surfactant, we write Eq. (7.20)

$$d\sigma = -[\Gamma_w d\mu_w + \Gamma_s d\mu_s] \tag{7.220}$$

From the Gibbs-Duhem expression at a constant T and P in the water-surfactant solution,

$$n_w d\mu_w + n_s d\mu_s = 0$$

or

$$d\mu_w = -\frac{n_s}{n_w} d\mu_s$$

Combining the above equations,

$$d\sigma = -\left[-\Gamma_w \left(\frac{n_s}{n_w}\right) + \Gamma_s\right] d\mu_s$$

From $d\mu_s = (\partial\mu_s/\partial x_s)_{T,P} dx_s$, and the fact that based on thermodynamic stability, $(\partial\mu_s/\partial x_s)_{T,P}$ is positive, then we can state that the addition of surfactant in the bulk liquid phase results in an increase of $d\mu_s$, that is $d\mu_s > 0$.

Provided the surfactant moves toward the interface and that generally n_s is orders of magnitude less than n_w, $\Gamma_s >> \Gamma_w(n_s/n_w)$. Then,

$$d\sigma \approx -\Gamma_s d\mu_s$$

Therefore, the addition of a small amount of a surfactant results in a decrease in interfacial tension.

7.7: Effect of Polyelectrolytes on Interfacial Tension Unlike surfactants, which move to the interface (at constant T and P), the addition of polyelectrolytes results in the accumulation in the bulk liquid phase and a low concentration at the interface. Show that the addition of polyelectrolytes may result in an increase in the interfacial tension.

Solution Let us use Eq. (7.220). Because $\Gamma_p << \Gamma_w(n_p/n_w)$, then one can write,

$$d\sigma \approx \left[\Gamma_w \left(\frac{n_p}{n_w}\right)_P\right] d\mu_p \tag{7.221}$$

From stability analysis $(\partial\mu_p/\partial x_p)_{T,P} > 0$, therefore, the interfacial tension increases as x_p increases. The magnitude of the increases is, however, small, unlike the decrease in interfacial tension by surfactants, where it can be very pronounced.

7.8: Effect of Charge on Vapor Pressure of Small Droplets Charge on aerosols and other small droplets has many fascinating aspects. The problem is very rich in Gibbs free energy analysis. Derive the basic expression to study the effect of charge on vapor pressure of small droplets and examine equilibrium and stability of charged droplets. Interestingly, many aspects of this problem are covered in the 1888 book by J. J. Thomson in relation to vapor pressure. The idea of electrons introduced by Thomson is also related to negative charge of particles; he called them "corpuscules," now we call them "electrons."

Solution In our derivation, which is different from the approach by J. J. Thomson, we first neglect the charge and start from Eq. (7.91),

$$\Delta G = W = -n\Delta\mu + \sigma\pi d^2 \tag{7.222}$$

where d is the diameter of the liquid droplet.

The number of molecules, n, in the droplet is given through $\rho^L V_n = \pi\rho^L d^3/6$ where $\rho^L V_n$ is the mass of the droplet,

$$n = \frac{N_A}{M}\frac{\pi}{6}\rho^L d^3 \tag{7.223}$$

where M is the molecular weight and N_A is the Avogadro number. Combining Eqs. (7.222) and (7.223),

$$\Delta G = -\frac{N_A}{M}\frac{\pi}{6}\rho^L d^3\Delta\mu + \sigma\pi d^2 \tag{7.224}$$

For a given $\Delta\mu(T, P)$ and $\rho^L(T, P)$, the critical droplet size (at constant T and P) is obtained by finding the extremum of ΔG, i.e. $(\partial\Delta G/\partial d)_{T,P} = 0$,

$$\boxed{d^* = \frac{4\sigma M}{N_A\rho^L\Delta\mu}} \tag{7.225}$$

For a droplet, the expression for $\Delta\mu$ is given by $\Delta\mu = kT\ln P/P_e$ [from approximation of Eq. (7.123)]. Combining the expression for $\Delta\mu$ and Eq. (7.225), we obtain (note that $R = N_A k$)

$$d^* = \frac{4\sigma M}{RT\rho^L \ln P/P_e} \tag{7.226}$$

An alternative form of Eq. (7.226) is given by (note that $v^L = M/\rho^L$)

$$\frac{P}{P_e} = \exp\left[\frac{4\sigma v^L}{RTd^*}\right] \tag{7.227}$$

Equation (7.227) is the Kelvin expression derived in Chap. 2. The above provides an alternative derivation, which clearly reveals that the equilibrium can be unstable.

We now include the effect of charge by adding $(q^2/d)(1/\varepsilon_0 - 1/\varepsilon)$, derived in Chap. 2, to Eq. (7.222). Note that a multiplier $1/4\varepsilon_0$ may be applied for consistency of units. Here q is the charge in the droplet, ε_0 is the gas dielectric constant, and ε is the liquid dielectric constant. The total change in free energy to form a droplet at constant T and P is given by

$$\Delta G = -\frac{N_A}{M}\frac{\pi}{6}\rho^L d^3 \Delta\mu + \sigma\pi d^2 + \frac{q^2}{d}\left(\frac{1}{\varepsilon_0} - \frac{1}{\varepsilon}\right) \tag{7.228}$$

For a given $\Delta\mu$ (constant T and P), the critical diameter of the droplet is obtained from $(\partial\Delta G/\partial d)_{T,P} = 0$.

$$\boxed{-\frac{N_A}{2v^L}\Delta\mu + \frac{2\sigma}{d^*} - \frac{q^2}{\pi d^{*4}}\left(\frac{1}{\varepsilon_0} - \frac{1}{\varepsilon}\right) = 0} \tag{7.229}$$

Using $\Delta\mu = kT\ln P/P_e$, the expression for P/P_e is given by

$$\boxed{\frac{P}{P_e} = \exp\left\{\frac{2v^L}{RT}\left[\frac{2\sigma}{d^*} - \frac{q^2}{\pi d^{*4}}\left(\frac{1}{\varepsilon_0} - \frac{1}{\varepsilon}\right)\right]\right\}} \tag{7.230}$$

Note the additional term due to the charge. Equation (7.230) implies that one can expect significant effect of charge at very small droplet sizes as we will see in the calculations below. The charge will drastically affect the state of stability and equilibrium of a small droplet. In the following we use Gibbs free energy analysis for the interpretation of droplet equilibrium and stability.

Let us consider a droplet with the following charges, $q = 0$, 10 e, and 100 e. Other relevant data are: $\sigma = 72$ dyne/cm, $\rho^L = 1$ g/cm^3, $\varepsilon = 80$, $\varepsilon_0 = 1$.

Figures 7.8a to 7.8d show the results of our calculations. We first use Eqs. (7.227) and (7.230) to obtain the equilibrium vapor pressure of an uncharged and charged droplet depicted in Fig. 7.8a. Note that charge has a significant effect on vapor pressure. The results in Fig. 7.8a show only equilibrium but one has to find out if the equilibrium is stable or unstable. Figure 7.8b depicts the change in Gibbs free energy from formation of an uncharged droplet for different values of $\Delta\mu$. Note a maximum in ΔG for all four $\Delta\mu$ values. For $(\Delta\mu/kT) = 0.25$, the value of the diameter is $d^* = 14.2$ nm. This implies that for $(\Delta\mu/kT) = 0.25$, any diameter larger than 14.2 nm will experience growth. A droplet with a diameter less than 14.2 nm for the same $(\Delta\mu/kT)$ may not grow; it may shrink. The diameters at the maximum of Gibbs free energy change for different values of $(\Delta\mu/kT)$ are the critical diameters. Let us take the diameter at the maximum of $(\Delta G/kT)$ for $(\Delta\mu/kT) = 0.5$ corresponding to $(P/P_e) = 1.65$. In Fig. 7.8a the diameter for this value of P/P_e is $d^* = 7.1$ nm, which is the same as the diameter at the maximum $\Delta G/kT$ in

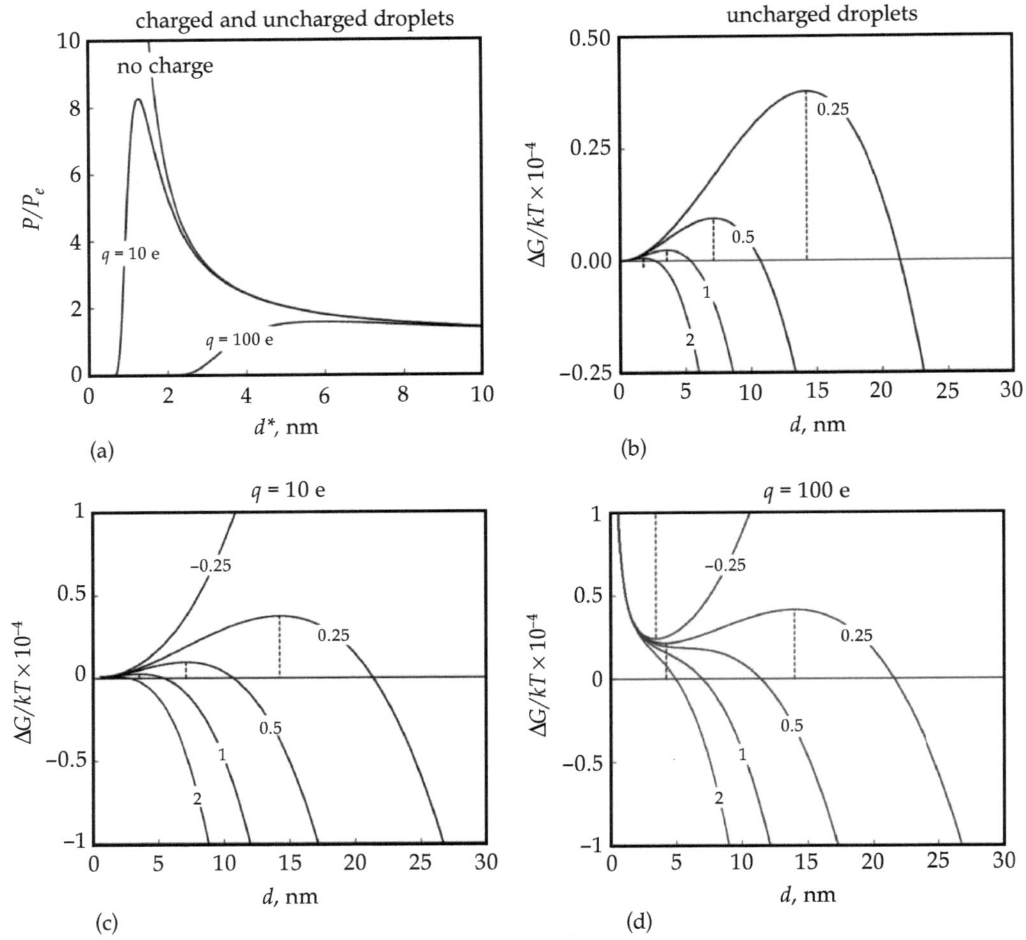

FIGURE 7.8 (a) Vapor pressure of charged and uncharged droplets; Gibbs free energy change of (b) uncharged and (c) $q = 10$ e and (d) $q = 100$ e charged droplets for various values of $(\Delta\mu/kT)$. The vertical dashed lines show the location of the maximum and/or minimum of Gibbs free energy change.

Fig. 7.8b. For $(\Delta\mu/kT) = 1$, one gets$(P/P_e) = 2.72$. The diameter at the maximum $\Delta G/kT$ for $(\Delta\mu/kT) = 1$ is 3.5 nm (in Fig. 7.8b). We obtain the same value for the diameter in Fig. 7.8a.

Now let us examine the effect of charge of $q = 10$ e and the results in Fig. 7.8c. For $(\Delta\mu/kT) = -0.25$, there is no maximum. This implies that all droplets show an increasing Gibbs free energy; they will shrink. For $(\Delta\mu/kT) = 0.25$, the right side of maximum results in growth, and the left side will have shrinkage. Similar behavior is expected for values of $(\Delta\mu/kT) = 0.5$, 1, and 2. In Fig. 7.8c, there is a critical size for $(\Delta\mu/kT) = 2$, but it is very small (not shown) in Fig. 7.8c.

The most interesting aspect is related to the plots in Fig. 7.8d. For $(\Delta\mu/kT) = -0.25$, that is, $(P/P_e) = 0.78$, there is a minimum in $(\Delta G/kT)$. The diameter corresponding to this minimum is around 3.4 nm. The implication of minimum is that whether we are on the right or the left side, for a diameter greater than or smaller than 3.4 nm, there will be shrinkage for $d > 3.4$ nm, and growth for $d < 3.4$ nm. In other words, the change is toward stable equilibrium. Without charge this behavior cannot be observed. There will be also a decrease in vapor pressure of the droplet, unlike the case of an uncharged droplet. For $(\Delta\mu/kT) = 0.25$, $(P/P_e) = 1.3$, there are three distinct regions. For diameters less than about 4.1 nm, the growth will be toward a stable equilibrium size of 4.1 nm. For diameters between $4.1 < d < 14$ nm, there will be shrinkage toward equilibrium value of 4.1 nm, and for $d > 14$ nm, there will be growth. For $(\Delta\mu/kT) = 0.5$, 1 and 2, there will be growth for all droplet sizes.

7.9: Interface in Weightless State Consider a spherical vessel of radius R at temperature T. The liquid and gas of a single-component fluid are at equilibrium within the vessel (see Fig. 7.9). In outer space, where gravity is virtually absent, the Helmholtz free energy change of the entire system is given by,

$$dA = \sigma^{LG} d\mathcal{A}^{LG} + \sigma^{LS} d\mathcal{A}^{LS} + \sigma^{GS} d\mathcal{A}^{GS} \tag{7.231}$$

where σ represents the surface energy density and $\mathcal{A}$ represents the surface area in Fig. 7.9a. The vessel is assumed to be larger than a few mm. First derive the above expression for change of Helmholtz free energy. Then, present working equations for the determination of the shape of the interface.

Solution We assign the Gibbs dividing surface to the three interfaces, such that $n^s = 0$ (i.e., $dn^L = -dn^G$) for all the three interfaces. The chemical potential of the phases and interfaces are the same at equilibrium. Based on thermal, mechanical, and chemical equilibrium, we write the following expressions,

$$dA^L = -P^L dV^L + \mu dn^L \tag{7.232}$$

$$dA^G = -P^G dV^G + \mu dn^G \tag{7.233}$$

The Helmholtz free energy of the total system is given by,

$$dA = dA^L + dA^G + dA^s \tag{7.234}$$

where A^s is the surface free energy of the system given by,

$$A^s = \sigma^{LG}\mathcal{A}^{LG} + \sigma^{LS}\mathcal{A}^{LS} + \sigma^{GS}\mathcal{A}^{GS} \tag{7.235}$$

$$dA^s = \sigma^{LG} d\mathcal{A}^{LG} + \sigma^{LS} d\mathcal{A}^{LS} + \sigma^{GS} d\mathcal{A}^{GS} \tag{7.236}$$

We have assumed that, as surface areas change, surface energy densities do not change. In the weightless state P^L and P^G in the phase do not change spatially (there is no gravity effect). An important consequence of constant P^L and P^G is that the interface curvature stays constant in the interface, except very close to the wall. From the Young-Laplace equation,

$$P^G - P^L = \frac{2\sigma^{LG}}{r} \tag{7.237}$$

(r is the interface radius), one can readily see that interface curvature does not vary. The volumes of the gas and the liquid phases add to the total volume $V = V^L + V^G$; therefore, $dV^L = -dV^G$; also, $dn^L = -dn^G$. Combining the above equations, we obtain Eq. (7.231). The ratio of liquid to gas volume x is expressed as,

$$x = \frac{V^L}{V^G} \tag{7.238}$$

We can also write

$$V^L + V^G = \frac{4\pi R^3}{3} \tag{7.239}$$

Therefore

$$V^L = \frac{x}{1+x}\frac{4\pi R^3}{3} \tag{7.240}$$

$$V^G = \frac{1}{1+x}\frac{4\pi R^3}{3} \tag{7.241}$$

The relation between x and h (see Fig. 7.9a) is expressed by

$$\left(\frac{h}{R}\right)^2\left(1 - \frac{1}{3}\frac{h}{R}\right) = \left(\frac{4}{3}\right)\frac{x}{1+x} \tag{7.242}$$

x and h/R can be used interchangeably in the analysis. Various possible shapes of the gas-liquid interface should be examined for the surface energy (see Fig. 7.9). There are many other possibilities. In the following, we will consider flat and convex surfaces. Expressions for the surface energies are then established. The lower surface energy is the more probable shape.

Flat Interface The volume of the liquid phase when the interface is flat (see Fig. 7.9a) is given by,

$$V^L = \Psi(\beta)\frac{4\pi R^3}{3} \tag{7.243}$$

Note that β is the angle shown in Fig. 7.9a. The expression for the shape factor $\Psi(\beta)$ is given by Eq. (7.110). Combining Eqs. (7.240) and (7.243),

$$\Psi(\beta) - \frac{x}{1+x} = 0 \tag{7.244}$$

Given V^L, or x, β can be obtained from Eq. (7.243).

We then write $\mathcal{A}^{LS} + \mathcal{A}^{GS} = 4\pi R^2$, and,

$$\mathcal{A}^{LG} = 4\pi R^2 \sin^2\beta \tag{7.245}$$

$$\mathcal{A}^{LS} = 2\pi R^2[1 - \cos\beta] \tag{7.246}$$

$$\mathcal{A}^{GS} = 2\pi R^2[1 + \cos\beta] \tag{7.247}$$

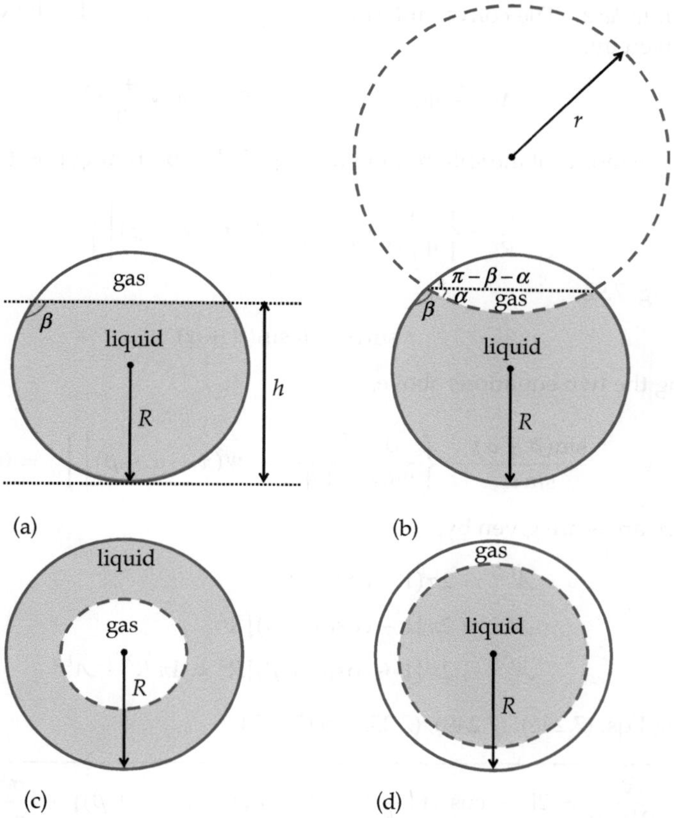

FIGURE 7.9 Some interface configurations in weightless state.

Substitution in Eq. (7.235) gives,

$$\frac{A^s}{\pi R^2 \sigma^{LG}} = 4\sin^2\beta + 2[1 - \cos\beta]\frac{\sigma^{LS}}{\sigma^{LG}} + 2[1 + \cos\beta]\frac{\sigma^{GS}}{\sigma^{LG}} \tag{7.248}$$

The surface energy densities are related by the Young equation,

$$\sigma^{LG}\cos\theta = \sigma^{GS} - \sigma^{LS} \tag{7.249}$$

where θ is the contact angle which is known.

Combining the above two equations, one obtains,

$$\boxed{\frac{A^s}{\pi R^2 \sigma^{LG}} = 4\sin^2\beta + 2\cos\theta(-1 + \cos\beta) + 4\frac{\sigma^{GS}}{\sigma^{LG}}} \tag{7.250}$$

Convex Interface The convex interface is depicted in Fig. 7.9b. The volume of the gas phase is given by,

$$V^G = \Psi(\pi - \alpha - \beta)\frac{4\pi}{3}R^3 + \Psi(\alpha)\frac{4\pi}{3}r^3 \tag{7.251}$$

where r is the radius of the sphere shown in Fig. 7.9b. Combining Eqs. (7.241) and (7.251),

$$\frac{r}{R} = \left\{\frac{1}{\Psi(\alpha)}\left[\frac{1}{1+x} - \Psi(\pi - \alpha - \beta)\right]\right\}^{\frac{1}{3}} \tag{7.252}$$

Based on Fig. 7.9b,

$$r \sin\alpha = R\sin(\beta + \alpha) \tag{7.253}$$

Combining the two equations above,

$$\frac{\sin(\beta+\alpha)}{\sin\alpha} - \left\{\frac{1}{\Psi(\alpha)}\left[\frac{1}{1+x} - \Psi(\pi - \alpha - \beta)\right]\right\}^{\frac{1}{3}} = 0 \tag{7.254}$$

The surface areas are given by,

$$\mathcal{A}^{LG} = 2\pi(1 - \cos\alpha)r^2 \tag{7.255}$$

$$\mathcal{A}^{LS} = 2\pi[1 - \cos(\alpha + \beta)]R^2 \tag{7.256}$$

$$\mathcal{A}^{GS} = 2\pi[1 + \cos(\alpha + \beta)]R^2 = 4\pi R^2 - \mathcal{A}^{LS} \tag{7.257}$$

Combining Eqs. (7.236), (7.249), (7.255) to (7.257),

$$\boxed{\frac{A^s}{\pi R^2\sigma^{LG}} = 2[1 - \cos\alpha]\left(\frac{r}{R}\right)^2 + 2\cos\theta[1 + \cos(\alpha+\beta)] + 4\frac{\sigma^{LS}}{\sigma^{LG}}} \tag{7.258}$$

The more probable configuration can be obtained by comparing Eq. (7.250) and (7.258); the lower surface energy is the more probable configuration.

Configurations in Figs. 7.9c and 7.9d will be addressed in Prob. 7.23.

7.10: Effect of Line Tension on Contact Angle In a recent work we have measured the effect of drop size on contact angle of oil/water/substrate system. We first use a drop of size 10 μL (petroleum fluid). As we increase the size of the drop to 30 μL a decrease in contact angle is measured, from about 80° for a drop of 10 μL to about 40° for a size of 30 μL and higher. Derive the following expression that relates the contact angle of a small drop $\theta_{\mathfrak{r}}$ to the contact angle θ given by Eq. (7.115),

$$\cos\theta_{\mathfrak{r}} = \frac{\sigma_s - \sigma_i}{\sigma} - \frac{\kappa}{\sigma\mathfrak{r}} = \cos\theta - \frac{\kappa}{\sigma\mathfrak{r}}$$

In the above expression $\mathfrak{r}$ is the radius of the drop base ($\mathfrak{r} = r\sin\theta_{\mathfrak{r}}$), and κ is the line tension. Note that as $\mathfrak{r}$ increases the second term in the right side vanishes.

Solution Add $2\pi\kappa d\mathfrak{r}$ to the expression for $d\phi_t$ in Eq. (7.112) to account for line tension. Note that the total volume of the system volume V and volume of drop V_n remain constant. Therefore, both $PdV = 0$, and $dV_n = 0$. P is the pressure outside the drop. The total volume is constrained in the Helmholtz free energy derivation. The combination of $d\phi_t = 0$ and $dV_n = 0$ lead to the above expression.

Various authors including Weijs et *al.* (2011) have discussed line tension effect on contact angle. While thermodynamics stability analysis shows that the surface tension (surface energy density) is positive, line tension sign cannot be inferred from thermodynamics stability analysis. In shale porous media due to small size of nanopores, the effect of size on contact angle may be important.

7.14 Problems

7.1 Show that for a one-component system

$$u^s = \sigma - T\frac{d\sigma}{dT}$$

where u^s is the superficial internal energy per unit area of the interface.

7.2 Show that the temperature dependence of the interfacial tension can be used to determine the interfacial entropy and then the total surface energy.

7.3 Using the data given in the table below, compute the adsorption "inventory" of ethanol at the interface relative to water as a function of the bulk concentration of ethanol.

x_2 (mol fraction of ethanol)	Surface Tension (dynes/cm)	Partial Pressure of Ethanol (mm Hg)
1.0	21.93	5.8767
0.9	22.93	5.5957
0.8	23.26	5.3602
0.7	23.97	5.1500
0.6	24.67	4.9530
0.5	25.43	4.7827
0.4	26.43	4.6136
0.3	27.60	4.4460
0.2	29.97	4.1704
0.1	36.72	3.5113
0.0	71.97	0

Temperature = 25°C

7.4 Consider a binary mixture of nitrogen and water in saturated gas and liquid phases at 17°C. Measured data show that as the pressure is increased from 1 to 3 atm, the interfacial tension between the equilibrium phases decrease about 0.2 dyne/cm. Use the interfacial tension change to calculate the relative adsorption of nitrogen at the interface.

7.5 Derive the expressions

$$U = TS - P''V + \sum_{i=1}^{n_c} \mu_i N_i$$

$$A = -P''V + \sum_{i=1}^{n_c} \mu_i N_i$$

for the internal energy and the Helmholtz free energy of a composite system with a curved interface. Also derive another form of the above functions given by

$$U = TS - P'V' - P''V'' + \gamma\mathcal{A} + \sum_{i=1}^{n_c} \mu_i N_i$$

$$A = -P'V' - P''V'' + \gamma\mathcal{A} + \sum_{i=1}^{n_c} \mu_i N_i$$

Obtain the expression of the Gibbs free energy of a composite system with a curved interface given by

$$G = U - TS + P''V = A + P''V$$

7.6 Two students were arguing on the condition for the minimum of internal energy U at equilibrium in a system of two homogeneous bulk phases, and in a heterogeneous system consisting of two homogenous bulk phases and an interface. One of the students argued that there is a difference in invariability of volume V in the two systems and stressed that in a system consisted of two bulk phases V should be constant while in a heterogeneous system with an interface, the invariability of boundaries defines the minimum of U. Which student do you side with?

7.7 Show that for a critical-size cluster, the cluster phase is in equilibrium with the other bulk phase. This equilibrium is of an unstable nature. What is the difference between the unstable and state equilibrium states?

7.8 Show that the driving force for condensation of a solute at constant pressure and temperature can be approximated by

$$\Delta\mu(c, T) = kT \ln\left[\frac{c}{c_e}\right]$$

where c is the actual concentration and c_e is the equilibrium concentration.

7.9 Show that both for a single component or a multicomponent mixture the reversible work of forming a spherical critical nucleus is given by

$$W^* = \frac{1}{3}\sigma\mathcal{A} = \frac{16\pi\sigma^3}{3(P_e - P)^2}$$

Also show that for both cases, the critical radius is given by

$$r^* = \frac{2\sigma}{(P_e - P)}$$

7.10 Derive the expression for the work of formation of lens-shaped clusters shown in Fig. 7.10c. The index h in the surface energy densities indicate the hydrate crystal.

The work of cluster formation for spherical and capped-shaped clusters was derived in the text (corresponding to Fig. 7.10a and 7.10b, below).

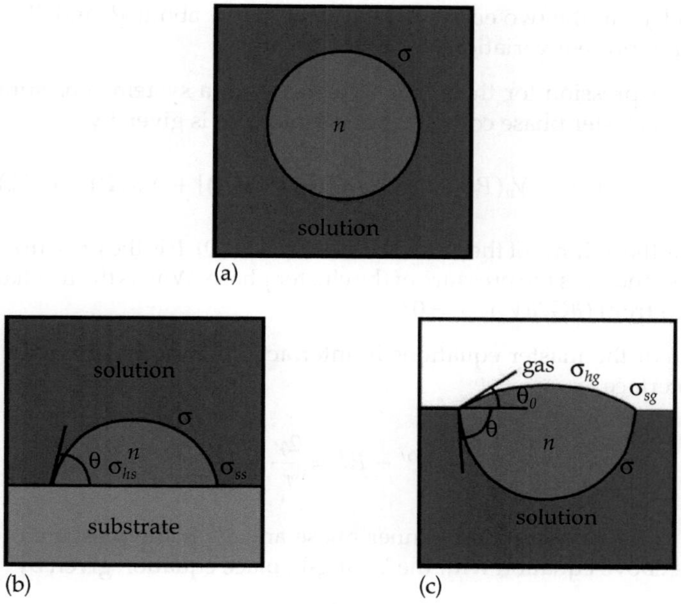

FIGURE 7.10 Various cluster shapes (adapted from Kashchiev and Firoozabadi, 2002).

Remarks You may need to first derive the expressions for contact angles θ and θ_0 given by

$$\cos\theta = \left(\frac{\sigma_{sg}^2 + \sigma^2 - \sigma_{hg}^2}{2\sigma_{sg}\sigma}\right)$$

$$\cos\theta_0 = \left(\frac{\sigma_{sg}^2 - \sigma^2 + \sigma_{hg}^2}{2\sigma_{sg}\sigma_{hg}}\right)$$

7.11 Based on Example 7.4, show that oil recovery from the solution-gas drive process is higher at higher rate of depletion. This is mainly true in lab scale.

7.12 The expression for work in a two-phase system with a curved interface is given by

$$đW = -P'\mathrm{d}V' - P''\mathrm{d}V'' + \gamma\mathrm{d}\mathcal{A} + \left[\frac{\partial\gamma}{\partial r}\right]\mathcal{A}\mathrm{d}r \tag{7.259}$$

where γ is the interfacial tension, $\mathcal{A}$ is the interface area, and r is the radius of the interface.

Derive the following equation from the above expression:

$$đW = -P'\mathrm{d}V' - P''\mathrm{d}V'' + \sigma\mathrm{d}\mathcal{A} \tag{7.260}$$

Are V' and V'' in the two equations the same? How about P' and P''? Does the second expression represent variation of σ with curvature?

7.13 The expression for the Gibbs free energy of a system consisting of a large bulk phase and a cluster phase consisting of n molecules is given by

$$G = -V_n(P_n - P) + n[\mu(P_n) - \mu(P)] + N\mu(P) + \phi(V_n)$$

where V_n is the volume of the cluster (based on $N^s = 0$), P is the pressure of the continuous bulk phase, and P_n is the pressure of the cluster phase. Why is the mechanical equilibrium established from $(\partial G/\partial V_n)_{n,P_n} = 0$?

7.14 One of the master equations in interfacial thermodynamics for a system with a curved interface is

$$P' - P'' = \frac{2\gamma}{r} + \left[\frac{\partial \gamma}{\partial r}\right]$$

where P' is the pressure of the inner phase and P'' is the pressure of the outer phase. Relate the above equation with the Young-Laplace equation given by

$$P' - P'' = \frac{2\sigma}{r}$$

Does r stand for the same physical parameter in the above two questions?

7.15 We derived the expressions for the thermodynamic driving force at constant temperature for condensation from vapor and evaporation from liquid. Derive similar expressions for the driving force at constant external pressure.

7.16 Consider pure liquid n-C_5 at $T = 450$ K and $P = 0.6\ P_e$ (note that the liquid is supersaturated). Plot the work of formation of a bubble versus the number of molecules in the bubble. What is your estimation of the critical bubble size? (Assume $\sigma = 1$ dyne/cm.) Repeat the same calculations for $P = 0.8\ P_e$.

7.17 Calculate the critical-bubble radius and the number of molecules in a critical-size bubble for n-pentane and benzene at the following conditions.

n-Pentane

$T = 146.1°C$, $P = 1.0$ bar; $= 3.28$ dyne/cm

$T = 162.0°C$, $P = 12.8$ bar; $= 2.00$ dyne/cm

Benzene

$T = 225.3°C$, $P = 1.0$ bar; $= 4.66$ dyne/cm

$T = 240.6°C$, $P = 14.7$ bar; $= 3.19$ dyne/cm

7.18 In the classical nucleation theory, some authors have suggested the following expression for the work of cluster formation in a binary mixture (Reiss, 1950; Doyle, 1961):

$$\Delta G(n_1, n_2) = n_1 \Delta\mu_1 + n_2 \Delta\mu_2 + \mathcal{A}\sigma \tag{7.261}$$

In the above equation, n_1 and n_2 are the number of molecules of species 1 and 2 in the cluster and $\Delta\mu_i = \mu_i^L - \mu_i^G$. Derive the above expression from our general expression given by Eq. (7.179) and discuss the assumptions.

For the critical nucleus where the variations in ΔG with respect to n_1 and n_2 should vanish, Lewis and Randall (1961) derive

$$\Delta\mu_i + \frac{2\sigma\bar{V}_i}{r^*} = 0 \tag{7.262}$$

where $\bar{V}_i$ is the partial molar volume of component i. Doyle's (1961, 1981) derivation includes an additional term to account for the effect of composition on the interfacial tension,

$$\Delta\mu_i + \frac{2\sigma\bar{V}_i}{r^*} + \frac{3(1-x_i)v}{r^*}\frac{d\sigma}{dx_i} = 0 \tag{7.263}$$

where v is the average molar volume and x_i is an average mol fraction. Note that σ is a function of n_1 and n_2. Provide your derivation.

Remarks There can be substantial inconsistency between Eqs. (7.262) and (7.263). Wilemsky (1984) has used Eqs. (7.262) and (7.263) to calculate the composition of a critical-size droplet for a mixture of ethanol and water at 293.2 K. Equation (7.262) gives a mol fraction of 0.06 for ethanol and a corresponding interfacial tension of 43.7 dyne/cm while the results from Eq. (7.263) are ethanol mol fraction = 0.19, and $\sigma = 30.7$ dyne/cm. Wilemski's calculations demonstrate that the agreement between theory and experiments for the nucleation of water-rich binary liquid droplets is much improved when Eq. (7.262) is used.

7.19 Show that the Young equation (which relates contact angles to various surface energy densities) is not applicable to the case when the contact angle is zero.

7.20: Work of Cluster Formation in 2D Nucleation and the Specific Edge Energy Derive the following expression for work of cluster formation in 2D (Fig. 7.11).

$$W = 2\pi r\kappa - \pi r^2\left(\frac{\Delta\mu}{a} - \Delta\sigma\right)$$

where a is the area per molecule, r is the radius, κ is the specific edge energy, $\Delta\mu$ is the driving force, and $\Delta\sigma \equiv \sigma + \sigma_{ns} - \sigma_s$. Detailed derivations are provided in Anklam and Firoozabadi (2005). Note that the specific edge energy and the line tension are not the same (Kashchiev,2000), although shown by the same symbol.

7.21: Derivation of Gibbs Adsorption Equation for Edge Energy in 2D Derive the expression relating the line energy change to the absorbed amount per length.

$$d\kappa = -\sum_{i=1}^{n_c} \tau_i^E d\mu_i$$

where $\tau_i^E = N_i^E/L$. The above expression is analogous to the Gibbs adsorption equation in a plane (Anklam and Firoozabadi, 2005).

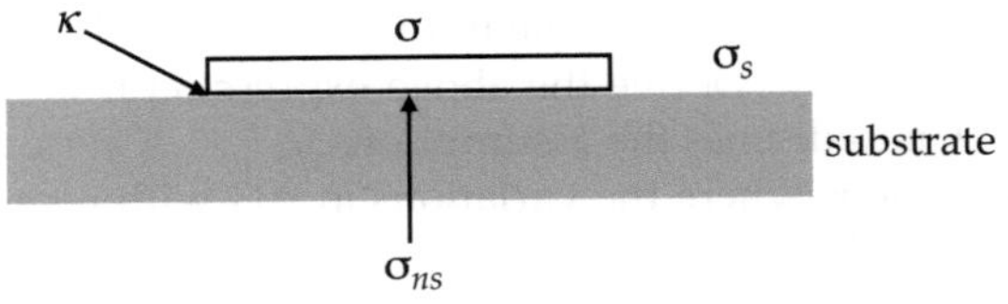

FIGURE 7.11

7.22: Critical Nucleus Size in 2D and Its Interpretation Use the work of cluster formation in Prob. 7.20 to derive the expression for critical radius.

$$r^* = \frac{\kappa}{\left(\frac{\Delta\mu}{a} - \Delta\sigma\right)} \tag{7.264}$$

Derive also the expression for the work to form a critical-2D nucleus given by,

$$W^* = \frac{\pi \kappa^2 a}{(\Delta\mu - a\,\Delta\sigma)} \tag{7.265}$$

What is the implication for a large $\Delta\sigma$?

7.23 Derive the following expressions for surface energy in weightless state of the following two configurations.

(a) Gas bubble: $A^s/(\pi R^2 \sigma^{LG}) = 4[1/(1+x)]^{\frac{1}{3}} + 4\sigma^{LS}/\sigma^{LG}$ (See Fig. 7.9c.)
(b) Liquid droplet: $A^s/(4\pi R^2 \sigma^{LG}) = 4[x/(1+x)]^{1/3} + 4\sigma^{GS}/\sigma^{LG}$ (See Fig. 7.9d.)

Symbols are provided in Example 7.9.

7.24: Rayleigh Limit for a Charged Droplet When charged droplets are smaller than certain limit, they will fly off in different directions. This is not the case in uncharged droplets, which may grow or shrink. In other words, when there is evaporation from a charged droplet, as it becomes smaller, there will be a size at which it becomes unstable and will break up into smaller droplets. Derive the following limiting diameter of a charge droplet which flies off in different directions, that is, the Raleigh limit.

$$d^{*3} = \frac{2q^2}{\pi\sigma}\left(\frac{1}{\varepsilon_0} - \frac{1}{\varepsilon}\right)$$

The derivation of the Rayleigh limit was first published by Lord Rayleigh in 1882.

7.15 References

Alberty R. A., Legendre transforms in chemical thermodynamics, *Chemical Reviews*, 1994. 94: pp.1457–1482.

Anklam, M. R., and A. Firoozabadi, An interfacial energy mechanism for the complete inhibition of crystal growth by inhibition adsorption. *Journal of Chemical Physics*, 2005. 123: p. 144708.

Anklam, M. R., and A. Firoozabadi, Driving force and composition for multicomponent gas hydrate nucleation from supersaturated aqueous solutions. *Journal of Chemical Physics*, 2004. 121(23): pp. 11867–11874.

Debenedetti, P. G., and H. Reiss, Reversible work of formation of an embryo of a new phase within a uniform macroscopic mother phase. *Journal of Chemical Physics*, 1998. 108(13): pp. 5498–5505.

Defay, R., and I. Prigogine, Surface tension and adsorption, 1966.

Doyle, G. J., Self-nucleation in the sulfuric acid-water system. *Journal of Chemical Physics*, 1961. 35(3): pp. 795–799.

Doyle, G. J., Response to the comment on self-nucleation in the sulfuric acid-water system. *Journal of Chemical Physics*, 1981. 75(3): pp. 1585–1586.

Drelich, J., Miller, and J. D. J. Hupka, The effect of drop size on contact-angle over a wide-range of drop volumes. *Journal of Colloid and Interface Science*, 1993. 155: pp. 379–385.

Everett, D. H., Manual of symbols and terminology for physicochemical quantities and units, appendix II: Definitions, terminology and symbols in colloid and surface chemistry, *Pure and Applied Chemistry*, 1972. 31: pp. 577–638.

Kashchiev, D., *Nucleation: Basic Theory with Applications*. 2000, Butterworth Heinemann, Boston. 2000.

Kashchiev, D., and A. Firoozabadi, Nucleation of gas hydrates. *Journal of Crystal Growth*, 2002. 243: pp. 476–489.

Kirkwood, J. G., and F. P. Buff, The statistical mechanical theory of surface tension. *Journal of Chemical Physics*, 1949. 17(3): pp. 338–343.

Laaksonen, A., and R. McGraw, Thermodynamics, gas-liquid nucleation, and size-dependent surface tension. *Europhys Letters*, 1996. 35(5): pp. 367–372.

Laaksonen, A., McGraw, R., and H. Vehkamäki, Liquid-drop Formation and Free-energy Surfaces in Binary Homogeneous Nucleation Theory. *Journal of Chemical Physics*, 1999. 111(2): pp. 2019–2027.

Larson, M. A., and Garside, Solute clustering and interfacial tension. *Journal of Crystal Growth*, 1986. 76: pp. 88–92.

Larson, M. A., and Garside, Solute clustering in supersaturated solutions. *Chemical Engineering Science*, 1986. 41(5): pp. 1285–1289.

Lewis, G. N., and M. Randall, *Thermodynamics*. 2nd ed. (revised by Pizter, K. S., and Brewer, L.), 1961. New York, McGraw-Hill, 723.

McGraw, R., and A. Laaksonen, Interfacial curvature free energy, the kelvin relation, and vapor-liquid nucleation rate. *Journal of Chemical Physics*, 1997. 106(12): pp. 5284–5287.

Nishioka, K., and I. Kusaka, Thermodynamic formulas of liquid phase nucleation from vapor in multicomponent systems. *Journal of Chemical Physics*, 1992. 96(7): pp. 5370–5376.

Ono, S., and S. Kondo, Molecular Theory of Surface Tension in Liquids, Handbuch der Physik, (1960) Bd. 10, Struktur der Fliissigkeiten (Springer, Berlin, Gottingen, Heidelberg).

Oxtoby, D. W., and D. Kaschiev, A general relation between the nucleation work and the size of the nucleus in multicomponent nucleation. *Journal of Chemical Physics*, 1994. 100: pp. 7665–7671.

Lord Raleigh, F. R. S., On the equilibrium of liquid conducting masses charged with electricity. *Philosophical Magazine Series 5*, 1882. 14(87): pp. 184–186.

Reiss, H., The kinetics of phase transitions in binary systems, *Journal of Chemical Physics*, 1950. 18(6): pp. 840–848.

Santiso, E., and A. Firoozabadi, Curvature dependence of surface tension in multicomponent systems. *AIChE Journal*, 2006. 52(1): pp. 311–322.

Thomson, J. J., *Applications of Dynamics to Physics and Chemistry*. 1888, London: Macmillan and Co.

Tolman, R. C., The effect of droplet size on surface tension. *Journal of Chemical Physics*, 1949. 17(3): pp. 333–337.

Weijs, J. H., A. Marchand, B. Andreotti, D. Lohse, J. H. Snoeijer, Origin of line tension for a Lennard-Jones nanodroplet. *Physics of Fluids*, 2011. 23, p. 022001.

Wilemski, G., Composition of the critical nucleus in multicomponent vapor nucleation. *Journal of Chemical Physics*, 1984. 3(80): pp. 1370–1372.

Wilemski, G., Revised classical binary nucleation theory for aqueous alcohol and acetone vapors. *Journal of Chemical Physics*, 1987. 91: pp. 2492–2498.

Wilemski, G., Some issues of thermodynamic consistency in binary nucleation theory. *Journal of Chemical Physics*, 1988. 88(8): pp. 5134–5136.

CHAPTER 8

Thermodynamics of Thin Liquid Films

There are ample evidences that a thin liquid film forms on the surface of various crystals below the melting temperature. Figure 8.1 shows an example of a film of water on ice in relation to ice skating. Thin liquid film may affect oil recovery from low salinity water injection (Myint and Firoozabadi, 2015a).

In Chap. 7, we assumed zero thickness for the interface between two fluid phases, but excess properties were assigned to the interface. The thermodynamic description of thin liquid films on solid surfaces is based on film's thickness. The thickness of the film affects agglomeration of solid particles through capillarity (Anklam et al., 2008).

Thermodynamic description of a thin liquid film is based on the concept of disjoining pressure Π, which is defined as the excess normal pressure in a liquid film $\Pi(h) = P_n^F - P^L$, where h is the film thickness, P_n^F is the normal pressure at the interface on the side of the film, and P^L is the pressure in the bulk liquid phase connected to the thin liquid film. Another key expression is the augmented Young equation $\gamma^{LG} \cos\theta = \gamma^{GS} - \gamma^{LS} - \Pi h$; θ is the contact angle and γ represents the surface energy density (GS and LS represent gas-solid, and liquid-solid, respectively). We will also define film tension, γ^F. Another important expression derived in this chapter is $\gamma^{LG}(\cos\theta - 1) = \Pi(h)h + \int_h^{\infty} \Pi(h')\mathrm{d}h'$, which connects disjoining pressure and contact angle; LG index represents liquid-gas. We also derive the expressions for film thermodynamic functions when the liquid film thickness is accounted for. All these derivations are central to thermodynamics of thin liquid films.

Before embarking on our thermodynamic derivations, we would like to examine the examples of the thin liquid film of water on the surface of hydrates and ice. Premelting of ice was first discussed in 1840 by Faraday (Faraday, 1860). Figure 8.2 shows the decrease in thickness of the water layer as the temperature decreases. In a recent work, Jiménez-Ángeles and Firoozabadi (2014) have computed the thickness of the liquid water film on the surface of ice and hydrates by molecular dynamics simulations. Ice and hydrate crystals were selected at a temperature below the crystallization temperature in contact with methane at a pressure of 100 bar and a temperature 1°C below the melting point. A thin layer of water was computed to form on the surface of ice and gas hydrates (see Fig. 8.3). The thickness of the liquid film on the hydrate surface is computed as 1.4 nm, and the thickness of water film on the surface of ice is 1 nm. The computed thickness on the surface of ice is in line with measurements presented in Fig. 8.2.

FIGURE 8.1 Ease of skating on ice due to film of water.

This chapter includes one example and a few problems. The example relates to the effect of size on the melting temperature of gold particles. Incorporation of a thin liquid film on the surface of gold particles allows proper modeling of melting of gold nanoparticles.

In the following derivations, we will follow a recent work by Myint and Firoozabadi (2015b) and consider a liquid droplet resting on a flat surface. The solid surface is part of

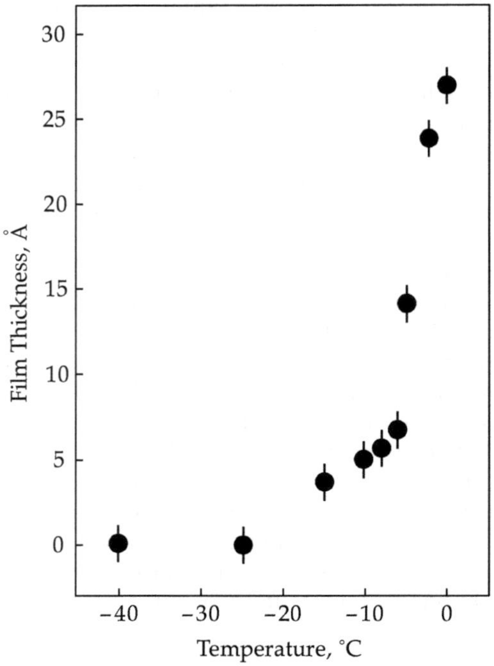

FIGURE 8.2 The thickness of liquid-like layer on ice surface as a function of temperature (adapted from Bluhm et al. 2002).

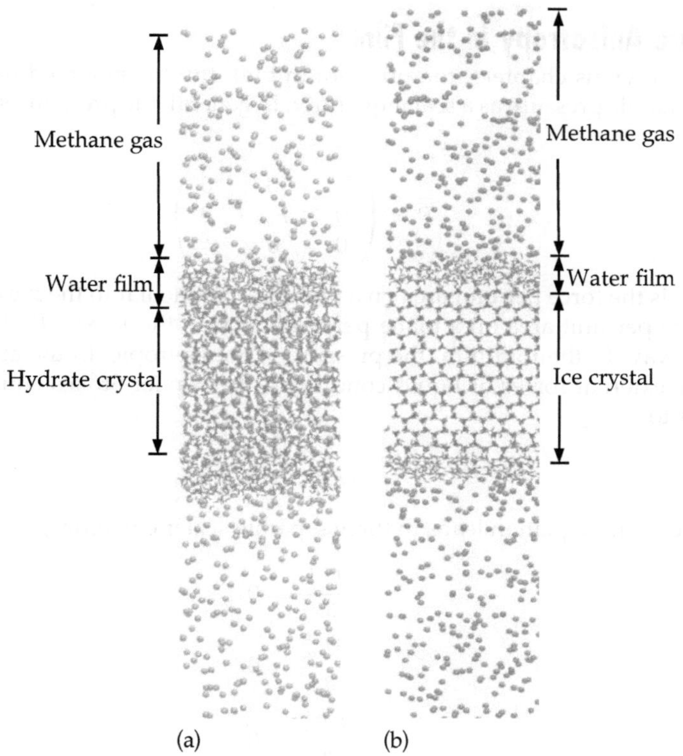

FIGURE 8.3 (a) Methane-hydrate/methane-gas, and (b) ice/methane-gas interface at 1°C below the melting temperature of hydrate and water. $P = 100$ bar. The simulation box used in the calculations has dimensions of around 3.5 nm × 3.5 nm × 14 nm (adapted from Jiménez-Ángeles and Firoozabadi, 2014).

a nondeformable container that encloses the droplet and surrounding gas phase. Figure 8.4 shows the droplet on the surface and the spreading of a thin liquid film. The contact angle and the film thickness are also shown in Fig. 8.4. When liquid is strongly wetting, the droplet may vanish and spread over the whole surface. In our derivations, we assume the film thickness to be uniform. We first examine the pressure in the liquid film.

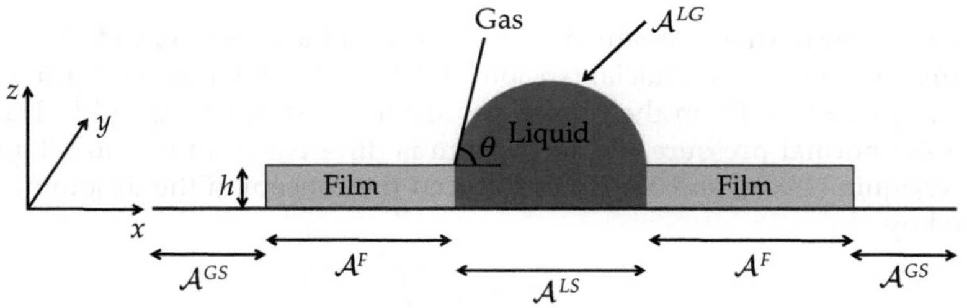

FIGURE 8.4 Sketch of liquid droplet and the thin film. Sizes are not scaled (adapted from Myint and Firoozabadi, 2015b).

8.1 Pressure Anisotropy in the Film

In all the previous chapters, the pressure in bulk fluid is assumed to be equal to the thermodynamic pressure as a scalar quantity. In general, the pressure is a tensor defined by,

$$\vec{\vec{P}} = \begin{pmatrix} P_{xx} & 0 & 0 \\ 0 & P_{yy} & 0 \\ 0 & 0 & P_{zz} \end{pmatrix} \tag{8.1}$$

where P_{xx} is the force per unit area on a plane perpendicular to the x-axis. Similarly, P_{yy} is the force per unit area on a plane perpendicular to the y-axis. P_{zz} is also defined in the same way. In the thin film, the pressure is not isotropic. In the absence of gravity and other external forces at liquid conditions, the Navier-Stokes equation in the film simplifies to

$$\nabla.\vec{\vec{P}}^F = 0 \tag{8.2}$$

The above vector equation leads to three separate scalar equations,

$$\frac{\partial P_{xx}^F}{\partial x} = 0 \tag{8.3}$$

$$\frac{\partial P_{yy}^F}{\partial y} = 0 \tag{8.4}$$

$$\frac{\partial P_{zz}^F}{\partial z} = 0 \tag{8.5}$$

We designate P_{zz}^F as P_n^F since it gives the pressure normal to the film-gas interface. For uniform thickness film P_n^F does not have any spatial dependence, while $P_{xx}^F(z) = P_{yy}^F(z)$ is a function of only z. Symmetry requires that P_{xx}^F and P_{yy}^F be independent of x and y consistent with Eqs. (8.3) and (8.4). We represent P_{xx}^F and P_{yy}^F by a single pressure $P_t^F(z)$, where the subscript indicates the pressure along directions tangent to the film-gas interface.

The interfacial balance of momentum normal to the static gas-liquid interface or gas-film interface yields the Laplace equation,

$$P_n^F - P^G = \frac{2\gamma}{r} \tag{8.6}$$

where P_n^F is the normal pressure on the film side of the interface, and P^G is the gas phase pressure, and γ is the interfacial tension, and $1/r$ is the interface curvature for spherical shapes. $P_n^F(z=h) = P^G$ in the film ($r \to \infty$); in the droplet, $P_n^F = P^L$. The implication is that the normal pressure, P_n^F in the film is different from the bulk liquid pressure, P^L. Derjaquin (1987a, and 1987b) introduced the concept of the disjoining pressure, Π, defined by,

$$\boxed{\Pi(h) = P_n^F - P^L} \tag{8.7}$$

The disjoining pressure decreases sharply with increase in thickness of the film (Hirasaki, 1991; Yeh et al., 1999).

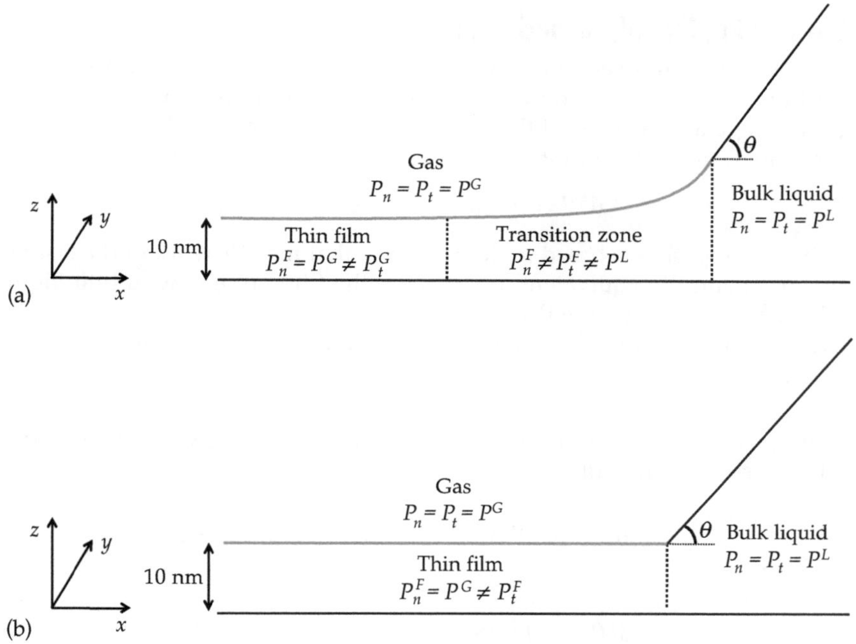

FIGURE 8.5 Two-dimensional cross-section of the system near the liquid-film-gas contact line at a scale comparable to h. The curvature of the liquid-gas interface is not noticeable at this scale. The pressure is anisotropic in the film and in the transition zone. The model approximates the real behavior in (a) with the idealized picture in (b) (adapted from Myint and Firoozabadi, 2015b).

The disjoining pressure $\Pi(h)$ is therefore, a reflection of the difference between the two bulk pressures. If the film thickness is uniform, the disjoining pressure is constant; it is given by Eq. (8.7). Let us consider Fig. 8.5a, in which there is a gradual transition from the bulk liquid phase to the thin liquid film. In that case, instead of Eq. (8.7), the disjoining pressure expression can be combined with the Young-Laplace equation, $P_n^F - P^G = 2\gamma/r$ to obtain,

$$P^L - P^G = \frac{2\gamma}{r} - \Pi(h) \tag{8.8}$$

Note that according to the shape of the interface in Figs. 8.4 and 8.5a,the pressure, P^L is higher than P^G. Equation (8.8) is the augmented Young-Laplace equation. When thickness h is high, such as in bulk like phases Eq. (8.8) reduces to the conventional expression. Figure 8.5b is an idealization of Fig. 8.5a. Figure 8.5a is related to Eq. (8.8), and Fig. 8.5b is related to Eq. (8.7) as already mentioned.

We combine thermodynamics of interfaces with no interface volume from Chap. 7 and thermodynamics of thin liquid films to derive the expression for thermodynamic functions of a thin liquid film. The thermodynamic functions of thin films are generally derived based on film tension (to be defined later). Next, we derive the conditions of equilibrium in a system with a film and start with expressions for film work and energy.

8.2 Thin Liquid Film Work and Energy

The liquid film comprises a volume, (FV) and interfaces. The interfaces for the film of Fig. 8.4 are shown by FG (on the gas side), and FS (on the solid side). Note that we neglect change in side areas of the film in pressure-volume work.

The work done on the film to change its area by $d\mathcal{A}^F$ is given by,

$$đW = \gamma^{FG} d\mathcal{A}^F + \gamma^{FS} d\mathcal{A}^F - P^L h d\mathcal{A}^F \tag{8.9}$$

The above expression is based in defining h so that the PV work in the last term of Eq. (8.9) is mechanically equivalent to replacing the film volume with bulk liquid (Myint and Firoozabadi, 2015b). See Prob. 8.6.

The work done on the film to change its thickness by dh is given by

$$đW = -P_n^F \mathcal{A}^F dh \tag{8.10}$$

Now we write the expression for internal energy of the film, which is composed of two interfaces, and the film volume.

$$dU^{FG} = T^F dS^{FG} + \gamma^{FG} d\mathcal{A}^F + \sum_{i=1}^{n_c} \mu_i^F dN_i^{FG} \tag{8.11}$$

$$dU^{FS} = T^F dS^{FS} + \gamma^{FS} d\mathcal{A}^F + \sum_{i=1}^{n_c} \mu_i^F dN_i^{FS} \tag{8.12}$$

The change in internal energy of the film volume is

$$dU^{FV} = T^F dS^{FV} - P_n^F \mathcal{A}^F dh - P^L h d\mathcal{A}^F + \sum_{i=1}^{n_c} \mu_i^F dN_i^{FV} \tag{8.13}$$

In the above expressions, T^F is the temperature of the film, S is the entropy, μ_i^F is the chemical potential of component i in the film, and N_i is the number of mols of component i. The differential of the internal energy of the film, as a whole, is

$$dU^F = dU^{FV} + dU^{FG} + dU^{FS} = T^F dS^F - P_n^F \mathcal{A}^F dh + (\gamma^{FG} + \gamma^{GS} - P^L h) d\mathcal{A}^F + \sum_{i=1}^{n_c} \mu_i^F dN_i^F \tag{8.14}$$

where $S^F = S^{FV} + S^{FG} + S^{FS}$ and $N_i^F = N_i^{FV} + N_i^{FG} + N_i^{FS}$.

The Helmholtz free energy of the film, $A^F = U^F - T^F S^F$, and therefore,

$$dA^F = -S^F dT^F - P_n^F \mathcal{A}^F dh + (\gamma^{FG} + \gamma^{FS} - P^L h) d\mathcal{A}^F + \sum_{i=1}^{n_c} \mu_i^F dN_i^F \tag{8.15}$$

8.3 Equilibrium Conditions

We will examine the equilibrium conditions for the system in Fig. 8.4 based on Helmholtz free energy. The Helmholtz free energy of the whole system is given by,

$$A = A^L + A^G + A^F + A^{LG} + A^{SG} + A^{LS} \tag{8.16}$$

where the superscripts, L, G, and F represent bulk liquid and gas phases, and the film respectively. The rest are the interfaces. At equilibrium,

$$\mathrm{d}A = \mathrm{d}A^L + \mathrm{d}A^G + \mathrm{d}A^F + \mathrm{d}A^{LG} + \mathrm{d}A^{SG} + \mathrm{d}A^{LS} = 0 \tag{8.17}$$

From Chap. 1 and 7,

$$\mathrm{d}A^L = -S^L \mathrm{d}T^L - P^L \mathrm{d}V^L + \sum_{i=1}^{n_c} \mu_i^L \mathrm{d}N_i^L \tag{8.18}$$

$$\mathrm{d}A^G = -S^G \mathrm{d}T^G - P^G \mathrm{d}V^G + \sum_{i=1}^{n_c} \mu_i^G \mathrm{d}N_i^G \tag{8.19}$$

$$\mathrm{d}A^I = -S^I \mathrm{d}T^I + \gamma^I \mathrm{d}A^I + \sum_{i=1}^{n_c} \mu_i^I \mathrm{d}N_i^I \quad I = LG, LS, GS \tag{8.20}$$

In Eq. (8.20), I represents the interface. At Helmholtz free energy minimum (the system is isothermal and the whole system is inside a container of constant volume), from thermal equilibrium

$$\mathrm{d}T^L = \mathrm{d}T^G = \mathrm{d}T^F = \mathrm{d}T^I = 0, \quad I = LG, LS, GS \tag{8.21}$$

From material balance,

$$\mathrm{d}N_i^L = -\mathrm{d}N_i^G - \mathrm{d}N_i^F - \sum_I \mathrm{d}N_i^I \quad i = 1, \ldots, n_c \tag{8.22}$$

Combining Eqs. (8.15), and (8.17) to (8.22),

$$\mathrm{d}A = -P^L \mathrm{d}V^L - P^G \mathrm{d}V^G - P_n^F A^F \mathrm{d}h + (\gamma^{FG} + \gamma^{FS} - P^L h)\mathrm{d}A^F + \sum_I \gamma^I \mathrm{d}A^I$$

$$+ \sum_{i=1}^{n_c} (\mu_i^F - \mu_i^L)\mathrm{d}N_i^F + \sum_{i=1}^{n_c} (\mu_i^G - \mu_i^L)\mathrm{d}N_i^G + \sum_I \sum_{i=1}^{n_c} (\mu_i^I - \mu_i^L)\mathrm{d}N_i^I = 0 \tag{8.23}$$

Since N_i^G, N_i^F, and N_i^I are independent of each other,

$$\boxed{\mu_i^L = \mu_i^G = \mu_i^F = \mu_i^I} \quad i = 1, \ldots, n_c \quad I = LG, LS, GS \tag{8.24}$$

The equality of chemical potentials relates to chemical equilibrium. We proceed to derive the condition of mechanical equilibrium, which is not trivial.

8.3.1 Mechanical Equilibrium

The condition for mechanical equilibrium is established from the remaining terms of Eq. (8.23),

$$-P^L \mathrm{d}V^L - P^G \mathrm{d}V^G - P_n^F A^F \mathrm{d}h + (\gamma^{FG} + \gamma^{FS} - P^L h)\mathrm{d}A^F + \sum_I \gamma^I \mathrm{d}A^I = 0 \tag{8.25}$$

Note that we have implemented the condition for thermal and chemical equilibrium in Eq. (8.25). Here, unlike in Chap. 1, chemical equilibrium only relates to the condition of equal chemical potentials and equal temperatures in all phases and interfaces. In Eq. (8.25), dV^L and dV^G are related to each other as discussed in Chap. 7 in relation to a cap. Some other terms are also dependent.

The system is constrained by constant volume and constant surface area of the container: $d(V^L + V^G + V^F) = 0$ and $d(\mathcal{A}^F + \mathcal{A}^{LS} + \mathcal{A}^{GS}) = 0$. The volume of the film is given by $V^F = h\mathcal{A}^F$. Use of these two constraints in Eq. (8.25) gives,

$$(P^G - P^L)dV^L + (P^G - P_n^F)\mathcal{A}^F dh + \left[(\gamma^{FG} + \gamma^{FS} - P^L h) - \gamma^{GS} + P^G h\right]d\mathcal{A}^F + (\gamma^{LS} - \gamma^{GS})d\mathcal{A}^{LS} + \gamma^{LG} d\mathcal{A}^{LG} = 0 \tag{8.26}$$

Because of the complicated nature of differentials and their dependency on each other, we use the virtual work associated with virtual displacement. The concept of virtual work is used in analytical mechanics (Lanczos, 1986) and was used by Gibbs and others (Gibbs, 1948; Johnson, 1959; and Li and Newmann, 1991).

Let us consider the sketch in Fig. 8.6 and follow four virtual displacements. The displacements normal to the liquid-gas interface, film-gas interface, liquid-film-gas contact line, and film-gas-solid contact line are represented by δn^{LG}, δh, δn^{LFG} and, δn^{FGS}, respectively (see Fig. 8.6). $\delta n^{LG} > 0$ when the displacement is directed away from the droplet; $\delta h > 0$ if h increases; $\delta n^{LFG} > 0$ if the displacement is directed away from the bulk liquid; and, $\delta n^{FGS} > 0$ if the displacement is directed away from the film. Based on these conventions, δn^{LG} and δh, in Fig. 8.6 are negative; δn^{LFG} and δn^{FGS} are positive. Li and Newmann (1991) have a similar approach as in the analysis of Myint and Firoozabadi (2015b), but only consider three displacements instead of four (see Prob. 8.7). We include displacement δn^{FGS}.

The virtual work associated with δn^{LG} is given by,

$$-\int_{\mathcal{A}^{LG}} P^L \delta n^{LG} d\mathcal{A}^{LG} - \int_{\mathcal{A}^{LG}} P^G(-\delta n^{LG}) d\mathcal{A}^{LG} + \int_{\mathcal{A}^{LG}} 2\gamma^{LG} H \delta n^{LG} d\mathcal{A}^{LG} = \int_{\mathcal{A}^{LG}} (P^G - P^L + 2\gamma^{LG} H)\delta n^{LG} d\mathcal{A}^{LG} \tag{8.27}$$

In the above expression, H (i.e., $1/r$) is the curvature, and the first two integrals on the left side represent the pressure-volume work done on the liquid and gas phases, respectively. The third term represents the interfacial work due to displacement, δn^{LG}. The interfacial work can be understood from Fig. 8.7.

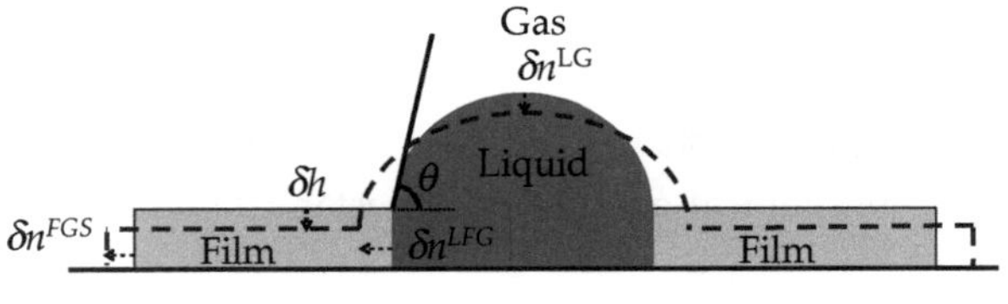

FIGURE 8.6 The four virtual displacements (δn^{LG}, δh, δn^{LFG}, and δn^{FGS}) of droplet and film (adapted from Myint and Firoozabadi, 2015b).

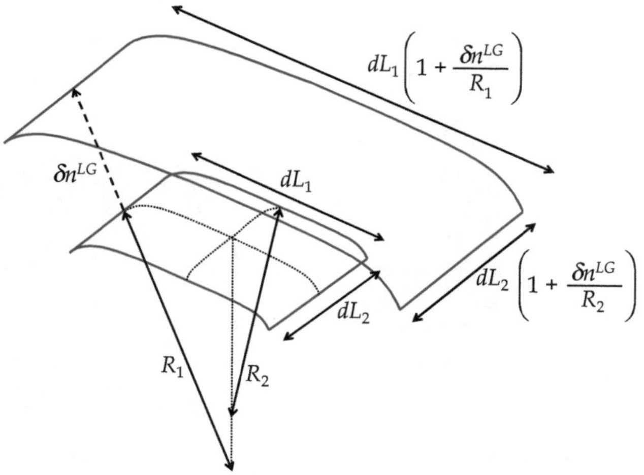

FIGURE 8.7 Sketch of a differential element of the liquid-gas interface (defined by radii R_1 and R_2) and the resultant shape of the element after the virtual displacement δn^{LG} is applied. Before undergoing the displacement, the area of the element is approximately $\mathrm{d}L_1\mathrm{d}L_2$. After the displacement, the element becomes deformed so that its area is approximately $\mathrm{d}L_1(1+\delta n^{LG}/R_1)\mathrm{d}L_2(1+\delta n^{LG}/R_2)$, which simplifies to $[1+\delta n^{LG}/(R_1+R_2)]\mathrm{d}L_1\mathrm{d}L_2$ if the quadratic term in δn^{LG} is neglected. Thus, the displacement δn^{LG} maps an area $\mathrm{d}A^{LG}$ of the interface to an area $[1+\delta n^{LG}/(R_1+R_2)]\mathrm{d}A^{LG}$ (adapted from Myint and Firoozabadi, 2015b).

The work associated with δh is given by,

$$-P_n^F\mathcal{A}^F\delta h - P^G\mathcal{A}^F(-\delta h) = \left(P^G - P_n^F\right)\mathcal{A}^F\delta h \tag{8.28}$$

The work associated with δn^{LFG} is given by,

$$\begin{aligned}
&\int_{L^{LFG}}\gamma^{LS}\delta n^{LFG}\mathrm{d}L^{LFG} + \int_{L^{LFG}}\gamma^{FG}(-\delta n^{LFG})\mathrm{d}L^{LFG} + \int_{L^{LFG}}\gamma^{FS}(-\delta n^{LFG})\mathrm{d}L^{LFG}\\
&\quad + \int_{L^{LFG}}\gamma^{LG}\cos\theta\,\delta n^{LFG}\mathrm{d}L^{LFG} - \int_{L^{LFG}}P^L h\,\delta n^{LFG}\mathrm{d}L^{LFG}\\
&\quad - \int_{L^{LFG}}P^L h(-\delta n^{LFG})\mathrm{d}L^{LFG}\\
&\quad = \int_{L^{LFG}}[-(\gamma^{FG}+\gamma^{FS}-P^L h)+\gamma^{LG}\cos\theta+\gamma^{LS}-P^L h]\\
&\quad \times\,\delta n^{LFG}\mathrm{d}L^{LFG}
\end{aligned} \tag{8.29}$$

In the above expression, $\mathrm{d}L^{LFG}$ is the differential element of the liquid-film-gas contact line. Figure 8.8 provides a schematic view of the droplet and thin system elements. In Eq. (8.29), the first term on the left side is the interfacial work from the change in the liquid-solid interfacial area, $\mathcal{A}^{LS}$. The interfacial work of the first term is coupled with the interfacial work done on the film in changing, $\mathcal{A}^F$ of the second and third terms. The fourth term on the left is the work done on the liquid-gas interface; $\gamma^{LG}\cos\theta$ is the

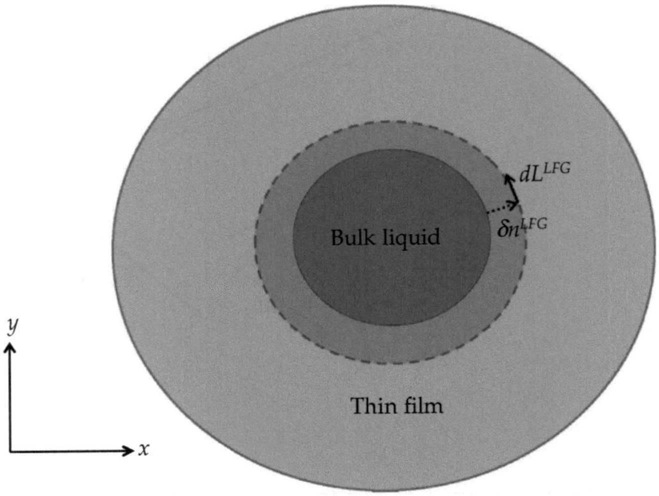

FIGURE 8.8 Overhead view of the system, where for clarity the droplet is being depicted spherical and the thin film as being cylindrical. The change in the area A^{LS} of the liquid-solid interface due to the virtual displacement δn^{LFG} is $\int_{L^{LFG}} \delta n^{LFG} \mathrm{d}L^{LFG}$, which is represented by the annulus between the dashed curve and the bulk liquid. This annulus is also minus the change in the area A^F of the film (adapted from Myint and Firoozabadi, 2015b).

force per unit length of the liquid-film gas contact line in the direction of displacement, δn^{LFG}. The last two terms on the left are the pressure-volume work (in differential volume element, $h\delta n^{LFG}\mathrm{d}L^{LFG}$) of the bulk liquid and the film volume, respectively.

The work associated with δn^{FGS} is given by,

$$\begin{aligned}
&\int_{L^{FGS}} \gamma^{FG} \delta n^{FGS} \mathrm{d}L^{FGS} + \int_{L^{FGS}} \gamma^{FS} \delta n^{FGS} \mathrm{d}L^{FGS} \\
&\quad + \int_{L^{FGS}} \gamma^{GS} (-\delta n^{FGS}) \mathrm{d}L^{FGS} - \int_{L^{FGS}} P^L h \delta n^{FGS} \mathrm{d}L^{FGS} \\
&\quad - \int_{L^{FGS}} P^G h (-\delta n^{FGS}) \mathrm{d}L^{FGS} \\
&\quad = \int_{L^{FGS}} \left[(\gamma^{FG} + \gamma^{FS} - P^L h) - \gamma^{GS} + P^G h \right] \delta n^{FGS} \mathrm{d}L^{FGS} \qquad (8.30)
\end{aligned}$$

In Eq. (8.30), $\mathrm{d}L^{FGS}$ is the length of a differential element of the film-gas-solid line.

Mechanical equilibrium requires that the sum of the right side of Eqs. (8.27) to (8.30) to be zero. Because each of these work terms are independent from each other, therefore, each of them should be zero. As a result, we obtain four conditions for mechanical equilibrium.

From Eq. (8.27), we obtain,

$$P^L - P^G = 2\gamma^{LG} H \qquad (8.31)$$

This is the mechanical equilibrium condition for a droplet from Chap. 2. It is the Young-Laplace equation.

From Eq. (8.28), we obtain,

$$P^G = P_n^F \tag{8.32}$$

which states the conditions of local mechanical equilibrium at the gas-film interface.
From Eq. (8.29),

$$\gamma^{FG} + \gamma^{FS} - P^L h = \gamma^{LG}\cos\theta + \gamma^{LS} - P^L h \tag{8.33}$$

From Eq. (8.30),

$$\gamma^{FG} + \gamma^{FS} - P^L h = \gamma^{GS} - P^G h \tag{8.34}$$

8.3.2 Augmented Young Equation

Combining Eqs. (8.7) and Eqs. (8.32) to (8.34), one obtains the augmented Young equation,

$$\boxed{\gamma^{LG}\cos\theta = \gamma^{GS} - \gamma^{LS} - \Pi h} \tag{8.35}$$

Note that the augmented Young equation is based on mechanical equilibrium. A different form of the augmented Young equation by Amirfazli (2004) is given in Problem 8.7.

8.4 Film Tension

In Eq. (8.14), the work contributions to the change in Internal free energy of the film is

$$đW = -P_n^F \mathcal{A}^F \mathrm{d}h + (\gamma^{FG} + \gamma^{FS} - P^L h)\mathrm{d}\mathcal{A}^F \tag{8.36}$$

Combining Eqs. (8.32) and (8.36), one can write,

$$\boxed{đW = -P^G \mathcal{A}^F \mathrm{d}h + (\gamma^{FG} + \gamma^{FS} - P^L h)\mathrm{d}\mathcal{A}^F} \tag{8.37}$$

Many authors (Toshev and Ivanov, 1975; Ivanov and Toshev, 1975; de Feijter et al., 1978; Rusanov and Schekin, 2005; Iota et al., 2004; Hirosaki, 1991; Amifazli, 2004; and Hess, 1994) start from a system sketched in Fig. 8.9a and express the work related to a thin film by

$$đW = -P^G \mathrm{d}V^F + \gamma^F \mathrm{d}\mathcal{A}^F \tag{8.38}$$

where γ^F is the film tension.

In the above equation $\gamma^F \mathrm{d}\mathcal{A}^F$ is the work done on the film to change its area by $\mathrm{d}\mathcal{A}^F$. The term, $P^G \mathrm{d}V^F$ represents the pressure-volume work. Using $V^F = h\mathcal{A}^F$, Eq. (8.38) can be written as (for the system sketched in Fig. 8.9a),

$$\boxed{đW = -P^G \mathcal{A}^F \mathrm{d}h + (\gamma^F - P^G h)\mathrm{d}\mathcal{A}^F} \tag{8.39}$$

Note that there are some basic differences between the films sketched in Figs. 8.4 and 8.9a, despite the fact that both are connected to the bulk liquid phase. The pressure of

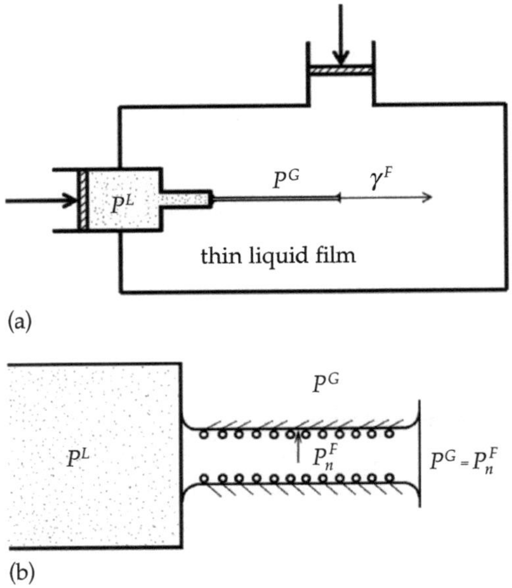

FIGURE 8.9 (a) Complete system that shows the thin-liquid film and the film tension, (b) thin-liquid film and the bulk liquid phase from which the thin film is made of.

the gas phase is higher than the liquid phase in Fig. 8.9a. The opposite is true in Fig. 8.4. The film is stabilized by surfactants in Fig. 8.9b. The film is rested on the solid surface in Fig. 8.4. The film is surrounded by the gas phase on both of its parallel interfaces in Fig. 8.9a. From Eqs. (8.37) and (8.39),

$$\boxed{\gamma^F = \gamma^{FG} + \gamma^{FS} + (P^G - P^L)h} \tag{8.40}$$

Combing the definition of disjoining pressure and Eq. (8.40), one obtains,

$$\boxed{\gamma^F = \gamma^{FG} + \gamma^{FS} + \Pi h} \tag{8.41}$$

8.5 Thermodynamic Functions of the Thin Liquid Film

The differential internal energy of the thin liquid film, in terms of film tension, can be written as,

$$dU^F = T^F dS^F - P^G dV^F + \gamma^F dA^F + \sum_{i=1}^{n_c} \mu_i^F dN_i^F \tag{8.42}$$

Equation (8.42) is obtained by substituting the film work from Eq. (8.38) in Eq. (8.14). The film's internal energy from Eq. (8.42) is then,

$$U^F = T^F S^F - P^G V^F + \gamma^F A^F + \sum_{i=1}^{n_c} \mu_i^F N_i^F \tag{8.43}$$

Helmholtz free energy is then readily obtained by using $A^F = U^F - T^F S^F$,

$$A^F = -P^G V^F + \gamma^F \mathcal{A}^F + \sum_{i=1}^{n_c} \mu_i^F N_i^F \tag{8.44}$$

For all thermodynamic systems, whether homogeneous or inhomogeneous, the Gibbs free energy is $\sum_{i=1}^{n_c} N_i \mu_i$. That is, the Gibbs free energy of the film is

$$\boxed{G^F = \sum_{i=1}^{n_c} N_i^F \mu_i^F} \tag{8.45}$$

The relation between A^F and G^F is then established by combining Eqs. (8.44) and (8.45),

$$\boxed{G^F = A^F + P^G V^F - \gamma^F \mathcal{A}^F} \tag{8.46}$$

The expression for enthalpy is also readily obtained by using $H^F = G^F + T^F S^F$,

$$H^F = T^F S^F + \sum_{i=1}^{n_c} \mu_i^F N_i^F \tag{8.47}$$

Some authors add an additional term to the expression for Gibbs free energy and write

$$G^F = \gamma^F \mathcal{A}^F + \sum_{i=1}^{n_c} N_i^F \mu_i^F \tag{8.48}$$

The additional term $\gamma^F \mathcal{A}^F$, then will not allow computation of molecular structure in the closed systems at constant temperature and constant pressure.

Taking the derivative of Eq. (8.43) and comparing the result with Eq. (8.42) (and $V^F = h\mathcal{A}^F$) yields the Gibbs-Duhem equation,

$$\boxed{\mathrm{d}\gamma^F = -\frac{S^F}{\mathcal{A}^F}\mathrm{d}T^F + h\mathrm{d}P^G - \sum_{i=1}^{n_c} \frac{N_i^F}{\mathcal{A}^F}\mathrm{d}\mu_i^F} \tag{8.49}$$

Next, we discuss a useful relation between contact angle and film thickness.

8.6 Relations between Disjoining Pressure and Contact Angle

Derjaguin (1987a and 1987b) originally introduced the relation between the disjoining pressure in the film and the contact angle,

$$\gamma^{LG}(\cos\theta - 1) = \Pi(h)h + \int_h^\infty \Pi(h')\mathrm{d}h' \tag{8.50}$$

For large droplets, like the one in Fig. 8.4, Πh can be neglected compared to the integral in the above equation. Therefore,

$$\gamma^{LG}(\cos\theta - 1) \approx \int_h^\infty \Pi(h')\mathrm{d}h' \tag{8.51}$$

We are interested to derive the above expression. In our derivation, we closely follow the work by Rusanov (1975). Combining Eqs. (8.33) and (8.41),

$$\gamma^F = \gamma^{LG}\cos\theta + \gamma^{LS} + \Pi h \tag{8.52}$$

The excess entropy and excess mols of component i in the film are defined by,

$$S^{\text{excess}} = S^F - s^L \mathcal{A}^F h \tag{8.53}$$

$$N_i^{\text{excess}} = N_i^F - c_i^L \mathcal{A}^F h \tag{8.54}$$

where s^L is the entropy density ($s^L = S^L / V^L$) of the bulk liquid. Similarly, $c_i^L = N_i^L / V^L$. Substituting the above two expressions in Eq. (8.49), we obtain

$$\mathrm{d}\gamma^F = -\frac{S^{\text{excess}}}{\mathcal{A}^F}\mathrm{d}T + h\mathrm{d}P^G - \sum_{i=1}^{n_c}\frac{N_i^{\text{excess}}}{\mathcal{A}^F}\mathrm{d}\mu_i - h\left(s^L\mathrm{d}T + \sum_{i=1}^{n_c} c_i^L \mathrm{d}\mu_i\right) \tag{8.55}$$

The superscript F has been dropped on T and μ_i since we have shown that equilibrium requires spatial uniformity of T in different phases and interfaces. From the Gibbs-Duhem equation for the bulk liquid, the quantity in parentheses is equal to $\mathrm{d}P^L$. Therefore,

$$\mathrm{d}\gamma^F = -\frac{S^{\text{excess}}}{\mathcal{A}^F}\mathrm{d}T + h\mathrm{d}\Pi - \sum_{i=1}^{n_c}\frac{N_i^{\text{excess}}}{\mathcal{A}^F}\mathrm{d}\mu_i \tag{8.56}$$

At constant temperature and chemical potentials, the above expression reduces to

$$\mathrm{d}\gamma^F = h\mathrm{d}\Pi \tag{8.57}$$

Applying integration by parts, we have

$$\begin{aligned}\gamma^F(h) &= \gamma^F(h \to \infty) + \Pi(h)h + \int_h^\infty \Pi(h')\mathrm{d}h' \\ &= \gamma^{LG} + \gamma^{LS} + \Pi(h)h + \int_h^\infty \Pi(h')\mathrm{d}h'\end{aligned} \tag{8.58}$$

Note that we have used $\gamma^F(h \to \infty) = \gamma^{LG} + \gamma^{LS}$. Combining Eqs. (8.52) and (8.58), one obtains Eq. (8.51).

8.7 Example and Theory Extension

8.1: Melting Point of Gold Nanoparticles Certain properties such as melting may change significantly with size in the nano-scale. Figure 8.10 presents data of gold nano-particles from Buffat and Borel (1976). We use two different plots of data in Fig. 8.10; one of reduced melting point temperature of particles versus particles radius, and the other

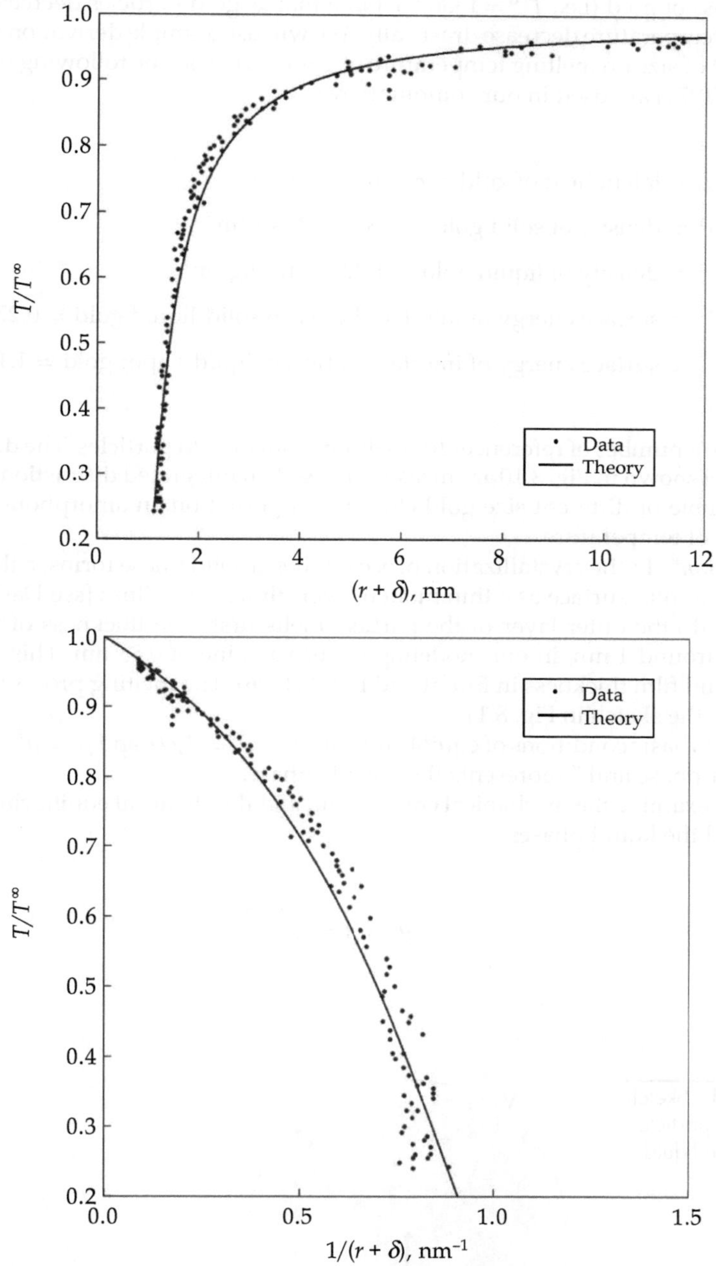

FIGURE 8.10 Normalized melting point of gold particles as a function of size (T^{∞} is the melting point for flat interface).

with inverse of particle radius. The melting point temperature of gold is 1336 K for a large mass of gold (i.e., $T^{\infty} = 1336$ K). Note that as gold particles decrease in size, their melting temperature decrease drastically. We will use a simple derivation that can model the effect of size on melting temperature of gold particles. The following properties from Samble (1971) are used in our computations:

$$L \equiv \text{latent heat of gold} = 6.276 \times 10^4 \text{ J/kg}$$
$$\rho^S \equiv \text{density of solid gold} = 1.84 \times 10^4 \text{ kg/m}^3$$
$$\rho^L \equiv \text{density of liquid gold} = 1.728 \times 10^4 \text{ kg/m}^3$$
$$\gamma^{SL} \equiv \text{surface energy of interface between solid-liquid gold} = 0.27 \text{ J/m}^2$$
$$\gamma^{LV} \equiv \text{surface energy of interface between liquid-vapor gold} = 1.135 \text{ J/m}^2$$

There are a number of references for melting point of gold particles. The data from Buffat and Borel shown in Fig. 8.10 are measured based on integrated diffraction intensity from an ensemble of different size gold clusters supported on an amorphous substrate as a function of temperature.

Solution In the crystallization process when a solid phase forms, a thin liquid often covers the solid surface as a thin molten layer (thin liquid film); [see Dash et al. (2006)]. As a result, the outer layer of the particle melts first. The thickness of the thin liquid layer is around 1 nm; in our modeling we use a value of 0.83 nm. This value is in line with liquid film thickness in Buffat and Borel (1976). The melting process formulation is based on the sketch in Fig. 8.11.

The two basic conditions of equilibrium are $P' - P'' = 2\gamma/r$ and $\mu' = \mu''$ where $'$ denotes the inner phase and $''$ represents the outside phase.

Let us examine the mechanical equilibrium and the chemical equilibrium between the solid and the liquid phases:

$$P^S - P^L = \frac{2\gamma^{SL}}{r} \tag{8.59}$$

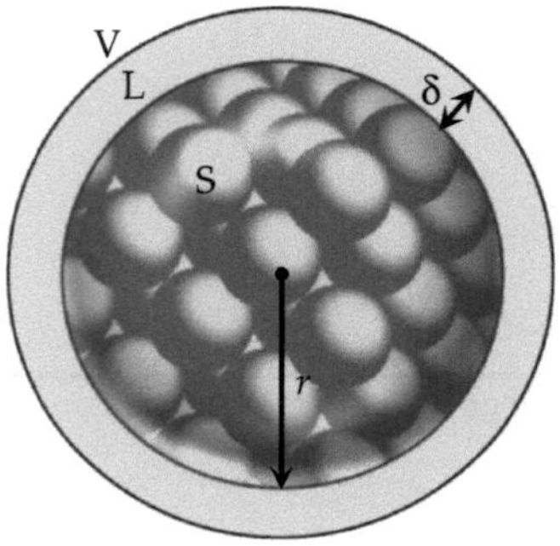

FIGURE 8.11 Sketch of a gold particle with a thin liquid film.

The Gibbs-Duhem equation is applied to phases S and L:

$$s^S \mathrm{d}T - v^S \mathrm{d}P^S + \mathrm{d}\mu^S = 0 \tag{8.60}$$

$$s^L \mathrm{d}T - v^L \mathrm{d}P^L + \mathrm{d}\mu^L = 0 \tag{8.61}$$

where s and v represent the molar entropy and volume, respectively.

Combining $d\mu^S = d\mu^L$ and Eqs. (8.60) and (8.61) provides

$$(s^S - s^L)\mathrm{d}T = \frac{\mathrm{d}P^S}{c^S} - \frac{\mathrm{d}P^L}{c^L} \tag{8.62}$$

where c represents the molar density ($c = 1/v$).

We can also write the Young-Laplace equation to relate pressure of vapor and liquid phases,

$$P^L - P^V = \frac{2\gamma^{LV}}{(r+\delta)} \tag{8.63}$$

The external pressure P^V is constant; therefore, $\mathrm{d}P^V = 0$. Eqs. (8.59) and (8.63) can provide

$$\mathrm{d}P^S = 2\mathrm{d}\left(\frac{\gamma^{SL}}{r}\right) + 2\mathrm{d}\left(\frac{\gamma^{LV}}{r+\delta}\right) \tag{8.64}$$

$$\mathrm{d}P^L = 2\mathrm{d}\left(\frac{\gamma^{LV}}{r+\delta}\right) \tag{8.65}$$

The latent heat of fusion is related to entropy of fusion

$$L = T(s^L - s^S) \tag{8.66}$$

Combining Eqs. (8.62) and (8.64) to (8.66), using specific properties instead of molar properties, ($c = \rho/M$, where M is the molecular weight), we obtain

$$\left(\frac{LM}{T}\right)\mathrm{d}T = -\frac{2}{\rho^S}\left[\mathrm{d}\left(\frac{\gamma^{SL}}{r}\right) + \mathrm{d}\left(\frac{\gamma^{LV}}{r+\delta}\right)\right] + \frac{2}{\rho^L}\mathrm{d}\left(\frac{\gamma^{LV}}{r+\delta}\right) \tag{8.67}$$

Making the assumption that L, δ, and densities are independent of temperature, and integrating from $r = \infty$ to finite r, we obtain

$$\boxed{\ln\left(\frac{T}{T\infty}\right) = -\frac{2}{\rho^S LM}\left[\frac{\gamma^{SL}}{r} + \frac{\gamma^{LV}}{r+\delta}\left(1 - \frac{\rho^S}{\rho^L}\right)\right]} \tag{8.68}$$

Note that without a thin liquid layer (i.e., when δ is very large), the second term in the right-hand side disappears. Without the second term, only nano-particles with a large diameter (10 nm) can be modeled. More accurate modeling requires allowance for variation of surface energy densities with size and temperature. Allowance should be made also for temperature dependency of latent heat and densities. Note that in this problem, we have assumed an isotropic pressure in the thin liquid film. Despite this assumption, the model is predictive, as Fig. 8.9 shows. There is good agreement between data and calculations based on Eq. (8.68).

8.8 Problems

8.1 Use Fig. 8.12 to derive the expression for the film tension given in Eq. (8.40). Note that the pressure in the film is assumed isotropic.

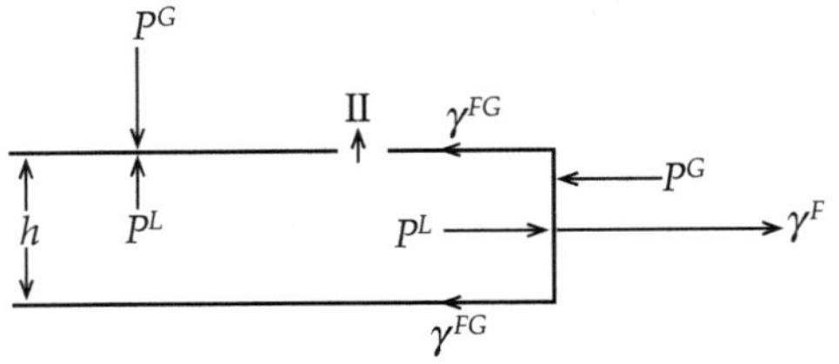

FIGURE 8.12

8.2 Consider the pressure in the liquid phase in the geometry sketched in Fig. 8.13. Write the general expression that relates the pressure in the gas and in the liquid phase.

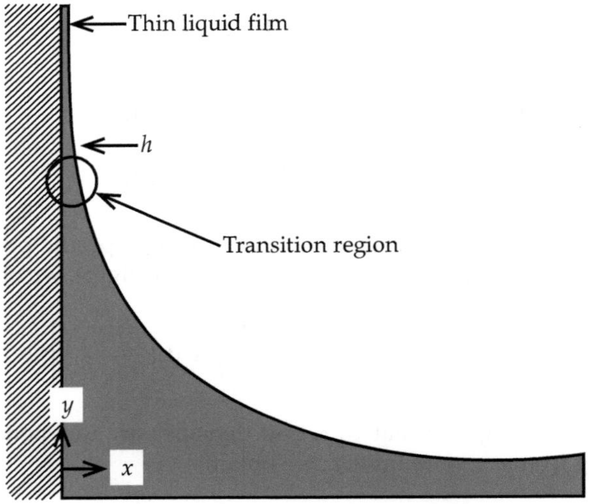

FIGURE 8.13

8.3 Consider the interfacial layer between a gas phase and a liquid phase sketched in the left in Fig. 8.14. The figure on the right shows the tangential pressure in the interfacial layer.

Explain the basis for the pressure sketched in the figure on the right.

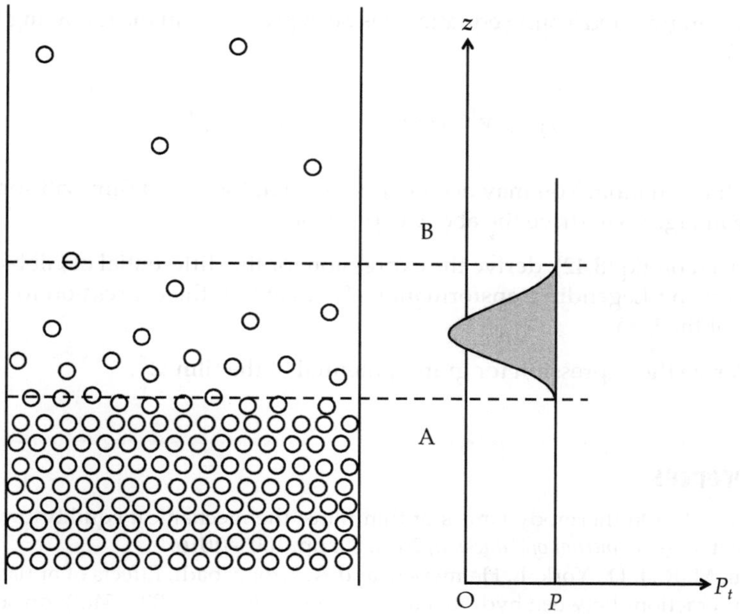

FIGURE 8.14

8.4 Use Eq. (8.26) and geometrical relations similar to the one in relation to a cap in Chap. 7 to derive Eqs. (8.31) to (8.34).

8.5 The difference between melting and crystallization temperature can be very significant (see the discussion of the observation below). The difference between condensation from vapor to liquid and vaporization from liquid to vapor can also be significant.

Discuss the reason for the differences.

NOTE *Hydrate crystallization temperature measurements at constant pressure is a strong function of rate of cooling. The melting, on the other hand, at a constant pressure is independent of heating rate. The difference between the two temperatures can be as high as 10 to 20°C in a typical measurement. The crystallization temperature in methane hydrates can be as low as 5°C. The melting temperature is, say 20°C. The melting (sometimes called dissociation) temperature is close to the equilibrium temperature. (See Fig. 4 of Li et al., 2010.)*

8.6 There are subtleties in Eq. (8.9). First derive the following expression:

$$\gamma^{FG} + \gamma^{FS} = -\int_0^h \left[P_z(z) - P^L\right] dz - \int_h^Z \left[P_z(z) - P^G\right] dz \tag{8.69}$$

(see Fig. 8.4). Then derive Eq. (8.9).

8.7 The augmented Young equation has been presented in the following form by Amirfazli (2004),

$$\gamma^{LG}\cos\theta = \gamma^{FS} + \gamma^{FG} - \gamma^{LS}. \tag{8.70}$$

Derive the equation. You may need to assume that the liquid film will spread the whole surface in Fig. 8.6 to drive the above expression.

8.8 Based on Eq.(8.42), derive the expression for the differential of Gibbs free energy of the film using Legendre transformation. Then, obtain the expression for the Gibbs free energy of the film.

8.9 Derive the expression for grand potential of the film Ω^F.

8.9 References

Amirfazli, A., On thermodynamics of thin films: the mechanical equilibrium condition and contact angles. *Journal of Adhesion*, 2004. 80. pp. 1003–1016.

Anklam, M. R., J. D., York, L. Helmerich, and A. Firoozabadi, Effects of antiagglomerants on the interactions between hydrate particles. *AIChE Journal*, 2008. 54(2): pp 562–574.

Bluhm, H., D. F. Ogletree, C. S. Fadley, Z. Hussain, and M. Salmeron, The premelting of ice studied with photoelectron spectroscopy. *Journal of Physics: Condensed Matter*, 2002. pp. I.227–I.233.

Buffat, P. and J-P. Borel. Size effect on the melting temperature of gold particles. *Physical Review A*, 1976. 6 (13): pp. 2287–2298.

Dash, J. G., A. W. Rempel, and J. S. Wettlaufer, The physics of premelted ice and its geophysical consequences. *Reviews of Modern Physics*, 2006. 78: pp. 695–741.

de Feijter, J. A., Thermodynamics of Thin Liquid Films, *Thin Liquid Films: Fundamentals and Applications*. Marcel Dekker, New York, 1988. pp. 1–48.

de Feijter, J. A., J. B. Rijnbout, and A Vrij, Contact angles in thin liquid films. I. thermodynamic description. *Journal of Colloid and Interface Science*, 1978. 64: pp. 258–268.

Derjaguin, B. V., Some results from 50 years' research on surface forces. *Progress in Colloid and Polymer Science*, 1987a. 74: pp. 17–30.

Derjaguin, B. V., N. V. Churaev, and V. M. Muller, *Surface Forces*. 1987b, New York: Springer US.

Faraday, M. Note on regelation. *Proceedings of the Royal Society of London*. 1860, 10: pp. 440–450.

Gibbs, J. W., *The Collected Works of J. Willard Gibbs:* In Two Volumes, 3rd ed. 1948, New Haven: Yale University Press.

Hess, P. C., Thermodynamics of thin fluid films. *Journal of Geophysical Research*, 1994. 99: pp. 7219–7229.

Hirasaki, G. J., *Thermodynamics of Thin Films and Three-Phase Contact Regions: Interfacial Phenomena in Petroleum Recovery*. Marcel Dekker, New York. 1991b. pp. 24–76.

Hirasaki, G. J., Wettability: Fundamentals and Surface Forces. *SPE Formation Evaluation*. 1991a. pp. 217–226.

Ivanov, I. B., and B. V. Toshev, Thermodynamics of thin liquid films II. Film thickness and its relation to the surface tension and the contact angle. *Colloid and Polymer Science*, 1975. 253: pp. 593–599.

Iyota, H., Krustev, R., and H.-J. Müller, Thermodynamic studies on thin liquid films. I. General Formulation. *Colloid and Polymer Science*, 2004. 282: pp. 1329–1340.

Jiménez-Ángeles, F., and A. Firoozabadi, Induced charge density and thin liquid film at hydrate/methane gas interfaces. *Journal of Physical Chemistry C*, 2014. 118: pp. 26014–26048.

Johnson, R. E., Conflicts between Gibbsian thermodynamics and recent treatments of interfacial energies in solid-liquid-vapor systems. *Journal of Physical Chemistry*, 1959. 63: pp. 1655–1658.

Lanczos, C., *The Variational Principles of Mechanics*, 4th ed. 1986, New York: Dover Publications.

Li, D., and A. W. Neumann, Thermodynamics of contact angle phenomena in the presence of a thin liquid film. *Advances in Colloid and Interface Science*, 1991. 36: pp. 125–151.

Li, X., L. Negai, and A. Firoozabadi, Anti-agglomeration in cyclopentane hydrates from bio- and co-surfactants. *Energy Fuels*, 2010. 24: pp. 4937–4943.

Myint, P. C., and A. Firoozabadi, *Thermodynamics of Flat Thin Liquid Films*. (In review), 2015b.

Myint, P. C., and A. Firoozabadi, Thin liquid films in improved oil recovery from low-salinity brine. *Current Opinions in Colloid and Interface Science*. 2015a. 20(2): pp. 105–114.

Rusanov, A. I., On the thermodynamics of phase equilibria in thin films. *Journal of Colloid and Interface Science*, 1975. 53: pp. 20–30.

Rusanov, A. I., and A. K. Shchekin, Local mechanical equilibrium conditions for interfaces and thin films of arbitrary shape. *Molecular Physics*, 2005. 103: pp. 2911–2922.

Sambles, J. R., An electron microscope study of evaporating gold particles: The kelvin equation for liquid gold and the lowering of the melting point of solid gold particles. *Proceedings of the Royal Society of London A*, 1971. 324: pp. 339–351.

Toshev, B. V., and I. B. Ivanov, Thermodynamics of thin liquid films I. Basic relations and conditions of equilibrium. *Colloid and Polymer Science*, 1975. 253: pp. 558–565.

Yeh, E. K., J. C. Newman, and C. J. Radke, Equilibrium configurations of liquid droplets on solid surfaces under the influence of thin-film forces Part I. thermodynamics, *Colloids and Surfaces A: Physicochemical and Engineering Aspects*, 1999. 156: pp. 137–144.

Index

Y

Printed by Libri Plureos GmbH in Hamburg,
Germany

9 780071 843256